INTRODUCTORY OCEANOGRAPHY

Sixth Edition

INTRODUCTORY OCEANOGRAPHY

HAROLD V. THURMAN
Mount Saint Antonio College

Macmillan Publishing Company
New York

Collier Macmillan Canada, Inc.
Toronto

Maxwell Macmillan International Publishing Group
New York Oxford Singapore Sydney

Cover photo: © Kurt Amsler, 1989 Telephoto.

The greatest beauty and diversity of marine life is surely found in the warm waters of the tropical coral reef. Paradoxically, this ecosystem can only exist in nutrient-poor waters. To add to the paradox, although the beauty of the coral reef has a soothing effect on the human psyche, with each human contact, the reef is degraded.

Senior Editor: Robert A. McConnin
Production Editor: Mary Harlan
Art Coordinator: Ruth A. Kimpel
Production Buyer: Janice E. Wagner
Cover Designer: Brian Deep

This book was set in Usherwood.

Copyright © 1991 by Macmillan Publishing Company, a division of Macmillan, Inc.

Previous editions copyrighted 1988, 1985, 1981, 1978, 1975 by Merrill Publishing Company.

Printed in the United States of America

All rights reserved. No part of this book may be reproduced or transmitted in any form or by any means, electronic or mechanical, including photocopy, recording, or any information storage and retrieval system, without permission in writing from the Publisher.

Macmillan Publishing Company
866 Third Avenue, New York, New York 10022

Collier Macmillan Canada, Inc.

Library of Congress Cataloging-in-Publication Data

Thurman, Harold V.
    Introductory oceanography/Harold V. Thurman.—6th ed.
        p.   cm.
    Includes bibliographical references and index.
    ISBN 0-675-21317-7 (hard)
    1. Oceanography.   I. Title.
GC16.T45   1991
551.46—dc20   90-13281
                            CIP

Printing: 1 2 3 4 5 6 7 8 9   Year: 1 2 3 4

PREFACE

In the previous editions of *Introductory Oceanography,* I have tried to present an unbiased view of the controversial issues related to marine science. With this edition, I find it necessary to espouse the position that as a result of human activities we may be experiencing, or soon will experience, an increased greenhouse effect. Although most people grasped that burning fossil fuels at increasing rates would produce ever-worsening problems, we somehow also envisioned the oceans gobbling up the excess carbon dioxide and depositing it on the sea floor.

Experts disagree as to whether the greenhouse effect is increasing and what the ocean's role is in the process. But it is clear that the atmosphere and the coastal oceans are being altered. We must act to reduce the risk of further degradation.

Because fifteen years of research have confirmed the initial warning that human activities are resulting in the depletion of ozone in the upper atmosphere, we now tend to give greater credibility to the present projections of an increasing greenhouse effect also resulting from those activities.

This text deals with questions of how the increased greenhouse effect is related to the interactions between the atmosphere and the oceans. Current research on the role of the oceans in this and other important environmental processes is presented. The emphasis of the text is still on marine processes, but the issue of interactions among humans, the ocean, and the atmosphere is covered in greater depth.

A major focus of this revision is the improvement of the illustrations. Many new line drawings have been included, and others have been redrawn. Full color photographs and illustrations have been added throughout the book.

A number of improvements have been made in the text. Chapter 1, "History of Oceanography," has been revised to expand the coverage of events from the late nineteenth century to the present. New renderings of figures and condensed coverage on the formation of the atmosphere are highlights of the revision of Chapter 2.

Chapters 3 and 4 on marine geology have been significantly revised. Formerly, Chapter 3 described the provinces of the ocean basins without discussing their origin, while Chapter 4 discussed their origin in terms of global plate tectonics. Now Chapter 3, "Global Plate Tectonics," precedes the discussion of marine provinces in Chapter 4. The provinces of the ocean floor are now introduced in terms of their origin as it relates to global plate tectonics. Chapters 3 and 4 have been reorganized and updated to provide the widest possible range of coverage of global plate tectonics and its role in producing the features of the ocean floor.

In Chapter 5, "Marine Sediments," the distribution of oceanic sediments is now based on sediment type—calcareous ooze, siliceous ooze, and abyssal clay—rather than organized by ocean basin.

In the chapters on physical oceanography, the following significant changes have been made. Chapter 6, "Properties of Water," now includes the hydrologic cycle and places increased emphasis on the role of water in determining the earth's climate. This effort culminates in the special feature, "Hurricanes—Nature's Safety Valves." Chapter 7, "Air-Sea Interaction," presents the El Niño–Southern Oscillation events and the greenhouse effect as examples of the interrelatedness of the atmosphere and the oceans and briefly discusses the effects of human activities on both environments. Chapter 8, "Ocean Circulation," has an expanded explanation of westward intensification, and the discussion of circulation in the various oceans is more concise. New line drawings in Chapter 9, "Waves," clearly illustrate the conditions of the "sea" area of wave generation, wave interference patterns, tsunami creation, and world coastal wave energy resources. The discussions of tide generation and tidal power in Chapter 10 have been simplified.

Chapter 11 includes a discussion of the role of government policy in producing overdevelopment of our coastal regions. Francis P. Shepard's classification of coasts has been added, and the present rise in sea level and the potential future rise are discussed in terms of possible human contributions. Coverage of barrier islands has been expanded, and the discussion of U.S. coasts now emphasizes present rates of erosion.

Increased attention to anthropogenic effects is found in Chapter 12, with emphasis on the Columbia River and Chesapeake Bay. An addition is the effect of estuarine input on the plankton communities of the southeastern U.S. shelf waters. A special feature, PROBES: Marine Science Steps Forward, describes how marine investigation has increased our understanding of the fisheries of the Bering Sea shelf waters.

Within the chapters on marine biology, a comparison of benthic and pelagic food chains has been added to Chapter 14. The discussion of cetaceans in Chapter 15 now includes more on taxonomy, behavior, and use of sound. In Chapter 16, the discussions of coral reefs and deep-sea vent and seep communities have been expanded.

Chapter 17, "Exploitation of Marine Resouces," includes the negative effects of human interaction with the ocean, as well as discussions of petroleum, mineral resources, fisheries, and mariculture. Chapter 18, "Marine Pollution," includes the areas of plastics, petroleum, sewage, radioactive waste, halogenated hydrocarbons, and mercury.

Acknowledgments

Many individuals have offered advice and assistance for previous editions of this work, including Ernest Angino, University of Kansas; James Conkin, University of Louisville; William T. Fox, Williams College; David M. Karl, University of Hawaii; C. Ernest Knowles, North Carolina State University; Ted Loder, University of New Hampshire; James M. McWhorter, Miami Dade Community College; Larry L. Malinconico, Southern Illinois University; Edward D. Stroup, University of Hawaii; and Mel Zucker, Skyline College.

I want to extend a special thank-you for the valuable suggestions made by those who reviewed the manuscript for the sixth edition of *Introductory Oceanography*: Marsha Bollinger, Winthrop College; Richard C. Brill, Honolulu Community College; John Ehleiter, West Chester University; William J. Frazier, Columbus College; William Hoyt, University of Northern Colorado; Jack Hyde, Tacoma Community College; Ronald Johnson, Old Dominion University; Albert M. Kudo, University of New Mexico; James E. Mackin, SUNY–Stony Brook; John Mylroie, Mississippi State University; Carl Nelson, University of San Diego; Robin Newbold-Sturn, San Juan Capistrano; Peter Nielsen, Keene State College; Carl H. Oppenheimer, University of Texas; Harold R. Pestana, Colby College; Paul Pinet, Colgate University; Robert G. Read, Carmel Valley; Jill Singer, SUNY–Buffalo; Yin S. Soong, Millersville University; and Robert W. Tank, Lawrence University.

Over the course of this revision, many members of the production staff contributed to the successful completion of this ambitious revision project. I particularly wish to thank Mary Harlan, who oversaw most of the project and kept it moving along on schedule. Jean Brown did an excellent job of copy editing the manuscript, and Ruth Kimpel successfully coordinated the significant improvement in the art work.

HVT

CONTENTS

3

GLOBAL PLATE TECTONICS   45

4

MARINE PROVINCES   79

5

MARINE SEDIMENTS   107

6

PROPERTIES OF WATER   137

7

AIR-SEA INTERACTION   157

INTRODUCTORY OCEANOGRAPHY

1
HISTORY OF OCEANOGRAPHY

With each passing year, human interaction with the ocean increases. The fishing effort is stepped up, and the search for petroleum and minerals intensifies. With this increasing rate of ocean exploitation, we are obliged to increase our knowledge of its effect on the marine environment. The following is a summary of events that have brought us to our present state of knowledge, and a discussion of how the pressing need for increased knowledge of the oceans may be met through the application of newly developed technology.

EARLY HISTORY

Early Human Interaction with the Oceans

If human beings evolved in Africa some two million years ago, as present evidence indicates, it is obvious that the vast extent of the oceans was no barrier to their movement from this early home to all parts of the habitable world. When European explorers set out at the end of the fifteenth century to see what lay beyond the horizon, they discovered that cultures unknown to them had preceded them to far-off places, from Tierra del Fuego in the south to the island of Greenland in the north.

Humankind probably first viewed the ocean as a source of food. At some later stage in the development of civilization, vessels were built to move upon the ocean's surface, thereby making it a wide avenue over which distant societies could begin to interact.

The first westerners that are known to have developed the art of navigation were the **Phoenicians**, who as long ago as 2000 B.C. were investigating the Mediterranean Sea, the Red Sea and the Indian Ocean. The first recorded circumnavigation of Africa in 590 B.C. was made by the Phoenicians, who had also sailed as far north along the European coast as the British Isles.

Contributions of the Greeks and Romans

The Greek view of the world in 450 B.C. is seen in the map of **Herodotus** (fig. 1–1). It shows the Mediterranean Sea surrounded by three continents: Europe,

Asia, and Libya. They are, in turn, surrounded by continuous seas.

The Greek astronomer-geographer **Pytheas** sailed northward to Iceland in 325 B.C. and worked out a simple method for determining latitude. It involved determining the angle between two lines of sight—one on the northern horizon and the other the North Star (see Appendix 1X).

Eratosthenes (276–192 B.C.), librarian at Alexandria, is the first known determiner of the world's circumference. His value for a meridional circumference was 40,000 km (24,840 mi) which compares well with a figure of 40,032 km (24,875 mi) that has been determined by today's more precise methods (fig. 1–2).

The Roman **Strabo** (63 B.C.–A.D. 24) observed volcanic activity in what is now Italy and concluded that, as a result of this activity, the land periodically sank and rose, causing the sea to invade and recede from the continents. He also recognized the role of streams in eroding the continents and depositing sediment in the seas.

In approximately A.D. 150, **Ptolemy** produced a map of the world that represented Roman knowledge at the time. He introduced lines of latitude and longitude on his map and included the continents of Europe, Asia, and Africa. The seas that surrounded these continents were considered to be similar to the Mediterranean—surrounded by as yet unknown landmasses. This map was later a great incentive to explorers keen to seek out the unknown lands beyond the seas.

The Middle Ages

After the fall of the Roman Empire and the rise of Christianity, the Mediterranean was dominated by the Moslem **Arabs**. During their dominance, the Arabs took advantage of the seasonal wind changes called the monsoons to trade across the Indian Ocean and became the dominant navigators in the Mediterranean region.

The post-Roman Empire concept of world geography degenerated considerably in southern Europe. Many believed the earth was represented by the map of a sixth-century navigator, **Cosmas**. This map showed the earth to be a flat rectangle 10,000 by 20,000 km (6200 by 12,400 mi).

To the north, the nautical inactivity of the southern Europeans was offset by the vigorous exploration of the **Vikings** of Scandinavia. Late in the ninth century, a warming of the climate freed the North Atlantic of ice. Vikings that had been banished from the mainland expanded to the west. They conquered Iceland,

Preceding page: The *JOIDES Resolution,* the drilling ship of the Ocean Drilling Program. This modern drilling ship has replaced the *Glomar Challenger* in the important work of sampling the floors of the world's oceans. Photo courtesy of Ocean Drilling Program.

FIGURE 1–1 The World of Herodotus.

The world according to Herodotus, 450 B.C. From the *Challenger* report, Great Britain, 1895.

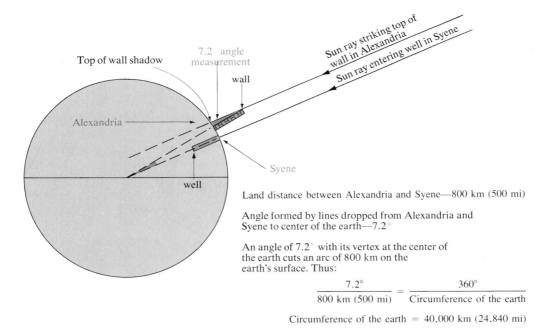

Land distance between Alexandria and Syene—800 km (500 mi)

Angle formed by lines dropped from Alexandria and Syene to center of the earth—7.2°

An angle of 7.2° with its vertex at the center of the earth cuts an arc of 800 km on the earth's surface. Thus:

$$\frac{7.2°}{800 \text{ km (500 mi)}} = \frac{360°}{\text{Circumference of the earth}}$$

Circumference of the earth = 40,000 km (24,840 mi)

FIGURE 1–2 Eratosthenes' Determination of the Circumference of the Earth (circa 200 B.C.).

Eratosthenes used the simple geometric principle that when two parallel lines are cut by a third straight line, corresponding equal angles are formed, to determine quite precisely the meridional circumference of the earth.

looting the Christian population and driving them away. From there they sailed west to Greenland. In 995, **Leif Eriksson** sailed westward from Greenland, discovered what was then called **Vinland**, and wintered in that part of North America that is now New-foundland (fig. 1–3).

By the beginning of the thirteenth century, the climate began to cool, and the northern Atlantic once again became choked with ice throughout most of the year. This deterioration in climate led to the isolation of the Greenland colonies and marked an end to the Viking expansion.

Age of Discovery

During the thirty-year period from 1492 to 1522, known as the **Age of Discovery**, the Western world came to a full realization of the vastness of the earth's water-covered surface.

After his well-known difficulties in originating the voyage, **Columbus** set sail with 88 men and three ships on August 3, 1492. Sailing from the Canary Islands, their first glimpse of the new world came on October 12, 1492. Columbus had reached the Bahama Islands—probably landing at **San Salvador Island**.

FIGURE 1–3   The Vikings Reach the New World.
The expansion of Viking influence spread from Europe and probably reached Iceland by 860. Erik the Red was banished from Iceland and sailed to Greenland in 981. He returned to Iceland in 984 and led the first wave of Viking colonists to Greenland. Bjarni Herjolfsson sailed from Iceland to join the colonists but sailed too far south and is thought to be the first Viking to have viewed what is now called Newfoundland. Bjarni did not land but continued trying to find his way to Greenland. He eventually did land at the settlement Herjolfsness, named for his father, who waited for him there. In 995, Leif Eriksson, son of Erik the Red, bought Bjarni's ship and set out for the land Bjarni had seen to the southwest. Leif set up a camp at Tickle Cove in Trinity Bay and named the land Vinland after the grapes that were found there.

The culmination of this period of discovery is marked by the circumnavigation of the globe by **Ferdinand Magellan**'s ship, *Victoria*. Sailing from Spain, Magellan sailed south through the Atlantic Ocean and west across the Pacific Ocean to the Philippines. After Magellan was killed by natives in the Philippines, **Sebastian del Caño** completed the circumnavigation. The *Victoria* arrived back in Spain on September 6, 1522 (fig. 1–4).

Spain maintained dominance of the seas until 1588, when it attempted an invasion of the British Isles. After the British defeated the **Spanish Armada**, they began a period of empire building that led to their domination of the seas until early in the twentieth century.

THE FORMATIVE YEARS OF MARINE SCIENCE

Captain James Cook

To help them maintain their marine superiority, the English undertook many voyages of discovery with scientific objectives. One of their most capable navigators in the conduct of such voyages was **Captain James Cook** (1728–79) (fig. 1–5). Between 1768 and 1779, Cook undertook three voyages. During these voyages he searched for, but did not find, Terra Australis, the supposed southern continent. Cook, armed with **John Harrison's chronometer**, made the first accurate maps of many regions in the southern oceans. This chronometer was the first timepiece capable of keeping accurate time aboard a ship at sea; this capability was necessary to accurately determine longitude (*see* Appendix IX). Cook discovered the South Georgia Islands, South Sandwich Islands, and Hawaiian Islands. Captain Cook is credited with conquering the disease scurvy by providing a nutritious diet for his men. This disease was later found to result from a deficiency of vitamin C. His ships led the way in sampling subsurface temperatures, measuring winds and currents, sounding, and collecting important data on coral reefs.

Benjamin Franklin and the Gulf Stream

An American contemporary of Captain Cook was an early contributor to greater understanding of the ocean's surface currents. **Benjamin Franklin**, deputy postmaster general for the colonies, determined that it took mail ships coming from Europe by a northerly route two weeks longer to reach the colonies than it did ships that came by a longer, more southerly

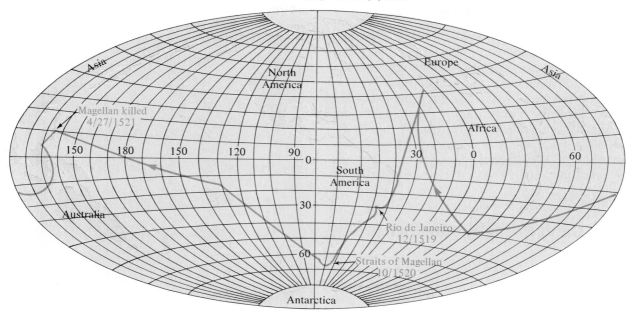

FIGURE 1–4 Voyage of Magellan.

Searching for a western route to the East Indies, Magellan culminated the Age of Discovery by beginning a voyage that ended for him with death in the Philippines. Juan Sebastian del Caño returned to Spain in the *Victoria*, one of five ships that began the voyage, to complete the first circumnavigation of the earth. Base map courtesy of National Ocean Survey.

route. His nephew **Timothy Folger**, who was a Nantucket whaling captain, had intimate knowledge of the warm **Gulf Stream** because whales were commonly found near the edges of this current. He drew Franklin a chart of the Gulf Stream (fig. 1–6). This chart was printed in 1769 and distributed to captains of the mail packets so they could shorten their voyages by avoiding the Gulf Stream, which flowed east across the North Atlantic.

FIGURE 1–5 Captain James Cook, RN (1728–79).
Courtesy U.S. Navy.

FIGURE 1–6 Chart of the Gulf Stream Compiled by Benjamin Franklin in 1777.
Courtesy of U.S. Navy.

Matthew Fontaine Maury

An even greater contribution to safe navigation and the gathering of marine data was made by **Matthew Fontaine Maury** (1806–73) (fig. 1–7). This career officer in the U.S. Navy was placed in charge of the Depot of Naval Charts and Instruments after suffering an injury early in his career. Maury helped organize the first **International Meteorological Conference** in Brussels in 1853 for the purpose of establishing uniform methods of making nautical and meteorological observations at sea. This standardization greatly increased the dependability of such data, which Maury summarized in *The Physical Geography of the Sea* (1855). Maury is often referred to as the father of oceanography.

Darwin, Coral Reefs, and Biological Evolution

Early in the nineteenth century, **Charles Darwin** (1809–82) entered the scientific scene (fig. 1–8). His interest in investigating the whole of nature led him to sign on as naturalist on the voyage of the *H.M.S. Beagle,* which lasted more than four and one-half years (fig. 1–9). Young Darwin saw many wonderful things on that voyage. He was greatly fascinated with the beautiful coral reefs. After returning to England, Darwin published, in 1843, *The Structure and Distribution of Coral Reefs,* which correctly described how the process of subsidence of oceanic islands produces the progressive change in reef form, which begins with the fringing reef and ends with an atoll. Of course, Charles Darwin is best known for his *On the Origin of Species,* published in 1859. In this publication Darwin presented his thoughts on the biological evolution of organisms through natural selection and revolutionized biological thought.

The Rosses, Edward Forbes, and Life in the Sea

The first major controversy in marine science developed around the work of **Sir John Ross**, his nephew **Sir James Clark Ross**, and **Edward Forbes**—all British naturalists/scientists. Sir John Ross took bottom measurements and collected animals from the ocean floor at depths of 1.8 km (1.1 mi) in Baffin Bay, Canada, during the years 1817–18. He used a device of his own design called a *deep-sea clamm.* His nephew undertook a similar mission in the Antarctic, where he recovered bottom dwellers to depths of as much as 7 km (4.3 mi). Sir James observed that the animals he recovered were the same species his uncle had recovered in the North Atlantic and that they were

FIGURE 1–7   Lieutenant Matthew F. Maury.
A photograph of an engraving by Lemuel S. Punderson after a daguerrotype, autographed and inscribed by Lt. Maury. Released Naval Historical Center photograph.

sensitive to temperature increase. This discovery led him to the conclusion that the waters making up the deep ocean must be of uniformly low temperature.

During this same time, Edward Forbes was studying the vertical distribution of life in the oceans. He came to the conclusion that plant life is limited to the sunlit zone near the surface. Forbes also found that the animal concentration was greater near the surface and decreased with increasing depth until only a small trace of life, if any, would remain in the deepest waters. After his death, his followers came to the conclusion that no life could exist in the deep ocean because of the high pressure and the absence of light and oxygen in this environment. Ignoring the findings of the Rosses in the high latitudes of the northern and southern oceans, they persisted in their belief that no life existed in the abyssal ocean.

THE OCEAN AS A LABORATORY

The *Challenger* Expedition

The controversy over the distribution of life in the ocean was one of the factors that helped stimulate interest in the first large-scale voyage with the ex-

FIGURE 1–8 Charles R. Darwin (1809–82).
Photo from The Bettmann Archive.

press purpose of studying the marine environment. In 1871, the Royal Society recommended that funds be raised from the British government for an expedition to investigate the following subjects:

1. physical conditions of the deep sea in the great ocean basins,
2. the chemical composition of seawater at all depths in the ocean,
3. the physical and chemical characteristics of the deposits of the sea floor and the nature of their origin,
4. the distribution of organic life at all depths in the sea, as well as on the sea floor.

A staff of six scientists under the direction of C. Wyville Thompson left England in December 1872 aboard the H.M.S. *Challenger* (fig. 1–10). A 2306-ton corvette that had been refitted to conduct scientific investigation, the *Challenger* returned in May 1876 after having traversed large portions of the Atlantic and Pacific oceans. The achievements of the *Challenger* expedition, which covered 127,500 km (79,223 mi), included 492 deep-sea soundings, 133 bottom dredges, 151 open-water trawls, and 263 serial water-

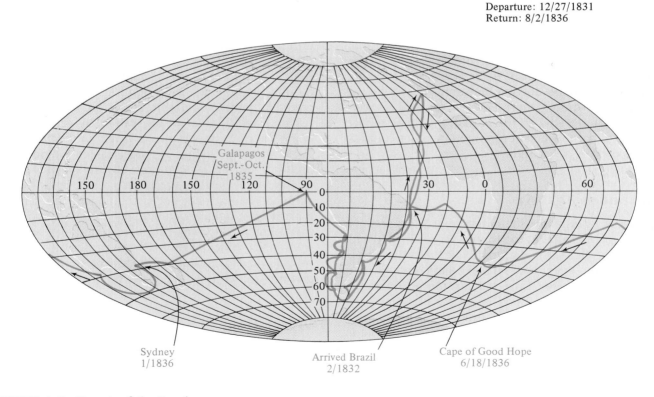

FIGURE 1–9 Voyage of the *Beagle*.
Sailing as naturalist aboard H.M.S. *Beagle,* Charles Darwin gathered the evidence that enabled him to develop his theory of biological evolution through natural selection. Base map courtesy of National Ocean Survey.

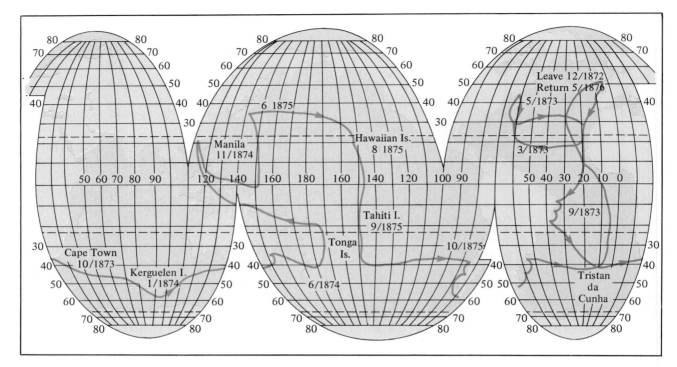

FIGURE 1–10   Voyage of H.M.S. *Challenger*, December 1872–May 1876.
The route traveled by H.M.S. *Challenger* during the first major oceanographic voyage. Base map courtesy of National Ocean Survey.

temperature observations. The more outstanding aspects of the voyage were the netting and classification of 4717 new species of marine life and the measurement of a water depth of 8185 m (26,850 ft) in the Mariana Trench. Seventy-seven samples of ocean water collected during the *Challenger* expedition were analyzed in 1884 by C. R. Dittmar. This first refined analysis of ocean water contributed greatly to our understanding of ocean salinity.

Alexander Agassiz

Alexander Agassiz, the son of the great Swiss scientist Louis Agassiz, came to the United States in 1846 with his parents (fig. 1–11). He graduated from Harvard in 1855, became superintendent of a large copper mining operation in 1866, and was a multimillionaire by 1875. Agassiz was a friend of C. Wyville Thompson and his successor, **Sir John Murray**, in the preparation of the *Challenger* reports. He had a prominent role in the processing of the *Challenger* collection in 1876. Although he contributed greatly of money and effort in the development of U.S. ocean study, Agassiz got on the wrong end of a couple of controversies—both of which he initiated. He took a position opposed to

Darwin's theory of coral reef development, and when **Carl Chun** concluded that there was an extensive midwater plankton community (tiny floating animals), Agassiz vigorously opposed this view. Even though he was not always right, Alexander Agassiz is credited by those who were at the forefront of marine research as it entered the twentieth century with being the most important individual in the field.

He was the first to use steel cables for deep-sea dredging on his 1877 voyage aboard the *Blake* and was the driving force that brought oceanography recognition as a science. Agassiz initiated the scheme that first brought oceanographic research strong financial and institutional support. Noted for his ability to organize efficient and effective research voyages, he also developed ingenious devices to improve the quantitative value of biological samples recovered.

Victor Hensen and Marine Ecology

In the late nineteenth century there was great concern about the fluctuation of fish abundance. Although grand international committees were formed to investigate the problem, one individual of great talent, whose broad goal was to understand the ecology of the oceans, learned more about how life in the oceans functions than the large groups focusing on "why do fish populations fluctuate?" The German scientist **Victor Hensen** coined the word "plankton" in 1887 and developed quantitative methods for its study. In 1889 most biologists expected to find more microscopic organisms (plankton) in the tropics than in colder waters. Hensen's Plankton Expedition aboard the *National* in that year showed the reverse was true. Later work showed that this was due to the lack of nutrients in warm surface waters as a result of thermal stratification: colder waters maintained higher nutrient levels because of storm-induced mixing. We will expand on this very important condition of marine ecology in later chapters. Hensen was the first investigator to take a broad ecological approach to the study of life in the oceans.

Polar Oceanography Begins with the Voyage of the *Fram*

One of the most unusual of voyages was that undertaken by **Fridtjof Nansen** aboard the ship *Fram*. On June 24, 1893, 13 men provisioned for five years set sail from Oslo, Norway. It was their plan to take the *Fram* to the New Siberian Islands, let it freeze into the sea ice, and have the moving ice carry them across the North Pole. They got frozen into the ice before they reached the New Siberian Islands, and the voyage was under way (fig. 1–12). They missed the North Pole by 394 km (244 mi) but survived the trip well. On August 13, 1896, the little *Fram* broke free of the ice near Spitzbergen after drifting a total of 1658 km (1028 mi) during her three-year entrapment.

The drift of the *Fram* proved that no continent existed in the Arctic Sea and that the ice that covered the polar area throughout the year was not of glacial origin; it was, instead, a freely moving pack ice accumulation that had formed directly on the ocean surface. During the voyage, the crew had found that the depth of this ocean was in excess of 3000 m (9840 ft) and that there was a rather surprising body of relatively warm water with temperature 1.5°C (35°F) between the depths of 150 and 900 m (about 500 to 3000 ft). Nansen correctly described this water as being a mass of Atlantic water that had sunk below the less saline Arctic water.

Nansen's realization of the need for more-accurate measurements of salinity and temperature led him to develop the Nansen bottle, once widely used for sampling the ocean depths. His observations of the direction of ice drift relative to the wind direction helped **V. Walfred Ekman**, a Scandinavian physicist, to develop the mathematical explanation for this phenomenon and the relationship of wind to the ocean currents it drives. Nansen, who was awarded the Nobel Peace Prize in 1922, Ekman, and other Scandinavian scientists led the way in the development of the field of physical oceanography in the early twentieth century.

FIGURE 1–11 Alexander Agassiz (1835–1910).
Contemporary illustration of Professor Agassiz with friends gathering natural history specimens at night near Newport, R.I. Photo from The Bettmann Archive.

FIGURE 1-12 Voyage of the Fram.

The course of the *Fram* after becoming frozen in ice near the New Siberian Islands, September 21, 1893. On August 13, 1896, the little ship was released by the ice off Spitsbergen.

TWENTIETH-CENTURY OCEANOGRAPHY

Early in the twentieth century, major institutions that concentrated on marine studies came into existence. The first in the United States was Scripps Institution for Biological Research, which began in 1903 at La Jolla, California. It gradually grew into **Scripps Institution of Oceanography** of the University of California, San Diego. In 1930, **Woods Hole Oceanographic Institution** was established at Woods Hole, Massachusetts, with a grant from the Rockefeller Foundation and grew rapidly. The last of the three leading U.S. institutions of marine research to come into existence was the Lamont Geological Observatory. It was founded at Torrey Cliffs, Palisades, New York, in 1949. Lamont originally concentrated on marine geology but has expanded its scope now as the **Lamont-Doherty Geological Observatory**; it investigates the full range of marine studies.

Some of the investigations and technological developments that were particularly significant in the addition of new knowledge during the early twentieth century are discussed in the next sections.

Voyage of the *Meteor*

In 1925, the German ship *Meteor* undertook a two-year voyage to refine the knowledge of circulation in the South Atlantic Ocean. This required a series of 310 stations at which 9383 pairs of temperature and salinity values were determined from the surface to the ocean floor (fig. 1-13). The ship also carried two **acoustic depth recorders**, which were used to make soundings every 2 to 3 nautical miles. The soundings gave new detail on the configuration of the ocean floor, making it easier to understand the patterns of water circulation observed.

The chief scientist on the expedition, **Georg Wüst**, published the paper that introduced the four-layer structure of circulation still recognized in the Atlantic Ocean and discussed in chapter 8.

The Expansion of Oceanography

Not until World War II did the study of the oceans begin a rapid expansion. The military supported many studies of the transmission of sound in the ocean, as well as other physical properties of ocean water important to aiding the war effort. Out of this beginning, a large ocean study program developed in the United States. It initially involved mostly scientists from Scripps, Woods Hole, and Lamont, but as the scope of the program increased, many other institutions developed oceanography schools. Although oceanographers have always been innovative in the development of devices to study the ocean, one of the first big-ticket items to grow out of this period of ex-

FIGURE 1–13 Voyage of the *Meteor*.

This 1925 cruise crisscrossed the South Atlantic, where marine data were sparse, in an effort to make possible a comprehensive study of the water masses and circulation of the Atlantic Ocean. From Wüst, G., 1935. The stratosphere of the Atlantic Ocean. Scientific results of the German American Expedition of the Research Vessel *Meteor* 1925–27, Vol. VI, sec. 1. W. J. Emery (trans. and ed.). Amerind Pub. Co., New Delhi, 1978.

panded ocean study was the *Floating Instrument Platform* (FLIP), developed in the 1960s. This strange-looking device (fig. 1–14), designed to provide a stable platform for instruments deployed in the open ocean, was just the first of many wonderful machines that have been developed in the quest to increase our knowledge of the oceans.

Large-Scale Oceanographic Research

The need for cooperative efforts was recognized in the **Deep Sea Drilling Project** (DSDP), which was being carried out under the leadership of the Scripps Institution of Oceanography and through the cooperation of a number of educational institutions that had developed oceanographic programs. In November 1963, the director of the National Science Foundation announced the establishment of a national program for sediment coring. Scripps Institution of Oceanography and three other major oceanographic institutions—the Rosenstiel School of Atmospheric and Oceanic Studies at the University of Miami, Florida; Lamont-Doherty Geological Observatory of Columbia University; and Woods Hole Oceanographic Institution in Massachusetts—united to form the Joint Oceanographic Institutions for Deep Earth Sampling (JOIDES). They were later joined by the departments of oceanography of the University of Washington, Texas A&M, the University of Hawaii, Oregon State University, and the University of Rhode Island.

A ship capable of drilling the ocean bottom while floating 6000 m (3.7 mi) above at the ocean surface

FIGURE 1–14 Floating Instrument Platform (FLIP).

FLIP, a 108-m (454-ft) Floating Instrument Platform, designed and developed by Marine Physical Laboratory at Scripps Institution of Oceanography. In the vertical position (inset) it gives scientists an extremely stable platform from which to carry out underwater acoustic and other types of oceanographic research. FLIP has no motive power of its own and must be towed to a research site in the horizontal position. Once on station, the ballast tanks are flooded and the vessel flips vertically, leaving 17 m (56 ft) of the platform above water. The work completed, water is forced from the tanks and FLIP resumes the horizontal position. Courtesy of Scripps Institution of Oceanography, University of California, San Diego.

THE *SEA BEAM/ARGO-JASON* REMOTE SYSTEM

A new remote underwater observation system has been designed to replace manned submersibles: a dual vehicle system, *Argo-Jason* is used in conjunction with the *Sea Beam* acoustical mapping system.

Currently, detailed deep-sea observation by the manned submersible *Alvin* can be carried out to depths of 3500 m (11,480 ft) using a slow and expensive procedure that has three phases.

1. Using sonar to make accurate maps of the ocean floor.
2. Towing an unmanned camera sled to obtain more detailed reconnaissance profiles of the ocean floor.
3. After identifying prime targets for detailed observation and sampling, sending down the manned submersible.

Following the lead of the offshore oil industry, which has found unmanned remote machines much more cost-effective than manned submersibles, scientists believe such systems will become increasingly more important in deep-ocean research.

While the preceding three-step procedure may enable the exploration of tens of square kilometers per month, the use of the *Sea Beam/Argo-Jason* system will allow the exploration of hundreds of square kilometers per month (fig. 1A). The system will work in the following fashion: *Sea Beam,* a complex sonar system connected to a computer and operated on a tow ship, can make instantaneous bathymetric maps of the ocean floor (fig. 1B) over which the *Argo-Jason* system will pass. The data from adjacent paths of the *Sea Beam* data can be converted by the computer into a three-dimensional model of the ocean floor. This image will aid in determining at what depth *Argo* should be towed.

Argo will be towed by a coaxial cable and will contain an array of sonar and television camera systems that will give observing scientists on the ship (and even on land) an instant wide view of the strip of ocean floor it passes over. Each television image will cover 4 acres; acoustical images, created by bouncing sound waves off the ocean floor, will include 150 acres of sea floor.

Jason (a tethered, self-propelled vehicle) will be "garaged" under *Argo.* When a particularly interesting area is identified on the *Argo* images, *Jason* can be sent to investigate further. *Jason* will have two mechanical arms and stereo color television cameras.

Sea Beam is already perfected and in use. *Argo* provided the first images of the *Titanic* since her sinking in 1912—video images during her first sea trials on September 1, 1985, *Jason*—in the prototype form of *Jason Junior*—pro-

FIGURE 1A (opposite) Sea Beam/Argo-Jason.
Sea Beam, with its computerized system, makes very accurate sonar maps that can be converted to three-dimensional models of the ocean floor. When the *Argo-Jason* system reaches the ocean floor that has been modeled, the shipboard operator will have a view of the ocean floor as if he or she were inside the remote system. Both *Argo* and *Jason* will use television eyes to guide the shipboard operator across the ocean floor. *Argo* will send out sonar signals to help determine its position on the three-dimensional model created by *Sea Beam.* Data courtesy Woods Hole Oceanographic Institution.

Position and sensor information

MOBILE UNIT

Uplink/Downlink control communication

SATELLITE

Ship, sled and rover controls

WOODS HOLE

JASON

ARGO

Support staff

SHIPBOARD

R.V. KNORR
KNORR
Seabeam 16-channel floor mapping

ARGO
Sled and rover hangar

ARGO video montage

JASON rover

JASON video view

ARGO
Forward-scanning sonar

ARGO
Side-scanning sonar

vided additional images of the *Titanic* in July 1986. Highly successful tests of this version of *Jason* provided pictures from inside the sunken ship (fig. 1C).

Dr. Robert Ballard (fig. 1D) of Woods Hole Oceanographic Institution, a driving force behind the use of manned submersibles and the more advanced remote systems such as *Argo-Jason,* says, "The ultimate challenge will be to integrate the entire *Sea Beam/Argo-Jason* system into a single operational reality."

The development of a more efficient means of deep-ocean exploration is a welcome event. However, even with the use of such remote systems that allow more rapid investigation of the ocean floor, it will be many years before we know as much about this surface as we do the far side of the moon.

FIGURE 1B The *Sea Beam/ Argo-Jason* Remote System.

A computer-produced map of the ocean floor made by the sophisticated *Sea Beam* mapping system shows portions of two small volcanic peaks up to 207 m (679 ft) high. The ocean floor mapped is about 3300 m (10,824 ft) deep at a location approximately 2900 km (1798 mi) east of Tahiti. Courtesy of Scripps Institution of Oceanography.

FIGURE 1C R.M.S. *Titanic* Is Photographed.

R.M.S. *Titanic,* unseen by human eyes since it sank 73 years ago 600 km (373 mi) south of Newfoundland, is at least visible in television and photographic images. The ship was discovered during sea trials of the new deep-sea investigative instrument *Argo* on September 1, 1985. Photos *A* and *B* were taken by a towed photographic unit, ANGUS (Acoustically Navigated Underwater Survey System), that takes high-resolution still photographs. *A,* anchor chains, capstans, and windlasses. *B,* loading cranes on the bow of the *Titanic. C,* electric winch on boat deck photographed by *Jason Jr. D, Jason Jr.* photographed from *Alvin. Jason Jr's.* "garage" on *Alvin* is in foreground. Courtesy of Woods Hole Oceanographic Institution.

FIGURE 1C *Continued*

A.

B.

C.

D.

FIGURE 1D Dr. Robert Ballard.

Dr. Robert D. Ballard in the personnel sphere of the research submersible *Alvin.* Courtesy of National Geographic Society.

had to be designed specifically to carry on this program. The *Glomar Challenger* was constructed and launched March 23, 1968 (fig. 1–15). The first leg of the Deep Sea Drilling Project commenced on August 11 of that year. The *Glomar Challenger* was 122 m (400 ft) long and 20 m (66 ft) in beam, and its displacement was 10,500 tn loaded. The vessel was a self-sustained unit capable of remaining at sea for 90 days.

Initially the oceanographic research program was financed by the U.S. government, but in 1975 it became international. Financial and scientific support is now received from West Germany, France, Japan, the United Kingdom, and the USSR. In 1983 the Deep Sea Drilling Project became the **Ocean Drilling Program** (ODP), with the broader objective of also drilling the continental margins. This new phase of operations is supervised by **Texas A&M University.** Accompanying this change in emphasis was the decommissioning of *Glomar Challenger* after a highly successful 15-year role in which it made possible a major advancement in the field of earth science—confirmation of sea-floor spreading, a process in which new ocean floor is being created along submarine mountain ranges. Following in the wake of the *Glomar Challenger* as the O.D.P. drill ship is the much larger *JOIDES Resolution,*

which conducted its first scientific cruise in January of 1985.

The **Geochemical Ocean Sections (GEOSECS)** program has been operating since 1972. This project involves sampling and analyzing water from all oceans to study ocean circulation patterns, mixing processes, and biogeochemical cycling. The GEOSECS program is discussed further in chapter 6.

Technology Invades the Deep Ocean

Putting humans into the marine environment to observe it directly has always been a dangerous business. The effects of increased pressure on body cavities and gases dissolved in body tissues limit the depth and duration of dives where the diver interacts directly with the marine environment. The record for such dives in the ocean is 457 m (1500 ft), but the researchers in this field believe humans will eventually be able to work for extended periods of time at depths of more than 600 m (1970 ft). Table 1–1 provides the dates of some significant events in the history of human progress toward deep-sea descent.

To the present, the use of **submersibles** that take humans deep into the ocean housed in chambers in which the pressure remains the same as at the ocean

A.

B.

FIGURE 1–15 Old and New *Challengers.*

A, H.M.S. *Challenger.* For propulsion, the *Challenger* relied on her sails and auxiliary steam power of 1234 hp. With a length of 61 m (200 ft) and displacement of 2300 tn, she shortened sails and used steam power to remain over a water station. Study of the samples recovered by this vessel marked the beginning of deep-sea marine geology. Illustration from the *Challenger* report, Great Britain, 1895. *B,* D/V *Glomar Challenger* could produce 8800 continuous or 10,000 intermittent hp for propulsion and for operating the drilling equipment. To remain over the drill site, she used dynamic positioning that could move the vessel in any direction. With a length of 122 m (400 ft) and displacement of 10,000 tn, the achievements of this vessel helped revolutionize geologic science. Photo courtesy of Victor S. Soleto, Deep Sea Drilling Project.

TABLE 1–1 Some Important Achievements in the History of Human Descent into the Ocean.

360 B.C.	Aristotle records the use of air trapped in kettles lowered into the sea by Greek divers in his *Problematum*.
330 B.C.	Alexander the Great descends in a glass berrylike bell during the siege of Tyre.
1620	Dutch inventor Cornelius van Drebbel tests first submarine in the Thames River. Drebbel descended to a depth of 5 m (16.5 ft) with King James I of England on board.
1690	Sir Edmond Halley descends in a lead-weighted diving bell. Air was replenished from barrels lowered into the Thames River.
1715	Englishman John Lethbridge tests a leather-covered barrel of air with viewing ports and waterproof armholes to depth of 18 m (60 ft). This design represented the first armored diving suit.
1776	Colonist David Bushnell's *Turtle* attempted the first military submarine attack on the British ship H.M.S. *Eagle*. The attempt failed, but the British moved their fleet to less accessible waters.
1800	Robert Fulton builds the submarine *Nautilus*. It was the first of several to bear this name and was powered by a hand-driven screw propeller when submerged.
1837	Augustus Sieve invents a prototype of the modern helmeted diving suit.
1913	A German company, Neufeldt and Kuhnke, manufactures an armored diving suit with articulating arms and legs.
1930	William Beebe and Otis Barton descend in a bathysphere to a depth of 923 m (3027 ft) off Bermuda.
1943	Jacques-Yves Cousteau and Emile Gagnan invent the fully automatic, compressed air aqualung.
1960	Jacques Piccard and Donald Walsh descend to a depth of 10,915 m (35,801 ft) in the bathyscaphe *Trieste*.
1962	Hannes Keller and Peter Small make an open-sea dive from a diving bell to a depth of 304 m (1000 ft) using a special gas mixture. Small dies during decompression.
1962	Albert Falco and Claude Wesly live for one week under conditions of saturation in Cousteau's Continental Shelf Station off Marseilles, France.
1964	*Sealab I*—U.S. Navy team led by George F. Bond stays 11 days at 59 m (193 ft) off Bermuda.
1966	Submersibles *Alvin, Aluminaut,* and *Cubmarine,* with the aid of unmanned remote-controlled vehicle *CURV,* locate and recover an atomic bomb from water over 800 m (2,624 ft) deep off Palomares, Spain.
1969	Armored diving suit, *Jim,* is built to operate at depths below 304 m (1000 ft)
1969–70	*Tektite I* and *II* accommodate 12 scientific teams for from 14 to 60 days at a depth of 15 m (50 ft) off the U.S. Virgin Islands.
1976	Jaques Mayol dove to 100 m (328 ft) while holding his breath for 3 minutes and 40 seconds.
1981	Diving scientists at Duke University simulate a dive to 686 m (2250 ft) in a pressure chamber.

surface (1 atmosphere) has been very successful as a means of direct observation of the deep-sea environment. In 1934, zoologist **William Beebe** descended to a depth of 923 m (3027 ft) off Bermuda in a tethered *bathysphere* to observe deep-sea life. After World War II, the Swiss scientist **Auguste Piccard** designed his *bathyscaphe* (deep boat), *Trieste,* which was bought by the U.S. Navy. This vessel, which operates much like a balloon with a gondola, submerged to the record depth of 10,915 m (35,801 ft) in the Mariana Trench off Guam on January 23, 1960 (fig. 1–16).

Two submersibles that are now widely used in deep-sea research are untethered vessels. *Alvin* can descend to a depth of 4000 m (13,120 ft), and *Sea Cliff II* can descend to a depth of 6000 m (19,680 ft).

A newly developed **Remote Underwater Manipulator (RUM III)**, about the size of a compact car, will provide deep-ocean research capability previously available only through the costly use of manned submersibles (fig. 1–17). Weighing 363 kg (800 lbs) in water, it is light enough to maneuver across the soft-

est sediments using its canvas tracks. Operated from a computerized console aboard ship, it will be towed by steel-armored coaxial cable. RUM III has a manipulator arm that can lift samples and experimental equipment weighing up to 100 kg (220 lbs), thrusters for positioning above and on the ocean floor, and television cameras. Two side-looking sonar instruments will be able to map the terrain of the ocean floor. RUM III began sea trials early in 1986 and can work at depths of 6000 m (20,000 ft).

Another remote system similar to RUM III is the **Sea Beam/Argo-Jason** system that is discussed in the special feature, "The Sea Beam/Argo-Jason Remote System," on page 12.

Remote Sensing

Remote sensing from satellites is also becoming an important tool in ocean studies. Instruments aboard satellites can measure the temperature, ice cover,

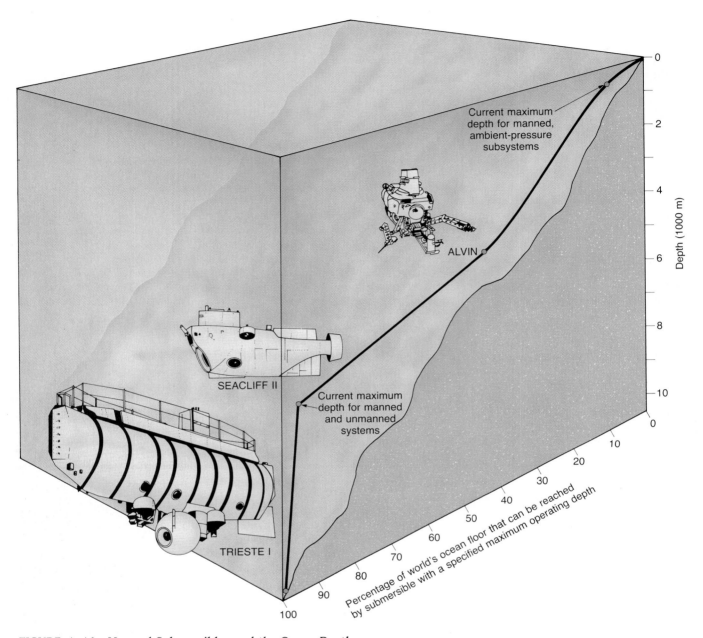

FIGURE 1–16   Manned Submersibles and the Ocean Depths.
This figure shows the percentage of ocean floor that is reachable by three submersibles of importance in the history of ocean exploration. *Alvin*, which is the first submersible to be used extensively in marine research, can explore about 44 percent of the ocean floor with its maximum operating depth of 4000 m (13,120 ft). *Sea Cliff II*, with its operations limited to depths of 6000 m
(20,000 ft) extends exploration to 96 percent of the ocean floor. The 1960 dive of *Trieste* to a depth
of 10,915 m (35,801 ft) shows it to have the capability of exploring any part of the deep ocean.
However, given its immobility, it is not practical to use it for deep-sea exploration.

color, and topography of the ocean surface as well as
give us much information about the topography of
the ocean floor. Figure 1–18 shows *Seasat A,* the first
truly oceanographic satellite, launched in 1978. Unfortunately it shorted out after three months but still

provided enough data to prove the great benefit to be
derived from remote satellite sensing of the ocean's
surface. The change in sea-surface temperature
throughout the world ocean during the lifetime of
Seasat A is shown in figure 1–19.

A.

B.

FIGURE 1-17 RUM III.
A. diagram of the RUM III showing *(A)* thrusters, *(B)* manipulator arm shown picking up a magnetometer capsule, *(C)* canvas tracks, and *(D)* counterweights to help stabilize RUM III when heavy objects are lifted. *B.* RUM III during sea trials in January 1986. Photo supplied by The Marine Physical Laboratory of the Scripps Institution of Oceanography.

FIGURE 1-18 *Seasat A.*
This satellite, which pioneered the remote sensing of the oceans, was launched July 7, 1978. It shorted out on October 10, 1978, but provided sufficient data to prove the value of satellite remote sensing to the study of the oceans. The ghostly image below *Seasat* is the H.M.S. *Challenger,* which pioneered the study of the oceans from the ocean surface in 1872. Courtesy Jet Propulsion Laboratory, NASA.

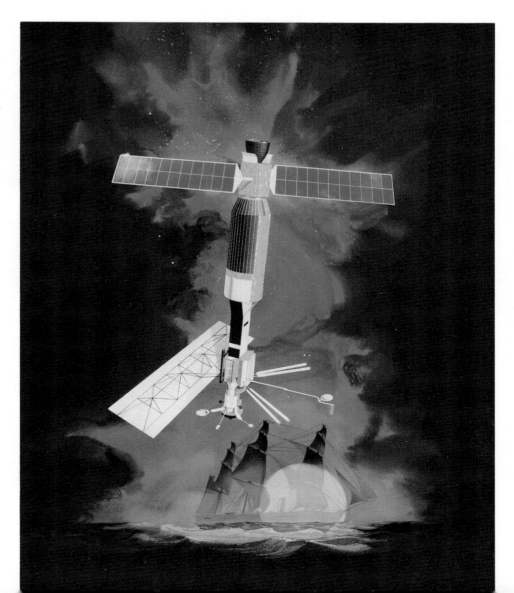

FIGURE 1–19   World Ocean Surface Temperature (SST) Change during the Lifetime of *Seasat*.

The top image shows SST averaged over a 14-day period just before *Seasat* shorted out. The middle image shows SST averaged over a 12-day period just after it was launched. The bottom image shows the change in SST that occurred between the times of the two upper images. Courtesy Jet Propulsion Laboratory, NASA.

SUMMARY

After human beings evolved, they migrated from a probable early home in Africa to all parts of the earth before the time of written history. Many of these migrations seem to have required long ocean voyages about which we know nothing.

We do know that in the western world, the Phoenicians were the first great navigators. Later the Greeks, Romans, and Arabs made significant contributions that lay hidden behind the curtain of the Middle Ages for over one thousand years. During this time, the Vikings expanded their influence across the North Atlantic and discovered North America in 995. The Age of Discovery renewed the western world's interest in exploring the unknown. It began with the voyage of Columbus in 1492 and ended in 1522 with the circumnavigation of the earth by a voyage initiated by Ferdinand Magellan.

Captain James Cook, who sailed in the late eighteenth century, was one of the first navigators to sail the oceans with the primary purpose of learning their natural history. He also made the first accurate maps, using the chronometer developed by John Harrison. Matthew Fontaine Maury was the first American to make a large-scale contribution to the study of the oceans with publication of *The Physical Geography of the Sea* in 1855. By the mid-nineteenth century, the English biologists Charles Darwin, Sir John Ross, Sir James Ross, and Edward Forbes had made significant contributions to increasing our knowledge of life in the oceans.

The voyage of H.M.S. *Challenger* from December 1872 through May 1876 represented the first major full-scale oceanographic voyage. A feeling for the significance of this exploration can be gained by contemplating the fact the 4717 new species of marine life were identified during the voyage. During the late nineteenth century, Alexander Agassiz contributed to the development of oceanographic study in the United States, and Victor Hensen developed the foundations for the study of marine ecology. An unusual voyage aboard the *Fram,* which spent three years entrapped in Arctic ice, was initiated by the Norwegian explorer Fridtjof Nansen in 1893. It was with this voyage that the first knowledge of the oceanography of the Arctic Ocean was gained. From 1925 through 1927, during a period of 25 months, the German ship *Meteor* crisscrossed the South Atlantic Ocean, obtaining the first large-scale depth record with an echo sounder. This voyage first revealed the true ruggedness of the ocean floor.

Recent oceanographic investigations have been numerous. In the United States, they were conducted early by government agencies and three pioneering oceanographic institutions—Scripps Institution of Oceanography, Woods Hole Oceanographic Institution, and Lamont-Doherty Geological Observatory. They have since been joined by many educational institutions. A major focus of this widespread involvement in marine research has been the Deep Sea Drilling Project and the subsequent Ocean Drilling Program. The Geochemical Ocean Sections (GEOSECS) program has provided a worldwide chemical analysis of ocean water.

The use of submersibles such as *Alvin* made it possible for oceanographers to see the phenomena of the deep ocean at first hand. In the future, much of the deep-ocean exploration will likely be done by use of less expensive and safer unmanned remote systems such as Remote Underwater Manipulator III and the *Sea Beam/Argo-Jason.*

The use of remote sensors mounted on airplanes and satellites is beginning to show promise of helping us rapidly expand our knowledge of the oceans.

QUESTIONS AND EXERCISES

1. Construct a time line that includes the major events of human history that have resulted in a greater understanding of our planet in general and the oceans in particular.

2. List the Mediterranean cultures that appear on your time line in the probable order of their dominance in marine navigation.

3. Describe the method used by Pytheas to determine latitude in the Northern Hemisphere.

4. State the plane geometry relationship used by Eratosthenes to calculate the circumference of the earth. Determine the circumference of the earth from the proportion stated in figure 1–2.

5. How did the Greek concept of geography provided by Herodotus differ from that of the Roman geographer Ptolemy?

6. While the Arabs dominated the Mediterranean region during the Middle Ages, what were the most significant events taking place in northern Europe?

7. List the important events from the fall of Constantinople through the Age of Discovery.

8. List some of the major achievements of Captain James Cook.

9. What was Matthew Fontaine Maury's major contribution to an increased knowledge of the oceans?

10. What are two of Charles Darwin's contributions to increasing scientific knowledge?

11. What was the controversy that developed as a result of the observations of the Rosses and Edward Forbes?

12. List some major achievements of the voyage of H.M.S. *Challenger.*

13. What were the major contributions of Alexander Agassiz and Victor Hensen to the progress of marine science?

14. What new knowledge of the Arctic polar region did the voyage of the *Fram* provide?

15. The voyage of which ship first revealed the true relief of the ocean floor?

16. List the names and locations of three pioneering oceanographic research institutions.

17. Name the major ocean research programs designed to drill into the ocean floor and study ocean circulation.

18. Discuss, giving your own perceptions, the barriers to human exploration of the deep ocean, and compare the limitations that might affect the use of tethered versus untethered submersibles; manned versus unmanned remote submersibles.

19. List the features of the oceans that might be studied remotely by use of satellite-borne sensors.

REFERENCES

Allmendinger, E. 1982. Submersibles: Past-Present-Future. *Oceanus* 25:1, 18–35.

Bailey, H. S., Jr. 1953. The voyage of the *Challenger. Scientific American* 188:5, 88–94.

Borgese, E. M., and Ginsberg, N. 1988. *Ocean Yearbook* 7. Chicago: The University of Chicago Press.

Brewer, P., ed. 1983. *Oceanography: The present and future.* New York: Springer-Verlag.

Mowat, F. 1965. *Westviking.* Boston: Atlantic-Little, Brown. Oceanography from space. 1981. *Oceanus* 24:3, 1–75.

Ryan, P. R. 1986. The *Titanic* revisited. *Oceanus* 29:3, 2–17.

Sears, M., and Merriman, D., eds. 1980. *Oceanography: The past.* New York: Springer Verlag.

The *Titanic*: Lost and found. 1985. *Oceanus* 28:4, 1–112.

SUGGESTED READING

Sea Frontiers

Baker, S. 1981. The continent that wasn't there. 27:2, 108–14.
  A history of the search for Terra Australis Incognita, the unfound southern land.

Browning, M. A. 1973. Stick charting. 19:1, 34–44.
  A description of the methods of navigation used by the Marshallese and the stick charts that served as maps.

Charlier, R. H., and Charlier, P. A. 1970. Matthew Fontaine Maury, Cyrus Field, and *The physical geography of the sea.* 16:5, 272–81.
  A biography of Maury with emphasis on his accomplishments as superintendent of the Department of Charts and Instruments. His role in laying the first transatlantic cable and a discussion of his most famous work, *The physical geography of the sea,* are included.

Denzel, J. F. 1976. Edward Forbes and the birth of marine ecology. 22:1, 16–32.
  The studies conducted by Edward Forbes in the Mediterranean Sea and seas around the British Isles and his division of living communities within the seas into eight zones are discussed.

Gaunt, A. 1975. Marking time for three hundred years. 21:6, 322–30.
  An information-rich history of Greenwich Observatory near London, England.

Heidorn, K. C. 1974. The weather luck of Christopher Columbus. 20:5, 302–11.
  A discussion of Columbus's luck in arriving in the Caribbean during hurricane season in 1492 without encountering a storm. Also described are his later encounters with storms.

Hennesey, H. 1982. Full circle for *Vema*. 28:2, 92–95. The history of the famous research vessel *Vema.*

Rice, A. L. 1972. H.M.S. *Challenger*—midwife to oceanography. 18:5, 291–305.
  A summary of the voyage of H.M.S. *Challenger* that was undertaken in December 1872 and laid the foundation for most branches of marine science.

Schuessler, R. 1984. Ferdinand Magellan: The greatest voyage of them all. 30:5, 299–307.
A brief history of the voyage initiated by Magellan to circumnavigate the globe.

Van Dover, C. 1987. *Argo* Rise: Outline of an oceanographic expedition. 33:3, 186–194.
A description of the first scientific use of the *Argo* system to survey hydrothermal vent activity on the East Pacific Rise off the coast of Mexico.

Scientific American

Bailey, H. S. 1953. The voyage of the *Challenger*. 188:5, 88–94
A summary of the accomplishments of the English oceanographic expedition of 1872.

Herbert, S. 1986. Darwin as a geologist. 254:5, 116–123.
Darwin's early interest in the developing science of geology aided him in the development of his theory of biological evolution.

Revelle, R. 1969. The ocean. 221:3, 54–65.
A discussion of how humans use the ocean and the geological creation of its present basins.

2
THE ORIGIN OF THE EARTH, ITS OCEANS, AND LIFE IN THE OCEANS

It can be a wonderful and rewarding experience to learn about the world's oceans. The wonder is heightened by adding this new knowledge to a general understanding of the place of the oceans in the cosmos. A great deal of speculation is involved in these considerations, but science has not made advances by restricting itself to that which is already known. We should, however, begin our discussion with those aspects of the universe we can readily observe and describe. Then we will discuss some theories of the origin of the universe and our planet.

THE UNIVERSE WE SEE

As astronomers looked into the sky surrounding earth, they gradually developed an understanding of earth's position in what is now called the solar system (fig. 2–1). The **solar system** is composed of the sun and its nine planets, which occupy a disc-shaped portion of the universe about 13 billion km (8 billion mi) in diameter. This structure seems enormous, but it represents a very minute portion of the total universe. Our sun is one of perhaps 100 billion billion, or 10^{20}, stars that constitute the known universe. In terms of distances with which we are familiar, the solar system is incomprehensibly isolated. The distance from the sun to the nearest star is 44 million billion km (27 million billion mi). This distance is usually expressed as 4.490 light-years. A **light-year** is a unit of measure used by astronomers to describe distances within the universe and is equal to the distance traveled by light at a speed of 300,000 km/s (186,000 mi/s) during a period of one year—almost 10 trillion (9.8×10^{12}) km.

While the planets revolve around the sun, the entire system is moving at about 280 km/s (174 mi/s) around the center of the **Milky Way**, the **galaxy** or the group of stars to which the solar system belongs. Estimates of the dimensions of the Milky Way indicate that it is a disc-shaped accumulation of stars about 100,000 light-years in diameter and from 10,000 to 15,000 light-years thick near the center, thinning gradually toward the edges. Within the Milky Way are some 100 billion (10^{11}) stars, tens of millions of which very probably possess families of planets, millions of which may be inhabited by intelligent creatures, according to the estimates of some astronomers.

All the stars you see with your naked eye belong to the Milky Way (fig. 2–2). If you have exceptional

vision and live in the Northern Hemisphere, you may see a hazy patch of light in the constellation **Andromeda** that represents another galaxy within the universe. This galaxy is some 1.5 million light-years away. It appears that galaxies are exceedingly numerous, and within a distance of a billion light-years from our galaxy there are at least 100 million galaxies large enough to be observed by our telescopes.

By observing light energy that radiates from these distant galaxies, astronomers have been able to determine that most are moving away from us.

The velocity with which the galaxies recede from our system is also proportional to their distance from it (fig. 2–3). Velocities of over 250,000 km/s (155,250 mi/s) have been calculated for the most distant galaxies as they recede from the Milky Way. This is in excess of four-fifths of the speed of light!

One may be left with the impression that all of the galaxies are being repelled by our solar system, but a more reasonable consideration would be that galaxies are moving away from one another as would fragments that result from an explosion. The universe appears to be rapidly expanding, its component galaxies moving ever farther apart. If all of these galaxies are moving away from one another in a manner similar to that of fragments created by explosion, we might ask if they all belonged to one large mass. We, of course, do not know but evidence in support of this condition continues to increase. If that is the case, the time required for the galaxies to have reached their present distribution would have been about 15 billion years.

ORIGIN OF THE EARTH

As insignificant as our planet is in the overall composition of the cosmos, it is nonetheless of interest to us to learn what we can of its origin. Although no complete understanding of the origin of the earth has yet been developed, we can outline some of the probable sequences that led to its present form. As galactic matter revolved around its center of rotation, the stars began to form as individual glowing masses due to the concentration of space particles under the force of gravity. These concentrations may have occurred in local eddies and over a vast period of time taken their present form. Many of the resulting stars may have developed assemblages of planets that revolve around them.

Such a general sequence may have occurred with the formation of our sun, and in its early stages it may have had a size at least equal to the diameter of the solar system that we observe today. As this large

Preceding page: Cape Kiwanda, Oregon. Photo by Ray Atkeson.

A.

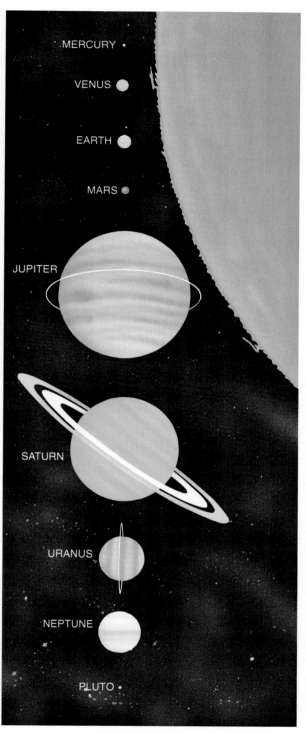

B.

FIGURE 2–1 The Solar System.

A, the orbits of the planets of the solar system are drawn to scale. *B,* relative sizes of the sun and the planets are shown. The distance scale is not maintained in *B.* Reprinted by permission from Tarbuck, E. J., and Lutgens, F. K., *Earth science,* 5th ed. (Columbus: Merrill, 1988), figs. 19.1 and 19.2.

FIGURE 2–2   The Milky Way.
The sun and its planets are located about two-thirds the distance from the center of the Milky Way to the edge. The Milky Way completes a rotation every 200 million years. To maintain its position in the Milky Way, the solar system travels around the center of the galaxy at a velocity of about 280 km/s (174 mi/s). Reprinted by permission from Tarbuck, E. J., and Lutgens, F. K., *Earth science,* 5th ed. (Columbus: Merrill, 1988), fig. 20.17.

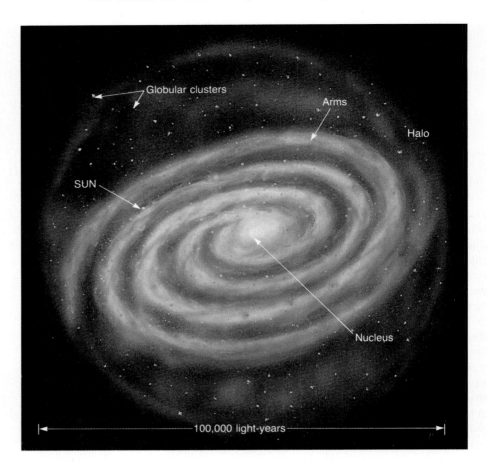

Constellation in which galaxy may be seen	Velocity km/s (mi/s)	Distance from sun (light-years)	Shift of absorption lines	
			Violet	Red
Virgo	1,200 (745)	43,000,000		
Corona Borealis	21,500 (13,351)	728,000,000		
Hydra	61,000 (37,881)	1,960,000,000		

FIGURE 2–3   Galaxies—Velocity, Distance, and Spectral Shift of Absorption Spectrum Lines.
Most galaxies are moving away from the Milky Way. This figure gives the characteristics of three such galaxies. A relationship can be seen among the velocity of the galaxy, its distance, and the observed amount of shift toward the red end of the light spectrum. Radiation from distant galaxies is absorbed so that dark lines (gaps) appear in the violet part of the spectrum. The greater the distance to the galaxies, the greater the shift of the dark lines toward the red end of the spectrum.

accumulation of gas and space dust began to contract under the force of gravity, a small percentage of it was left behind in smaller eddies with features similar to the large eddies in which the stars developed in association with the rotation of the galaxies.

This remaining material flattened into a disc that became increasingly more dense. Because of the increase in density of this disc, it became gravitationally unstable and broke into separate small clouds. These were protoplanets, which later consolidated into the present planets.

Protoearth was a huge mass, perhaps 1000 times greater in diameter than the earth today and 500 times more massive. The heavier constituents of the planet earth, as with all the other **protoplanets**, were moving toward the center to form the heavy core surrounded by lighter materials. Similarly, the satellites we observe around the planets, such as our moon, began to form.

During this early formation of the protoplanets and their satellites, the sun eventually began to "shine" as it condensed into such a large mass with such high temperature that forces were set up within its interior, creating energy through a process known as **hydrogen burning**, the first step in the fusion reaction. When temperatures reach tens of millions of degrees, hydrogen atoms, in the presence of carbon, are converted to helium atoms, and huge amounts of energy are released. In addition to the light that is so obviously emitted by the sun, there is an intense emission of ionized, electrically charged, particles. In the early stages of the development of our solar system, this emission of ionized particles, called the **solar wind**, served to blow away the nebular gas that remained from the formation of the four inner planets and their satellites.

As the protoplanets were being warmed by the solar radiation, their "atmospheres" of hydrogen and helium began to boil away. The combination of emission of ionized solar particles and the internal warming of the planets resulted in a drastic decrease in the size of the planets closer to the sun. Since the gas envelope that surrounded most of these planets was possibly composed mostly of hydrogen and helium (we may consider this to be the composition of the earth's first atmosphere), the energy that was available easily allowed atoms of small size to escape from the gravitational fields of the planets. Much of the atmosphere lost by the inner "terrestrial" planets was swept up by Jupiter, Saturn, and Uranus, the huge planets beyond Mars. As the protoplanets continued to contract, the heat produced within the planet as a product of spontaneous disintegration of atoms (radioactivity) became more intense.

The evidence indicates that the earth may have became molten, allowing the heavier components to concentrate in the **core**. The lighter of the materials segregated according to their densities in concentric spheres around this core. Surrounding the core, containing abundant iron and nickel, is a zone of material composed of heavy iron and magnesium silicates—the **mantle**. Subsequently, a thin **crust** formed at the surface.

On the basis of data gained from the observation of seismic waves passed through the earth's interior, scientists have constructed a model of the earth that shows a solid **inner core** with a radius of approximately 1216 km (755 mi) composed of iron and nickel and other heavy elements. The solid core gradually changes to a liquid **outer core** with a thickness of approximately 2270 km (1409 mi). Surrounding the core is a 2885-km (1790-mi) thick mantle. The mantle is composed of silicates of iron and magnesium and is overlain by a very thin crust made up of silicates of iron, magnesium, and aluminum. The crust ranges in thickness from 4 to 60 km (2.5 to 37 mi) and is separated from the mantle by a distinct density change called the **Mohorovicic Discontinuity (Moho)**, after the Yugoslavian seismologist who discovered its existence (fig. 2–4). It should be noted that the internal structure of the earth appears to be much more complex than might be gathered from this brief discussion of the broad features of this structure.

ORIGIN OF THE ATMOSPHERE AND OCEANS

The solidification of the earth's crust marks the beginning of **geologic history**, and evidence indicates that this event occurred about 4.6 billion years ago (fig. 2–5). At that time the earth had lost all but a small fraction of its original gas envelope because of the heat radiating from it as it cooled.

The Atmosphere Forms

Where did the material to make up the present atmosphere and ocean come from? Although the earth's surface had become relatively cool by this stage in its development, there was still much volcanic activity according to the geologic record. As volcanic gases bubbled up from the earth's upper mantle, they formed a new atmosphere as soon as the earth had cooled enough to retain one. The atmosphere produced by this **outgassing** was far different

FIGURE 2–4 The Earth's Structure.

The cross section shows the three major subdivisions of the earth: core, mantle, and crust. The inner core, outer core, and mantle are drawn to scale, but the thickness of the crust is exaggerated about five times. Reprinted by permission from Tarbuck, E. J., and Lutgens, F. K., *The earth: An introduction to physical geology,* 3d ed. (Columbus: Merrill, 1990), fig. 1.11.

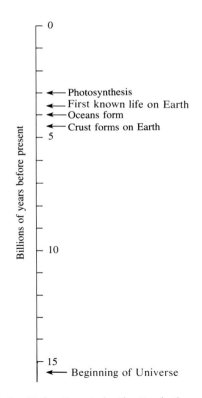

FIGURE 2–5 Major Events in the Evolution of the Earth.

from the one that the earth had lost and from the one it presently has. It contained little free oxygen and was composed of large percentages of **carbon dioxide** and **water vapor.** As the earth cooled to the point at which it could retain these gases, a vast envelope of clouds surrounded the planet, causing the earth to reflect about 60% of all the sunlight that was directed toward it. Much of the energy it absorbed from the sun remained in the thick envelope of gases.

Water vapor released from the mantle of the earth was captured in the thick envelope and was forced to remain there as a vapor until sufficient cooling of the earth's crust would allow it to condense and begin to accumulate as a liquid. Possibly by 4 billion years ago (b.a.) cooling of the crust allowed the water vapor to condense and fall under the force of gravity to the earth's surface. Finally, with further decrease in the surface temperature, permanent accumulation of water began in depressions on the earth's irregular surface (fig. 2–6).

The Oceans Evolve

These accumulations were the initial oceans on planet earth, and it is assumed that this water has been permanently in existence since its formation. The primary crystalline rocks that made the earth's crust were weathered and eroded. These processes were aided by the high concentration of **carbonic acid** (H_2CO_3) resulting from the combination of carbon dioxide and water molecules in the atmosphere. The inorganic compounds of which the rocks were composed were dissolved in the waters that fell relent-

FIGURE 2–6 Formation of the Earth's Oceans.

A, widespread volcanic activity released H_2O and smaller quantities of CO_2, Cl_2, N_2, and H_2, which produced a water vapor atmosphere that also contained carbon dioxide (CO_2), methane (CH_4), and ammonia (NH_3). *B,* as the earth cooled, the water vapor (1) condensed and (2) fell to the earth's surface. There it accumulated to (3) form the oceans.

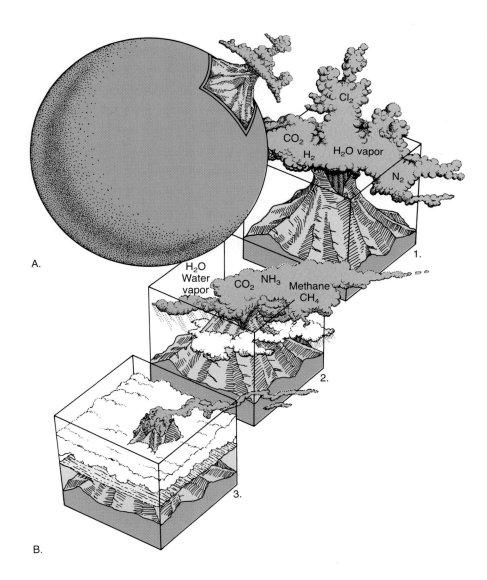

lessly upon the rock surfaces and were carried into the newly forming oceans.

The process of cooling the earth's surface, which allowed the water vapor to condense and accumulate to form the earth's first oceans, was accelerated by water's property of high heat capacity and latent heat of vaporization. The heat energy that prevented the water from accumulating on the earth's surface was being absorbed by that massive cloud of water vapor and carbon dioxide that surrounded the earth's solid surface. As the molecules of water absorbed heat at the earth's surface and moved through the cloud to its outer edges, where they were frozen into ice crystals, large quantities of heat were released into outer space.

THE SOURCE

We have obviously been proceeding under the assumption that the materials that make up the atmosphere and the ocean came from within the earth's interior and were brought to the surface by the process of volcanism. Is this a valid assumption? Let us consider the water of the oceans. The material that is brought to the earth's surface as lava must have its source in the lower crust or the upper mantle of the earth. The mantle of the earth has been calculated to have a volume of 10^{27} cm^3 and a density that averages 4.5 g/cm^3. This amounts to a total mass for the mantle of 4.5×10^{27} g. The water of the oceans has a mass of 1.4×10^{24} g. Therefore, the mantle must have lost 0.031% of its mass as water to have produced the earth's oceans. (*See* calculations in appendix VII.)

For **meteorites** that contain silicates and are believed to resemble the composition of the earth's mantle, we can obtain a water content from which we can estimate a water content for the mantle. The average water content of these stony meteorites is about 0.5%, which is about 16 times the loss required from the mantle to account for the present oceans. If our assumption is correct, the mantle would contain more than enough water to create the oceans, and therefore the mantle could very adequately serve as a source for the waters of the earth's oceans.

An Alternate Theory—Ocean Water from Space

Although some oceanographers can't believe he is serious, **Louis Frank**, one of the world's top space scientists, believes the ocean's water may have come to

earth as cometlike balls of ice about 9 m (30 ft) in diameter. Frank believes they enter the earth's atmosphere at a rate of 20 per minute.

The theory evolved from the need to explain some features on images of the earth's atmosphere produced from data obtained from an ultraviolet photometer aboard the *Dynamic Explorer I,* a polar-orbiting satellite that reaches an altitude of 23,350 km (14,500 mi). The image is of **dayglow** created by the absorption and reradiation of solar energy by atomic oxygen 290 km (180 mi) above sea level. The dayglow, which makes the earth look like a yellow ball, had a number of short-lived dark spots (fig. 2–7). These dark "holes" in the atmosphere were 48 km (30 mi) across and existed for up to 3 minutes. Because of the absorption spectra of these dark spots, they can be explained only by the cometlike balls of ice breaking up and vaporizing 1600–3200 km (1000–2000 mi) above the earth's surface.

At the observed rate of occurrence, the earth would receive 0.0025 mm (0.0001 in.) of water per year. Four billion years of such bombardment would give us enough water to fill the oceans to their present volume.

THE CHEMICAL BALANCE SHEET

Because chemical weathering of the primary rocks that composed the original earth's crust was made possible by the appearance of water, the process of **chemical weathering** needs to be considered in some detail. As a result of chemical weathering, the elements that were contained in the **primary crystalline rock** were freed, dissolved in the water, and carried off to accumulate in the ocean. As the water carried these dissolved **elements** and **compounds** into the ocean, it also carried a great load of particulate material that had been weathered and eroded from the crystalline rocks and deposited this material as sediment. The components of the primary crystalline rock that were freed by chemical weathering must be found dissolved in that water in the ocean or as components of the earth's atmosphere. Some of the components will remain chemically bound with the sediments that have been carried into the sea by the rivers (*see* fig. 2–8).

The estimates that have been made for these various volumes will not be exact, but the estimate for the ocean is probably one of the most accurate, since the ocean volume is relatively well known and its composition is comparatively uniform because it is

FIGURE 2–7 Image of Earth's Dayglow.

Recorded by ultraviolet photometer aboard the satellite *Dynamic Explorer I.* The dark spot or "hole," thought to be caused by the vaporization of cometlike balls of ice, is shown in the inset. The circular emission at the top is the northern auroral oval. Courtesy L. A. Frank, University of Iowa.

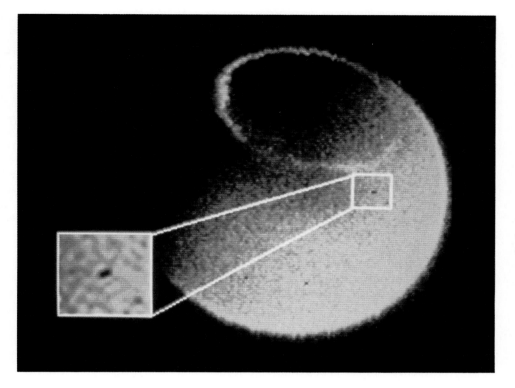

well mixed. The estimates are less accurate for crystalline rock, which is quite heterogeneous, and for the sediments, which are of various types. Nevertheless, geochemists have drawn up balance sheets for the various elements and those calculated by different individuals working on the problem agree fairly well. For those elements which are most common in the crystalline rocks, the geochemical equations balance well within the limits set by the accuracy of the estimates.

Excess Volatiles

There are some components that do not balance in any of the attempts that have been made to do so. These elements are much more abundant in the atmosphere, the ocean, and the sediments than can be accounted for by the volume of crystalline rocks which have been estimated to have undergone chemical weathering. All of these substances that are found in excessive amounts compared to the products of chemical weathering are volatile at or slightly above the average temperatures found at the earth's surface and are therefore called **excess volatiles.** Most abundant of the gaseous substances that make up the excess volatiles are water vapor and carbon dioxide; also important are quantities of chlorine, nitrogen, sulfur, hydrogen, and fluorine. (*See* table 2–1.)

The total mass of crystalline rock components freed by chemical weathering must be found as components of the atmosphere, oceans, or sediments.

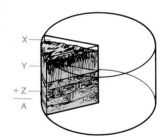

When considering one element represented by the dark wedge in the "pie" to the left, (A), the sum of the masses of that element found in the atmosphere, (x), oceans, (y), and sediments, (z), should equal the mass weathered from the crystalline rock, A = x + y + z

FIGURE 2–8 Geochemical Balances.

TABLE 2–1   Excess Volatiles. The volatile elements and compounds that are found in the atmosphere, ocean, and sedimentary rocks in far greater amounts than have been made available by chemical weathering of primary crystalline rocks are called *excess volatiles*. They are believed to have been released from the interior of the earth by volcanic activity. Note the absence of free oxygen.

	Mass of Elements or Compounds in 10^{20} g					
	H_2O	C as CO_2	Cl	N	S	H, B, Br, Ar, F, etc.
Found in:						
Atmosphere and ocean	14,600	1.5	276	39	13	1.7
Buried in sedimentary rocks	2,100	920	30	4	15	15
Total	16,700	921.5	306	43	28	16.7
Supplied by:						
Weathering of crystalline rocks	130	11	5	0.6	6	3.5
Excess volatiles (unaccounted for by rock weathering)	16,570	910.5	301	42.4	22	13.2

Courtesy of the Geological Society of America. After Rubey, 1951

Volcanic activity is presently producing these gases and venting them into the earth's atmosphere. A great percentage of the gases that surface as a result of volcanic activity may simply be recycled volatiles that have been dissolved in groundwater and carried down from the earth's surface, but it has been determined that a small percentage of the total volatiles is always composed of new gases that have for the first time been released from the crystalline rock of the earth's crust. When molten magma cools, solid minerals are formed, and some of the gaseous constituents of the magma are freed. These gases, freed during the crystallization process, represent the new material coming to the earth's surface from the mantle.

We have geological evidence in the form of fossilized algae, assumed to have grown in a marine environment, that the oceans have existed on the earth's surface for well over 3 billion years. If we consider the rate of discharge of excess volatiles by volcanoes to be equal to the average rate of production of excess volatiles throughout the last 3 billion years, we find that enough water vapor would have been produced to fill the oceans to more than 100 times their present volume. We must remember that much of this water that comes to the surface is water that is being recycled; it does not represent new water released from the crystallization of magma. Only 1% of this water would require such an origin to account for the present volume of the earth's oceans from this source.

DEVELOPMENT OF THE OCEANS AND THEIR BASINS

When the earth began to form its solid crust about 4.6 billion years ago, the mass of the oceans was zero, while at present it is 1.4×10^{24} g. We do not know what the rate was at which volatiles escaped the earth's mantle via volcanic activity and hydrothermal springs throughout this period of 4.6 billion years. If we assume it to be at least equal to the present rate, we can visualize the formation of the earth's oceans as a gradual process.

Does continual outgassing mean that the oceans are gradually covering more and more of the earth's surface? Not necessarily, since the area that the oceans cover is directly determined by the volume that the basin in which they form is able to accept. The crust, which we now describe as continental, may not have been present during the initial solidification of the earth's crust and may well have formed gradually as the oceans themselves were formed. Since the volume of the ocean basins is directly related to the difference in thickness between the oceanic and the continental crusts, and if we assume that the continental material has gradually accumulated, we can see that the capacity of the oceans' basins could have gradually increased along with the volume of water that was being produced over this period of time. Thus, it is quite possible that the oceans have had a distribution very similar in total surface area to that of the present oceans for a long period of geologic time and that the major change in the oceans' character has been an increase in depth. Throughout the geologic time during which the oceans and continents were increasing their volumes, the ocean basins were getting deeper to accommodate the increasing volume of water that was being transferred from the underlying mantle to the earth's surface. In fact, the area covered by the oceans may be decreasing as a result of continued growth of the continents. This process will be discussed further in chapter 3.

Considering the history of ocean salinity, we might certainly ask if the oceans have possessed a relatively uniform salinity throughout their history or if they are getting more or less saline. By far the most important component of salinity is the chlorine ion, Cl⁻, which is produced by volcanic action, the same process that produces the water vapor forming the oceans. So our considerations boil down to one question: Has the relative amount of water vapor to chlorine ion production remained constant throughout geologic time, or has it varied? Although it is impossible to develop a definitive answer to this question, we have no indication that there has been any major fluctuation in this ratio throughout geologic time. We must then consider, on the basis of the present evidence, that the oceans' salinity has been relatively constant, while their volume and the volume of the crystalline rock that makes up the continental crust has increased throughout the 4.6 billion years since the earth's crust first began to form.

A CHEMICAL BACKGROUND

Before going further in our discussion, we need to define the following terms. An **element** is one of a number of substances each of which is comprised entirely of like particles, atoms, that cannot be broken into smaller particles by chemical means. The **atom** is the smallest particle of an element that can combine with similar particles of other elements to produce compounds. The periodic table of elements, appendix III, lists the elements and describes their atoms. A **compound** is a substance containing two or more elements combined in fixed proportions. A **molecule** is the smallest particle of an element or compound that, in the free state, retains the characteristics of the substance.

As an illustration of these terms, consider **Sir Humphrey Davey's** use of electrolysis to break the compound water into its component elements, hydrogen and oxygen. Atoms of the elements hydrogen (H) and oxygen (O) combine in the proportion 2 to 1, respectively, to produce molecules of the compound water (H_2O). As an electric current is passed through the water, the molecules dissociate into hydrogen atoms that collect near the cathode (negatively charged electrode) and oxygen atoms that collect near the anode (positively charged electrode). Here they combine to form the diatomic gaseous molecules of the elements hydrogen (H_2) and oxygen (O_2). Because there are twice as many hydrogen atoms as oxygen atoms in a given volume of water, twice as many molecules of hydrogen gas (H_2) as oxygen gas (O_2) are formed. Further, because under identical conditions of temperature and pressure the volume of gas is proportional to the number of gas particles (molecules) present, two volumes of hydrogen gas are produced for each volume of oxygen.

A Look at the Atom

Building upon earlier discoveries, the Danish physicist **Niels Bohr** (1884–1962) developed his theory of the atom as a small "solar system" in which a positively charged nucleus takes the place of the sun and the "planets" that orbit around it are represented by negatively charged electrons. Although this theory has since been diagrammatically altered, it is still commonly used to demonstrate the arrangement of electrons and nuclear particles in the atom.

Bohr's earliest concern was with the atom of hydrogen, which he considered to consist of a single positively charged **proton** in its nucleus orbited by a single negatively charged **electron**. Since the mass of the electron is only about 1/1840 the mass of the proton, it will be considered negligible in our coming discussion of atomic masses.

According to Bohr, the number of protons, or units of positive charge in the nucleus, coincides with the **atomic number** of the element. Thus hydrogen, with an atomic number of 1, has a **nucleus** containing a single proton. Helium, the next heavier element, having an atomic number of 2, contains two nuclear protons, and so forth. The atomic number will also indicate the number of electrons in a normal atom of any element, because this number is equal to the number of protons in the nucleus (fig. 2–9).

An **isotope** is an atom of an element that has a different **atomic mass** than other atoms of the same element. As we will later see in some detail, chemical properties of atoms are determined by the electron arrangement that surrounds the nucleus. This arrangement, in turn, is determined by the number of protons in the nucleus. Because there are isotopes that have different atomic masses but identical chemical characteristics, we suspect that the proton is not the only nuclear particle that can influence the atomic mass of an atom. There must be a nuclear particle that has no effect on the atom's chemical properties, in that it does not affect the electron structure surrounding the nucleus (fig. 2–10).

Nuclear research has discovered the additional particle, the **neutron**, which was postulated by **Rutherford** in 1920 and was first detected by his associate **James Chadwick** in 1932. It has a mass very similar to that of the proton but no electrical charge. This characteristic of being electrically neutral has made it one of the particles most utilized by physicists. These scientists continue to explore the atomic nu-

FIGURE 2–9 Bohr-Stoner Orbital Models for Atoms.

Each atom is composed of a positively charged nucleus with negatively charged electrons around it. The nucleus, which occupies very little space, contains most of the mass of the atom. The atomic number is equal to the number of protons (positively charged nuclear particles) in the nucleus of an atom of an element. Note that the first shell holds only two electrons. Other shells can hold no more than eight electrons when they are in the outer position.

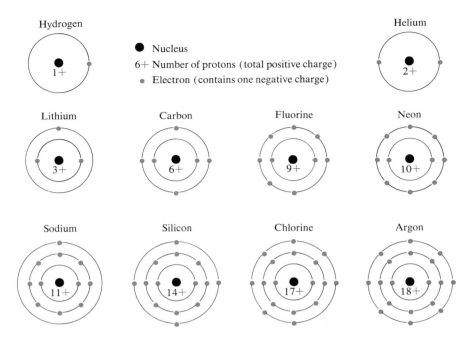

cleus and to make new discoveries of nuclear particles. We will not consider these particles, as knowledge of them is not necessary for our understanding of the chemical nature of atoms.

We should summarize here that an atom of a given element can be changed to an ion of that ele-

ment by adding or taking away one or more electrons. It is changed to an isotope of the element by adding or taking away one or more neutrons. Adding or taking away one or more protons will change the atom to an atom of a different element.

We can now divide the atom into two parts—the **nucleus,** which contains neutrons and protons, and the **electron cloud** surrounding the nucleus that is involved in chemical reactions. It is not possible to determine the precise location of electrons in this cloud at any instant, but it is possible to estimate the most probable position of an electron in this cloud. We will picture these regions in which we would more likely find particular electrons as concentric spheres, or shells, that surround the nucleus (*see* fig. 2–9).

Chemical Bonds

In considering the chemical reactions in which atoms are involved, we will be concerned primarily with the distribution of electrons in the outer shell. With some exceptions, when atoms combine to form compounds, they combine in one of two ways:

1. **covalent bonding**—sharing electrons in the outer shells of the atoms
2. **ionic bonding**—some of the atoms lose electrons from their outer shell, and others gain electrons in their outer shell

This latter process produces an **ion,** an electrically charged atom that no longer has the properties of a neutral atom of the element it represents. A positively charged ion, a **cation,** is produced by the loss of elec-

FIGURE 2–10 Neutrons as Nuclear Particles.

Isotopes are atoms of an element that have different atomic masses. The hydrogen atom ($_1H^1$) accounts for 99.98% of the hydrogen atoms on earth. It contains one nuclear particle, a proton. Deuterium ($_1H^2$) contains one neutron and one proton in its nucleus and combines with oxygen to form "heavy water" with a molecular mass of 20. Tritium ($_1H^3$) is a very rare radioactive isotope of hydrogen that has a nucleus containing one proton and two neutrons.

trons from the outer shell, the positive charge being equal to the number of electrons lost. A negatively charged ion, an **anion**, is produced by the gain of electrons in the outer shell of an atom, and its charge is equal to the number of electrons gained.

In a statement of great oversimplification, let us say it is the "desire" of an individual atom to assume the outer-shell electron content of the inert gases such as helium (2), neon (8), and argon (8) that makes it chemically reactive (see fig. 2–9). Normally, if an atom can assume this desired configuration by either sharing one or two electrons with an atom of another element or by losing or gaining one or two electrons, the elements are highly reactive. By contrast, the elements are less reactive if three or four electrons must be shared, gained, or lost to achieve the desired configuration.

We can describe the combining power of atoms of different elements by the concept of **valence**. Valence can be defined as the number of hydrogen atoms with which an atom of a given element can combine. Valence can be either ionic or covalent. Elements with lower valences of one or two combine chemically in a more highly reactive manner than do those with higher valences of three or four, or in certain instances, more than four. Although those elements with higher valences do not react as violently as low-valence elements, they have a greater combining power and can gather about them larger numbers of atoms of other elements than can the atoms with lower valence values.

An example of a covalent bond is the sharing of electrons by hydrogen and oxygen atoms in the water molecule. In the formation of this molecule, both hydrogen atoms and the oxygen atom assume the inert gas configuration they seek by sharing electrons (fig. 2–11A).

An example of a compound formed by ionic bonding is seen in the metallic salt sodium chloride (fig. 2–11B). In the formation of this compound, the sodium atom, which has one electron in its outer shell, loses this electron, forming a sodium ion with a positive electrical charge of one. The chlorine atom, which contains seven electrons in its outer shell, completes its shell by gaining an electron and becoming a chloride ion with a negative charge of one. These two ions are held in close proximity by an electrostatic attraction between the two ions of equal and opposite charge.

LIFE FORMS IN THE OCEANS

We did not include free oxygen in the excess volatiles that were produced at the earth's surface by volcanic

activity. If such a gas was produced, it is certain that it could not remain free in the earth's atmosphere because we find iron that occurs in two forms in volcanic rock—ferrous iron (FeO) and ferric iron (Fe_2O_3).

The ferrous iron atom displays a valence of two, while the iron atom in a molecule of ferric iron has a valence of three. The ferric state is more highly oxidized (contains more oxygen per iron atom) than the ferrous. The iron in volcanic rock is generally observed to be ferrous. Any oxygen that surfaced during volcanic activity would be used up converting ferrous iron to the ferric state, thus preventing free oxygen from being found as an excess volatile. As a result of this condition, we may consider that the earth's early atmosphere did not contain this gas that makes up almost 21% of our present atmosphere.

$$8 FeO + 2O_2 \longrightarrow 4 Fe_2O_3$$
Ferrous iron + Oxygen ⟶ Ferric iron

Ozone Formation

We depend on the oxygen in our atmosphere for protection from the ultraviolet radiation that is constantly bombarding our planet. Much of the energy represented by this ultraviolet radiation is exhausted in our upper atmosphere, converting the free oxygen (O_2) into ozone (O_3), thus allowing very little of the ultraviolet energy to reach the surface of the earth.

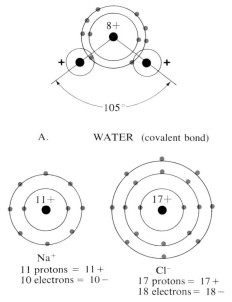

A. WATER (covalent bond)

Na⁺
11 protons = 11 +
10 electrons = 10 −

Cl⁻
17 protons = 17 +
18 electrons = 18 −

B. SODIUM CHLORIDE (ionic bond)

FIGURE 2–11 Covalent and Ionic Bonds.
Notice the contrast in the outer shells of the ions of sodium (Na⁺) and chlorine (Cl⁻) with the atoms of these elements in figure 2–9.

Free oxygen molecules (O_2) absorb ultraviolet light, and the energy taken up exceeds the bond strength. The bond breaks to give two oxygen atoms.

$$O_2 + \text{Ultraviolet light} \longrightarrow O + O$$

The oxygen atoms produced are very reactive and quickly combine with an oxygen molecule to produce a molecule of ozone.

$$O_2 + O \longrightarrow O_3$$

The First "Organic" Substances

Of course, if oxygen was missing from the primitive atmosphere, the ultraviolet radiation would have readily penetrated the atmosphere and reached the surface of the young oceans. These oceans contained the gases methane (CH_4) and ammonia (NH_3), products of the chemical combination of hydrogen (H_2), nitrogen (N_2), and carbon (C) made available by the great outgassing of carbon dioxide (CO_2).

Laboratory experiments have shown that exposing a mixture of hydrogen, carbon dioxide, methane, ammonia, and water to ultraviolet light and electrical spark will produce a large assortment of organic molecules (fig. 2–12). The exposure of the mixture to ultraviolet radiation causes photosynthesis, carried on in the absence of chlorophyll. **Photosynthesis** is an endothermic reaction by which energy from the sun is stored in the products of the photosynthetic reaction—the organic molecules. The term **organic molecules** is used in reference to molecules of substances such as sugars, amino acids, and fats produced by organisms. The production of organic substance in the shallow oceans that developed early in the earth's history must have resulted in a vast amount of "organic" material, as do the laboratory experiments, but this material did not yet represent life on the planet earth.

Carbon: The Organizer of Life

Let us now leave these general considerations for a moment and consider more specifically the element that seems to be the most important in allowing the build-up of the organic compounds we have just mentioned. This element is carbon. We even refer to the study of the chemistry of carbon compounds as **organic chemistry**. In studying the chemistry of carbon, we are constantly concerned with molecules which are several times larger than those with which

FIGURE 2–12 Synthesis of Organic Molecules.
The apparatus used by Stanley L. Miller in the 1952 experiment that resulted in the synthesis of the basic components of life, amino acids. A mixture of water vapor, methane, ammonia, and hydrogen were subjected to an electrical spark that provided the energy for synthesis. This mixture is thought to resemble the composition of the atmosphere and oceans that existed when "organic" molecules first formed on earth.

the inorganic chemist is commonly concerned. The greater percentage of inorganic molecules contains fewer than 10 atoms, while in organic chemistry, molecules that exceed 100 atoms per molecule are not uncommon.

This fact arises out of the **"combining power"** of the carbon atom, which has four electrons in its outer electron shell. It has a capacity of accepting atoms of other elements with a total valence of four. It is tetravalent.

To demonstrate how this ability gives carbon an unusually high combining power, let us compare the carbon atom with other common atoms with valences less than four. We will consider valence to be the number of hydrogen atoms with which an atom can combine. If we use chlorine, which occurs as a gas, we can see that a single molecule of chlorine combines with a single molecule of hydrogen to produce two molecules of **hydrogen chloride, HCl.**

$$Cl_2 + H_2 \longrightarrow 2HCl \qquad\qquad Cl-H$$

Oxygen, which also occurs as a gas, has a valence of two. It combines with two molecules of hydrogen in the formation of **water, H_2O.**

$$O_2 + 2H_2 \longrightarrow 2H_2O \qquad\qquad O\begin{smallmatrix}\diagup H\\ \diagdown H\end{smallmatrix}$$

Another gas common in our atmosphere, nitrogen, is trivalent. One molecule of nitrogen will combine with three molecules of hydrogen to produce the gas ammonia, NH_3.

$$N_2 + 3H_2 \longrightarrow 2NH_3$$

We return to the carbon atom, with a valence of four, and find that it will combine with two molecules of hydrogen to produce the gas methane, CH_4.

$$C + 2H_2 \longrightarrow CH_4$$

We can readily see that the higher the valence of an atom of an element, the greater the complexity of the molecules it can form. If we consider that carbon has a valence of four compared to the valence of one for chlorine, then carbon must have a much greater combining power than chlorine or any other element with a valence of one. Consider chlorine as it combines with the metal sodium, which also has a valence of one.

$$2Na + Cl_2 \longrightarrow 2NaCl$$

We can see that these two elements can combine to form only one compound, sodium chloride (common table salt).

By contrast, methane, an important constituent of the earth's early atmosphere, represents only the simplest of the group of carbon compounds called **hydrocarbons**—those containing only carbon and hydrogen atoms. Somewhat more complex are the **carbohydrates**, which are composed of carbon and water molecules. Sugars and starches belong to this group. Even more complex carbon compounds that are known as **lipids (fats)** and **proteins** may or may not have been present in those early oceans as products of inorganic photosynthesis.

Acidity and Alkalinity

As water vapor and carbon dioxide were spewed into the earth's atmosphere by the volcanic activity on the earth, they combined to form carbonic acid, H_2CO_3.

$$H_2O + CO_2 \longrightarrow H_2CO_3$$

This acid is very weak, but it is important because it falls with water to the earth's surface and aids the water in dissolving the mineral substances locked within the rocks that make up the landmasses. These mineral substances were later incorporated into living organisms. Carbonic acid also played a very important role in keeping even the very earliest oceans slightly alkaline, as life evolved to fit this medium.

Before discussing this condition further we will describe what is meant by acid. An **acid** is a compound containing hydrogen that, when dissolved in water, sets hydrogen ions (H^+) free and thus increases the number of such ions per unit of water mass. A strong acid is one whose molecules readily release hydrogen ions when they are part of a dilute solution. The term **alkaline** refers to the presence of compounds in water that dissociate to increase the concentration of hydroxide ions (OH^-). A strong **base** is a compound whose molecules readily release hydroxide ions in dilute solutions.

Both hydrogen and hydroxide ions are present at all times in water as the water molecules dissociate and reform.

$$H_2O \longrightarrow H^+ + OH^-$$

If they are produced in pure water by the water molecules themselves breaking up, they are always found in equal concentrations. The water is neutral; there is no excess of either the hydrogen or the hydroxide ion.

If we add a given amount of hydrochloric acid (HCl) to a volume of water, the water will become acidic because it will receive a high concentration of hydrogen ions as a result of the dissociation of the HCl molecules. If we add a strong base (alkaline substance) such as sodium hydroxide (NaOH) instead, there will be an excess of hydroxide ions (OH^-) as the NaOH molecules break up.

The acidity or alkalinity of a solution is indicated by a **pH scale** ranging from 0 to 14. A solution with a pH of 7 is neutral, such as pure water. The lower the pH of the solution falls below 7, the stronger its acidity, indicating an increasing concentration of hydrogen ions. The higher the pH rises above 7, the stronger the alkalinity because of an increasing content of hydroxide ions.

pH scale

0	7	14
Acid	Neutral	Alkaline

The introduction of carbon dioxide into the oceans as carbonic acid serves to maintain the slightly alkaline conditions of the oceans (fig. 2–13). When the carbon dioxide molecule combines with a water molecule, two of the electrons in the oxygen atom of the water molecule are shared with two of the electrons of the carbon atom from the carbon dioxide molecule. This situation leaves the hydrogen nuclei weakly attached to the water molecule as it becomes part of the H_2CO_3 molecule. One of these hydrogen nuclei may be lost and become a hydrogen ion (H^+), transforming the carbonic acid molecule into a negatively charged **bicarbonate ion, HCO_3^-**.

Should the bicarbonate ion lose its hydrogen ion, it then is transformed into a double-charged negative **carbonate ion, CO_3^{--}**. The arrows in figure 2–13 indicate the reaction can go in both directions. If for any reason the pH of the ocean rises, the reaction proceeds to the right, releasing H^+ and lowering the pH. Should the pH of the ocean drop, the reaction proceeds to the left and raises the pH by removing H^+. At the normal pH of ocean water, about 8.1, most of the carbon dioxide in it occurs as the bicarbonate ion (HCO_3^-). This process is termed **buffering**.

A solution must be slightly alkaline before complex organic combinations can form within it. Solutions that are either highly acid or highly alkaline lack the necessary tolerance for the vast range of bonding arrangements needed to form organic compounds. Carbon dioxide combined with water as carbonic acid did much to prepare the environment in which life was to begin. Most important of all is the fact that the carbon dioxide molecule contains that atom which became the organizer of these complex organic building blocks—the carbon atom.

We do not know precisely how the organic material took on the characteristics that allowed it to become living substance, nor do we know when this occurred. Through some process, however, the organic material became chemically self-reproductive, developed the ability to actively seek light, and began to grow toward a characteristic size and shape dictated by forces from within. These living entities were composed mostly of a substance with a chemical composition not greatly unlike that of the water in which they originated. We call this material **protoplasm**. Once organisms formed in the ocean, they could be protected from the lethal ultraviolet radiation by sinking only a few millimeters beneath the ocean surface.

Plants and Animals Evolve

These very earliest forms of life must have been **heterotrophic** organisms, those that depend on an external food supply. That food supply was certainly abundant in the form of molecules of the same nonliving organic material from which the first living single-celled organisms originated. The **autotrophs**, those organisms that do not depend on an external food supply but manufacture their own, eventually evolved. These first autotrophic forms may well have been organisms similar to some of our present-day bacteria. They may have been able to oxidize various inorganic compounds that would release energy that could be used to produce their own food internally. This process is called **chemosynthesis**. At some later date, the more complex single-celled plants probably evolved. Having developed a green pigment called **chlorophyll**, plants could capture the sun's energy and produce through **photosynthesis** their own food supply from the carbon dioxide and water that surrounded them.

The autotrophs had developed security in the form of a built-in mechanism to assure them of a food supply. Other forms, not so endowed, were at a considerable disadvantage because they had to search for this necessity of life. To meet this need, animals eventually emerged with their inherent characteristic of mobility and began to develop an awareness or a consciousness that gave them a much higher place in the biological community than that possessed by the plants.

Plants and animals developed in a beautifully balanced environment where the waste products of one filled the vital needs of the other; we can see this expressed chemically by looking at the processes of photosynthesis and respiration. In the photosynthetic process, energy is being captured and stored in the form of organic compounds. This is an **endothermic** chemical reaction. Respiration is just the opposite—an **exothermic** reaction. It is the reaction by which the energy stored in the foods produced by photosyn-

FIGURE 2–13 Maintenance of Ocean Alkalinity.

When the pH of ocean water drops (7.5), the reaction moves to the left to remove H^+ from the water, reversing the drop in pH. If the pH of ocean water rises (9.0), the reaction moves to the right to release H^+ into the water, bringing the pH back down.

$$\underset{\text{Water}}{H_2O} + \underset{\text{Carbon dioxide}}{CO_2} \longleftrightarrow \underset{\text{Carbonic acid}}{H_2CO_3} \longleftrightarrow \underset{\text{Bicarbonate ion}}{HCO_3^- + H^+} \longleftrightarrow \underset{\text{Carbonate ion}}{CO_3^{--} + H^+ H^+}$$

$$\underset{7.5}{\text{Low pH}} \longleftrightarrow \underset{9.0}{\text{High pH}}$$

thesis is extracted by the plants and animals to carry on their life processes.

Photosynthesis ⟶ *energy is stored (endothermic)*

$6H_2O + 6CO_2 + Energy \longleftrightarrow C_6H_{12}O_6 + 6O_2$
Water + Carbon + Energy   Sugar + Oxygen
       dioxide

Respiration ⟵ *energy is released (exothermic)*

The first evidence we have in the fossil record of life on earth is in the form of algae and bacterialike forms that have been preserved in rocks more than three billion years old. Every living organism that inhabits the earth today is the result of a process of **natural selection** that has been going on since these early times and by which various life forms (species) have been able to inhabit increasingly numerous niches within the earth environment. As these diverse life forms developed in an attempt to exist and adapt to various environments, they also modified the environments in which they lived.

As the plants emerged from the oceans to inhabit the terrestrial environment, they changed the appearance of the landscape from the harsh, bleak panorama we may envision as being typical of a lunar surface to the soft green hues that cover much of the land surface of the earth today. Other changes were manifested in the ocean itself as vast quantities of the hard parts of dead organisms began to accumulate on the bottom. These accumulations of calcium carbonate ($CaCO_3$) and silica (SiO_2) formed the **limestone** and **diatomaceous** deposits that we now see exposed on the continents as evidence of a marine origin. During the course of geologic time, vast amounts of carbon dioxide have been stored in the calcium carbonate deposits on the ocean floor, which have formed through chemical precipitation and the accumulation of the hard parts secreted by marine organisms such as corals, molluscs, and foraminifers.

Probably the most important environmental modification was that which involved the atmosphere. It was this change that made it possible for animals to develop. A byproduct of photosynthesis in the early oceans was the free oxygen that makes up 21% of our present atmosphere. Carbon dioxide, which at one time must have made up a large portion of our atmosphere, was removed by photosynthesis to produce the present concentration of 0.03% of the earth's present atmosphere. These changes in the atmosphere must have had a very significant effect on the climates that have evolved throughout the earth's history (fig. 2–14).

We also see, in **petroleum** and **coal** deposits, the remains of plant and animal life that were not com-

pletely oxidized but were buried in a reducing (oxygen-free) environment that allowed their energy to remain stored. In some instances, especially in the formation of coal, the dead plant matter accumulated in such quantities that the reducing environment was produced by the decay of the organic matter using up the available oxygen. These deposits provide us with over 80% of the energy used for domestic and industrial purposes today. So we, as animals, are not only dependent upon the present productivity of plants to supply the energy required by our life processes, but we also depend very heavily on the energy stored by plants during the geologic past.

SUMMARY

Our **solar system**, consisting of the sun and nine planets, belongs to the **galaxy** of stars we call the **Milky Way**. All the stars we see with the unaided eye belong to this galaxy; yet there are some 100 million galaxies thought to exist in the universe. Most seem to be moving away from us as though all were set in motion by a single large explosion, and the farther away a galaxy is, the faster it seems to be moving away.

The galaxies are thought to be accumulations of the debris from this explosion in large eddies, inside of which are smaller eddies where concentration of matter has produced stars and their associated planets. **Stars** result from accumulation of masses large enough to produce sufficient internal temperatures and pressures to set off processes causing them to "shine" as they give off energy at high rates. The **planets** were smaller, so they did not reach this state.

Protoearth was composed mostly of **hydrogen** and **helium**, but as it condensed and heated up, these elements were driven off into space. There are indications the earth became molten and then developed an atmosphere rich in **water vapor** and **carbon dioxide** produced by volcanic activity. **Methane** and **ammonia** were also present in significant quantities. As the earth's surface cooled sufficiently, the water vapor condensed and accumulated in depressions on the surface to give the earth its first oceans.

Study of atomic bonding has revealed two predominant types. Water molecules typically display **covalent bonds**, while **ionic bonds** are characteristic of metallic salts.

Life is thought to have begun in these oceans. As **ultraviolet radiation** fell on the oceans with their dissolved carbon dioxide, methane, and ammonia, inorganic molecules may have combined to produce carbon-containing molecules that are now formed

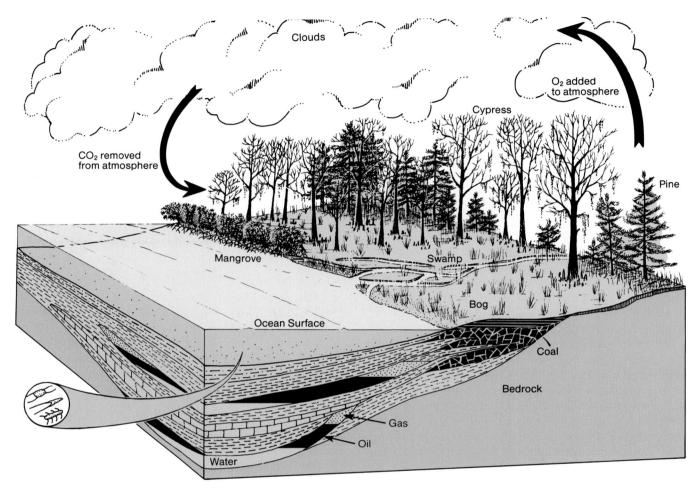

FIGURE 2–14   The Effect of the Evolution of Plants on the Earth Environment.
Subsequent to the origin of single-celled plants in the ocean, the earth's atmosphere began to be enriched with O_2 and its CO_2 content was decreased as a result of photosynthesis. As these plants and the microscopic animals that fed off them died and sank to the ocean floor, these remains were incorporated into the marine sediments. Some of the remains were buried before they could be completely decomposed and were ultimately converted to petroleum (oil and gas). When plants moved onto the continents, they evolved into the various forms that support the animals that subsequently evolved there. In the low-lying, highly productive wetlands, plant remains accumulated in the sediment to be converted to peat and ultimately coal.

naturally on earth only by organisms. As these chance combinations became more complex, some of the combinations that sank deep enough into the ocean to be protected from the lethal ultraviolet rays evolved into **bacteria**. Eventually some included **chlorophyll** in their makeup, and **plants** were the result. The **photosynthesis** of plants extracted carbon dioxide from the atmosphere and produced the first free **oxygen** at the earth's surface. This produced an oxygen-rich atmosphere in which **animals** as we know them could survive. Eventually, both plants and animals evolved into forms that could survive on the stark environment of the continents, producing the lush continental environments we enjoy today.

QUESTIONS AND EXERCISES

1. Construct a graph with velocity in thousands of km/s plotted along the vertical axis from 0 to 100 and light-years of distance plotted along the horizontal axis from 0 to 2 billion. On this base, plot the values given for the constellations Virgo, Corona Borealis, and Hydra in figure 2–3. Establish a line by connecting the three points. How much farther from us is a galaxy for each increase in velocity of ten thousand km/s?

2. Why is it theorized that the observed motions of galaxies were initiated by an explosion?

3. During its history, the earth may have possessed three distinctly different envelopes of gases, or atmospheres. Describe the atmosphere of protoearth, of the earth when the oceans first formed, and of the earth at present.

4. In what three components of the earth's makeup must we find the total mass of crystalline rock components that were eroded from the primary crystalline rocks of the earth's surface?

5. List the five most abundant excess volatiles. Why are they called excess volatiles?

6. Why does the fact that new water is continually being released to the atmosphere by volcanic activity not necessarily mean the oceans will progressively cover an increasing percentage of the earth's surface?

7. Locate sulfur (S) on the periodic table of the elements, appendix III. List its atomic number, atomic mass, and valence. Draw a diagram similar to those used for the atoms in figure 2–9 showing the number of protons, neutrons, and electrons in the most common isotype of the element.

8. List the two types of chemical bonds discussed in the text, describe the manner in which the outer shell of electrons is filled, and give an example of a compound containing each type of bond.

9. Why do we believe that free oxygen (O_2) was not released into the atmosphere by volcanic activity?

10. How does the presence of oxygen (O_2) in our atmosphere help reduce the amount of ultraviolet radiation that reaches the earth's surface?

11. How did the photosynthesis that produced the first organic molecules in the early oceans differ from plant photosynthesis? How does chemosynthesis differ from photosynthesis?

12. When we say carbon (C) has a greater combining power than oxygen (O), we are referring to chemical valence. How is valence related to combining power?

13. How are the pH scale, acidity-alkalinity, and relative H^+ ion and OH^- ion concentrations related?

14. How does the chain of reactions made possible by dissolving CO_2 in water help maintain a slightly alkaline condition in the oceans?

15. Describe some basic characteristics of living things.

16. Discuss photosynthesis and respiration, and explain their relationship to the chemical processes of endothermic and exothermic reactions.

17. As plants evolved on the earth, great changes in the earth's environment were produced. Describe some of these major changes caused by plants.

REFERENCES

Davis, K. S., and Day, J. A. 1961. *Water: The mirror of science*. Garden City, N.Y.: Doubleday.

Frank, L. A.; Sigwarth, J. B.; and Craven, J. D. 1986. On the influx of small comets into the earth's upper atmosphere—observations and interpretations. *Geophysical Research Letters*. 13:4, 303–310.

Glaessner, M. F. 1984. *The dawn of animal life: A biohistorical study*. Cambridge: Cambridge University Press.

Gregor, B. C., Garrels, R. M., Mackenzie, F. T., and Maynard, J. B., eds. 1988. *Chemical cycles in the evolution of the earth*. New York: John Wiley and Sons, Inc.

Holland, H. D. 1984. *The chemical evolution of the atmosphere and oceans*. Princeton, N.J. Princeton University Press.

Rubey, W. W. 1951. Geologic history of sea water—an attempt to state the problem. *Geological Society of America Bulletin*. 62:1110–19.

Strom, K. M., and Strom, S. E. 1982. Galactic evolution: A survey of recent progress. *Science*. 216:4546, 571–80.

SUGGESTED READING

Scientific American

Barghoorn, E. S. 1971. The oldest fossils. 224:5, 30–54.
The evidence that bacteria and algae were present on earth over 3 billion years ago is discussed along with evolutionary considerations.

Barrow, J. D., and Silk, J. 1980. The structure of the early universe. 242:4, 118–28.
Although the universe is inhomogeneous on the small scale of the solar system or a galaxy, it is very homogeneous on the scale of the universe as a whole.

Clarke, B. 1975. Causes of biological diversity. 223:2, 50–61.
The diversity that is observed within species and its relationship to natural selection are pursued. The physiological and social implications of the data are discussed.

Frieden, E. 1972. Chemical elements of life. 227:1, 52–64.
The roles of the twenty-four elements known to be essential to life are discussed, including some background on how they may have been selected from the physical environment in which life evolved.

Gott, J. R., and Gunn, J. E. 1976. Will the universe expand forever? 234:3, 62–79.
Based on data regarding the recession of galaxies, average density of matter, and chemical elements, the authors suggest expansion will be reversed.

Stebbins, G. L., and Ayala, F. J. 1985. The evolution of Darwinism. 253:1, 72–85. New advances in molecular biology and new interpretations of the fossil record add to the knowledge of evolution.

Vidal, G., 1984. The oldest eukaryotic cells. 250:2, 48–57.
Cells with a nucleus appear to have evolved as marine plankters 1.4 billion years ago.

Wilson, A. C. 1985. The molecular basis of evolution. 253:4, 164–175.
Mutations within the genes of organisms play an important role in evolution at the organismal level.

Woese, C. R. 1981. Archaebacteria. 244:6, 98–125.
A discussion of two classes of bacteria, archaebacteria and eubacteria.

3
GLOBAL PLATE TECTONICS

The fact that we live on a dynamic earth on which movement is the rule rather than the exception has long been accepted by geologists. Geologic processes responsible for the ever-changing landscape of the continents are believed to require long periods of geologic time to build such features as mountains. However, many geologists were not prepared until recent years to accept a dynamic definition of the earth that was of a much broader scope—the movement of continents. When geologists first spoke of the movement of continental masses across the earth's surface, they referred to this phenomenon as **continental drift.**

Marine geologists who studied the ocean floor and in the 1960s developed theories to explain why the continents are moving relative to one another used the term **sea-floor spreading.** The name for this theory is derived from evidence indicating that new oceanic crust and rigid upper mantle material is being added along a series of mountain ranges on the ocean floor and moving at right angles away from these mountainous ridges. The continents, which are thought to float on this denser material, are carried along on a moving layer, or plate, of denser rock.

A third term used to describe this entire process is **global plate tectonics.** Study of the process has indicated that there are a number of major plates into which this rock sphere, or **lithosphere,** can be divided. The interaction of these plates as they move builds the structural features of the earth's crust that we observe. Thus, **tectonics** refers to the building of the earth's crustal structure and is derived from Greek *tektonikos,* which means *to construct.*

The idea that the continents may be moving across the surface of the earth is not new. Such a possibility was suggested early in the nineteenth century, and the first major theory attempting to explain the movement was presented in 1912. With such a long history of awareness of the possibility of such movement, why has the acceptance of theories about the process been so long in coming?

DEVELOPMENT OF THE THEORY

Alfred Wegener (fig. 3–1) is considered by most scientists to be the pioneer of the modern continental drift theory. This German scientist originally was drawn to the concept in an attempt to explain the an-

Preceding page: The deep-sea research submersible *Alvin.* Photo by Rod Catanach, courtesy of Woods Hole Oceanographic Institution.

cient climates that were recorded in the rocks deposited in ocean basins and on landmasses of the past. Wegener's theory was published in 1912 and met a great deal of resistance from the scientific community.

Wegener considered that about 200 million years ago all of the continental mass of the earth was one large continent, **Pangaea.** About 180 million years ago the continent began to break up, and the various continental masses we know today started to drift toward their present positions. Because it was impossible to present evidence that would support a mechanism for such transportation, the theory did not receive wide acceptance.

Although many Southern Hemisphere geologists had accepted continental drift as a geological reality, it was not until the 1950s that geologists of the Northern Hemisphere began to give it serious attention. The impetus for the renewed attention arose from the study of the record of the earth's ancient magnetism recorded in rocks. In 1960, **Harry Hess** (fig. 3–2) developed the theory of mantle convection as a mechanism to drive sea-floor spreading. It was strongly supported by the work of geophysicist **Frederick Vine** in 1963. Vine showed that the parallel bands of paleomagnetic anomalies on either side of oceanic mountain ranges resulted from the combined effects of reversals of the polarity of the earth's magnetic field and the generation of new ocean floor along the axes of the mountain ranges. The details of these considerations will be discussed later in this chapter.

On the basis of this study of the earth's ancient magnetic fields, convincing arguments could be made for the fact that the continents had moved relative to one another. As study continued, more evidence was gathered to support this movement, and additional data suggested the mechanism by which the movement might have taken place. The confirmation of the spreading process was achieved during the early stages of the Deep Sea Drilling Project in 1968.

ISOSTASY

Because the relative densities of the earth's crust and mantle contribute to the mechanism of global plate tectonics, we will provide a brief discussion of that relationship.

The crust can be divided into two distinct types. The basaltic oceanic crust has an average density of about 3.0 g/cm^3. The continents are composed of a lighter granitic crustal material ranging in density

FIGURE 3–1 Alfred Wegener.
Shown waiting out the 1912–13 Arctic winter during an expedition to Greenland, where he made a 1200-km traverse across the widest part of the island's ice cap. Courtesy of Bildarchiv Preussischer Kulturbesitz, West Berlin.

FIGURE 3–2 Harry Hess.
The theory of sea-floor spreading was developed in 1960 by this geologist from Princeton University. Photo courtesy of Seeley G. Mudd Manuscript Library, Princeton University.

from 2.67 g/cm^3 near the surface to 2.8 g/cm^3 deep beneath the continental mountain ranges. The continental crust averages about 35 km (22 mi) in thickness but may reach a maximum thickness of 60 km (37 mi) beneath the highest mountain ranges.

The granitic continental blocks and basaltic oceanic crust "float" on the denser mantle beneath. This concept of crustal flotation, which may be compared with the flotation of ice on water, is called **isostasy** (fig. 3–3).

If we take a regularly shaped prism of ice with a density of 0.91 and float it on water with a density of 1.0, the ice will sink into the water until it displaces water equal in mass to the ice. When this displacement, called *buoyancy,* occurs, 91% of the ice mass will be submerged. Therefore, 91% of the length of the prism of ice will be beneath the surface of the water. Similarly, about 55% of the mass of the continents is found submerged in the mantle. Actually, we would expect about 91% of the continents to be submerged because they are, on the average, about 91% of the density of the upper mantle. However, where high heat flow in the mantle heats and reduces the density of the mantle beneath the continents, this low-density mantle provides *isostatic compensation.* In such situations as exist beneath the southwestern United States and eastern Africa, the low-density mantle material provides the buoyancy that would normally be provided by the deep root of continental crust.

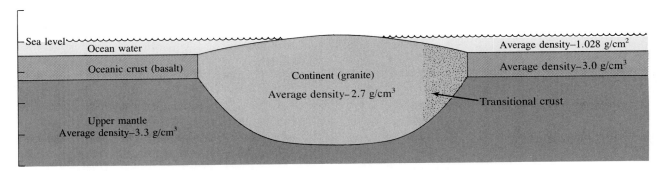

FIGURE 3–3 Isostasy.

Isostasy is a state of equilibrium reached by different components of the earth's crust and the mantle. So that the lithosphere (the rock sphere of the earth's structure), which contains the continental and oceanic crustal units as well as the upper mantle, may maintain a uniform average density, continental crustal units that extend up to 14 km (8.7 mi) above the floor of ocean basins must send deep roots into the mantle. Along the margins of continents that have been separated from other continental masses by sea-floor spreading, the separation process heats and converts granite to a composition intermediate between oceanic and continent crusts in composition and density. It is called transitional crust and is present on all Atlantic Ocean continental margins.

SUPPORTING EVIDENCE

Continental Jigsaw

Before we look at the bulk of scientific evidence that has been gathered in relation to plate tectonics, we should consider how well the continents do fit together. Attempts were made by many of the early investigators to arrange the continents in a manner that would achieve a reasonable fit and support their data. Most of these attempts used the existing shoreline as the margin of the continent.

Sir Edward Bullard, an English geophysicist, constructed a computer fit of all the continents in 1965. The arrangement that produced minimum overlaps and gaps was found to be formed by continents outlined by the 2000-m (6560-ft) depth contour. This contour represents a depth that is approximately halfway down the continental slope—a better marker for the edge of a continent than the coastline. (fig. 3–4).

CONTINENTAL GEOLOGY

To test the fit of the continents, we may compare the rocks along the margins of the two continents that were at one time united. We need to identify rocks of the same type and age on the continents along their common margin. Identification is not easy in some areas, since during the millions of years since their separation, younger rocks have been deposited and

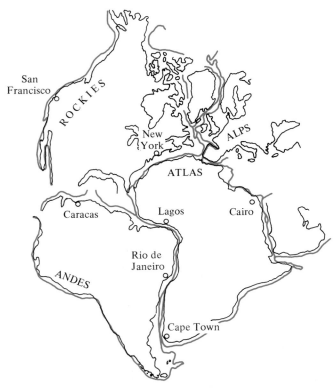

FIGURE 3–4   Computer Fit of Continents by Sir Edward Bullard.

Sir Edward Bullard's fit of the continents was attempted in 1965 after a convincing fit pattern had been achieved in 1958 by the Australian geologist S. Warren Carey without the aid of computers. Bullard's fit was in complete agreement with that of Carey. From *Continental drift* by Don. and Maureen Tarling, © 1971 by G. Bell & Sons, Ltd. Reprinted by permission of Doubleday & Co., Inc.

cover those rocks that might hold the key to the past history of the continents. However, there are many areas where such rocks are available for observation, and in these areas we can compare the ages by the use of (1) **fossils**, the remains of ancient organisms preserved in rocks, and (2) **radiometric dating** of the rocks.

The Fossil Record

The use of fossils became an important dating device in the early nineteenth century when it was realized that a particular layer of sedimentary rock could contain the remains of organisms that, as a group, were unique in time. Rocks laid down earlier and those laid down later would have a distinctly different assemblage of plant and animal fossils. Once these assemblages are recognized, a geologist can tell if rocks exposed in one area are younger or older than in another. The unique character of these assemblages is the result of evolution throughout geologic time. Appendix IV shows the general pattern of this evolution and the geologic time scale.

Relative Dating Dating by assemblages of organisms is referred to as **relative dating**. One can tell only whether or not an assemblage is the same age or younger or older than another, but not its actual age in years (fig. 3–5A). Such assemblages have been found abundantly in sedimentary rocks of the last 600 million years. Note that this is but a little over one-eighth of the time of earth's existence, which is 4.6 billion years. Therefore, this method is not effective in comparing the ages of the continental margins in those rare instances where they are composed of rocks older than 600 million years.

Absolute Dating Most of the rocks we find on the continents contain small amounts of radioactive ele-

FIGURE 3–5 Relative and Absolute Dating.
A, fossil assemblages used in relative dating can be found in outcrops L and R. Assemblages 3, 4, and 5 may be matched, telling us the rocks in segments of the outcrops formed at the same time. We could not, however, tell how many years ago this formation occurred when fossil evidence was the only dating tool we had. The concept of superposition states that younger sediments are laid down on older deposits. *B,* Outcrops L and R contain the radioisotope uranium 235. It has a half-life of 710 million years and decays to lead 207. Dating with radioisotopes is possible for many types of rocks but is more broadly possible with igneous rocks like the basalt and granite that make up most of the oceanic and continental components, respectively, of the earth's crust. At outcrop L the ratio of U^{235} to Pb^{207} of 1:1 means that half of the U^{235} atoms have decayed to Pb^{207}, so the rocks are one half-life of U^{235} old, or 710 million years. In outcrop R, the ratio of U^{235} to Pb^{207} is 3:1. This means that one-fourth of the original U^{235} atoms have decayed to Pb^{207}, so the rock is one-half of a half-life old, or 355 million years.

A.
RELATIVE DATING

B.
ABSOLUTE DATING

ments such as uranium, thorium, and potassium, which break down into atoms of other elements. Each radioactive **isotope** (atom of an element with an atomic weight different than that of other atoms of the element) has a specific **half-life**, the time it takes for one-half of the atoms present in a sample to decay to atoms of some other element. By comparing the quantities of the radioactive isotope with the quantities of decay products in given rocks and by knowing the rate of decay (half-life), the age of the rock may be determined within 2–3% of its actual age. Although in very old rocks this error can be quite significant, this method of age determination is a very powerful tool and is the first method that we have had of giving the actual age of rocks in years. We refer to such dating as **absolute dating**. As compared to the relative dating possible with the use of fossils, absolute dating not only tells us which rocks are younger or older, but also their actual age in years (fig. 3–5B).

Ancient Life and Climates

The fossil record of plants and animals in the sedimentary rocks can tell us much about the environments of the past. For instance, it is relatively simple to determine whether an organism lived in the ocean or on land. This distinction can normally be made also by the characteristics of the sedimentary rock itself. However, finer differences in environmental conditions may also be determined.

Upon examining the distribution of some fossil assemblages and other characteristics of rocks that give clues to the climate under which they were formed, we find some that could not have formed under the present climatic conditions. These anomalous locations of fossil assemblages and rock types may be explained by assuming that either (1) the factors that control climatic belts today are different than the factors that controlled climates in the past or (2) the factors controlling climate throughout geologic time have not changed. The second assumption, which seems more logical, lets us conclude that these assemblages and rock types representing sediments of the past must have been carried to their present positions by the movement of the continental masses.

The dominant factor controlling climatic distribution is latitudinal position on the rotating earth. Assuming that the earth's axis of rotation has not changed significantly throughout its history, we may conclude that latitudinal belts have possessed climatic characteristics that have not changed greatly during the evolution of life on earth.

Laurasia and Gondwanaland Modern reef-building corals are known to exist in an environment where the water is clear and shallow and its temperature does not fall below 18°C (64.2°F). Although we cannot be sure such conditions have been required throughout the long history of coral evolution, we can assume that the conditions required were similar.

Exactly the same species of coral are found in rocks 350 million years old in western Europe and eastern North America, as well as throughout the Alps and Himalayas, implying that they were in close proximity at that time. Other evidence indicates that from 350 to 250 million years ago a major ocean, the **Tethys Sea**, separated the large continent **Laurasia**, composed of North America, Europe, and Asia, to the north, from the large continent **Gondwanaland**, composed of South America, Africa, India, Australia, and Antarctica, to the south (fig. 3–6). The record also indicates that the two supercontinents periodically came into direct or close contact across the Tethys Sea. Both continents were joined in forming one landmass, **Pangaea**, about 200 million years ago (fig. 3–7). Pangaea was surrounded by the ocean **Panthalassa**, the ancient precursor of the Pacific Ocean.

Fossils from sediments that were laid down within a land environment aid in determining the latitudinal positions of the two large continents. From 350 to 285 million years ago there existed two distinct floral assemblages on the two supercontinents. The Laurasian floral assemblage included many species of tropical plants that were incorporated into the sediment to form the extensive coal beds mined throughout the eastern United States and Europe. These tropical flora imply a low-latitude position for Laurasia during that time. Throughout Gondwanaland, there existed an assemblage represented by a few species of plants that are thought to have grown in a cold climate, presumably at a high southern latitude. Supporting this belief are indications of glaciation in South America, Africa, India, and Australia, which at that time must have been very near the southern polar region.

Continental Paleomagnetism

Although we cannot reconstruct the relative positions of continents prior to 200 million years ago with as much accuracy, there is enough evidence from that time to the present to give us some idea of the paths followed by the continents after the breakup of Pangaea. The first clues to such movements came from the study of the magnetism of continental rocks.

Most igneous rocks contain some particles of magnetite (an iron oxide, Fe_3O_4) that become magnetized by and align themselves with the earth's magnetic

FIGURE 3–6 Fossil and Glacial Evidence Supporting Continental Drift.

A, a fossil coral assemblage that flourished in the Tethys Sea and continental floral assemblages A and B that developed to the north and south of the Tethys suggest Pangaea was divided into Laurasia to the north and Gondwanaland to the south during the time span covering 280 to 350 million years ago. The *B* flora are thought to be cold-climate forms. This belief is supported by evidence of glaciation. *B*, Laurasia and Gondwanaland may have been in the relative positions shown here during the interval 250 to 350 million years ago. The Tethys coral assemblage could have been deposited in the narrow Tethys Sea while the tropical Laurasian floral assemblage and the polar Gondwanaland floral assemblage were developing on the continents.

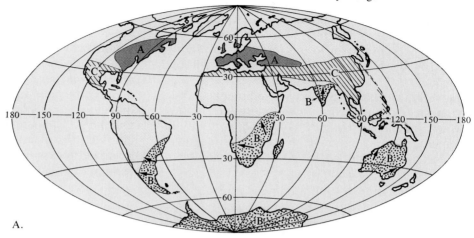

GLACIATION
250 – 300 million years ago
Arrows show direction of glacial movement.

A TROPICAL LAURASIAN FLORA
280–350 million years ago

B POLAR GONDWANALAND FLORA
280–350 million years ago

C TETHYS CORAL ASSEMBLAGE
350 million years ago

A.

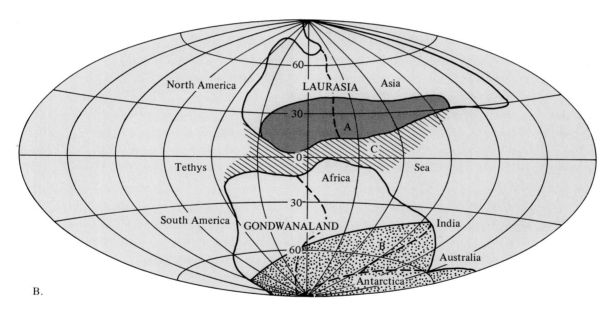

B.

FIGURE 3–7 Pangaea.

About 200 million years ago, near the beginning of the Mezozoic era, all the present continents were joined together into one continent—Pangaea. The ocean that surrounded Pangaea was called Panthalassa.

field at the time of the rocks' formation. Volcanic lavas such as basalt are high in magnetite content and solidify from molten material that possesses temperatures in excess of 1000°C (1832°F). As they cool below 600°C (1112°F), the **Curie point**, the magnetite particles become oriented in the direction of the earth's magnetic field, recording that field relative to the rock location permanently. If the earth's magnetic field changes subsequent to the formation of the igneous rock, the alignment of these particles will not be affected.

Magnetite is also deposited in sediments. While the deposit is in the form of a sediment surrounded by water, magnetite particles have an opportunity to align themselves once more with the earth's magnetic field. This alignment is preserved when the sediment is buried.

Although a number of rock types may be used for the study of the earth's paleomagnetism, the basaltic lavas and other igneous rocks high in magnetite content are best for such studies. The magnetite particles act as small compass needles, as shown in figure 3–8. They not only point in a north-south direction but also point into the earth at an angle relative to the earth's surface called the **magnetic dip**, or **inclination**, which is related to latitude. At the equator, the "needle" will not dip at all, but will lie horizontally. It will point straight into the earth at the magnetic north pole and straight out of the earth at the magnetic south pole. At points between the equator and the pole the angle of dip increases with increased latitude. It is this dip that is retained in magnetically polarized rocks. By measuring the dip angle, we can estimate the latitude at which the rock formed.

A.

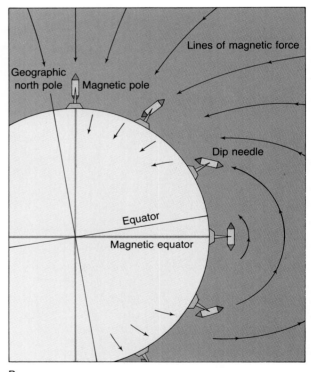

B.

FIGURE 3–8   Nature of Earth's Magnetic Field.

A, the earth's magnetic field consists of lines of force much like those a giant bar magnet would produce if placed inside the earth. Although the earth's magnetic poles seldom coincide with the geographic poles, over long periods of time, the average position of the magnetic poles can be considered to represent the position of the geographic poles. *B,* magnetite particles in newly formed rocks are aligned with the lines of force of the earth's magnetic field. This causes them to dip into the earth. The angle of dip increases uniformly from 0° at the magnetic equator to 90° at the magnetic poles. When the rocks solidify, the magnetite particles are frozen into position and become fossil compass needles that tell today's investigators about the strength and alignment of the earth's magnetic field when the rocks of which they are a part formed. The angle at which they dip into the earth also tells at what latitude they formed. Reprinted by permission from Tarbuck, E. J., and Lutgens, F. K., 1990, figs. 18.8 and 18.9.

Apparent Polar Wandering

Apparent Polar Wandering Because the present magnetic poles do not coincide with geographical poles related to latitude, it might be felt that determining latitude by magnetic dip would give some incorrect determinations. However, for the last few thousand years the average positions of the magnetic poles have coincided with those of the geographic poles. If we can assume that this is true for the past, we can determine average positions of the magnetic poles during a time interval and consider them to represent the geographic poles.

As these average positions for the magnetic poles are determined for rocks on the continents, it is found that their positions changed with time. It appears the magnetic poles were wandering. The wandering curve of the pole for North America shows an interesting relationship with that determined for Europe. Both curves had a similar shape, but for all rocks older than 70 million years the pole determined from North American rocks lies to the west of that determined by the study of European rocks. This difference implies that North America and Europe have changed position relative to the pole and relative to each other (fig. 3–9).

If there can only be one north magnetic pole at any given time and its position must be at or near the north geographic pole, a problem exists that can be solved only by moving the continents. The two wandering curves can be made to coincide by moving the two continents together as we go back in time. Now we have a single wandering curve, which shows the magnetic north pole to be much too far south during the interval of time from 200 million to 300 million years ago. Rotating the merged continents can bring the pole into the proper position. The fact that moving the continents was the only solution to this problem was very strong evidence in support of the movement of continents throughout geologic time.

Estimating the latitude of formation for rocks from many areas studied throughout the continents on the basis of paleoclimatic and paleomagnetic evidence produces similar results. The most logical explanation for the changes that have occurred throughout the past in both climate and magnetic dip of the rocks at a given location is that the continents have moved.

Magnetic Polarity Reversals

Magnetic Polarity Reversals Not only have the magnetic poles as determined from continental rocks seemed to wander throughout geologic time, but also the **polarity**, the direction of the magnetic field, seems to have reversed itself periodically. Reversals in polarity can be described in the following way. A compass needle that would point to the magnetic

A.

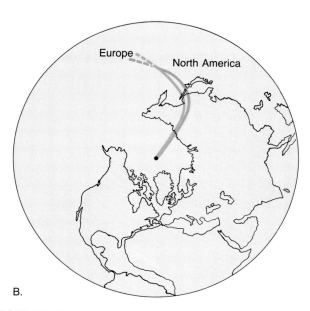

B.

FIGURE 3–9 Apparent Wandering Curves of North Magnetic Pole Determined from Rocks of North America and Europe.

A, apparent wandering curves of the north magnetic pole as determined from North America and Europe, going back in time from the present to 300 million years ago, follow divergent paths. *B,* if one moves the apparent position of the pole as determined from North America to a position that makes it coincide with the pole as determined from Europe, the direction and distance equal that required to close the Atlantic Ocean between the two continents. This still leaves the apparent north magnetic pole out in the North Pacific Ocean about 40° south of the geographic north pole. Again, movement of the continent is required to solve this problem. Rotation of Laurasia, which is formed by closing the Atlantic Ocean and joining Europe and North America, can bring the wayward apparent north magnetic pole to a position coincident with the north geographic pole. Reprinted by permission from Tarbuck, E. J., and Lutgens, F. K., 1990, fig. 18.10.

north today would, during a period of reversal, point south. It is not known why these reversals occur, but they have for the last 76 million years occurred once or twice each million years. The period of time during which a change in polarity occurs lasts for a few thousand years. It is identified on the basis of magnetic properties of rocks by a gradual decrease in the intensity of the magnetic field of one polarity until it disappears, followed by the gradual increase in the intensity of the magnetic field with the opposite polarity. The time during which a particular paleomagnetic condition existed can be determined by the radiometric dating of the igneous rock from which the paleomagnetic measurements were taken.

The earth's present magnetic field has been weakening for the last 150 years, and some investigators believe that the present polarity will disappear in another 2000 years.

MARINE GEOLOGY

We have so far considered only data obtained from the continents in considering the theory of moving continents. This continental evidence did not convince many geologists that the continents had moved. For many years no other data were available since

extensive sampling of the deep ocean bottom did not become technologically feasible until the 1960s.

The **Deep Sea Drilling Project**, which began in 1968, and the **Ocean Drilling Program**, which followed it in 1983, have added greatly to our knowledge of the ocean floor. The impetus for such programs was provided in great part by the need for observations of oceanic sediment and crust to check the theory of sea-floor spreading. Extensive geophysical studies, including those related to the paleomagnetic characteristics of the earth's crust, were first carried out. The results of these studies gave the planners of the Deep Sea Drilling Project clues as to where the drilling should be concentrated in order to gain the greatest amount of new knowledge.

Paleomagnetism

A detailed study of the magnetism of the Pacific Ocean floor by Scripps Institution of Oceanography identified narrow strips where magnetic properties of oceanic crust differed from those currently forming (magnetic anomalies). The **magnetic anomalies** ran parallel to the **Juan de Fuca Ridge**. Each anomaly represents a period when the polarization of the earth's magnetic field was reversed compared to today's. The anomalies were separated by bands of ocean crust that displayed present-day polarity.

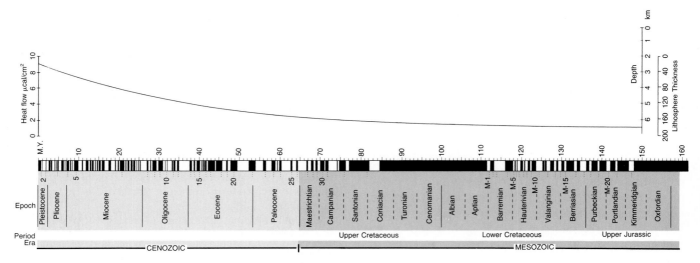

FIGURE 3–10 Magnetic Time Scale, Heat Flow, Ocean Depth, and Lithosphere Thickness.

Black bands represent periods when the magnetic polarity of the earth was the same as at present, and white bands represent periods when it was reversed. The curve shows that the ocean gets deeper and the lithosphere thicker with increasing age while the rate of heat flow to the earth's surface through the lithosphere decreases as the lithosphere ages. Although this curve could not be accurately applied to any particular region of the ocean basins, it indicates that, in general, heat flow decreases while ocean depth, thickness of the lithosphere, and age of the ocean floor increase with increasing distance from the spreading centers. Time is the independent variable for all of these changes. The geologic time scale is shown at the bottom (*see* Appendix IV).

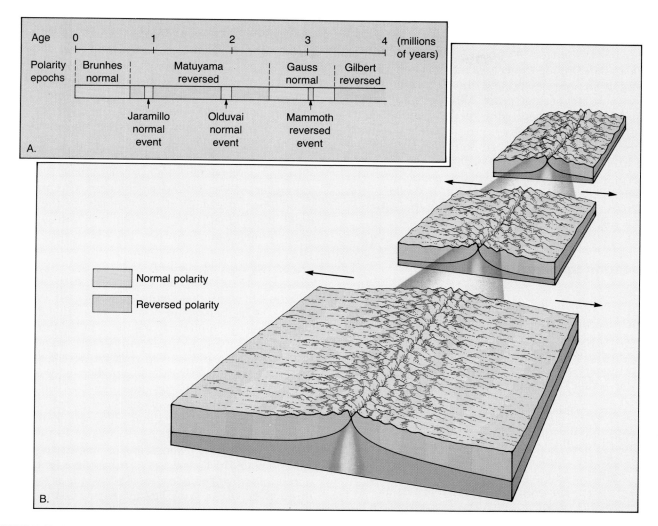

FIGURE 3–11 Magnetic Evidence in Support of the Sea-Floor Spreading Hypothesis.
A, time scale of the earth's magnetic field in the recent past. This time scale was developed by establishing the magnetic polarity for volcanic lavas of known ages. *B,* new sea floor records the polarity of the magnetic field at the time it formed. Hence it behaves much like a tape recorder, as it records each reversal of the earth's magnetic field. Data from Allan Cox and G. B. Dalrymple. Reprinted by permission from Tarbuck, E. J., and Lutgens, F. K., 1990, fig. 18.12.

Frederick Vine observed that the sequence of polarity changes on one side of an **oceanic ridge** (an oceanic mountain range) is identical to the sequence of reversals on the opposite side of the ridge. During dating of the reversal points on both flanks of the ridge, it became apparent that the rocks became older with increased distance from the ridge axis. The evidence indicated new oceanic crust was being formed at the oceanic ridges and moving away from them on opposite sides of the ridges. Having determined from continental studies the dates at which many of the more recent reversals occurred, it was possible to determine the age of the ocean floor at each strip boundary. Dividing the width by the number of years that polarity lasted produced the rate at which the ocean floor appeared to be moving away from the ridge (figs. 3–10 and 3–11).

Confirmation of the spreading of the oceanic crust away from the oceanic ridges required obtaining actual samples of the crust at various locations for radiometric dating. In addition, since sediment could not be laid down on crustal material until it formed at the oceanic ridges, it would be expected that the fossil assemblages observed in sediments immediately overlying the oceanic crust should be representative of organisms that existed at the time of the crustal formation. Also, it should be expected that the sediment thickness would be greater on older sea floor than on younger sea floor. A significant task of the Deep Sea Drilling Project was to check the age of the ocean bottom by drilling through sedimentary sections into the oceanic crust. Although all attempts to do this did not meet with success, enough data based on fossil assemblages immediately overlying the crust and radiometric age determinations of the crust itself confirmed that the oceanic floor is moving in the manner suggested by paleomagnetic data. The youngest oceanic crust was found at the axes of the oceanic ridges, with the age of the crust increasing at greater distances from the ridge crests.

Sea-Floor Spreading

New ocean floor appears to be forming at the oceanic ridges and rises. It then is carried away from the axes of such features, and volcanic processes fill the void with yet a younger strip of ocean floor. Thus, the axes of oceanic ridges and rises are referred to as **spreading centers**. Harry Hess was right, and we subsequently applied the term **sea-floor spreading** to his theory.

It is, however, more than the sea floor that is moving. Although the details are far from clear, the mechanism of spreading involves two fundamental units of the earth's structure confined to the uppermost 700 km (435 mi) (fig. 3–12). The **lithosphere** (rock sphere) is a relatively cool, rigid shell that includes the crust and upper mantle. The lithosphere is broken into a dozen or so lithospheric plates (fig. 3–13) that we observe moving across the earth's surface. Underlying the lithosphere is a high-temperature plastic layer within the mantle. This layer, called the **asthenosphere** (weak sphere), can flow slowly, allowing the rigid lithospheric plates resting at its upper surface to move.

The ultimate fate of a lithospheric plate, if mass is to be conserved, is its **subduction**—a process by which it descends beneath another plate and is ultimately resorbed into the mantle.

A Possible Mechanism for Spreading An absolute answer to the mechanism that might cause the formation of lithosphere at the oceanic ridges and its movement away from those ridges at right angles has not been determined. However, the lithosphere thickens from a few kilometers near the ridge to over 200 km (124 mi) beneath some continental regions. The increasing thickness of the lithosphere correlates well with increasing depth of the ocean and decreasing heat flow as the lithosphere ages (*see* fig. 3–10).

Within the earth, radioactive atoms are breaking down and releasing energy that must find its way to the surface as heat. It is conceivable that this heat moves to the surface through Hess's **convection cells** that carry the heat up in the regions of the oceanic spreading centers. If this is the case, there must be some regions on the earth where the cooler portions of the mantle descend to complete the convection cell.

Heat-flow measurements taken throughout the earth's crust show that the quantity of heat flowing to the surface along the oceanic ridges is as much as eight times in excess of the average for the earth's crust (1.25 μcal/cm^2/s). In addition, areas where ocean **trenches** are known to exist have heat flow as little as one-tenth the average.

The lateral movement that occurs between the regions of creation (ocean ridges and rises) and destruction of the lithosphere (trenches) may originate within the asthenosphere. These movements in the asthenosphere could carry the more rigid lithosphere, which contains the crust and upper mantle (fig. 3–14). It has been observed that the trenches do not fit the isostasy patterns characteristic of the continents, and it appears that there may be some force other than gravity pulling the lithosphere down in these regions. The descending convection cells might well provide that force.

The observed relationship of increasing lithospheric plate thickness, age of ocean floor, and ocean depth, along with a decrease in the rate of heat flow through the ocean floor with increasing distance from spreading centers can be explained by the convection process. The lithosphere at the spreading centers would be thin because of the proximity of high temperatures immediately beneath the ocean floor. This heat would also cause expansion of the upper mantle rocks, raising the oceanic ridges and rises to elevations well above those of the ocean floor on either side of the zones of upwelling heat. As the lithosphere moves away from the spreading centers toward zones of convection downwelling beneath the trenches, it will encounter decreasing rates of heat flow through the ocean floor from the underlying mantle. This would result in increasing thickness of the lithosphere as cooling asthenosphere at the base of the lithosphere is converted to rigid lithosphere. The thermal contraction of hot, fluffy asthenosphere into cold, brittle lithosphere accounts for the increasing depth of the ocean away from the spreading centers.

It was once believed that most of the heat driving global plate tectonics was generated primarily by radioactivity in the upper mantle. However, now there is evidence that a significant amount of heat is de-

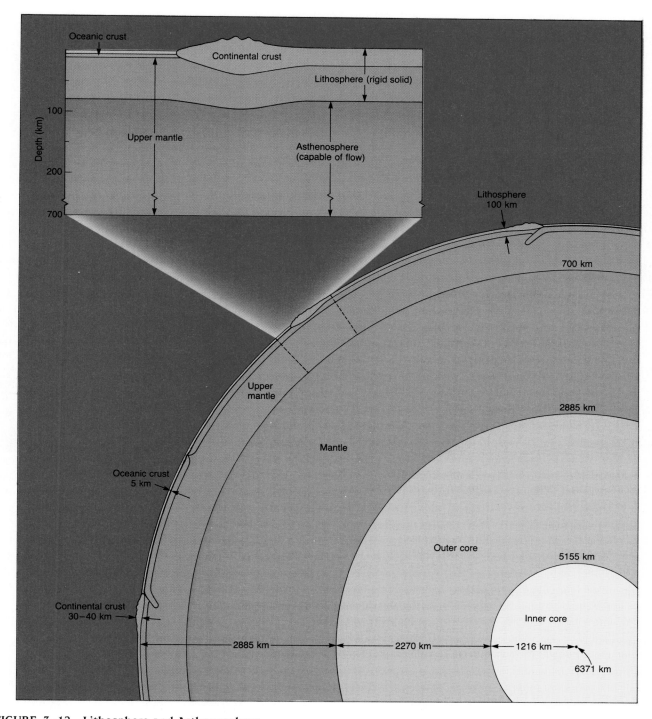

FIGURE 3–12   Lithosphere and Asthenosphere.
The lithosphere is the rigid outer shell of the earth's structure that includes the crust and cooler upper mantle. It is supported and allowed to move across the earth's surface by flow within the warmer, plastic underlying portion of the mantle called the asthenosphere. This weaker, possibly partially molten layer is thought to extend to a depth of approximately 700 km (435 mi). The three major subdivisions of the earth—core, mantle, and crust—are shown to clarify the position of the lithosphere and asthenosphere. Reprinted by permission from Tarbuck, E. J., and Lutgens, F. K., 1988, fig. 5.19.

Key

1 Eurasian Plate	9 African Plate
2 North American Plate	10 Indo-Australian Plate
3 Philippine Plate	11 Nazca Plate
4 Pacific Plate	12 Scotia Plate
5 Juan de Fuca Plate	13 Antarctic Plate
6 Cocos Plate	14 Arabian Plate
7 Caribbean Plate	15 San Andreas Fault
8 South American Plate	16 East African Rift Valley

Convergent boundaries

Divergent boundaries

Transform fault boundaries

Direction of plate movement

2.0   Average spreading rate (cm/yr)

FIGURE 3–13 Lithospheric Plates.
The lithospheric plates form at the axes of submarine mountain ranges (divergent boundaries) and dive back into the earth's interior beneath deep-ocean trenches (convergent boundaries). The divergent plate boundaries are offset by transform faults, along which the plates slide past one another in opposite directions. The lithospheric plates move at rates ranging from 1.5 cm (0.8 in.) per year for the American plates to more than 10 cm (5.4 in.) per year for the Pacific Plate.

rived from the core of the earth. If heat from the core is a significant source of energy driving mantle convection, a study of three-dimensional spherical models of convection in the earth's mantle shows that **upwelling** occurs as cylindrical plumes arising from the base of the mantle. Such features are known to exist, and their surface expressions are referred to as **hot spots**. The high rates of volcanic activity associated with hot spots have produced such features as the **Hawaiian Islands** and **Iceland**.

In these models, the **downwelling** occurs in sheets that form a ringlike pattern around the upwelling plumes. The overall pattern is similar to what we now see occurring in the Pacific Ocean. Because seismic data indicate the magma chambers that provide the lavas for the spreading centers extend no deeper than 350 km (220 mi) into the mantle, it may well be that they are passive features that result from the lithosphere being pulled apart by the subducting slabs at the edges of these sheets (fig. 3–15).

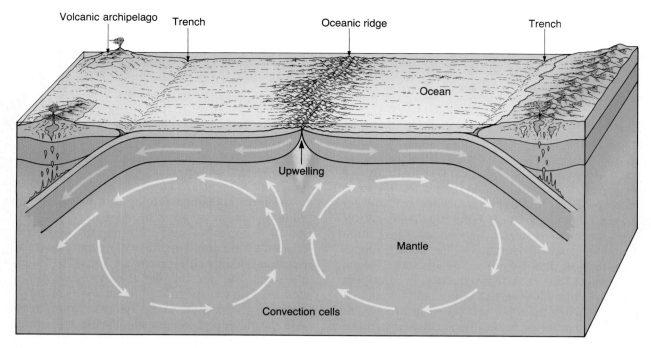

FIGURE 3–14 Movement in the Asthenosphere.
Lateral flow within the asthenosphere is a possible mechanism for the lateral movement of the lithosphere, including the crust. The asthenosphere is a low-strength plastic zone within the mantle where pressure is low enough and temperature sufficiently near the melting point of the material to allow it to flow slowly. Some form of thermal convection within the asthenosphere appears to create new lithosphere at the oceanic ridges and rises and may carry old lithosphere back into the mantle to be subducted beneath oceanic trench systems. Reprinted by permission from Tarbuck, E. J., and Lutgens, F. K., 1990, fig. 18.11.

FIGURE 3–15 Three-Dimensional Spherical Model of Convection in the Earth's Mantle with Half of the Heating Occurring from within the Mantle and Half from the Core/Mantle Boundary.

A, contours of vertical velocity of convection at a level midway between the base and top of the mantle. Red and yellow represent high and moderate rates of upwelling, respectively. Green represents no vertical motion, and light blue and dark blue represent low and high rates of downwelling, respectively. Note that upwelling appears as circles where the upwelling plume cuts through the spherical surface. Downwelling is seen as rings around the upwelling plumes where sheetlike downwellings intersect the sphere. *B,* in this longitudinal cross section of the model mantle, contours represent entropy values. Colors can be thought of as representing hot (red), warm (yellow), cool (cyan), and cold (blue). Red regions represent the centers of upwelling plumes that rise to the surface from the core/mantle boundary. Blue contours are found in the areas of convective downwelling. Images courtesy of D. Bercovici. From Bercovici, D., et al. "Three-Dimensional Spherical Models of Convection in the Earth's Mantle," *Science* 244, pp. 950–55, 26 May, 1989. Used with permission of the author and publisher. Copyright 1989 A.A.A.S.

PLATE BOUNDARIES

The process of sea-floor spreading sets up stresses within the lithosphere as plates move across the spherical surface of the earth and interact at relative velocities of up to 16 cm (5 in.)/yr. There are seven major and a half-dozen or more minor plates, some of which are shown in figure 3–13. Our first clues as to the locations of plate boundaries was that most tectonic events such as earthquakes and volcanic activity were confined to linear belts (fig. 3–16). Closer examination of these active regions identified three fundamental types of boundaries. The plate boundaries where new lithosphere is being added along oceanic ridges are called **constructive** or **divergent boundaries**. Boundaries where plates collide and one subducts beneath the other are called **destructive** or **convergent boundaries**. Lithospheric plates also move past one another along **transform** or **shear** boundaries (fig. 3–17).

Constructive Boundaries

Oceanic ridges and **oceanic rises** are the names given to the mountain ranges associated with oceanic spreading centers. Because deepening of the ocean floor is correlated directly with the amount of time the lithosphere has cooled, the rate of sea-floor spreading has a significant effect on the steepness of the flanks of the mountain ranges associated with the spreading.

The term **spreading rate** applies to the total widening rate of an ocean basin resulting from the motion of both plates away from a spreading center where they are created. These rates are shown for the major spreading centers in figure 3–13. The faster the spreading rate, the broader the mountain range associated with the spreading. These gently sloping features are the oceanic rises; the best example is the **East Pacific Rise** between the **Pacific Plate** and the **Nazca Plate**, where spreading rates are as high as 16.5 cm (6.5 in.)/yr. The **Mid-Atlantic Ridge**, with

FIGURE 3–16 World Distribution of Earthquakes over a Nine-Year Period.
Note that the pattern of earthquake distribution matches that of the lithospheric plate boundaries shown in figure 3–13. Data from NOAA. Reprinted by permission from Tarbuck, E. J., and Lutgens, F. K., 1988, fig. 5.11.

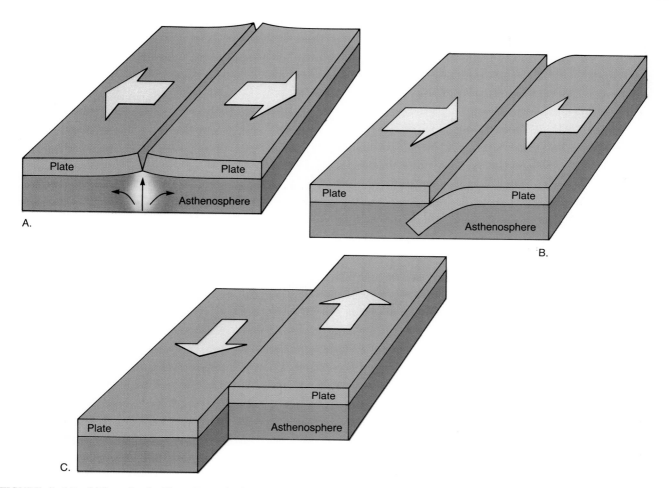

FIGURE 3–17 Lithospheric Plate Boundaries.
A, constructive (divergent) plate boundary. New material is added to plates at the spreading centers. The plates then diverge from one another as they move away from the spreading center. *B*, destructive (convergent) plate boundary. Lithospheric plates are destroyed when they converge at trenches and one subducts to be melted back into the mantle. *C*, transform (shear) plate boundary. Spreading centers are offset by transform faults along which plates slide past one another without the creation or destruction of lithosphere. Reprinted by permission from Tarbuck, E. J., and Lutgens, F. K., 1988, fig. 6.7.

spreading rates in the 2–3 cm (0.8–1.2 in.)/yr range, displays the steeper slopes characteristic of slow spreading (fig. 3–18). Project FAMOUS, the French-American Mid-Ocean Undersea Study, conducted in 1974, made observations in the rift valley running along the axis of the Mid-Atlantic Ridge. These observations provide a clue to the spreading process (fig. 3–19).

Fissures oriented parallel to the ridge axis range from hairline size near the center of the rift valley to more than 10 m (32.8 ft) in width near both margins of the valley. This evidence supports the theory that the plates are being continuously pulled apart rather than being pushed apart by upwelling of material be-

neath the ridges. Possibly, the upwelling of magma beneath the oceanic ridges is more that of filling in the void left by the separating plates of lithosphere. Whatever the mechanism of emplacement, a large mass of magma or viscous igneous material must exist near the surface beneath the ridges (fig. 3–20). This is manifested in the oceanic ridges and rises being elevated to mountainlike proportions above the deep-ocean basins. This process produces 20 km^3 (4.8 mi^3) of new ocean crust per year.

With new high-resolution acoustical devices such as *Sea Beam*, investigators have been able to identify offsets of between 2 and 10 km (1.2 and 6.2 mi) in the axis of the East Pacific Rise. These features are

(a)

(b)

0   100 km

FIGURE 3–18   Oceanic Ridges and Rises.
A, the Mid-Atlantic Ridge has a steeper and more irregular topography as a result of a low spreading rate. B, the high spreading rate of the East Pacific Rise has produced a more gently sloping and smoother bottom topography. Vertical exaggeration 50×. *The ocean basins: Their structure and evolution,* The Open University and Pergamon Press.

FIGURE 3–19   Rift Valley Fissures.
A, a fissure in the rift valley of the Mid-Atlantic Ridge photographed from the submersible *Alvin* during Project FAMOUS, 1974. Courtesy of Woods Hole Oceanographic Institution. B, a more readily observable fissure above sea level in the rift valley of Iceland. Here the Mid-Atlantic Ridge rises above the ocean surface as an island because of the high rate of volcanic activity in the Iceland hot spot. Courtesy of Br. Robert McDermott, S.J.

A.

B.

FIGURE 3–20 Magma Chambers and Their Relationship to Offset Spreading Centers.

A, location of Magma expedition that investigated the relationship between offset spreading centers and the distribution of magma chambers beneath them. The axis of the East Pacific Rise is outlined by the 3000-m depth contour. Map courtesy of Mark S. Burnett, Scripps Institution of Oceanography, University of California, San Diego. B, schematic cross section of the crust beneath the East Pacific Rise, based on gravity and seismic data. A mass of very hot rock that is for the most part not molten is found beneath the rise axis. The mushroom-shaped magma chamber that caps this mass of hot rock is fed by molten material rising from the upper mantle and is best developed where the ridge axis stands highest at the center of segments of the offset spreading

centers. The magma chamber, which reaches a maximum width of 4 km (2.5 mi), gradually disappears toward the ends of the segments and is absent beneath the overlapping offsets. Courtesy of K. C. Macdonald; University of California, Santa Barbara. C, false-color map of the MAGMA Expedition area of study with depths indicated by the color scale to the left. This map was prepared using data from Sea Beam and SeaMARC II side-directed sonar systems. It has a vertical exaggeration of 6. The axial ridge is broken into three segments by two overlaps, which are identified by the latitude values 12°37' and 12°54'N. The axial magma chamber is present beneath the rest of the rise, but disappears at the two overlapping spreading centers. This map was produced by S. P. Miller using NECOR software and unpublished data of K. C. Macdonald and P. J. Fox and was provided courtesy of K. C. Macdonald; University of California, Santa Barbara.

called **overlapping spreading centers** because the ends of adjacent segments of the offset spreading center overlap. The offsetting overlaps have an interesting relationship to the mass of hot, partially molten rock found beneath the axis of the East Pacific Rise. Figure 3–20A shows the location of two overlapping spreading centers at 12°37′N and 12°54′N latitude. Figure 3–20C shows that the axial ridge is lower and narrower at the overlaps and highest and broadest at the center of the segments between two overlapping ends. The mass of hot, partially molten rock (fig. 3–20B) extends uninterrupted along the entire axis and bends to follow the offsets. However, there appears to be no **magma chamber** (reservoir of molten rock) at the offsets. A thin magma chamber about 4 km (2.5 mi) wide is best developed at the center of each segment, where its top is about 1.4 km (0.9 mi) below the ocean floor. There is a well-developed rift or summit **graben** at midsegment, where the magma chamber is present; the graben weakens and disappears toward the ends of the segments, where the offsets occur. These data clearly show that segmentation on a relatively small scale (some segments are only 20 km long) is a fundamental property of oceanic ridges and rises and is associated with segmentation of magma chambers beneath the axis of the spreading center.

Destructive Boundaries

Island Arc Trench System   Oceanic ridges, oceanic rises, and trench systems represent significant discontinuities in the earth's crustal structure and are characterized by earthquakes and volcanic activity. There is a significant difference in the earthquake activity of the two regions, however. Shallow quakes, usually less than 10 km (6 mi) in depth, are associated with the constructive ridge boundaries where the lithosphere is thin, and movements that cause earthquakes in the trench regions may occur at depths down to 670 km (415 mi).

The earthquake **focus**, the point at which the movement that causes the quake occurs, is usually shallow in the immediate area of the trench. Earthquake activity in the subduction zone deepens and reaches a maximum value of 670 km. The **Wadati-Benioff seismic zone** is a band 20 km (12.5 mi) thick that dips from the trench region under the overriding plate, within which all of the earthquake foci are located. The angle of dip is approximately 45°, becoming steeper at greater depths (fig. 3–21A). Because the descending lithosphere would become too ductile below depths of 70–100 km to produce earthquakes,

some other mechanism must be found to explain them to depths of 670 km. It is possible that these deep-focus earthquakes result from the changing of the mineral, olivine, to a denser form, spinel. Olivine is a mineral that makes up a large percentage of the descending lithospheric mantle. This process is considered a likely cause of deep-focus earthquakes because it should occur between the depths of 400 and 670 km.

This band within which the earthquakes originate must lie within the slab of the lithosphere that is descending. The fact that the lithosphere generated at the oceanic rises and ridges is subducted at the trenches is supported by the fact that no oceanic crustal material more than 175 million years old has been recovered from the ocean bottom. Continental crust, which is very unlikely to be subducted because of its low density, remains at the earth's surface, and continental rocks as old as 3.8 billion years have been found.

As the lithosphere is carried into the asthenosphere, it is heated by being subducted to greater depths. Water and other volatiles are freed, producing a low-density mixture that rises to the surface on the overriding plate side of the trenches to produce **basaltic** volcanoes; these volcanoes may become island arcs, which usually develop about 100 km (62 mi) above subducting lithosphere (fig. 3–21B).

Continental Arc Trench System   Should an oceanic plate subduct beneath a plate with a continent at its leading edge, the melting of the subducting oceanic plate occurs beneath the continent. Consequently, the rising basalt melt passes through and mixes with the granite of the continental crust. This results in a continental arc of volcanoes along the edge of the continent (fig. 3–21A). The volcanoes are composed of a rock called **andesite**, which is of a composition intermediate between those of basalt and granite and is thought to form from the mixing of the two basic crustal rock types.

If the spreading center producing the subducting plate is far enough from the subduction zone, an oceanic trench is well developed along the margin of the continent. The **Peru-Chile Trench** is an example. It is associated with the **Andes Mountains** continental arc of volcanoes from which the rock, andesite, takes its name. No trench is visible where the **Juan de Fuca Plate** subducts beneath the **North American Plate** (NAP) off the coasts of Washington and Oregon to produce the **Cascade Mountains** continental arc. Here, the **Juan de Fuca Ridge** is so near the NAP that the subducting lithosphere is less than 10 million years old and hasn't cooled enough to produce deep-

FIGURE 3–21 Effects of Plate Collisions.

A, when an oceanic lithospheric plate subducts beneath a plate with continental crust at its leading edge, an andesitic continental arc develops. The volcanic rock andesite is named for the Andes Mountains, where a continental arc exists. Andesite has a composition between that of granite and basalt. It is thought to form as melted basalt from the descending oceanic plate rises through the continental granite to produce the volcanic continental arc. The sloping belt in which earthquake foci are located is the Wadati-Benioff seismic zone. *B,* classic oceanic trench systems develop where oceanic lithosphere subducts beneath a plate with oceanic crust at its leading edge and a continent trailing some distance behind. A basaltic island arc-trench system develops. *C,* as two plates carrying continents near their leading edges converge, trenches do not develop because sediments that accumulated in the sea between the continents are too light to be subducted. The shortening of the earth's crust in such regions is accommodated by the folding of the sediments into mountain ranges such as the Appalachians, Alps, Himalayas, and Urals. The lower lithosphere and asthenosphere may be involved in low rates of subduction beneath the mountains. Reprinted by permission from Tarbuck, E. J., and Lutgens, F. K., 1988, fig. 6.11.

A.

B.

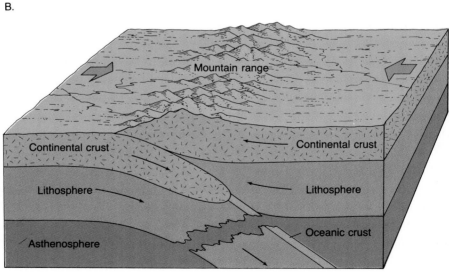

C.

ocean depths. Also, the large amount of sediment carried to the ocean by the **Columbia River** has filled what trench structure may have developed. Many of the Cascade volcanoes have been active within the last 100 years. Most recently, Mount St. Helens erupted in May of 1980 and killed 62 people (fig. 3–22).

Continental Mountain System If two lithospheric plates collide and both contain continental crust near their leading edges, the surface expression will be a mountain range composed of the folded sedimentary rocks derived from sediments deposited in the sea that previously separated the continental blocks. For the most part, these sedimentary rocks do not subside in a zone of subduction because of their relatively low density, having been derived from the continents. The oceanic crust itself may, however, subside beneath such mountains. The exact nature of

the subsidence of oceanic crustal material in these regions has not been determined (fig. 3–21C).

Some degree of folded mountain formation is frequently associated with plate collisions, but ocean trenches are absent in some areas. See "ophiolites" on page 68 for a discussion of these special features associated with destructive plate boundaries.

Shear Boundaries

Transform Faults As newly formed lithosphere moves away from the axis of an oceanic ridge, it travels perpendicular to the axis. On a spherical surface like that of the earth, all of this new lithosphere will converge toward two points. For instance, if the equator were an oceanic ridge, the points of convergence would be the North and South poles. Obviously, all this new matter cannot be accommodated at a single

FIGURE 3–22 Mount St. Helens.

A, tectonic features of the Mount St. Helens region. The mountains shown are volcanoes within the Cascade Mountain Range. *B,* the eruption of Mount St. Helens resulted from the subduction of the Juan de Fuca Plate beneath the North American Plate.

FIGURE 3–23   Transform Faults—Fracture Zones.

A, the axes of oceanic ridges are offset by transform faults due to the stresses resulting from the motion of rigid plates moving across the spherical surface of the earth. Earthquakes are common and of relatively great magnitude along the transform faults. They are called transform faults because they cut across the larger physiographic features of the ocean floor, oceanic ridges, and rises. Extending beyond the offset ridges are fracture zones, where earthquakes are rare because there is no relative movement on the opposite sides of the fracture zone. Fracture zones are scars of old transform fault activity and are maintained as significant topographic features because the age of the ocean floor on one side of the fracture zone is older than the ocean floor immediately opposite it on the other side of the fracture zone. Point X represents younger ocean floor than point Y. Therefore point Y has subsided more because of cooling and thermal contraction than the ocean floor at point X. Thus there is an escarpment along the fracture zone in Plate B that faces the older ocean floor on the side of the fracture zone where point Y is located. There is also an escarpment on Plate A, but it faces the opposite direction. *B,* the San Andreas Fault is a classic example of a transform fault that offsets an oceanic ridge or rise and forms a boundary between plates that are sliding past each other. Along this contact, the Pacific Plate is moving north relative to the North American Plate. Reprinted by permission from Tarbuck, E. J., and Lutgens, F. K., 1990, figs. 18.22 and 18.23.

point, and stresses are set up within the lithosphere. The stresses cause the lithosphere to break along lines perpendicular to the ridge axis. Movement along the breaks offsets the ridge axis, producing a **transform fault** (fig. 3–23). Along such faults one plate slides past another, producing shallow earthquakes in the thin lithosphere. An example of such a fault that has come ashore is the **San Andreas Fault,** which runs across California from the head of the Gulf of California to the San Francisco area.

Fracture Zones   Close observation of the axes of oceanic ridges and rises shows they do not move in a

continuous meandering path through the world's oceans, but are offset periodically by transform faults that run perpendicular to the axes. Continuing along the alignment of the transform fault beyond the offset axes are **fracture zones.** The transform faults represent active displacements of the axes; the fracture zones are evidence of past transform-fault activity. On opposite sides of the transform fault, the lithospheric plates are moving in opposite directions, whereas there is no relative plate motion occurring along a fracture zone. In other words, the transform faults serve as plate boundaries, while the fracture zones are embedded in a single plate (fig. 3–23).

OPHIOLITES

It is a well-accepted tenet of the global plate tectonic theory that new oceanic lithosphere created at spreading centers ultimately is *subducted* and returns to the underlying mantle. However, it seems that about one one-thousandth (0.001)% of the oceanic lithosphere does not return. Instead, it is *obducted* onto the edge of a continental mass (fig. 3A). These fragments of oceanic lithosphere embedded into the continents are called *ophiolites.* They characteristically consist of a thin layer of sedimentary rock that undoubtedly formed in the ocean, underlain by a sequence of crystalline rocks typical of ocean crust and the underlying mantle. Many of the ophiolites contain rich metallic ores with patterns of enrichment similar to those known to occur in the sediments and oceanic crust near hydrothermal vents on oceanic spreading centers.

With our present understanding of the sea-floor spreading process and its role in global plate tectonics, we have developed a reasonably clear view of how the units of an ophiolite assemblage may have formed. How they become obducted onto the continents is not so well understood.

In the idealized ophiolite sequence of rock types, there is a layer of marine sediments and an underlying series of crystalline rocks. Immediately be-

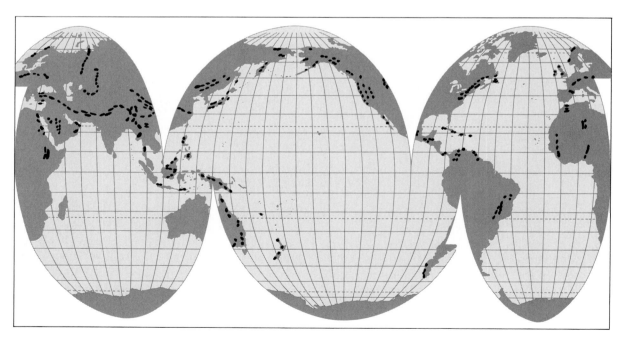

FIGURE 3A Ophiolite Distribution.

Distribution of ophiolites throughout the world. Ophiolites near the margins of the present continents where subduction is occurring are generally younger (less than 200 million years old) than those embedded deep within the continents. Some ophiolites are thought to be as much as 1.2 billion years old.

neath the sediments, one finds *pillow basalts* that flowed out on the ocean floor and show the pattern characteristic of submarine cooling. Descending through the lava flow, one encounters an increasing concentration of basaltic dikes that replace the flows completely about 1 km (0.62 mi) below the pillow lava surface (fig. 3B). It is believed that these *sheet dikes* form as vertical sheets of lava fill injected into narrow gaps formed as lithospheric plates are pulled away

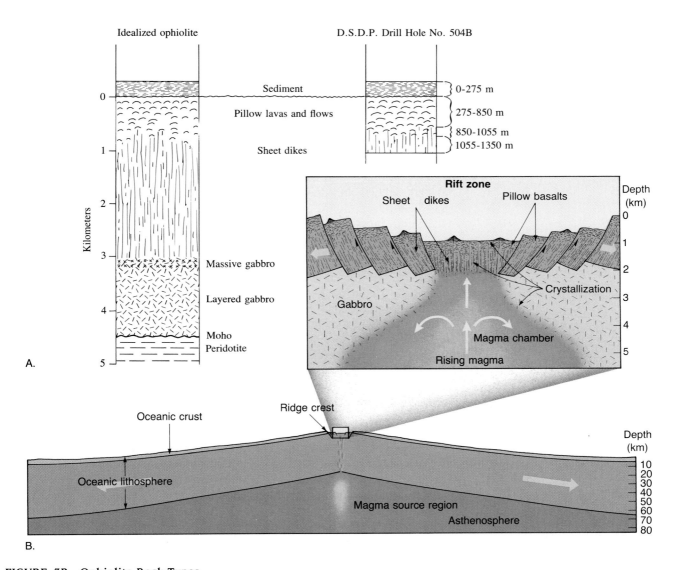

FIGURE 3B Ophiolite Rock Types.

A, the column on the left shows the sequence of rock types found in ophiolite complexes such as the Troodos Massif of Cyprus. On the right is the sequence of rock types encountered in the Deep Sea Drilling Project drill hole No. 504B, drilled to a total depth of 1350 m (4428 ft) beneath the ocean floor. It closely resembles that of the upper ophiolite complex and adds support to the belief that ophiolites are pieces of oceanic lithosphere that have been obducted onto the margins of continents. *B,* diagram showing how the pillow basalts, sheet dikes, and gabbros of an ophiolite might form. Reprinted with permission from Tarbuck, E. J., and Lutgens, F. K., 1990, fig. 19.17.

from the axis of spreading. At a distance of approximately 3 km (1.9 mi) beneath the pillow lava surface, the dikes give way to *massive gabbro,* a rock similar to basalt in composition but of a coarser texture owing to slower cooling. The gabbro may represent the cooling of magma near the roof of the magma chamber that was the source of the overlying basalts. The massive gabbro may be up to 300 m (984 ft) thick. It is underlain by *layered gabbro* that may have crystallized within the upper region of the magma chamber. Layered *peridotite,* a mantle rock composed mostly of the iron- and magnesium-rich minerals pyroxene and olivine, is found beneath the layered gabbro. The boundary between the gabbro and peridotite is thought to represent the *Moho,* or the contact between the ocean crust and the underlying mantle. The Moho is usually found about 4.5 km (2.8 mi) below the top of the pillow lavas.

The main features of the idealized ophiolite sequence observed on the margins of continents were partially encountered in place beneath the ocean floor. The Deep Sea Drilling Project drill hole No. 504B, located on the flank of the eastern Galápagos Spreading Center, was drilled to a depth of 1.35 km (0.84 mi) beneath the ocean floor in 3.46 km (2.15 mi) of water. The crustal rocks penetrated are thought to be just over 6 million years old and the sequence of rocks penetrated in the hole are as follows:

1. Sediment   0–275 m (0–902 ft)
2. Pillow lavas and flows   275–850 m (902–2788 ft)
3. Pillow lavas, flows, and sheet dikes   850–1055 m (2788–3460 ft)
4. Sheet dikes   1055–1350 m (3460–4428 ft)

Figure 3B shows how well this sequence resembles that observed in the ophiolite complexes.

One of the oldest known ore deposits is the Troodos Massif of the island of Cyprus (fig. 3C). It has been mined for copper since the time of the Phoenicians. The main sulfide ore bodies are found in the volcanics near the top of the ophiolite sequence, and the sediment section above the volcanics is rich in iron and manganese. Such a sequence of metal enrichment is known to be occurring now at oceanic spreading centers as a result of hydrothermal activity.

The difficult question of how the ophiolites become obducted onto the continents may have numerous answers. One of the more straightforward explanations (fig. 3D) is as follows:

1. A plate containing a continent has a section of oceanic lithosphere between the continent and a subducting plate of oceanic lithosphere.
2. The subducting oceanic plate contains water-bearing minerals such as zeolites and amphiboles. Subduction will heat these minerals and release water. The heated water converts the mantle rock, peridotite, to serpentine, which is much less dense than the surrounding peridotite. The serpentine rises and pushes up the overlying crust and mantle so that the oceanic crust is raised above sea level.
3. Continued compressional stresses may cause the oceanic crust and associated serpentine to break into layers that are obducted onto continental rock.

Continued study of ophiolites is needed to answer many questions that remain about the nature of all the processes involved in forming and obducting these ore-rich bodies along the margins of continents.

A.

B.

FIGURE 3C Ophiolite Mining.
A, location of the Troodos Massif ophiolite on the island of Cyprus in the eastern Mediterranean Sea. *B,* mining the Troodos Massif. The sulfide ore is the black material in the bottom of the pit. Photo courtesy of R. Koski, U.S.G.S.

A.

B.

C.

FIGURE 3D The Obduction of an Ophiolite onto a Continent.
A, subducting lithosphere contains water-bearing minerals. *B,* water-bearing minerals are heated and the hot water is released into the peridotite, which is altered to less dense serpentine. The expansion of the serpentine lifts the overlying mantle and oceanic crust. *C,* continued compressional stress breaks the uplifted segment of oceanic crust and pushes it on to the continent where it remains as an ophiolite structure.

Earthquakes with foci above a depth of 10 km are common along the transform faults, whereas the fracture zones are aseismic.

Because the age of the ocean floor on one side of a fracture zone is greater than that of the ocean floor on the opposite side, the older ocean floor will have subsided more from thermal contraction. This results in a vertical escarpment that faces the older ocean floor and increases in height with increased difference in the age of ocean floor on opposite sides of the fracture zone.

THE GROWTH OF OCEAN BASINS

So far in this chapter we have discussed some specific features associated with construction, transport, and destruction of tectonic plates in light of what they can tell us about the specifics of these processes. In this section we concentrate on the broader aspect of changes that have occurred since the breakup of Pangaea 175 million years ago. The study of the historical changes of shapes and positions of the continents and oceans is **paleogeography**, while the term **paleoceanography** applies to the study of the changes in physical and biological character of the oceans brought about by paleogeographical changes.

Figure 3–24A and B show the changing positions of the continents we know today during the last 150 million years for the Northern Hemisphere and Southern Hemisphere, respectively. Observing the changes in the Northern Hemisphere, it is clear that the major change is the separation of North America and South America from Europe and Africa to produce the Atlantic Ocean. During this process, Europe, Africa, and North America also moved a bit northward. Many investigators of plate motion observe that the evidence indicates rates of spreading have historically shown a pattern of higher rates at lower latitudes. This shows up in the pattern of formation of the North Atlantic Ocean. Europe and North America have separated at a much lower rate than have Africa and North America. It was not until 60 Ma (million years ago) that a noticeable separation showed up, and the island of Greenland appeared to be moving on the same plate as Europe. By 30 Ma, the north end of the Mid-Atlantic Ridge shifted to the east of Greenland, putting it back on the North American Plate. North America and South America were not fully connected by the Isthmus of Panama until about 5 Ma.

This event had a marked effect on the patterns of ocean circulation and biological species distribution.

In the Southern Hemisphere, the early stages of continental breakup show South America and a continent composed of India, Australia, and Antarctica separating from Africa. By 90 Ma, there is a clear separation between South America and Africa, and India had moved north away from the Australia-Antarctica mass, which began moving poleward. As the Atlantic Ocean continued to open, India moved rapidly north and collided with Asia about 35 Ma. Subsequent to this collision, the spreading center that ran between India and Australia foundered and was replaced by one that began to separate Australia from Antarctica. The scars representing shear-plate boundaries during the northward movement of the Indian Plate are the Chagos-Laccadive Ridge to the west and the Ninety East Ridge to the east. These features are clearly visible in figure 4–14.

The overall effect of global plate tectonic events over the past 180 million years is the creation of the new Atlantic Ocean, which continues to grow, and the division of Panthalassa into the Pacific Ocean and Indian Ocean. The fate of the Indian Ocean in terms of its changing size is questionable, but it is clear that the Pacific Ocean continues to shrink as oceanic plates produced within it subduct into the many trenches that surround it and continent-bearing plates bear in from east and west.

AGE OF THE OCEAN FLOOR

Much can be learned of the history of ocean basins from mapping the pattern of age distribution for the ocean crust. Observing figure 3–25, it can be seen that the Atlantic Ocean has the simplest and most symmetrical pattern of age distribution. As Pangaea was rifted by the incipient Mid-Atlantic Ridge, the first separation involved North America and Africa over 175 Ma. The Pacific Ocean has the least symmetrical pattern because of the large amount of subduction that surrounds it. The ocean floor east of the East Pacific Rise that is older than 40 Ma has been subducted, whereas ocean floor in the northwestern Pacific is over 170 Ma. The Rise itself has even disappeared under North America. The broader age bands on the Pacific Ocean floor are clear evidence that the spreading rate throughout this history has been greater in the Pacific Ocean than in the Atlantic or Indian oceans.

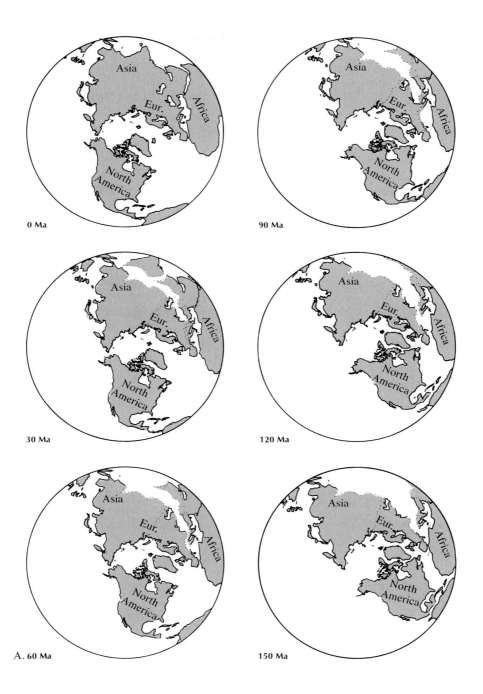

FIGURE 3–24A Paleogeographic Reconstruction—Northern Hemisphere.

These paleoceanic projections represent an attempt to determine the relative positions of the continents over the last 150 million years. Ages are quoted in Ma (millions of years ago) at intervals of 30 million years. *A,* Northern Hemisphere; *B,* Southern Hemisphere.

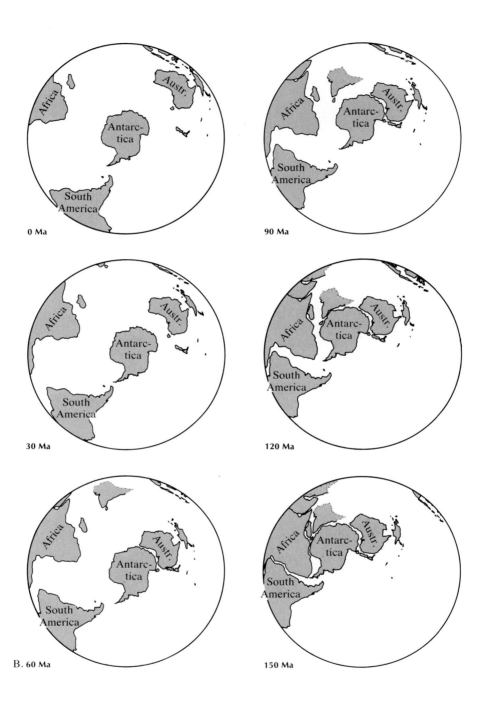

FIGURE 3–24B   Paleogeographic Reconstruction—Southern Hemisphere.

The projections are based on data taken from the first 60 legs of the Deep Sea Drilling Project,
conducted during a decade of deep-ocean drilling from August 1968 to August 1978. Courtesy of
P. L. Firstbrook, B. M. Funnell, A. M. Hurley, and A. G. Smith in association with Scripps Institution
of Oceanography, University of California, San Diego.

FIGURE 3–25   Relative Age of Oceanic Crust beneath Deep-Sea Deposits.

Notice that the youngest rocks (bright red areas) are found along the oceanic ridge crests and the oldest oceanic crust (brown areas) is located adjacent to the continents. When you observe the Atlantic basin, a symmetrical pattern centered on the Mid-Atlantic ridge crest becomes apparent. This pattern verifies the fact that sea-floor spreading generates new oceanic crust equally on both sides of a spreading center. Further, compare the widths of the yellow stripes in the Pacific basin with those in the South Atlantic. Because these stripes were produced during the same time period, this comparison verifies that the rate of sea-floor spreading was faster in the Pacific than in the South Atlantic. Reprinted from Tarbuck, E. J., and Lutgens, F. K., 1990, fig. 19.16, after *The Bedrock Geology of the World,* by R. L. Larson et al. Copyright © 1985 by W. H. Freeman.

KEY	
0–2 m.y.	58–66 m.y.
2–5 m.y.	66–84 m.y.
5–24 m.y.	84–117 m.y.
24–37 m.y.	117–144 m.y.
37–58 m.y.	144–208 m.y.

SUMMARY

Scientific investigation of evidence suggesting continents are moving relative to one another began with **Alfred Wegener** at the start of the twentieth century. The evidence gathered by the late 1960s indicated all the present continents were combined into one large continent, **Pangaea**, about 200 million years ago. Just before the formation and possibly also after the breakup of Pangaea, two large continents appear to have existed, **Laurasia** in the north and **Gondwana-land** to the south.

The discovery that the earth's **magnetic field** has changed **polarity** throughout the earth's history and that the record of these changes was permanently recorded in the rocks of the oceanic crust made it possible to develop a hypothesis of **sea-floor spreading** as the process by which the continents could be moved. According to the hypothesis, new rock was added to rigid **lithospheric plates** by intrusion and extrusion of molten material along submarine mountain ranges. These plates are composed primarily of dense mantle rocks with a thin crust of basaltic lava beneath the ocean basins. The continents "float" on these plates because they are composed of a less dense igneous rock, granite. As material is added to the plates near the axis of the oceanic ridges and rises, it moves away to make room for new material surfacing at the axis. Thus the crustal rocks beneath the oceans increase in age as one moves from the axis of the mountain ranges down their slopes into the deep-ocean basins. Continents located on a lithospheric plate would be carried passively along as the plate moved away from the oceanic ridge or rise. The sea-floor–spreading hypothesis was confirmed when the *Glomar Challenger* began the Deep Sea Drilling Project to sample the sediments and crust beneath the oceans in 1968. The movement of the lithosphere appears to be made possible by the existence of a plastic region of mantle material—the **asthenosphere**, immediately beneath the lithosphere—with temperatures near the melting point of the mantle.

As the lithospheric plates move away from the high-heat-flow oceanic ridges and rise toward low-heat-flow trenches, they cool and thicken. The age and depth of the ocean floor increase with increasing distance from the spreading centers.

As new mass is added to the lithosphere at the oceanic rises and ridges (**constructive** or **divergent plate boundaries**), the leading edges of the plates are subducted into the mantle at ocean trench systems or beneath continental mountain ranges such as the Himalayas (**destructive** or **convergent plate boundaries**). Additionally, there are plate boundaries called **transform faults**, along which oceanic ridges and rises are offset, and plates slide past one another at **shear plate boundaries**. Although the earth is over 4.6 billion years old, no rocks have been found on the ocean floor that are more than 175 million years old. It is becoming more clear that sinking at the leading edges of the plates rather than upwelling of new material at the ridges drives the convectionlike movement, and it is obvious that the process produces recycling of material between the lithosphere and the deeper mantle.

The **paleogeography**, or changing shape and location of ocean basins and continents, has produced changes in the physical and biological character of the oceans. The study of these changes is called **paleoceanography**. Since the breakup of Pangaea, which began 175 Ma, the Atlantic Ocean has come into existence and Panthalassa has been reshaped into the Pacific Ocean and Indian Ocean.

QUESTIONS AND EXERCISES

1. Why couldn't Alfred Wegener convince most scientists that the continents were indeed moving across the earth's surface?

2. What discovery made absolute dating of rock units possible? Why was this technique a significant improvement over relative dating?

3. What is the age of a rock that has a ratio of U^{235} to Pb^{207} of 1 to 3?

4. Describe the evidence found on the present continents that suggests Pangaea had broken into two continents, Laurasia and Gondwanaland, between 250 and 350 million years ago.

5. How does magnetic dip of magnetite particles found in igneous rocks tell us at what latitude they formed?

6. How does continental movement account for the two apparent wandering curves of the north magnetic pole as determined from Europe and North America?

7. What property of the magnetic record found in oceanic crustal rocks gave rise to the idea of sea-floor spreading?

8. Describe a possible mechanism responsible for moving the continents over the earth's surface. Include a discussion of the lithosphere and asthenosphere.

9. Describe the general relationships that exist among distance from the spreading centers of the oceanic ridges, heat flow, age of the oceanic crustal rock, and ocean depth.

10. Discuss the three types of plate boundaries, spreading centers of the oceanic ridges, trenches, and transform faults. Explain why earthquake activity is usually confined to depths less than 10 km (6 mi) along the spreading centers and transform faults while it may occur as deep as 670 km (415 mi) near the trenches. Construct a plan view and cross section showing each of the types of boundary and direction of plate movement.

11. Describe the relationship between fracture zones and transform faults, including a comparison of the earthquake activity along each.

12. When does evidence for the following events first show up on the paleogeographic reconstructions of figure 3–24?
    a. Greenland separates from Europe
    b. South America begins movement away from Africa
    c. India separates from Antarctica
    d. Australia separates from Antarctica

REFERENCES

Anderson, D. L. 1975. Accelerated plate tectonics. *Science* 187:1077–79.

Bercovici, D., Schubert, G., and Glatzmaier, G. A. 1989. Three-dimensional spherical models of convection in the earth's mantle. *Science* 244:4907, 950–54.

Bullard, Sir Edward. 1969. The origin of the oceans. *Scientific American* 221:66–75.

Burnett, M. S., Caress, D. W., and Orcutt, J. A. 1989. Tomographic image of the magma chamber at 12°50'N on the East Pacific Rise. *Nature* 339:206–208.

Carrigan, C. R., and Gubbins, D. 1979. The source of the earth's magnetic field. *Scientific American* 240–2:118–33.

Dietz, R. S., and Holden, J. C. 1970. The breakup of Pangaea. *Scientific American* 223:30–41.

Hamilton W. B. 1988. Plate tectonics and island arcs. *Bulletin Geological Society of America,* 100:1503–1526.

Hess, H. H. 1962. *History of ocean basins. Petrologic studies: A volume to honor A. F. Buddington,* Engel, A. E. J., Lames, H. L., and Leonard, B. F., eds., 559–620. New York: Geological Society of America.

Jordan, T. H. 1979. The deep structure of the continents. *Scientific American* 240–1:92–107.

Menard, H. W. 1986. The ocean of truth: A personal history of global tectonics. Princeton, N. J., Princeton University Press.

Olson, P., Silver P. G., and Carlson, R. W. 1990. The large-scale structure of convection in the Earth's mantle. *Nature,* 344:209–214.

Sclater, J. G., Parsons, B., and Jaupart, C. 1981. Oceans and continents: Similarities and differences in the mechanisms of heat loss. *Journal of Geophysical Research,* 86:11, 535–52.

Shepard, F. P. 1977. *Geological oceanography.* New York: Crane, Russak.

Tarbuck, E. J., and Lutgens, F. K. 1990. The earth: An introduction to physical geology, 3d. ed. Columbus: Merrill.

—————. 1988. Earth science. 5th ed. Columbus: Merrill.

Tarling, D., and Tarling, M. 1971. *Continental drift: A study of the earth's moving surface.* Garden City, N.Y.: Doubleday.

Vine, F. J., and Matthews, D. H. 1963. Magnetic anomalies over oceanic ridges. *Nature* 199:947–49.

SUGGESTED READING

Sea Frontier

Burton, R. 1974. Instant islands. 19:3, 130–36.
A description of the 1973 eruption on the island of Heimaey south of Iceland and its relationship to plate tectonics processes.

Dietz, R. S. 1976. Iceland—where the mid-ocean ridge bares its back. 22:1, 9–15.
A description of the rift zone of Iceland and its relationship to the Mid-Atlantic Ridge and sea-floor spreading.

—————. 1977. San Andreas: An oceanic fault that came ashore. 23:5, 258–66.
A discussion of the San Andreas Fault and its relationship to global plate tectonics.

—————. 1971. Those shifty continents. 17:4, 204–12.
A very readable and informative presentation of the crustal features and the possible mechanism of plate tectonics.

Emiliani, C. 1972. A magnificent revolution. 18:6, 357–72.
A discussion of advances in studying earth science from the sea, including climate cycles, plate tectonics, and the economic potential of resources lying within marine sediments.

Mark, K. 1974. Earthquakes in Alaska. 20:5, 274–83.
A discussion of the origin, nature, and effects of earthquakes in Alaska.

—————. 1972. Ocean fossils on land. 18:2, 95–106.
A discussion of the significance of marine fossils found in rocks now many miles inland is centered on the work of an early American paleontologist, James Hall.

Rona, P. 1984. Perpetual seafloor metal factory. 30:3, 132–141.
A discussion of how metallic mineral deposits form in association with hydrothermal vents located on oceanic ridges and rises.

Scientific American

Bonatti, E., and Crane, K. 1984. Oceanic fracture zones. 250:5, 40–51.
The role of oceanic fracture zones in the plate tectonics process is related to the spherical shape of the earth.

Courtillot, V., and Vink, G. E. 1983. How continents break up. 249:1, 43–49.
The process of continental rifting and deformation is considered.

Dietz, R. S., and Holden, J. C. 1970. The breakup of Pangaea. 229:4, 30–41.
A description of the breakup of the hypothetical supercontinent and the movement of continents to their present positions. The possible mechanism of continental drift is explored.

Francis, P., and Self, S. 1983. The eruption of Krakatau. 249:5, 172–187.
An anlysis of the 1883 event based on the volcanic deposits and the timing of air and sea waves created by the eruption.

Gass, I. G. 1982. Ophiolites. 247:2, 122–31.
The relationship between ophiolite emplacements and global plate tectonic processes is discussed.

Hallam, A. 1975. Alfred Wegener and continental drift. 232:2, 88–97.
This article presents a history of Wegener's development of his theory of continental drift presented in 1912 and the reasons for its general rejection until the 1960s.

Heirtzer, J. R., and Bryan, W. B. 1975. The floor of the Mid-Atlantic Rift. 233:2, 78–91.
The data gathered by Project FAMOUS (French-American Mid-Ocean Undersea Study) are summarized. This investigation of the Mid-Atlantic Rift south of the Azores involved the use of the submersibles *Alvin, Archimede,* and *Cyana.*

Menard, H. W. 1969. The deep-ocean floor. 221:3, 126–45.
A summary of the dynamic effects of sea-floor spreading is presented with a description of related sea-floor features.

Molnar, P. 1986. The structure of mountain ranges. 255:1, 70–79. Folded mountain ranges and their relationship to tectonic plates are discussed.

Pollack, H. N., and Chapman, D. S. 1977. The flow of heat from the earth's interior. 237:2, 60–76.
Based on data determined from thousands of heat-flow measurements on continents and the ocean floor, this article describes the relationship between lithospheric thickness and the heat flow through it.

Tokosöz, M. N. 1975. The subduction of the lithosphere. 233:5, 88–101.
A discussion of how the subduction of the lithosphere is related to ocean trenches, island arcs, volcanism, and earthquakes.

4
MARINE PROVINCES

An understanding of the general bathymetric features of the ocean is important because these features are related ultimately to global plate tectonics as well as physical and biological phenomena to be discussed in following chapters. **Bathymetry** in oceanography is analogous to **topography** in geography.

The **hypsographic curve** showing the distribution of the earth's solid surface at elevations above and below sea level indicates that 71% of the earth's surface is to be found beneath the oceans (fig. 4–1). The mean elevation of the continents above the sea (840 m, or 2755 ft) and mean depth of the oceans (3800 m, or 12,460 ft) are the result of different densities of the continental and ocean crustal rocks. The ocean crust is denser.

TECHNIQUES OF DETERMINING BATHYMETRY

Needless to say, our greatest understanding of the ocean floor relates to the shallow shelf areas at the margins of the continents, over which the most important commercial fisheries have been located. They are presently being explored extensively because of their high potential for petroleum accumulation. Because of their economic importance and the fact that we have a greater knowledge of this region than any other ocean region, we will begin our description of the ocean floor with these marginal shelves.

However, before we discuss the various submarine provinces, we should discuss briefly the technique that has been most used to obtain data regarding submarine geological processes, **seismic surveying**. Such surveys are made by emitting a sound signal that will travel through the water, bounce off an object, and return to be picked up by a receiver. If the velocity at which sound will travel in water has been determined, the distance to the object is equal to the velocity times one-half the time required for the sound signal to make the trip from the surface to the object or ocean floor and back to the receiver (fig. 4–2).

If one wishes to know only the distance to an object in the water or the ocean floor, relatively weak high-frequency sound signals of the **echo sounder** can be used. Marine geologists, however, need to know the thickness of various rock units beneath the ocean floor, and for this purpose they use stronger low-frequency signals generated by explosions. These sound signals can penetrate the rocks beneath the ocean floor, reflect off the contacts between rock

Preceding page: Pillow lava observed at the FAMOUS site in the North Atlantic Ocean. *Alvin's* collecting basket is in the foreground. Photo courtesy of the Scripps Institution of Oceanography, University of California, San Diego.

FIGURE 4–1 Elevations of the Earth's Crust.

The horizontal dashed lines indicate that the average height of the continents is 840 m (2755 ft) above sea level and that the average depth of the oceans is 3800 m (12,460 ft) below sea level. The vertical dashed line marks the division between land and sea. Almost 30 percent or 150 million square kilometers of the earth's surface is above sea level. It is clear that the volume of the earth below sea level is many times greater than the volume of the continents above sea level. Reprinted with permission from Tarbuck, E. J., and Lutgens, F. K., 1990, fig. 1.14.

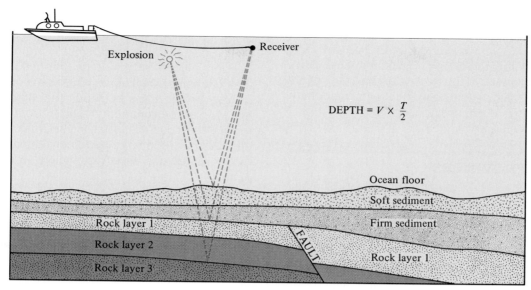

$$\text{DEPTH} = V \times \frac{T}{2}$$

FIGURE 4–2   Seismic Profiling.

Low-frequency sound that can penetrate bottom sediments is emitted by an explosion. It reflects off the boundaries between rock layers and returns to the receiver. The depth of each reflecting layer is equal to the velocity of sound travel *(V)* times one-half the time *(T)* required for the sound to travel from the source to the reflecting layer and back to the receiver.

units, and produce seismic reflection profiles such as those shown in this chapter and chapter 5.

A variation of this technique is **side-scan sonar**, which allows a survey ship to map the topography of the bottom along a strip of ocean floor up to 60 km (37 mi) wide. The map is developed with the aid of computers and sound emitters directed away from both sides of the ship. Figure 4–3 shows the British GLORIA system for deep-sea side-scanning sonar surveys.

FIGURE 4–3   GLORIA System for Side-scanning Sonar Surveys.

GLORIA is an acronym for Geological Long-Range Inclined ASDIC. ASDIC is an acronym for Acoustical Side Direction. The neutrally buoyant GLORIA is towed 300 m (985 ft) behind a research vessel. At water depths of 5000 m (16,400 ft), it can map a 60-km (37 mi)-wide strip of ocean floor. At shallower depths, the width of the strip is reduced.

Owing to the gravitational effects on the ocean surface, sea level rises or falls 4 m (13 ft) with each decrease or increase in ocean depth of 1000 m (3280 ft). Altimeters aboard *SEASAT A* detected these differences and provided data to produce a map of the ocean surface (fig. 4–4). The map reflects major bathymetric features of the ocean floor.

MARINE PROVINCES

As our knowlege of the bathymetry of the oceans grew, it became clear the ocean basins could be divided into two basic regions, the **continental margins** and **deep-ocean basins**. Figure 4–5 illustrates the formation of an ocean basin that resembles the present Atlantic Ocean very much. It shows that from the initial rifting of a continental mass, the deep-ocean basin forms with a shallow continental margin composed of rift blocks overlain by a thick mass of sediment carried to the oceans from the continent.

The ocean basin deepens and the continents subside as a result of the thermal contraction of the lithosphere as it cools with time. Despite this subsidence, the ocean floor rises abruptly to the surface of a thick mass of sediments eroded from the continent and deposited at its edge to form the continental margin. The weight of these sediments also contributes to the subsidence of the continent.

The continental margin of the type shown in figure 4–5 is referred to as an **Atlantic-type margin**, or **passive margin**, because there is little seismic activity associated with it. Along other continental margins, including the western margins of North America and South America, lithospheric plates converge and are destroyed through subduction. Here a high level of seismicity is manifested in earthquakes and volcanism. Along these **Pacific-type margins**, or **active margins**, the broad, thick wedge of sediment found along Atlantic-type margins may be greatly reduced and replaced by trenches and their associated features.

The deep-ocean basins include the spreading centers, the deepening basins on either side, trenches, and a vast variety of other fascinating features to be discussed on the following pages.

CONTINENTAL MARGIN

The surface of the continental margin sedimentary wedge typical of Atlantic-type margins is divided into the continental shelf, continental slope, and continental rise on the basis of depth and slope changes as one crosses the continental margin from the shoreline toward the deep-ocean basin. Because the continental margins are in large part depositional features, we will consider this aspect of their character in some detail.

Continental Shelf

Extending from the shoreline is a shelflike feature called the **continental shelf**. With few exceptions, coastal topography can be expected to extend beyond the shore and onto the continental shelf. The continental shelf is defined as a shelflike zone extending from the shore to a point at which a marked increase in slope occurs. This point where an increase in the rate of slope occurs is referred to as the **shelf break**, and the steeper slope beyond the break is known as the **continental slope** (shown in fig. 4–6).

The width of continental shelves may vary from a few tens of meters to about 1300 km (800 mi). The broadest shelf developments occur off the northern coasts of Siberia and North America in the Arctic Ocean and in the North and West Pacific from Alaska to Australia. The average width of the continental shelf is about 70 km (43 mi) and the average depth at which the break occurs is about 135 m (443 ft). Around the continent of Antarctica, however, the shelf break occurs at a depth of 350 m (1150 ft). The mean slope of the continental shelf is 0°07′, or 1.9 m/km (10 ft/mi). Sediment is transported across the continental shelf by current flow and mass wasting (submarine sediment slides), which are often generated by earthquakes.

Continental Slope

The continental slopes beyond the shelf break are features similar in relief to mountain ranges we find

FIGURE 4–4 (opposite)   Ocean Surface Mapped by *SEASAT*.
A, due to gravitational effects, the ocean surface rises over high volcanic peaks and drops over depressions such as trenches on the ocean floor. An altimeter on the *SEASAT* satellite that operated in 1978 was able to measure these changes in the ocean surface with sufficient precision to map major bathymetric features on the ocean floor. *B,* in the southern Indian Ocean and Pacific Ocean, this remote sensing of the ocean surface was able to identify some new bottom features and provide information that significantly altered the shape or location of previously identified features (circled).

TOPOGRAPHIC RELIEF FROM SEASAT ALTIMETER MEAN SEA SURFACE
JULY 7 – OCTOBER 10, 1978

PARKE, STAVERT
AND HUSSEY (1982)

NW – SE GRADIENT, METERS/DEGREE

-5   0   +5

A

SEA SURFACE HEIGHT ANOMALIES
SEASAT ALTIMETER DATA, JULY 7 – OCTOBER 10, 1978

Features newly discovered or significantly
modified SEASAT altimeter data

Plateau

Conrad
Rise

Kerguelen I.

Kerguelen Plateau

Elan
Bank

Seamount

Trough

Marion
Durfusne
Seamount

Rise

Seamount

Large rise on
geoid high

Seamounts

Louisville Ridge

PARKE, DIXON
AND HUSSEY
1982

METERS

-4   -3   -2   -1   -.5   0   .5   1   2   3   4

B

83

FIGURE 4—5   The Evolution of Deep-ocean Basins and Continental Margins.
A, a continent begins to rift apart above a region of magma development in the lithosphere. *B,*
rifted blocks drop down on either side of a rift valley above the spreading center. *C,* further spreading produces a narrow ocean, sediment derived from the continents begins to accumulate on the
downdropped rift blocks. *D,* a deep-ocean basin forms with continental margins produced through
continental sediment deposition at its margins. Adapted from Tarbuck, E. J., and Lutgens, F. K.,
1988, fig. 6.9.

on the continent. The break at the top of the slope
may be from 1 to 5 km (0.62 to 3 mi) above the deep-
ocean basin at its base. In areas where the slope descends into submarine trenches, even greater vertical
relief is measured. Off the west coast of South America the total relief from the top of the Andes Mountains to the bottom of the Peru-Chile Trench is about
15 km (9.3 mi).

Continental slopes vary in steepness from 1° to 25°
and average about 4°. Around the margin of the Pa-

cific Ocean, where the slope is associated with the
processes forming coastal mountain ranges and submarine trenches, the continental slopes are steeper
than in the Atlantic and Indian oceans. Slopes in the
Pacific Ocean average more than 5°, while those in
the Atlantic and Indian oceans are about 3°.

Submarine Canyons   The continental slope and,
less commonly, the continental shelf are cut by large
submarine canyons that resemble the largest of can-

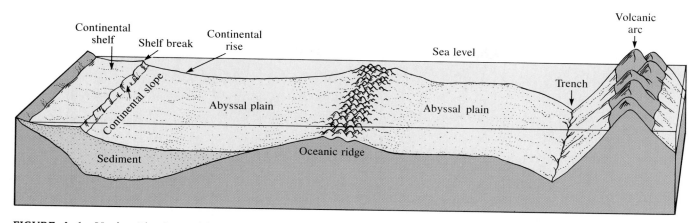

FIGURE 4–6   Marine Physiographic Provinces.

A longitudinal profile showing submarine physiographic provinces. To the left of the oceanic ridge is a continental margin typical of the Atlantic Ocean, with a well-developed continental shelf, slope, and rise at the surface of a thick sedimentary deposit. The features to the right are more typical of the Pacific Ocean, where trenches and associated volcanic arcs are common.

yons cut on land by rivers. Like the canyons cut by rivers, the submarine canyons have tributaries and steep V-shaped walls that expose a wide range of rock types varying greatly in geological age. Exposed in the walls of the canyons are rocks ranging from soft shales to quartzite and granite. The canyons are obviously erosional features; the problem is to explain how the erosion occurred.

Submarine canyons were one of the primary objects of study of the "father of marine geology," the late Francis P. Shepard, who began his informal studies of the ocean floor in the 1920s (fig. 4–7). Dr. Shepard continued to investigate coastal changes and the effect of volcanic activity on climate until his death at age 87 in 1985.

Recent work done with side-scan sonar indicates that the continental slope is dominated by submarine canyon topography along the Atlantic coast from Hudson Canyon to Baltimore Canyon (fig. 4–8). Smooth intercanyon areas are particularly rare on the middle and upper surfaces of the slope. There is also a correlation between the spacing of canyons and the continental slope gradient. No canyons are present where the slope is less than 3°. Canyons are from 2 to 10 km (1.2 to 6 mi) apart where the slope is between 3° and 5°, and they are 1.5–4 km (0.9–2.5 mi) apart where the slope gradient exceeds 6°.

Canyons confined to the continental slope are straighter and have steeper canyon floor gradients than those that extend into the continental shelf. This condition suggests that the canyons are initiated on the continental slope by some marine process and cut their way headward into the continental shelf as they age.

FIGURE 4–7   Francis P. Shepard (1898–1985), Father of Marine Geology.

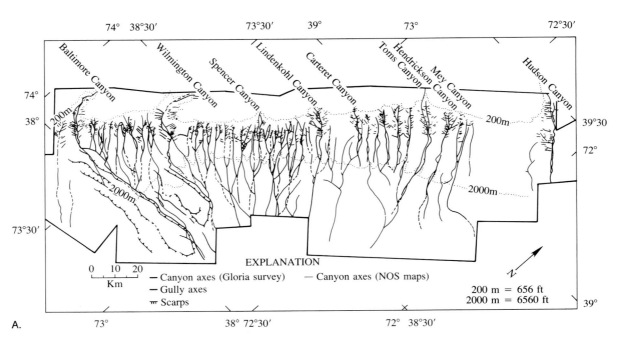

A.

FIGURE 4–8   Submarine Canyons.

A, submarine canyons along the Atlantic coast. There are no submarine canyons between Hudson Canyon and Mey Canyon where the slope gradient is less than 3°. The distance between adjacent submarine canyons decreases as the slope gradient increases. Between Lindenkohl Canyon and Baltimore Canyon, the gradient exceeds 6° for most of the continental slope. Where the slope is steepest, the submarine canyons are spaced as little as 1.5 km (0.9 mi) apart. *B,* erosional and depositional features associated with the formation of submarine canyons.

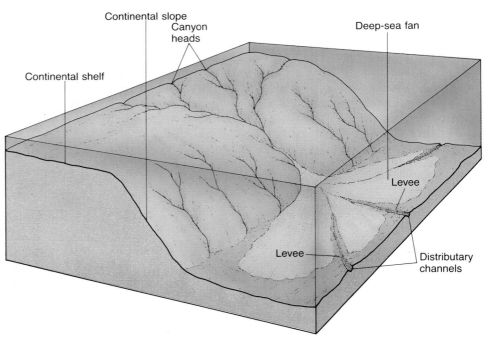

B.

Turbidity Currents Probably the most widely supported theory concerning the origin of submarine canyons is erosion by **turbidity currents,** flows of sediment-laden water that periodically move down the canyon. Turbidity flows resulting from river input near the heads of some Pacific-coast submarine canyons that extend well onto the continental shelf have been observed. However, turbidity currents in many canyons confined to the continental slope apparently are initiated after sediment moves across the continental shelf into the head of the canyon and accumulates there. The actual initiation of the current may be produced by an earth tremor or other disturbance that makes the mass collapse and start to move down the canyon under the force of gravity. As it moves, the sediment will mix with the water, producing a high-density mass that moves down the canyon, spreads out, and is deposited at the base of the continental slope as part of a **deep-sea fan** structure commonly found at the foot of a submarine canyon. The flow moves out across the fans somewhat as a river flow moves across the surface of a delta. A pattern of distributaries develops. They are bordered by levees that are deposited as the turbid mass overflows the banks of the relatively shallow **distributary channels** (fig. 4–8B).

The remote sonar recording of a turbidity current generated on August 26, 1976, by the discharge of mine tailings into Rupert Inlet, British Columbia (fig. 4–9), may be the only real-time observation of a marine turbidity current moving through its channel. The sonograms show acoustic sounding profiles across the submarine channel in Rupert Inlet before and during the event that lasted for about one and one-half hours.

Another line of evidence that indicates that turbidity currents move across the ocean bottom is associated with trans-Atlantic cable breaks that occurred on the continental shelf and slope south of Newfoundland after an earthquake on November 18, 1929 (fig. 4–10). These cables were broken in a pattern that indicated that the cables closest to the earthquake were broken first and those that crossed the shelf at greater depths and distances from the **epicenter** (point on ocean floor directly above earthquake) were broken later. This phenomenon could be explained by a dense flowing mass moving away from the epicenter down the slope and snapping the cables as it passed. The velocity required in this particular case would be approximately 27 km/h (16.8 mi/h), which seems quite high. The possibility that such a mass may move with this velocity makes it easier to understand how the submarine canyons may have been formed on the continental slope and continental shelf.

Continental Rise

Deep-sea fans accumulate as deposits at the mouths of submarine canyons (fig. 4–8B). The merging of these deep-sea fans along the base of the continental slope is partly responsible for the development of the **continental rise**.

The **Amazon Cone** is one of the largest deep-sea fans. It extends 700 km (434 mi) northeast of Brazil (fig. 4–11). Examination of the fan's distributary system using GLORIA long-range side-scan sonar revealed the following features. The main channel extends seaward from the **Amazon Submarine Canyon**. At a depth of about 2200 m (7200 ft), the channel splits into western and eastern systems that extend across the middle of the fan. Additional branching and development of meanders occur across the middle of the fan.

The onset of meandering of the distributary channels across the middle fan corresponds well to the decreased gradient of the middle fan surface as compared to that of the steeper upper fan. The meander patterns, as well as other features—abandoned meander loops, cutoffs, etc.—closely resemble in form and scale those found on the floodplains of mature river systems on land such as the lower Mississippi River.

A major factor in shaping the continental rise is the strong **western boundary undercurrent (WBUC)** or **slope current** that flows toward the equator at the base of the continental slope along the western boundaries of most ocean basins (fig. 4–10). This deep boundary current and the continental rise over which it flows have been extensively studied along the east coast of North America using submersibles, surface ships, and side-scan sonar.

The **Coriolis effect** (described in chapter 7) forces the WBUC, which originates in the Norwegian Sea and flows into the North Atlantic through deep troughs between Greenland and Scotland, to veer to the right and flow snugly against the base of the continental slope at all times. Picking up volcanic debris from Iceland and sediment from periodic turbidity flows, and scouring sediment from some portions of the continental slope and rise, the western boundary undercurrent, flowing at velocities that reach 40 cm/s (1 mi/h), creates a **benthic nepheloid layer (BNL)** of suspended sediment above the ocean floor (fig. 4–12). As the current is forced around bends in the continental slope, its velocity decreases and the rate of deposition increases, producing the deposits called **drifts or ridges,** shown in figure 4–10.

The surfaces of large drifts and ridges consist of **mud waves** that have an average distance between crests of 2–3 km (1.2–1.9 mi). These mud waves,

Sonograms

FIGURE 4–9   Rupert Inlet, British Columbia.

Top, bathymetry of submarine channel. Line T-T' is path followed by the Canadian Survey Ship *Vector* on August 26, 1976, when it recorded the occurrence of a turbidity current. The black circle denotes the approximate location of a bottom-moored current meter, and the diamonds labeled A, B, and C are locations of positioning transponders. Bottom, in sonogram A, the ship crossed the channel at about 12:53 P.M., and no turbidity current was flowing in the channel. Note the levees on either side of the main channel. The discrete spots in the water column above the bottom are thought to be fish. In B, the ship was over the channel at 1:03 P.M., and the turbidity current had begun to flow in the main channel. By 1:28 P.M. in C, the turbidity current had overspilled the levees. The turbidity current was not detectable 1.5 h after its onset. The steep wall is on the north side of the channel. The inclined vertical lines are the result of transmitted pulses from another sounder. Reprinted by permission from Hay, A. E., et al., "Remote acoustic detection of turbidity current surge," *Science* 217:4562, pp. 833–835, 27 August 1982. Copyright 1982 by the AAAS. Photos courtesy of Alex E. Hay.

FIGURE 4–10 Floor of the North Atlantic Ocean.

Arrows indicate part of North Atlantic Deep Water or western boundary undercurrent. Hexagon indicates cable breaks associated with the 1929 Grand Banks earthquake. North Atlantic panaroma
excerpted from *The World Ocean Floor* by Bruce C. Heezen and Marie Tharp, published by the
Lamont-Doherty Geological Observatory of Columbia University and sponsored by the U.S. Navy
through the Office of Naval Research. Copyright © 1977 by Marie Tharp.

FIGURE 4–11   The Amazon Cone.

The Amazon Canyon branches into two major distributary systems, the Western Levee Complex and the Eastern Levee Complex. They are composed of meandering channels bounded by levees. Note the large debris flows originating on the steep upper slope of the fan cone. Map courtesy of John E. Damuth and Roger D. Flood. Permission granted by the Geological Society of America.

shown in figure 4–13, are also found along much of the lower continental rise off the central Atlantic states and are known as **lower continental rise hills.**

Increased knowledge of continental-margin features and the processes that create them may be important in selecting ocean sites for toxic waste disposal and effective exploitation of oil, gas, and minerals.

DEEP-OCEAN BASIN

Beyond the continental slope, where depths generally exceed 4000 m (13,120 ft), lies the deep-ocean basin, which is entirely underlain by basaltic oceanic crust. Along the passive, Atlantic-type margins, the continental slope gradually descends to flat abyssal plains with slopes less than 1:1000. Continental slopes may descent into deep trenches along the more-active Pacific-type margins. Of course, the oceanic ridges and

rises running through the deep basins may raise bottom elevations to an average depth at their axes of about 2500 m (8200 ft). In the following paragraphs we will discuss some of the more interesting conditions of the deep ocean associated with these major elements of their framework.

Oceanic Ridges and Rises

Extending across some 65,000 km (43,365 mi) of the deep-ocean basin are continuous mountainous features, the oceanic ridges and rises which occupy about one-third of the basins and appear prominently on figure 4–14.

Hydrothermal Circulation at Spreading Centers   Near the axes of oceanic ridges and rises, where magma chambers may be less than 1 km (0.6 mi) beneath the ocean floor, seawater seeps down

FIGURE 4–12 Benthic Nepheloid Layer.

A, two light-scattering curves showing the nepheloid layer on the Blake-Bahama Outer Ridge. (See fig. 4–10). B, distribution of suspended material in the nepheloid layer in the Atlantic Ocean. Note the high concentrations found only in the western portions of the ocean overlying the continental rise. Compare the concentration pattern in the North Atlantic with the location of the western boundary undercurrent in figure 4–10. After Biscaye and Eittreim, 1977.

FIGURE 4–13 Mud Waves.

Echo sounding of mud waves taken along an east-west line across the Blake-Bahama Outer Ridge, a large depositional feature on the continental rise north of the Bahamas. The near-bottom bathymetric profile beneath the echo sounding gives a more accurate shape of the waves. Courtesy of Roger Flood, Lamont-Doherty Geological Observatory, Columbia University.

FIGURE 4—14   The Floor of the Oceans.

120°   140°   160°   180°   160°   140°   120°   100°   80°   60°

70°
60°
40°
20°
0°
20°
40°
60°
70°

LAPTEV SEA

NEW SIBERIAN ISLANDS

WRANGEL ISLAND

CHUKCHI SEA

CHUKCHI ABYSSAL PLAIN

CANADA ABYSSAL PLAIN

ELLESMERE

BAFFIN BAY

BAFFIN ISLAND

VICTORIA ISLAND

BROOKS RANGE

ALASKA

HUDSON BAY

NORTH AMERICA

BERING STRAIT

OKHOTSK SEA

BERING SEA

BERING ABYSSAL PLAIN

STANOVOY MTS.

KOLYMSKOYE MTS.

SAKHALIN

YELLOW SEA

Emperor Seamounts

Kurile Tr.

Japan Tr.

Aleutian Tr.

Mendocino F. Z.

Philippine Tr.

Mariana Tr.

Hawaiian Chain

Line Is.

Marquesas Is.

Tuamotu Is.

Society Is.

Cook Is.

Kermadec–Tonga Tr.

Eltanin F. Z.

East

Pacific

Rise

Mid. America Tr.

Galápagos Rift

Peru–Chile Tr.

SOUTH AMERICA

AUSTRALIA

GIBSON DESERT

SIMPSON DESERT

NULLARBOR PLAIN

ANTARCTICA

120°   140°   160°   180°   160°   140°   120°   100°

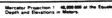
Editions Pierre Charron, 51 rue Pierre-Charron, 75008 Paris., Draeger, Imp.

Mercator Projection 1   40,030,000 at the Equator
Depth and Elevations in Meters.

Based on Bathymetric studies by
Bruce C. Heezen and Marie Tharp
of the Lamont Doherty Geological Observatory
Columbia University Palisades, New York, 10964
SUPPORTED BY THE UNITED STATES NAVY
OFFICE OF NAVAL RESEARCH

fractures in the ocean crust and is heated by the hot underlying magma. It then rises back toward the surface as hot water (fig. 4–15A). This rising water may exit into the ocean as **warm-water vents** (10°–20°C), **white smokers** (30°–330°C) containing white particles of barium sulfate, or **black smokers** with temperatures around 350°C (662°F) and containing black metal sulfides (fig. 4–15B). The entire volume of ocean water is recycled through this hydrothermal circulation system in about 3 million years, so the chemical exchange between ocean water and the basaltic crust has a major influence on the nature of ocean water. The deposition of metallic sulfides around these vents is likely a major source of continental sulfide ore deposits on earth as they may be transferred to the continents during subduction of the oceanic plate.

The oceanic hydrothermal vents support unusual biological communities first discovered in 1977 in the **Galápagos Rift** between South America and the Galápagos Islands (fig. 4–15C). Large clams, mussels, tube worms, and a large variety of other animals depend on bacteria that oxidize the hydrogen sulfide gas dissolved in the spring water as a source of energy to chemosynthetically produce food for the community.

Fracture Zones The bathymetry associated with the transform fault zones and the fracture zones that extend beyond the offset ridge and rise axes often displays great relief, and the length of fracture zones often exceeds thousands of kilometers. The **Mendocino Fracture Zone** in the North Pacific and the **Eltanin Fracture Zone** in the South Pacific are good examples (fig. 4–14). The physiographic relief near the fracture zones may extend through a width of up to 200 km (124 mi). The ocean floor on either side of a fracture zone may have depths differing by as much as 1500 m (4920 ft). For example, the ocean bottom on the southern side of the Mendocino Fracture Zone is more than 1000 m (3280 ft) deeper than that to the north. Therefore, this feature is sometimes referred to as the **Mendocino Escarpment**. Much of this difference in elevation can be attributed to the fact the ocean floor on the south side of the escarpment is as much as 40 million years older than that to the north and has undergone much more thermal contraction and deepening than the younger ocean floor to the north.

Although fracture zones are not known to be highly seismic, there is much evidence in the structure of fracture zones in the equatorial Atlantic and Indian oceans that they are tectonically active. The three fracture zones we will discuss are the **St. Paul Fracture Zone** and **Romanche Fracture Zone** in the equatorial Atlantic Ocean and the **Owen Fracture Zone** in the Indian Ocean. They are labeled on figure 4–14.

The St. Paul islets are peaks on a **transverse ridge** that runs parallel to the fracture zone. Charles Darwin collected peridotite, a rock typical of the mantle, when he sampled these small islands during the voyage of the H.M.S. *Beagle*. Significant recoveries of mantle rocks from the two Romanche Fracture Zone transverse ridges, which are separated by a deep trenchlike valley, and from the transverse ridge of the Owen Fracture Zone, shown in figure 4–16, indicate tectonic uplift of mantle material in these fracture zones. The uplift of mantle blocks may well result from compression across the fracture zone that occurs when the alignment of the offset spreading axes changes after the fracture zone has formed.

Intraplate Features

As newly created ocean floor moves away from the spreading centers, volcanic submarine peaks and islands are clearly affected in their development by the increasingly depth of the ocean floor that supports them during their millions of years of travel before they are subducted at a destructive plate boundary.

Seamounts and Tablemounts Although the features described below are abundant in all oceans, the evidence of volcanic activity on the bottom of the Pacific Ocean is particularly widespread—more than 20,000 volcanic features are found there. Some marine volcanic activity produces widespread gently sloping surfaces where large volumes of lava flowed out in broad sheets, solidified, and became **archipelagic aprons**. Such features commonly surround volcanic islands.

Volcanic peaks that extend more than 1 km (0.62 mi) above the abyssal floor are termed **seamounts**, unless they have tops flattened by erosion, in which case they are classified as **tablemounts, or guyots**. Tablemounts are less common than seamounts and are found in linear trends in the Pacific Ocean. The tops of most tablemounts in the Pacific Ocean are between 1800 and 3000 m (5904 and 9840 ft) below the present ocean surface.

Volcanic features on the ocean bottom that are less than 1 km (0.62 mi) above the ocean floor are called **abyssal hills**. They cover a large percentage of the entire ocean basin floor and have an average relief of about 200 m (656 ft); many such features are found buried beneath the sediments of abyssal plains.

A.

C.

B.

FIGURE 4–15   Hydrothermal Vents along Spreading Center Axes.
A, cross section across the axis of a spreading center to illustrate the hydrothermal circulation in
oceanic crust. Seawater percolates down along a broad zone of fractured crustal rock, is heated as
it approaches the underlying magma chamber and rises in a narrow zone near the axis of the
spreading center. The resulting vents usually lie within 200 m (656 ft) of the ridge or rise axis.
Reprinted with permission from Tarbuck, E. J., and Lutgens, F. K., 1988, fig. 10.17. *B,* black
smoker. During a 1979 expedition to 21°N on the East Pacific Rise, this "black smoker" spewing out
315°C (600°F) water was observed. In the foreground is *Alvin's* sample basket. *C,* large tube worms
observed at East Pacific Rise warm-water vent. The tubes are about 1 m (3.2 ft) long and the vestimentiferan worms that inhabit them have no mouth or digestive tract. Symbiotic chemosynthetic
bacteria oxidize the hydrogen sulfide gas dissolved in the vent water and use the energy released
to produce food for themselves and the worms. Photos courtesy of Scripps Institution of Oceanography, University of California, San Diego; Photo *B* by Dr. Fred N. Spiess.

FIGURE 4–16 Owen Fracture Zone.

This cross section of the Owen Fracture Zone shows the transverse ridge to the west of the Wheatly Deep (arrow). All dredges of the ridge recovered ultramafic mantle-type rocks and indicate the ridge resulted from uplift of mantle material rather than volcanism. Courtesy of Enrico Bonatti, Lamont-Doherty Geological Observatory of Columbia University.

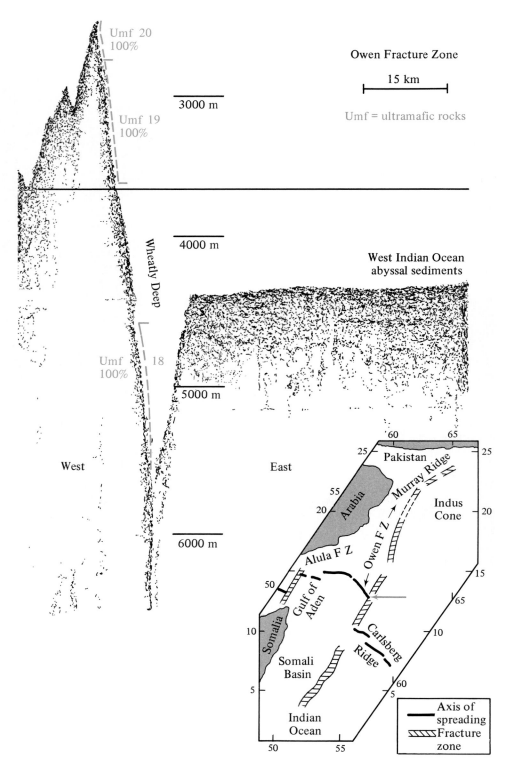

A theory related to the formation of such features in association with the oceanic ridge systems could account for the existence of many seamounts and tablemounts (fig. 4–17). Considering oceanic ridge systems on a worldwide basis, active oceanic volcanoes are characteristic of the axes of the ridges. Moving away from the ridge axis, the volcanic activity gradually decreases as the cracks in the earth's crust serving as conduits for rising lava near the axis of these ridges are sealed off as they move away. Active volcanoes over 30 million years of age are rare.

Let us consider the life history of a theoretical volcano that progressively moves away from an oceanic ridge, being carried by the lithospheric plate. A volcano may originate near the axis of the ridge and gradually increase in size as it moves downslope. By the time it is 10 million years old, if it has been a very active volcano, it may develop into an island that may exist for another 10 to 15 million years if the volcano remains active. The volcano will probably become dormant within 30 million years of its origin, and as it continues to move down the flank of the oceanic ridge, the top will be flattened by wave action and be ultimately submerged. The flat-topped structure will sink deeper and deeper beneath the ocean surface as it descends the slope of the ridge. It has now become a guyot or tablemount, most of which are at least 30 million years of age.

Coral Reef Development Although the development of a comprehensive theory of global plate tectonics was not to be developed until over 100 years later, Charles Darwin developed a theory on the origin of coral reefs that depended on the subsidence of volcanic islands. Darwin published his theory in the book *The Structure and Distribution of Coral Reefs* in 1842. Figure 4–18 illustrates Darwin's theory of reef development. Drilling into reefs before and after World War II provided the evidence to prove the validity of Darwin's theory.

There are three basic types of reefs—fringing, barrier, and atoll (fig. 4–18). **Fringing reefs** can develop at the margin of any landmass where the temperature, salinity, turbidity, and depth of the water are suitable for reef-building corals. The greatest concentration of live reef will be on the seaward margin, which is more protected from sediment and salinity changes due to runoff. Without some change in sea level relative to the landmass during the existence of the fringing reef, it is not likely that the other two types of reefs we have mentioned, the barrier reef or the atoll, will form.

Barrier reefs are linear or circular developments separated by a lagoon from the continent or island on whose margin they are growing. It is thought that barrier reefs form when the substratum on which a fringing reef is built begins to subside. As the island or margin of the continent subsides, the reef maintains its position at the optimum water depth by growing upward. Studies of reef growth rates indicate most have grown 3 to 5 m/1000 yr (9.8 to 16.4 ft/1000 yr) during the recent geologic past. There is some evidence, however, that reefs in the Caribbean have grown at rates of over 10 m/1000 yr (32.8 ft/1000 yr).

The most outstanding example of a barrier reef is the **Great Barrier Reef** along the northeast coast of Australia. It is 150 km (93 mi) wide and more than 2000 km (1242 mi) long. Smaller barrier reefs are found around the island of Tahiti and other islands in the western Atlantic and Pacific oceans.

Atolls may form either on a subsiding continental shelf or around the margin of a sinking volcanic island in the open ocean. Most atolls are on sinking volcanic peaks, and the greatest number occur in the equatorial Pacific. Atolls may take a circular or irregular shape and surround a lagoon that is usually 30 to 50 m (98 to 164 ft) deep at its deepest part. The fringing reef will generally have many channels that allow circulation between the lagoon and the open ocean. The reef flat is commonly broader on the windward side of the atoll, and channels are more numerous on the leeward side (fig. 4–19).

FIGURE 4–17 Formation of Seamounts and Tablemounts.
Volcanic activity develops near the crest of the ridge because of tensional fractures that develop parallel to the ridge. These fractures serve as conduits for lava but usually seal within 30 million years.

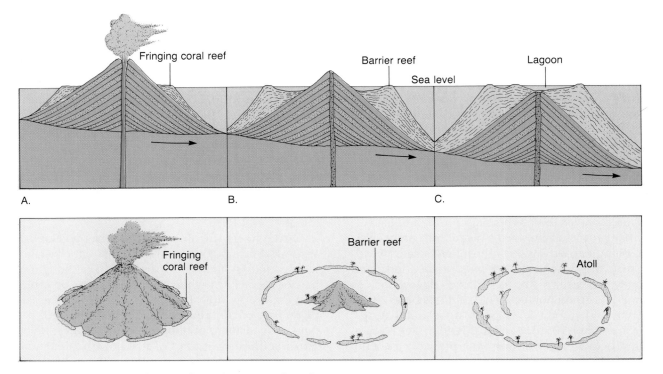

A.   B.   C.

FIGURE 4-18 Darwin's Theory of Development of Reef Types.

A volcanic island forms in warm-water latitudes. *A,* a fringing coral reef forms in the shallow, sunlit waters adjacent to the island shore. *B,* as the dormant volcano gradually subsides due to thermal contraction of the lithosphere, the reef maintains its position near the surface by growing upward. This produces a barrier reef where the actively growing reef is separated from the volcanic island by a well-developed lagoon. *C,* continued subsidence submerges the volcano, but the reef may continue to grow upward to produce an atoll with active reef growth confined mostly to the outer edge of the reef. Deposition of reef debris may produce low-lying islands inside the protective ring of coral reef. Within the atoll ring lies a broad, shallow lagoon. Reprinted by permission from Tarbuck, E. J., and Lutgens, E. K., 1988, fig. 10.19.

Not only is the subsidence of the substrate of coral reefs recorded in the reef development, but in the case of the large Great Barrier Reef, the northward horizontal movement of the Australian continent is clearly recorded as well. The Great Barrier Reef thins and the basal reef material becomes younger to the south. Figure 4–20 shows that the reef began forming about 30 million years ago when the northern end of the reef reached a latitude near the Tropic of Capricorn and temperate climatic conditions were replaced by sufficiently tropical conditions to allow for coral reef development.

Pacific Ocean Hot Spots Many northwest–southeast-trending island chains are located on the Pacific Plate. The most intensely studied of these is the chain of islands between Midway Island and Hawaii. There are no active volcanoes on Midway Island; volcanic activity appears to have ceased 20 million years ago. The ages of the islands decrease toward Hawaii, where the only active volcanism in the chain exists. It is manifested in the volcano Kilauea, which is less than 1 million years old. An active submarine volcano, Loihi, currently exists southeast of Hawaii and may either become part of that island or create a new island (fig. 4–14).

The explanation for the existence of this **nematath**, a chain of islands and seamounts with a progressive increase in age toward the northwest, is that a **hot spot**, a relatively stationary upwelling (plume) of magma, exists beneath the Pacific Plate in the region of the island of Hawaii. As the plate moves across this hot spot, volcanic eruptions produce islands that subsequently become dormant, erode, and subside as they move off to the northwest (fig. 4–21).

Continuing northwest from Midway Island, submerged seamounts have the same alignment as the island chain from Midway to Hawaii. At about 188°W longitude and 30°N latitude, an abrupt change in the alignment of the seamounts occurs. From this point on, the alignment is slightly west of true north. The age of the seamounts at this juncture of the two trends is thought to be between 30 and 40 million years, whereas the oldest seamounts in the northward-trending **Emperor Seamount Chain** formed some 70 million years ago. If the Emperor Seamount Chain and the **Hawaii seamounts and islands** were generated by the same hot spot, there must have been an abrupt change in the direction of the move-

FIGURE 4–19 (opposite) Atolls.
View from space of a group of atolls in the Pacific Ocean. Photo courtesy of NASA.

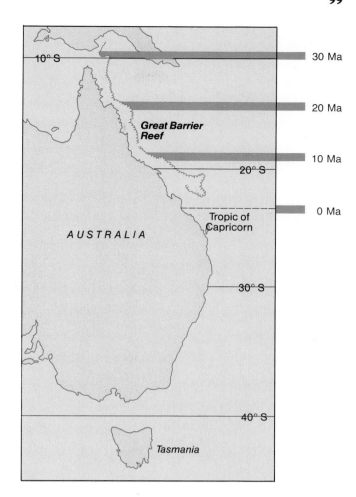

FIGURE 4–20 Australia's Great Barrier Reef and Global Plate Tectonics.

The Great Barrier Reef extends for more than 2000 km (1240 mi) along the northeastern coast of Australia. It extends between latitudes 9° and 24°S. The reef first began to develop about 30 million years ago when northern Australia moved into tropical waters from colder Antarctic waters to the south. The southern end of the Great Barrier Reef has moved into tropical waters only recently.

ment of the Pacific Plate between 30 and 40 million years ago.

The **Line Islands** to the south of the Hawaiian Islands and other northwest–southeast-trending volcanic island chains—the **Marquesas Islands**, **Tuamotu Islands**, **Society Islands**, and **Cook Islands**—may have formed more or less at the same time. Although there are indications these chains of islands may have a hot spot origin, the evidence is not as clearly developed as for the Hawaiian Islands.

The origin of hot spots is still controversial. The evidence presented by the Emperor Seamount–Hawaiian Island Chain would lend support to a stationary lower mantle source resulting from heating the mantle from below; the evidence found at the

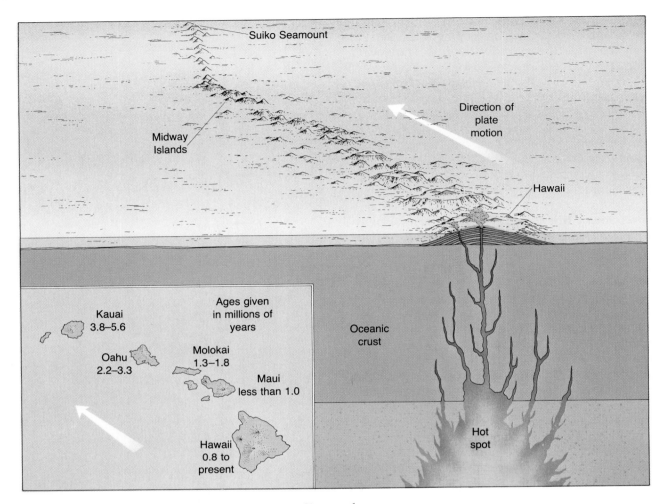

FIGURE 4–21   Hawaiian Islands–Emperor Seamount Nematath.
The chain of islands and seamounts that extends from Hawaii to the Aleutian trench results from
the movement of the Pacific Plate over an apparently stationary hot spot. Radiometric dating of the
Hawaiian Islands shows that the volcanic activity decreases in age toward the island of Hawaii.
Reprinted with permission from Tarbuck, E. J., and Lutgens, F. K., 1990, fig. 18.28.

other linear island chains mentioned does not support such an origin. Some investigators believe that
hot spots are associated with rifts in the lithosphere
resulting from cooling at the surface and plate interactions. Whatever the origin, it is now clear that hot
spots are not as stationary as once believed and move
around at rates of at least 1 to 2 cm (0.4 to 0.8 in.)
per year.

Hot Spots and Flood Basalts   Large volumes of
basalt that may exceed 2 million km^3 (480,000 mi^3)
are present on the margins of continents and appear
to have been produced at rates of about 1 km^3 (0.24

mi^3) per year over a relatively short interval of about
1 or 2 million years (fig. 4–22). These deposits are
called **flood basalts**. When hot spot heat plumes
reach the lithosphere, they form broad heads that
may range from 1000 to 2000 km (620 to 1240 mi) in
diameter. The temperature of the mantle within the
plume head is from 100° to 200°C (180° to 360°F) hotter than the surrounding mantle and produces a
broad doming at the earth's surface. If such a mushroom-shaped plume head is beneath a continent
when it rifts apart, the decompression will allow a
large amount of the hot mantle to melt and bleed out
at the surface to produce the flood basalt deposits.

FIGURE 4–22 Flood Basalts.

Four occurrences of paired flood basalts are plotted on the map. The age in Ma of their formation is indicated in parentheses adjacent to the still-active hot spots that produced them. The flood basalt deposits are believed to have formed when the continents on which the deposits are found first rifted apart over the hot spot. Associated with the Iceland hot spot are the Greenland-Hatton Bank flood basalts (off Scotland). The Parana and Etendeka flood basalts of South America and southwest Africa, respectively, were produced by the Tristan da Cunha hot spot. The Reunion hot spot east of Madagascar produced the Deccan Traps of India and the Saya de Malha Bank, a submarine plateau east of the Seychelles Islands. In the South Indian Ocean, the Marion-Prince Edward hot spot produced the Karoo (Africa) and Antarctic flood basalts.

Subduction Zones

Around the margin of the Pacific Ocean we see ocean trenches as dominant physiographic features of the ocean bottom. It is in these regions that the older lithospheric material is being subducted as new lithosphere is produced along the ridges and rises. The Pacific trenches range in depth from just under 7 km (4 mi) to just over 11 km (6.8 mi) in the Mariana Trench (table 4–1).

The Wadati-Benioff seismic zone mentioned in the previous chapter marks the position of a subducting slab of lithosphere, but the slab sinks more steeply

TABLE 4–1 Dimensions of Trenches.

Trench	Depth (m)	Mean width (km)	Length (km)
Peru-Chile	8,100	100	5,900
Aleutian	7,700	50	3,700
Middle America	6,700	40	2,800
Mariana	11,022	70	2,550
Kurile	10,500	120	2,200
Kermadec	10,000	40	1,500
Tonga	10,000	55	1,400
Philippine	10,500	60	1,400
Japan	8,400	100	800

than the angle of slope of the Wadati-Benioff zone. This allows the "hinge," or bend, in the subducting plate to move backward with time as the overriding plate advances. It is this process that has allowed the Pacific Ocean to become smaller over time as the purely oceanic lithospheric plates produced within it subduct beneath the plates carrying continents—North American, South American, Eurasian, and Indo-Australian plates.

Mariana Island Arc

Seaward Migration of Island-Arc Systems
When we think of plates colliding at subduction zones, it is easy to envision them as zones of compression. However, due to the seaward migration of "hingelines," tensional stresses are much more commonly observed. This may well result from the fact that the thick, massive plate that is already subducted pulls the subducting plate over the "hingeline" with a force that greatly exceeds any other force that may contribute to plate motion.

Figure 4–23 shows an example of the existence of tensional stress within the Mariana Island-Arc subduction system. Keeping in mind that an active volcanic island arc develops along an arc that is approximately 100 km (62 mi) above the top of the subducting plate, the subducting plate is at present 100 km beneath the **active arc** containing Pagan Island. About 200 km to the west is the original island arc for the Mariana Trench, which is now a **remnant arc**, the West Mariana Ridge. The volcanic peaks on this ridge are old and dormant, and those that once were islands have been submerged by subsidence. It appears that the 100-km depth to the top of the subducting plate has migrated 200 km eastward relative to the West Mariana Ridge.

Between the two volcanic arcs lies a **back-arc spreading center**. Here we have a small-scale sea-floor spreading system with its rift valley, hydrothermal circulation, and even fracture zones. This is clear evidence of the tensional forces at work within the island arc–subduction zone system. Although the origin of back-arc spreading centers is also controversial, it is widely believed they form as a result of the rapid seaward migration of the trench and subduction "hinge" that lies seaward of the trench. This pulling apart of the arc system produces linear fractures in a thin volcanic crust, allowing the upwelling of underlying magma. As long as the seaward migration continues at a sufficient rate, new ocean floor is created within the rift valley of the spreading center.

A Strange Kind of Seamount
A previously unknown feature of subduction zones was observed during Ocean Drilling Program operations in the Mariana fore-arc in 1989 (fig. 4–23). The fore-arc is that part of the overriding plate that lies seaward of the volcanic arc. What were thought to be volcanic seamounts in a belt 50 to 150 km (31 to 93 mi) from the trench axis turned out to be serpentine, a mudlike material that forms when water mixes with peridotite, the rock characteristic of the earth's mantle. Apparently, seawater entered the mantle from above through fractures in the fore-arc crust and from below from water trapped in subducting sediments. Because of its low density, the serpentine flowed up to the surface of the fore-arc block, carrying blocks of peridotite and basalt (possibly from the subducting ocean crust) to produce volcanolike seamounts near the seaward front of the fore-arc structure.

SUMMARY

Most of the knowledge we have gained about the nature of the ocean floor has been obtained using **echo sounders, seismic surveys,** and **side-scan sonar.** Seismic surveys can reveal the topography of the ocean floor and the location of rock unit boundaries beneath it. Side-scan sonar is used to make strip maps of ocean-floor topography along the course of the survey ship.

Extending from the shoreline of continents are gently sloping **continental shelves** that extend out to sea for an average distance of 70 km (43 mi). Sediment is transported across the continental shelves by currents and mass wasting (submarine sediment slides). When the shelves reach an average depth of 135 m (443 ft), they steepen into a feature called the **continental slope.** The slopes continue to the deep-ocean floor with steepness ranging from 1° to 25° and averaging 4°. Cutting deep into the slopes and reaching well onto the continental shelves, in some cases, are **submarine canyons.** The origin of the canyons is unknown, but they may be caused in part by periodic flows that are a mixture of sediment and water called **turbidity currents.** Turbidity currents deposit their sediment loads at the base of the continental slope to produce **deep-sea fans** or **cones** through a series of meandering **distributary channels,** which branch out from the mouth of the submarine canyon. The merging of these fans produces a gently sloping **continental rise** at the bases of some continental slopes.

FIGURE 4–23 Mariana Island-Arc Subduction System.

Location map, detail strip map, and cross section of the Mariana subduction system. The remnant arc (West Mariana Ridge) represents an ancient dormant volcanic arc separated from the present active arc (Mariana Islands) by a back-arc spreading center. All of these features are products of subduction initiated at the Mariana Trench. As the trench migrates seaward, the region of active volcanism that produces island arcs above the 100-km depth of the descending slab of lithosphere also migrates seaward. The tensional stresses associated with the seaward migration of the trench fracture the thin oceanic crust of the overriding plate and initiate the back-arc spreading. Continued migration maintains the back-arc spreading, which continues to increase the distance between the remnant arc and the active arc—also migrating seaward and maintaining a position above the 100-km depth of the descending Pacific Plate. The seamountlike structures on the leading edge of the fore-arc are produced by rising mudlike serpentine produced when water combines with peridotite (mantle rock) in the overriding plate at temperatures above 100°C. Courtesy of Peter Lonsdale, Scripps Institution of Oceanography, University of California, San Diego.

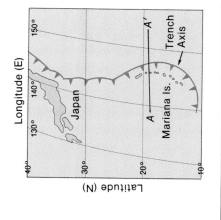

103

A major factor in distributing sediment along the continental slope is the **western boundary undercurrent (WBUC)** or slope current that flows toward the equator along the western boundary of ocean basins. The sediment suspended and carried above the bottom by these currents creates a **benthic nepheloid layer**. When the current slows to make turns along the base of the continental slope, it deposits sediment onto features called **drifts** or **ridges** that become prominent features of the continental rise.

The continental rises gradually become flat, extensive, depositional features of the deep-ocean basin called **abyssal plains**. These plains are often penetrated by volcanic peaks. Along the axes of oceanic ridges and rises, seawater percolates down through fractures to be heated and returned to the surface as **hydrothermal vents** with temperatures up to 350°C (662°F). This process has produced most of the earth's **metal sulfide ore deposits**, and new biological communities were discovered in 1977. These **vent biocommunities** depend on chemosynthesis by hydrogen sulfide–oxidizing bacteria rather than the photosynthetic plankton that support most marine life.

Fracture zones exhibit great amounts of relief, ranging from the 1000-m (3280-ft) cliff of the Mendocino Escarpment, which results from the differential cooling of different-aged lithosphere on opposite sides of the fracture zone, to the deep valley between two **transverse ridges** of the **Romanche Fracture Zone**.

Intraplate features that illustrate the effect of increasing depth of the ocean floor with its increasing age are a **nematath** with the alignment and age progression of the **Emperor Seamount–Hawaiian Island Chain** produced by a single **hot spot** and the progression of coral reefs through stages from **fringing reef**, **barrier reef**, and **atoll**. Scattered throughout much of the ocean floor are volcanic peaks called **seamounts** if they extend more than 1 km (0.62 mi) above the ocean floor and **abyssal hills** if they don't. Submerged peaks that once had their peaks flattened by wave erosion are **tablemounts** or **guyots**.

Some interesting features related to subduction zones are illustrated by the structure of the **Mariana Island Arc** subduction system. Here, the seaward migration of the trench has produced a **remnant arc** and **active arc** separated by a **back-arc spreading center**. The incorporation of seawater into the mantle rock at the subduction zone has produced strange seamountlike structures composed of **serpentine** on the seaward half of the **fore-arc** platform.

QUESTIONS AND EXERCISES

1. Discuss the slope and width of the continental shelf and the areas in which it is best developed.

2. Describe the most widely accepted theory explaining the development of submarine canyons on the continental slope. Include a discussion of the physical evidence that lends support to the theory.

3. What features observed on the Amazon Cone are also found on the floodplains of mature rivers?

4. What are the sources of sediment that become suspended in the nepheloid layer?

5. Why is the nepheloid layer best developed along the western margins of the ocean basins?

6. Describe the general process of hydrothermal circulation associated with spreading centers and their relationship to mining ore deposits and chemosynthetically supported biocommunities.

7. Explain the reasons for the existence of a 1000-m escarpment on the Mendocino Fracture Zone and the transverse ridges present in the St. Paul, Romanche, and Owen fracture zones.

8. Explain why seamounts and guyots increase in age and depth with increased distance from the oceanic ridges and rises. Discuss the length of time they probably were active volcanoes.

9. Discuss the possible relationship of coral reef development to sea-floor spreading. Include the effects of both vertical and horizontal motions.

10. How are the alignments and age distribution patterns of the Emperor Seamount and Hawaiian Island chains explained by the hot spot theory?

11. Describe the process thought to be responsible for the development of the remnant arc, active arc, and back-arc spreading center of the Mariana Arc subduction system.

REFERENCES

Biscaye, P. E., and Eittreim, S. L. 1977. Suspended particulate loads and transports in the nepheloid layers of the abyssal Atlantic Ocean. *Marine Geology* 23:155–72.

Bonatti, E. 1978. Vertical tectonism in oceanic fracture zones. *Earth and Planetary Science Letters* 37, 369–79.

Damuth, J. E.; Kolla, V.; Flood, R. D.; Kowsmann, R. O.; Monteiro, M. C.; Gorini, M. A.; Palma, J. J.; and Belderson, R. H. 1983. Distributary channel meandering and bifurcation patterns on the Amazon deep-sea fan as revealed by long-range side-scan sonar (GLORIA). *Geology* 11:2, 94–98.

Davies, P. J., Symonds, P. A., Feary, D. A., and Pigram, C. J. 1987. Horizontal plate motion: A key allocyclic factor in the evolution of the Great Barrier Reef. *Science* 238:4834, 1697–99.

Green, H. W., and Burnley, P. C. 1989. A new self-organizing mechanism for deep-focus earthquakes. *Nature* 341:6224, 733–37.

Hamilton W. B. 1988. Plate tectonics and island arcs. *Geological Society of America Bulletin* 100:10, 1503–27.

Hay, A. E.; Burling, R. W.; and Murray, J. W. 1982. Remote acoustic detection of a turbidity current surge. *Science* 217:4562, 833–835.

Hollister, C. D.; Flood, R. D.; and McCave, I. N. 1978. Plastering and decorating the North Atlantic. *Oceanus* 21:1, 5–13.

Molnar, P., and Stock, J. 1987. Relative motions of hotspots in the Pacific, Atlantic and Indian oceans since late Cretaceous time. *Nature* 327:6123, 587–91.

Shepard. 1973. *Submarine geology.* New York: Harper and Row.

Sverdrup, H. U.; Johnson, M. W.; and Fleming, R. H. 1942. Renewal 1970. *The oceans: Their physics, chemistry, and general biology.* Englewood Cliffs, N.J.: Prentice-Hall.

Tarbuck, E. J., and Lutgens, F. K. 1990. *The earth: An introduction to physical geology.* 3d ed. Columbus: Merrill.

————. 1988. *Earth science.* 5th ed. Columbus: Merrill.

Twichell, D. C., and Roberts, D. G. 1982. Morphology, distribution and development of submarine canyons on the United States Atlantic continental slope between Hudson and Baltimore canyons. *Geology* 10:8, 408–12.

Weirich, F. H. 1984. Turbidity currents; Monitoring their occurrence and movement with a three-dimensional sensor network. *Science* 224:4647, 384–87.

SUGGESTED READING

Sea Frontiers

Mark, K. 1976. Coral reefs, seamounts, and guyots. 22:3, 143–49.
A discussion of the role of global plate tectonics in explaining the distribution of seamounts, guyots, and the evolution of coral reefs.

Schafer, C., and Carter, L. 1986. Ocean-bottom mapping in the 1980s. 32:2, 122–30.
The use of SeaMARK (Seafloor Mapping and Remote Characterization), a side-scan sonar device, in mapping the continental margin off the coast of Labrador is described.

Shepard, F. P. 1975. Submarine canyons of the Pacific. 21:1, 2–13.
A description of canyons around the Pacific margin and discussion of their possible origin.

White, R. S., and McKenzie, D. P. 1989. Vulcanism at Rifts. 261:1, 62–77.
New information about the size, depth, and length of magma chambers beneath spreading center rifts is discussed.

Scientific American

Emery, K. O. 1969. The continental shelves. 221:3, 106–25.
The nature of the continental shelves and the effect of the advance and retreat of the shoreline across them as a result of glaciation are discussed.

Heezen, B. C. 1956. The origin of submarine canyons. 195:2, 36–41.
Theories explaining the origin of submarine canyons are presented along with data on the location and nature of such features.

Herbert, Sandra. 1986. Darwin as a geologist. 254:5, 116–23.
Before publication of *The Origin of Species,* Charles Darwin considered himself to be primarily a geologist. This article outlines his contributions in this field, with special attention being given to his theory of coral reef formation.

Menard, H. W. 1969. The deep-ocean floor. 221:3, 126–45.
A summary of the dynamic effects of sea-floor spreading is presented with a description of related sea-floor features.

5
MARINE SEDIMENTS

Over half of the rocks exposed at the surface of the continents above the present shoreline were formed by the lithification of sediment laid down in past ocean environments. This relationship between the continents and oceans had puzzled geologists until global plate tectonics theory explained the mechanism for incorporating ocean sediments into the continents: folded mountain ranges formed when two plates collided and compressed the sediments against the margin of the existing continent. Geologists once believed they could examine the sediments in the deep-ocean basins, which they assumed had existed since the earth's oceans first formed, and find an undisturbed record of the earth's history. Knowledge of the plate tectonics process has removed that dream, but much of that ancient record is still available in ancient marine sediments exposed in continental mountain ranges.

Although it seems unlikely that marine sediments more than 180 million years old will be found in the present ocean basins, the enthusiasm of marine geologists for examining them has not dimmed. There is still much to learn about this shorter period of history. Clues to past climates, movements of the ocean floor, changes in ocean circulation patterns, and nutrient supply for marine plant populations are embedded in the sedimentary deposits throughout the ocean basins.

SEDIMENT TEXTURE

Sediment texture, which is determined primarily by grain size, is indicative of the energy condition under which a deposit is laid down. The **Wentworth scale** presented in table 5–1 classifies particles in categories ranging from boulders to colloids. Between these extremes are cobbles, pebbles, granules, sand, silt, and clay-sized particles. Deposits that are laid down in areas where wave action is strong (areas of high energy) may be composed primarily of the larger particles, cobbles, and boulders. The deposition of clay-sized particles occurs in areas where the energy level is low and the current speed is minimal.

Sediments composed of particles primarily in the same size classification are well sorted. A beach sand is usually a well-sorted deposit. Glacial deposits sometimes found on the continental shelf are an ex-

Preceding page: Pteropod, *Corolla spectabilis*. The only opaque part of this swimming gastropod is the conical internal calcium carbonate shell, which is an important component of some marine sedimentary deposits. This organism inhabits the surface waters of the oceans. Photo courtesy of James M. King, Graphic Impressions.

ample of a poorly sorted deposit because they contain particles ranging in size from the colloid to boulders that have been carried by glaciers and dropped out as the glaciers melted.

As sediment is carried from the source, it increases in maturity. This change results from the association with moving water, which has the capacity to carry particles of a certain size away from a deposit and leave behind other larger particles. With continued reworking by wave and current action, clay content of the sediment is reduced. As the duration of transport increases, particles of sand size or larger become more rounded through chemical weathering and abrasion. Increasing **sediment maturity** is indicated by (1) decreasing clay content, (2) increasing degree of size sorting, and (3) increasing rounding of the grains within the deposit.

The poorly sorted glacial till, which contains relatively large quantities of clay-sized and larger particles that have not been well rounded, is an example of an immature sedimentary deposit. The beach sand we used as an example of a well-sorted sediment contains no clay-sized material and is usually composed of well-rounded particles that have undergone considerable transportation. The beach sand is an example of a mature sedimentary deposit. Figure 5–1 illustrates the nature of sediments of varying degrees of maturity based on the criteria.

SEDIMENT TRANSPORT

The sediment carried to the margins of the ocean by the rivers and glaciers flowing from the continents either settles out to fill estuaries, becomes a part of a delta, or is spread across the continental shelf by currents and mass wasting (submarine sediment slides). Some sediment is carried beyond the shelf to the deep-ocean basin. The greatest volume of sediment transport in the ocean is achieved by the longshore currents caused by breaking waves at the margins of the continent (*see* chapter 9). Lower-energy currents distribute the finer components of the sediment along the margin of the continental shelf and even into the deep-ocean basin. Glacial transport and iceberg rafting of sediment is important in subpolar oceans.

Figure 5–2 shows the relationship between average horizontal current velocity and the erosion, transportation, and deposition of particles ranging in size from 0.001 to 100 mm (0.0000394 to 3.94 in.). As might be expected, the curve that separates transportation of particles from their deposition shows that larger particles will settle out at higher velocity than smaller ones.

TABLE 5–1   Wentworth Scale of Grain Size for Sediments.

	Particle	Minimum Diameter (mm)
	Boulder	256
	Cobble	64
	Pebble	4
	Granule	2
SAND	Very coarse sand	1
	Coarse sand	1/2
	Medium sand	1/4
	Fine sand	1/8
	Very fine sand	1/16
SILT	Coarse silt	1/32
	Medium silt	1/64
	Fine silt	1/128
	Very fine silt	1/256
CLAY	Coarse clay	1/640
	Medium clay	1/1024
	Fine clay	1/2360
	Very fine clay	1/4096
	Colloid	

Wentworth (1922), after Udden (1898)

Maturity increases ⟶
Degree of sorting increases
Clay content decreases
Rounding of sand particles increases ⟶

— Clay particle

FIGURE 5–1   Sediment Maturity.

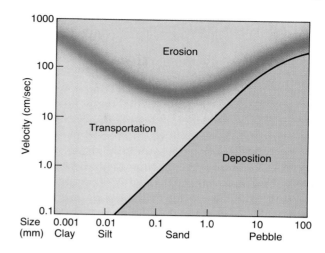

FIGURE 5–2   Hjulstrom's Curve.
Horizontal current velocity vs. erosion—deposition of sediment by particle size. Reprinted with permission from Tarbuck, E. J., and Lutgens, F. K., *The earth: An introduction to physical geology,* 3d ed. (Columbus: Merrill, 1990), fig. 10.15.

This causes a great cohesive force in clay sediments. Overcoming this cohesive attraction and picking up these particles once they are deposited requires a high-velocity current.

Particles less than 20 μm (0.008 in.) in diameter can be carried far out over the open ocean by prevailing winds. Such particles either settle out as the velocity of the wind decreases or find their way into the ocean during precipitation. They commonly serve as nuclei around which raindrops and snowflakes form. Some particles that reach high altitudes can be carried very rapidly by the jet stream that exists in the midlatitudes.

Studies have shown that the amount of windborne particles has increased since 1965. This may be the result of the increasing aridity of the major sources of these particles, the deserts of China and Africa. Dust from China is carried by the prevailing westerlies across the Pacific and has been recorded from the Bering Strait in the north to as far south as Antarctica. The trade winds transport Sahara dust across the Atlantic Ocean to the Caribbean Sea.

CLASSIFICATION OF MARINE SEDIMENT

The fundamental classification of marine sediment particles is based on the origin of the particle. There are four basic sources for marine sediment—rock (**lithogenous**), organic material (**biogenous**), compounds dissolved in water (**hydrogenous**), and outer space (**cosmogenous**).

Observing the curve that separates the process of erosion from transportation, we can see that it takes higher-velocity currents to *erode* (pick up and carry away) larger particles from the sand-sized particles up. The surprising fact that tiny clay-sized particles require a higher-velocity current than that needed to erode the larger sand-sized particles results from the fact that clay-sized particles are flat rather than round like typical sand-sized particles. Thus they have a greater surface area per unit of mass in contact with one another when they are laid down in a deposit.

Lithogenous Sediment

Common Rock-forming Minerals We will first direct our attention to the composition of sediments derived from rock, either from the continental masses or from volcanic activity in the open ocean. Since the sources of lithogenous sediment are the rocks of the earth's crust, we will be interested in identifying the most abundant compounds in these rocks. We call the elements and compounds that make up the rocks **minerals**. Although over 2000 minerals exist, we need to recognize fewer than two dozen of these to be able to identify those that make up most of the earth's crust. As we discuss these minerals, we will refer to their chemical composition. Table 5–2 lists the elements that are the most abundant constituents of the common rock-forming minerals and therefore the most abundant elements in the earth's crust. These are the only ones we will need to consider as we discuss the chemical composition of the common rock-forming minerals. In some common rock-forming minerals, ions of metals combine with a negatively charged ion composed of one atom of silicon and four atoms of oxygen, the **silicate tetrahedron** (fig. 5–3).

Those minerals composed of silicate tetrahedra and ions of iron and magnesium are called **ferromagnesian minerals**. They are dark or black and have higher densities than other rock-forming minerals. The most important ferromagnesian minerals are **olivine**, **augite**, **hornblende**, and **biotite**. Because of different crystalline structures of these minerals, they have different physical appearances. Olivine usually occurs in greenish granular masses, while augite and hornblende are usually black and cleave, or break cleanly, into blocky forms. Biotite has excellent cleavage in one direction along a plane where sheets of

molecules are poorly bonded together. As a result, the sheets of biotite, black mica, can be separated along very smooth, shiny surfaces. The density of these iron- and magnesium-containing minerals ranges from 2.8 to 3.4 g/cm^3.

Three other common rock-forming minerals, which do not contain iron and magnesium and are therefore known as **nonferromagnesian minerals**, are varieties of **feldspar, muscovite,** and **quartz**. All the varieties of feldspar are aluminum silicates that contain varying amounts of potassium, sodium, or calcium. Muscovite is similar to biotite in composition, except that iron and magnesium are excluded, and quartz is composed exclusively of the silicate tetrahedra with-

TABLE 5–2 Relative Abundance of Elements in Earth's Crust.

Element	Approximate Percentage by Weight
Oxygen (O)	46.6
Silicon (Si)	27.7
Aluminum (Al)	8.1
Iron (Fe)	5.0
Calcium (Ca)	3.6
Sodium (Na)	2.8
Potassium (K)	2.6
Magnesium (Mg)	2.1
All others	1.5
Total	100

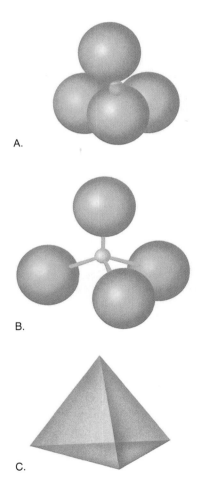

A.

B.

C.

FIGURE 5–3 Three Representations of the Silicon-Oxygen Tetrahedron.

A, the four large spheres represent oxygen atoms and the dark sphere represents a silicon atom. The spheres are drawn in proportion to the radii of the atoms. *B*, an expanded view of the tetrahedron using rods to depict the bonds that connect the atoms. *C*, diagrammatic representation using four points to represent the positions of the oxygen atoms. Reprinted by permission from Tarbuck, E. J., and Lutgens, F. K., *The earth: An introduction to physical geology,* 3d ed. (Columbus: Merrill, 1990), fig. 2.12.

out metallic ions. The density of these light-colored nonferromagnesian minerals ranges from 2.6 to 3.1 g/cm^3.

Classification of Igneous Rocks

Igneous rocks, those that solidify from a molten mass, are classified on the basis of texture (mainly the size of the crystals of which the rock is composed) and mineral composition. The texture of igneous rocks is determined by how rapidly they cool. Rocks extruded at the earth's surface during volcanic activity cool rather rapidly and have a fine-grained or even glassy texture if the cooling occurs so rapidly that crystals do not have a chance to form. They are **extrusive rocks**. A typical example of fine-grained igneous rock is **basalt**, which is characteristic of the oceanic crust and is the most common igneous rock in volcanic islands rising above the deep-ocean basins.

Coarse-grained igneous rocks form as a result of slow cooling at great depth. They are intruded masses that do not reach the surface of the earth's crust before they solidify and therefore cool more slowly because of the layer of insulating rock between them and the earth's atmosphere. The most common such rock is **granite**, which is the rock type characteristic of the continents. These are **intrusive rocks**.

Figure 5–4 shows the most common varieties of igneous rocks, of both intrusive and extrusive origin. **Granite** and **rhyolite**, its volcanic equivalent, are light-colored rocks on one end of the classification scheme that are composed primarily of nonferromagnesian minerals. This composition accounts for the fact that these rocks are light in color and of relatively low density. Such rocks are sometimes referred to as **sialic** because they characteristically contain high percentages of the elements silicon and aluminum.

Basalt and its coarse-grained equivalent, **gabbro**, are near the opposite end of the scale and are characterized by dark color and higher density. Because they have a relatively large ferromagnesian mineral content, the rocks at this end of the scale are referred to as **simatic**. This term is derived from the elements silicon and magnesium. The rock **peridotite** is com-

FIGURE 5–4 Igneous Rock Classification.

A, igneous rocks. This table shows the most common types of igneous rocks. The coarse-grained ones cool slowly deep beneath the earth's surface, and granite, the typical intrusive rock of the continents, is the most common variety. The fine-grained rocks, which are associated with volcanic activity, cool rapidly at or near the earth's surface. The most common variety is basalt, which is characteristic of the oceanic crust. The rocks on the left of the table are light colored and contain mostly nonferromagnesian minerals. Moving to the right, ferromagnesian minerals replace nonferromagnesian minerals, and the rocks become darker in color. *B,* examples of igneous rocks: (A) granite, (B) andesite, (C) basalt. Photo from Gary Cawthon.

posed almost entirely of ferromagnesian minerals; it is relatively rare at the earth's surface but is representative of the composition of the mantle.

The common rock-forming minerals that make up igneous rocks form at temperatures and pressures well above those of atmospheric conditions at the earth's surface. They are therefore chemically unstable and begin to weather into more stable compounds when exposed to the lower temperatures and pressures of atmospheric conditions. Physical and chemical processes of weathering free the individual grains of minerals from the parent rock mass, after which they are transported and deposited by whatever means of transportation might be available. Most of this material ends up in major river systems that carry it to the ocean basins.

The various rock-forming minerals have different degrees of stability under atmospheric conditions, quartz being the most stable and undergoing the slowest alteration. Where granitic masses are available as a source, quartz will commonly provide the majority of the coarse sand-sized particles in beach deposits. Feldspars and ferromagnesian minerals alter to clays and oxides of iron.

The weathering of basalt and other ferromagnesian rocks will produce primarily clays and iron oxides. In addition to the particles that result from the decomposition of igneous rocks, other substances that will be dissolved in the ocean water are also produced. These dissolved substances include ions of calcium, carbonate, iron, magnesium, silicon, potassium, sodium, and oxygen.

Biogenous Sediment

The chemical composition of organically produced particles in marine sediment will usually be either **calcium carbonate** ($CaCO_3$) or **silica** (SiO_2). These are the compounds most used by organisms to secrete protective external coverings. Contributing most of the silica component of biogenous particles, which rarely exceed 100 μm in diameter, are microscopic plants, **diatoms**, and animals, **radiolarians**. **Foraminifers**, close relatives of radiolarians, are responsible for most of the calcium carbonate component, but other significant calcareous contributors are the larger pteropods and microscopic algae, **coccolithophores** (fig. 5–5).

Coral reef material is also in the biogenous classification. Composed of the calcareous ($CaCO_3$) skeletal structure secreted by corals and algae, coral reefs are massive sedimentary rock structures.

We will not discuss coral reefs here because they are discussed fully in chapters 4 and 16. Another in-

teresting form of calcareous biogenous deposit, stromatolite deposition, is discussed later in this chapter.

Hydrogenous Sediment

Manganese Nodules Potentially of economic importance are the manganese nodules (polymetallic nodules) found on the deep-ocean floor. Since the voyage of the *Challenger,* we know them to be relatively abundant in all of the major ocean basins. They consist mainly of MnO_2 and Fe_2O_3, with the average manganese dioxide content ranging around 30% and the average iron oxide content around 20% by weight. Also occurring in economically significant concentrations in the nodules are copper, cobalt, and nickel. Although these concentrations are usually less than 1% they can exceed 2% by weight (fig. 5–6).

Since iron and manganese both occur in hydrogenous deposits in concentrations greater than those they reach in igneous rocks, there exists a problem of explaining the concentration of these elements in the marine deposits. This problem has not yet been successfully solved, but it has been concluded that there are three possible sources for manganese and iron. It is thought that they are derived from (1) the weathering of volcanic material produced by volcanic activity on the ocean floor, (2) concentration in hydrothermal spring water found at the axes of oceanic ridges and rises, and (3) runoff that carries the iron and manganese minerals as soluble compounds from the continents.

There seems to be little variation in the chemistry of deep-ocean water throughout the world ocean. Studies on the relationship between manganese nodules and organisms suggest that the presence of microorganisms, primarily bacteria, may be a determining factor in the formation of nodules. The Manganese Nodule Project (MANOP), which began in 1977, uses sediment traps to catch particles falling to the ocean floor and is carrying out seabed experiments designed to explain how nodules form and why they have the observed range of composition.

Phosphorite Phosphorus in the form of P_2O_5 is found abundantly as a precipitate in nodules and in the form of a thin crust on the continental shelf and banks at depths above 1000 m (3280 ft). Concentrations of phosphates in such deposits commonly reach 30% by weight. The deposits seem to be associated with areas of upwelling water rich in nutrients. A study of phosphate nodules on the Peruvian continental margin, an area of intense upwelling, found nodules forming preferentially where the oxygen minimum zone of the water column intersects the

A.

C.

B.

D.

FIGURE 5–5   Microscopic Skeletons from the Deep Sea.

A, magnified 160 times by scanning electron microscopy, the skeletons are foraminifers and radiolarians found in sediment cores taken at Site 55 in the Caroline Ridge area of the west Pacific
Ocean by the Deep Sea Drilling Project. They were recovered in water 2843 m (9315 ft) deep, with
further penetration of 130 m (426 ft) into the ocean bottom. Rounded globose forms are foraminifers, whose skeletons are calcareous. Forms with reticulate skeletons are radiolarians and are siliceous. In life, these tiny organisms lived near the ocean surface, but upon death their skeletons
fell to the sea bed to become entombed in the sediments. By carefully studying them, scientists
can determine how long ago they lived and can learn much about the history of the ocean. Courtesy of the Deep Sea Drilling Project, Scripps Institution of Oceanography, University of California,
San Diego. *B,* coccoliths magnified about 10,000 times. Photo courtesy of Deep Sea Drilling Project,
Scripps Institute of Oceanography. *C,* foraminifer, *Orbulina universa,* is magnified 300 times. Photo
courtesy of Dr. Howard Spero. *D,* pteropod, *Corolla spectabilis.* Photo courtesy of James M. King,
Graphic Impressions, P. O. Box 21626, Santa Barbara, CA 93121.

A.

B.

C.

FIGURE 5–6 Hydrogenous Sediment.

A, photograph of the surface of a manganese nodule, magnified 156 times by a scanning electron microscope. The small tube and dome structures were built by organisms for shelter, according to Dr. Jimmy Greenslate at Scripps Institution of Oceanography. He has studied 71 nodules from the Pacific Ocean and found biologically derived structures both buried inside them and covering large portions of their surfaces. Nodules may owe their existence to organisms that participate in their formation. *B,* manganese nodules on the Pacific Ocean sea floor are formed as chemical precipitates. They contain quantities of manganese, iron, copper, nickel, and cobalt, making harvesting of the lumps a probability in the future. Courtesy of Scripps Institution of Oceanography, University of California, San Diego. *C,* MANOP Bottom Lander designed for geochemical experiments on the sea floor to learn how nodules form and why they have the observed range of composition. Courtesy R. F. Weiss, O. H. Kirsten, and R. Ackermann, Scripps Institution of Oceanography, University of California, San Diego.

ocean floor at depths from 800 to 1000 m (2624 to 3280 ft). The nodules are growing slowly (1 to 10 mm/1000 yrs, or 0.0394 to 0.394 in./1000 yrs). Instead of growing in all directions, as do manganese nodules, the phosphorite nodules grow down into the sediment. Their tops are older than their bottoms. The source of phosphorus and the other elements

found in the nodules is the water trapped in the pores between sedimentary particles.

Glauconite This generally greenish hydrous silicate has a complex and variable composition which includes ions of potassium, magnesium, and iron. Glauconite may form by submarine weathering of biotite, although the exact mode of formation is not known. Glauconite, which is found to depths of 2500 m (8200 ft) but more commonly on topographically high areas near the coastline, is considered to form only in the marine environment. Forming as casts in organic remains and as encrustations, glauconite is more commonly found in grains of silt or sand size in mud and sand deposits. Deposits in which the glauconite is sufficiently abundant are frequently referred to as **green sands** or **green muds** owing to the coloration that results from the presence of glauconite.

Carbonates   The two most important carbonates in marine sediment are **aragonite** ($CaCO_3$) and **calcite** ($CaCO_3$). Aragonite and calcite have the same chemical composition but different crystalline structures. Aragonite is the least stable form and over a period of time changes to calcite.

During the geologic past, deposits of limestone ($CaCO_3$) in the marine environment appear to have been widespread. Most contain fossil evidence indicative of a biogenous origin, while others contain no fossils or characteristics that give any indication they formed organically. There appear to be very few places where nonbiogenous carbonates are presently forming. The **Bahama Banks** is the largest region where significant carbonate precipitation is presently occurring, although deposits are also forming on the Great Barrier Reef, in the Persian Gulf, and in other local areas in the low latitudes. Carbonate precipitation may occur directly from the water through physiochemical precipitation when water conditions are right. The most common forms of physically precipitated carbonate are aragonite crystals less than 2 μm (0.0008 in.) in length and **oolites**, onionlike spheres that develop around a nucleus and reach diameters of less than 2 mm (0.08 in.).

Since ocean water is essentially saturated with calcium (Ca^{++}), the only factor preventing precipitation of calcium carbonate (the most common carbonate) is the absence of the carbonate ions (CO_3^{--}). **Carbon dioxide** (CO_2) is important in carbonate chemistry. To see how it influences the precipitation of carbonates, we need to look at the various forms in which CO_2 is stored in ocean water. The reactions presented in chapter 2 and below tell us that when carbon dioxide combines with water in the ocean, carbonic acid (H_2CO_3) is formed. Carbonic acid may dissociate and produce a hydrogen ion (H^+) and a bicarbonate ion (HCO_3^-). On further dissociation of the bicarbonate ion, an additional hydrogen ion may be formed along with carbonate ion (CO_3^{--}).

$$H_2O + CO_2 \rightleftharpoons H_2CO_3 \rightleftharpoons H^+ + HCO_3^- \rightleftharpoons H^+ + H^+ + CO_3^{--}$$

| Water | Carbon dioxide | Carbonic acid | Hydrogen ion | Bicarbonate ion | Hydrogen ions | Carbonate ion |

Low pH ←————————————————→ High pH
7.5   9.0

The presence of carbon dioxide and the carbonate ion in ocean water is mutually exclusive. Carbon dioxide is present in free form only at pH values below that of the average ocean pH, 8.1. The concentration of carbonate ions increases with increased pH, so it is necessary to raise the pH of ocean water to make CO_3^{--} available for combination with Ca^{++}.

Removal of carbon dioxide raises the pH, so we might look to areas where photosynthesis is occurring at rates that remove significant amounts of carbon dioxide from the water for areas of carbonate precipitation. Heating water will also reduce its ability to dissolve carbon dioxide and decrease carbon dioxide concentration. Both factors are at work in areas where precipitation is known to occur. For instance, in the Bahama Banks, water that is moving north from the deep flow of the Florida Straits moves up over the shallow bank, where it is heated and exposed to high plant productivity. Both changes, along with lower pressure in near-surface waters, tend to reduce carbon dioxide content of the water and enhance precipitation of carbonates.

Other Hydrogenous Minerals   Other minerals found on the ocean bottom are secondary **zeolites** and **clays** commonly associated with deposits of the deep-ocean basin. The most likely process by which these minerals—zeolite (phillipsite) and **montmorillonite** (clay)—form is through the chemical alteration of volcanic material on the ocean bottom.

Zeolites in general are hydrated silicates of aluminum, calcium, sodium, and potassium. They have the peculiar property of gradually losing their water of hydration and possess the ability to substitute various metals for the alkaline metal originally present in their structures. **Phillipsite** is a white to reddish mineral that helps give the abyssal red clay deposits, particularly in the Pacific Ocean, their color. Although it may contain no sodium, phillipsite always contains considerable potassium. The total zeolite concentration in deep-ocean sediment frequently exceeds 50%.

Cosmogenous Sediment

Particles that originate in space occur in marine sediment as either nickel-iron spherules that are magnetic or as silicate chondrules composed of olivine and augite and some of their ferromagnesian relatives called **pyroxenes** (fig. 5–7). Such particles had been thought to be fragments remaining from

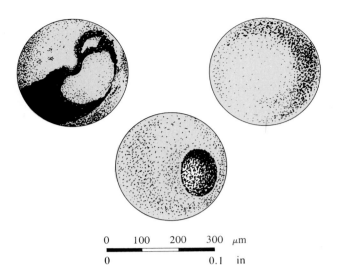

FIGURE 5–7 Cosmogenous Sediment.

meteorites. However, recent studies indicate they form in the asteroid belt as sparks produced when asteroids collide. **Iron spherules** are generally about 30 μm (0.012 in.) in diameter, while **silicate chondrules** group in diameters of 30 μm or 125 μm (0.012 or 0.05 in.). The overall size range of these cosmic particles falls between 10 and 640 μm (0.004 and 0.25 in.).

There is some evidence that the large comet or asteroid that many scientists think struck the earth 65 million years ago and caused the extinction of dinosaurs and many other species may have impacted earth in the Caribbean Sea. A bed of ejecta material .5 m (1.6 ft) thick on Haiti and evidence of erosion and deposits of a giant tsunami in marine sediments north of Colombia and on Cuba are just the right age to have been formed by such an impact. Seismic investigations reveal a tenuous crater beneath the ocean sediments off the coast of Colombia that is 300 km (186 mi) in diameter. Whether the crater itself is still present beneath the ocean sediments, the deposits that have been observed certainly make the Caribbean Sea the leading candidate for the location of such an impact.

The special feature, "Evidence for the Oceanic Impact of a Meteorite," on page 118, describes the evidence from marine sediments of a meteorite striking the earth's oceans.

TYPES OF SEDIMENTARY DEPOSITS

As a result of marine sediment deposition, the relative amounts of lithogenous, biogenous, hydrogenous, and cosmogenous particles found in marine sedimentary deposits will vary considerably. The two largest categories into which marine sedimentary deposits can be divided are **neritic deposits**, found near the continents, and **oceanic deposits**, characteristic of the deep-ocean basins. These two broad categories, the subgroups into which they are divided, and the processes determining their formation are discussed in the following paragraphs.

Neritic sediment refers to deposits with a wide range of particle size that are composed primarily of lithogenous particles derived from the continents. For the most part, such material enters the ocean at or near the shore. It accumulates rapidly on the continental margin. Neritic sediment also contains biogenous, hydrogenous, and cosmogenous particles. These constitute only a minor percentage of the total sediment mass because of the rapid deposition rate of lithogenous particles near the continents. Neritic sediment is distributed along the continental shelf by waves, currents, and mass wasting. It is transported down the continental slope to be deposited on the continental rise by sediment slumps and turbidity currents. Here the sediment is further distributed by deep boundary currents.

Oceanic sediment consists of the very fine-grained accumulations that form at a slower rate on the deep-ocean basin floor. It is possible for a greater variety of oceanic sediment types to accumulate because the rate of deposition of lithogenous particles is greatly reduced from that observed on the continental margins. Although the lithogenous component dominates the sediment found in most of the deeper basins of the ocean floor, biogenous and hydrogenous components are abundant in many areas of the ocean bottom.

NERITIC SEDIMENTS (SEDIMENTS OF THE CONTINENTAL MARGIN)

Owing to the rise in sea level that occurred with the melting of glaciers at the end of the last ice advance, many rivers of the world today deposit their sediment in estuaries rather than carry it onto the continental shelf as they did at various times during the geological past (fig. 5–8). The continental margin of the Atlantic coast of North America is an area where most river sedimentation is presently taking place in estuaries. In such regions the sediments that cover the continental shelf, **relict sediments**, may range from 3000 to 7000 years in age. Relict sediments currently cover about 70% of the world's continental shelf.

Of the 20 billion MT of sediment carried to the earth's continental margins by rivers each year, almost 80% is provided by runoff from Asia. Figure

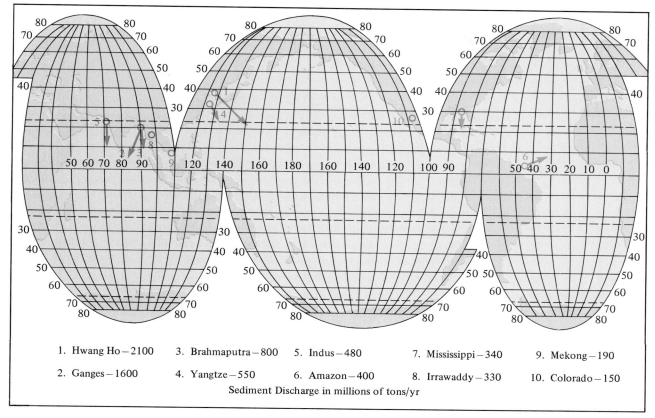

1. Hwang Ho — 2100   3. Brahmaputra — 800   5. Indus — 480   7. Mississippi — 340   9. Mekong — 190

2. Ganges — 1600   4. Yangtze — 550   6. Amazon — 400   8. Irrawaddy — 330   10. Colorado — 150

Sediment Discharge in millions of tons/yr

FIGURE 5–8 Sediment Discharge from Major Rivers.
Base map courtesy of National Ocean Survey.

5–8 shows that of the ten most important rivers contributing to this total, seven drain the Asian continent. The Hwang Ho, Yangtze, and Mekong empty their sediment into the marginal seas of the western Pacific Ocean. The Ganges, Brahmaputra, and Irrawaddy flow into the Bay of Bengal between India and Burma at the north end of the Indian Ocean. Also flowing into the Indian Ocean via the Arabian Sea is the Indus River.

The sediment of the Colorado River remains near the northern end of the Gulf of California. The Mississippi River is building a major delta on the continental shelf of the Gulf of Mexico, while the Amazon flows into the equatorial Atlantic. These two large rivers are typical of the long, low-gradient rivers that feed sediment onto the broad continental shelves of passive continental margins.

Sediment that reaches the shore is sorted by current and wave action that generally leaves the sand to be transported along the shore to form beach deposits. The finer particles, silt- and clay-sized, are carried farther out on the shelf and deposited in areas where wave energy has decreased sufficiently. Where sediments are being deposited on the continental shelf, the rate at which they accumulate may be greater than 10 cm/1000 yr (4 in./1000 yr). Relative to the deposition of other types of sediments in the marine environment, this is a very high rate of accumulation that allows burial of sediment before it has time to be acted upon chemically by the ocean water. Therefore, these deposits usually contain materials that have not been significantly altered since they entered the marine environment. They are commonly gray to greenish in color owing to lack of oxidation.

Turbidites

Although it seems unlikely that wave action and ocean current systems could carry coarse material beyond the continental shelf into the deep-ocean basin, there is evidence that much neritic sediment has been deposited at the base of the continental slope forming the continental rise. These accumulations thin gradually toward the abyssal plains. Such deposits are called **turbidites** and are thought to have been deposited by turbidity currents that periodically move down the continental slopes through the submarine canyons, carrying loads of neritic material that spread out across the continental rise (fig. 5–9).

EVIDENCE FOR THE OCEANIC IMPACT OF A METEORITE

Analysis of piston cores taken in the Antarctic Basin about 1400 km (870 mi) west of Cape Horn indicates that an asteroid meteorite struck the area 2.3 million years ago. This is the only known impact of a meteor into a deep-ocean basin (fig. 5A).

The evidence is in the form of three types of particles—all of which are thought to represent meteoric material. The particles are a nonterrestrial basalt that shows signs of impact stress, vesicular particles that formed from melted meteoric material, and a few pieces of metal. The basaltic particles and vesicular particles have similar compositions (fig. 5B). Commonly the basaltic particles are embedded in or are in contact with vesicular material. This indicates that vesicular material, while in a plastic state, collided with the solid basaltic particles in the atmosphere during impact.

FIGURE 5A Evidence of Meteorite Impact in Antarctic Basin.

The impact of an asteroid ranging in diameter between 0.5 and 1.0 km (0.31 and 0.62 mi) occurred after an approach from a direction a little south of west according to the distribution of Ir-rich particles recovered in piston cores labeled on the map. The greatest concentrations of Ir-rich particles occurs in cores 13-3 and 14-4 which are starred. Map courtesy of Frank T. Kyte, University of California, Los Angeles.

Iridium concentration in 2.3-million-year-old sediments (nanograms/cm^2)

10–15   15–50   50–200   200

A.

B.

FIGURE 5B   Iridium-Rich Meteoric Particles.

A, scanning electron micrograph (SEM) of a vesicular particle. Up to 50% of the volume of these particles is pore space. Bar length is 100 μm. *B,* SEM of a vesicular particle with a basaltic inclusion. Sharp boundary between basalt and vesicular material indicates the solid, rigid basalt collided with a plastic mass of vesicular material as they were propelled through the atmosphere. Bar length is 100 μm. SEMs courtesy of Frank T. Kyte, University of California, Los Angeles.

The metallic element iridium (Ir) is found in greater concentrations in meteoric material than on earth, so layers of sediment that contain unusually high concentrations of Ir are usually accepted as evidence of meteoric impact if there is no evidence of certain types of volcanic events that might have produced them.

Based on the Ir concentrations shown in figure 5A, the amounts and distribution pattern suggest that the meteorite that hit the Pacific Ocean was between 0.5 and 1 km (0.31 and 0.62 mi) in diameter. It hit the ocean without producing a crater in the ocean floor.

This event occurred near the start of the Pleistocene ice age. Although it surely didn't cause the climatic change, it could have contributed to a more rapid decrease in global temperature resulting from the meteoric dust and water that the event could have sent into the stratosphere.

Reference: Kyte, F. T., Lei, Z., and Wasson, J. T. 1988. New evidence on the size and possible effects of a late Pliocene oceanic asteroid impact. *Science* 241:4861, 63–64.

Turbidites are characterized by **graded bedding**. This means that each deposit that is laid down has coarser material at its base and decreasing particle size toward the top of the layer. This pattern results from particles of different sizes settling out of a moving water mass as the velocity decreases. The coarser material settles out first; the finer material stays in suspension longer.

During the millions of years that sediments accumulated at the margins of continents, the continental shelf, slope, and rise developed into a sedimentary wedge more than 10 km (6.2 mi) thick that contains over 75% of the sediment to be found on the ocean floor. Less than 25% of marine sediment will be found in the deep-ocean basins that cover in excess of 80% of the ocean floor.

Glacial Deposits

Poorly sorted deposits containing particles of all sizes, from boulders to clay, may be found in the high-latitude portions of the continental shelf. These **glacial deposits** were laid down by melting glaciers that covered the continental shelf area during the **Pleistocene Epoch**, when glaciers were more widespread than they are today and sea level was lower. Glacial deposits are still forming around the continent of Ant-

FIGURE 5–9 Turbidites and Graded Bedding.

A, turbidity currents flow down submarine canyons and deposit sediment as deep-sea fans. *B,* examination of the deep-sea fans shows that they are composed of a series of deposits called turbidites. Each turbidite represents the sediment deposited by one turbidity current event. Each turbidite displays graded bedding in which the sediment particles grade from coarse at the bottom to fine at the top. Reprinted by permission from Tarbuck, E. J., and Lutgens, F. K., *The earth: An introduction to physical geology,* 3d ed. (Columbus: Merrill, 1990), figs. 19.7 and 19.8.

A.

B.

FIGURE 5–10 Stromatolites in the Bahamas.
Subtidal oolitic stromatolites on the crest of an oolitic tidal bar on Eleuthera Bank, Bahamas. Most are 0.5 to 1.0 m (1.6 to 3.3 ft) high. They are colonized by forms of algae that produce a mucous scum which traps sediment. Photo courtesy of Jeff Dravis. Permission granted by Exxon Production Research Company and American Association for the Advancement of Science.

arctica and around the island of Greenland by **ice rafting**, a process by which rock particles trapped in glaciers are carried out to sea by icebergs that break away from glaciers as they push into the coastal ocean. As the icebergs melt, the lithogenous particles are released and settle to the ocean floor.

Stromatolites

Stromatolites are interesting calcareous sedimentary structures that are often dome-shaped (fig. 5–10). To produce these deposits, **algal mat** mucous secretions trap fine oolitic calcium carbonate sediment. Other forms of algae produce long filaments that are responsible for binding the carbonate particles together. When the **filamentous algae** die, the filaments become calcified and cement the oolitic particles into the stromatolite structure.

Commonly observed in rocks derived from sediments deposited in ancient oceans, stromatolites have long been known to be forming in Hamelin Pool of **Shark Bay**, Western Australia. In this location, the hypersalinity of the water is believed to protect the algae from grazing gastropods. It had been thought that stromatolites could form only in such an environment. However, they have also been observed forming on **Eleuthera Bank**, Bahamas (fig. 5–11), where normal marine salinity exists.

It has been suggested that migrating bars of oolitic sand that periodically cover and uncover the stromatolites, as well as a veneer of unconsolidated oolitic sand shifting in response to tidal currents, sufficiently limit gastropod grazing to allow the stromatolites to form in the Eleuthera Bank occurrence.

Blake Plateau

Just beyond the narrow continental shelf off the coast of Florida and extending northward to the Cape Hatteras area is a special feature of the continental margin that is receiving very little sediment at present, except for the hydrogenous accumulations of phosphorite and manganese nodules. The floor of the **Blake Plateau** lies at the base of the **Florida-Hatteras Slope** at a depth of about 700 m (2296 ft). The plateau is actually a terrace that begins to form south of Cape Hatteras and increases to a width in excess of 300 km (186 mi) off the Florida coast. This gently sloping surface, which slopes at a rate similar to that of the average continental shelf, reaches a depth of about 1000 m (3280 ft). The ocean floor then descends at slopes of up to 30° along the **Blake Escarpment** into the deep ocean basin. The presence of this terrace is usually ascribed to the fact that the **Gulf Stream** flowing northward against the Florida-Hatteras Slope has sufficient velocity to prevent the deposition of sediment, therefore preventing the growth of a broad continental shelf. Also contributing to the development of the plateau is a fundamental lack of sediment. It is free of the influence of major rivers and is too far north for coral reef development. Extensive accumulations of manganese nodules are found on the outer Blake Plateau, and phosphorite nodules occur in a belt along its inner margin (fig. 5–12).

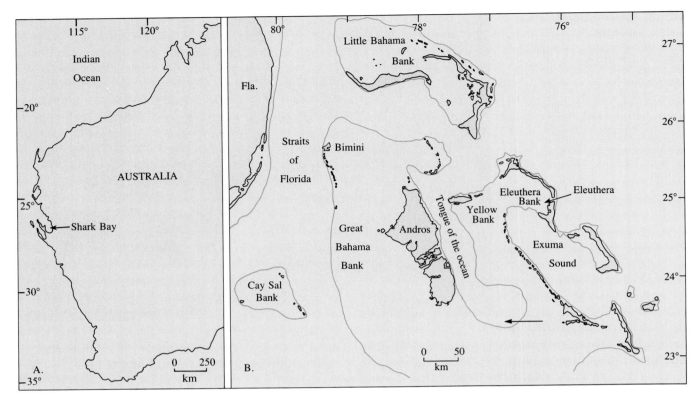

FIGURE 5–11   Locations of Two Regions of Present-Day Stromatolite Development.
A, map of western Australia showing Shark Bay. Arrow denotes approximate location of Hamelin Pool stromatolites. Restricted circulation with the open ocean produces hypersaline conditions that prevent algal grazing by gastropods. *B,* map of the Bahamas showing the locations of Eleuthera Bank and southern end of Tongue of the Ocean locations of stromatolite development (arrows). The physical movement of oolitic sediment helps to protect the algal mats from mollusk grazing in this environment where water salinity is normal. Map courtesy of Jeff Dravis. Permission granted by Exxon Production Research Company and American Association for the Advancement of Science.

OCEANIC SEDIMENTS (SEDIMENTS OF THE DEEP-OCEAN BASIN)

As was previously discussed, most of the neritic sediment is deposited on the continental margin as part of the continental shelf, continental slope, or continental rise. Since the continental rise is a transitional feature between the continental margin and the deep-ocean basin, we may well conclude that neritic material is deposited in significant quantities in the deep-ocean basin as part of the deep-sea fan deposits making up the continental rise. This is true, but when we consider the total surface area of the deep-ocean basin, we find that more commonly no neritic material will be found in the sedimentary section. Covering most of the deep-ocean floor are two types of oceanic sediment deposits—abyssal clay and oozes.

Abyssal Clay

Accumulating at a rate of approximately 1 mm/1000 yr (0.04 in./1000 yr), **abyssal clays** cover most of the deeper ocean floor. They are commonly red-brown or buff in color owing to phillipsite and oxidized iron. Abyssal clay consists predominantly of clay-sized particles derived from the continents. The particles are carried by winds or ocean currents to the open-ocean regions and, along with some cosmic and volcanic dust, settle slowly to the ocean floor.

Significant components of the abyssal clay are the minerals montmorillonite and phillipsite, which are thought to form from the interaction of ocean water with volcanic and other lithogenous particles.

Oozes

At somewhat shallower depths, biogenous material makes up a significant portion of the sediment that reaches the ocean bottom. It consists primarily of the minute shells of microscopic plants and animals that accumulate on the ocean floor. If sediment contains 30% or more skeletal material by weight, that deposit is called an **ooze**. Oozes may be classified as **calcareous** ($CaCO_3$) or **siliceous** (SiO_2), depending on the chemical composition of the dominant type of skeletal material making up the deposit. Oozes can form only where the deposition of material other than the remains of plants and animals occurs at a very low rate.

More specific names may be given to oozes to indicate the types of organic remains most abundant in the deposits. **Diatom oozes** are siliceous oozes that contain mostly the remains of diatoms. Other descriptive names commonly used on this same basis are **radiolarian ooze** (siliceous), **foraminiferan ooze** (calcareous), and **pteropod ooze** (calcareous). The predominant type of ooze is **Globigerina ooze**, which is especially widespread in the Atlantic and Pacific oceans. *Globigerina* is an abundant and widespread genus of foraminifer. Pteropods are snails with thin shells and a foot modified for swimming.

The rate of accumulation of biogenous material on the ocean floor depends upon three fundamental processes—productivity, destruction, and dilution. **Productivity** of the planktonic forms that contribute most of the biogenous material in oceanic sediments is greater in the areas of upwelling along the margins of continents and associated with the diverging water masses in equatorial regions.

Destruction of skeletal remains occurs primarily through solution. The ocean is undersaturated with silica at all depths. So for siliceous shells to last until they are incorporated in the sediment, they must be thick, as thin fragments will be dissolved in the ocean water before they can settle out and be covered by the sediment. About 80% of the silica reaching the ocean floor is thought to be dissolved back into the water before it is protected by being covered.

The solubility of calcium carbonate increases with increased CO_2 content of the water and increased pressure. Near the surface, where temperatures are high, there is little CO_2 in the water and the water does not dissolve much $CaCO_3$. At greater depths, the temperature decreases and pressure increases. This lower temperature and higher pressure increase the capacity of the water for dissolving carbon dioxide; this, in turn, increases the water's ability to dissolve calcium carbonate. Below 4500 m (14,760 ft) calcium carbonate dissolves quite readily. Carbonate oozes are generally rare below a depth of 5000 m (16,400 ft). The depth at which calcium carbonate is dissolved at the same rate it is arriving from above is the **calcium carbonate compensation depth**. It may be as deep as 6000 m (19,680 ft) in parts of the Atlantic

FIGURE 5–12 Blake Plateau.
The relationship of the Blake Plateau to the Gulf Stream can be seen as the mighty current sweeps along the Florida-Hatteras Slope.

Ocean, but above 3500 m (11,480 ft) in regions of low biological productivity in the Pacific Ocean. About 50% of the total calcium carbonate reaching the deep-ocean floor is dissolved into the ocean water before it is protected by being covered with sediment. Essentially all of this dissolving of deposited calcium carbonate occurs below a depth of 3500 m. Therefore, even though calcium carbonate particles may reach the deep-ocean floor, they may not survive to be incorporated into the sediment.

Dilution is illustrated by the fact that calcareous oozes are not found on the continental margins, where rapid rates of deposition of lithogenous sediment prevail. Even in areas of upwelling where biological productivity is high, oozes do not occur on the continental shelf because the concentration of biologically derived silica or calcium carbonate remains below 30% by weight as a result of the high rate of deposition for lithogenous materials. The rates of deposition for biogenous sediment range from 1 to 15 mm/1000 yr (0.04 to 0.6 in./1000 yr), depending on the net effect of productivity and destruction. Figure 5–13 shows the relationships among calcium carbonate compensation depth, sea-floor spreading, and productivity in determining what type of sediment will accumulate on the ocean floor.

Manganese Nodules

Although what is known about the formation and distribution of manganese nodules is greatly exceeded by what is not known, we do know they are found only in deep-ocean environments where other sediments accumulate at rates of less than 7 mm/1000 yr (0.3 in./1000 yr). The nodules grow very slowly, adding layers from 1 to 200 mm (0.04 to 8 in.) thick every million years.

There appears to be a correlation between the biological productivity of overlying waters and the metallic composition and growth rates of manganese nodules. Beneath unproductive areas near the center of the subtropical oceans, the nodules are rich in iron and cobalt extracted from the bottom waters, and they grow slowly (less than 5 mm/million yr, or 0.2 in./million yr). Beneath waters of moderate levels of biological productivity, such as those north and south of the equatorial Pacific Ocean, nodules have increased concentrations of copper, nickel, and manganese. It is believed that these metals are transported to the bottom in the remains of plants and animals that settle from above. They are transferred to the nodules through a process that is not yet fully understood. In this environment nodules grow at rates of from 5 to 10 mm/million yr (0.2 to 0.4 in./million yr). Beneath very productive waters, another process that is not fully understood produces nodules so rich in manganese that the economically important copper, nickel, and cobalt are diluted to very low levels of concentration. These nodules grow very rapidly. From 10 to 200 mm/million yr (0.4 to 8 in./million yr) may be added. Figure 5–14A shows the pattern of copper concentration in Pacific Ocean nodules. The nodules with the greatest economic potential are those that form under conditions of moderate levels of biological productivity in the overlying waters. They are rich in copper and nickel.

An interesting finding resulting from the study of small nodules in the high-copper-nickel belt north of the equator at 140° W longitude and 11° N latitude was that the cores of these nodules were composed of molds of biogenic material: coccoliths and radiolarians. All of the $CaCO_3$ and SiO_2 of which these forms were originally composed has been replaced by a form of manganese oxide. It must be emphasized that the fundamental reason for these composition and distribution patterns is unknown. Future research conducted as part of the Manganese Nodule Program (MANOP) may help identify the processes that affect the composition and growth rates of manganese nodules.

Cobalt-Rich Ocean Crusts

Cobalt is an important strategic metal that is not mined in the United States. It is required to produce dense, strong alloys with other metals for use in high-speed cutting tools, powerful permanent magnets, and jet engine parts.

Major enrichments of cobalt in the earth's crust are confined to central southern Africa and deep-ocean nodules and crusts. Considering the unstable political situation in Africa, the United States has been looking to the ocean floor as a more dependable source. However, the political and economic uncertainties associated with the United Nations Law of the Sea Convention have made mining interests back away from the mining of deep-sea nodules beyond the 200-mile-wide Exclusive Economic Zone (EEZ). In 1981 **cobalt-rich manganese crusts** were found on the upper slopes of islands and seamounts that lie within the EEZs of the United States and its allies. The cobalt concentrations in these crusts are on the average half again as rich as in the richest African ores and at least twice as rich as in deep-sea manganese nodules.

The greatest concentrations seem to be associated with depths between 1000 and 2500 m (3280

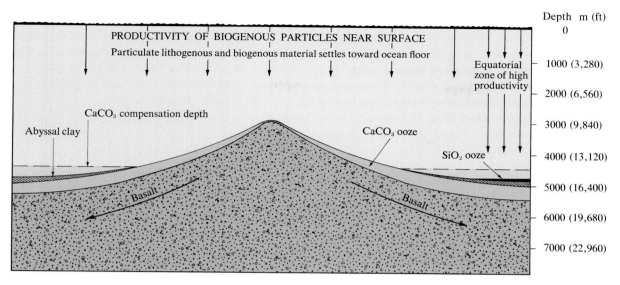

FIGURE 5–13   Sea-Floor Spreading and Sediment Accumulation.

New basaltic oceanic crust is forming at ridges that run across the ocean floor and rise to depths of about 3000 m (9840 ft). As the crust moves away from the ridges and sinks to greater depths, the type of sediment that accumulates on it changes. Until it sinks below the $CaCO_3$ compensation depth, the typical sediment that forms is $CaCO_3$ ooze. Below this depth $CaCO_3$ is dissolved, and abyssal clays dominate. If the crust passes beneath a region of high biological productivity, such as the Pacific equatorial region, oozes could again accumulate.

FIGURE 5–14   Metal-rich Hydrogenous Deposits.

A, copper content of nodules—percent of copper by weight. The areas of high copper content (greater than 1%) form under conditions where biological productivity in overlying waters reaches moderate levels. Nodules found here are also enriched in nickel and manganese. Blank areas are nodule-free because of the deposition of lithogenous particles near the continents and calcareous ooze along the equator and on the East Pacific Rise exceeds 7 mm (0.28 in.) per thousand years. Courtesy of G. Ross Heath, University of Washington. *B,* cobalt-rich manganese crusts. Crusts composed primarily of iron and manganese oxides also contain up to 2.5% cobalt. This high cobalt concentration in the crusts, which are up to 2 cm (0.8 in.) thick, has aroused interest in their economic potential. Drawing courtesy of U.S. Geological Survey.

A.

B.

and 8200 ft) on the flanks of islands and seamounts in the central Pacific Ocean. Additional sampling needs to be done to assure that crust deposits are extensive enough to provide a dependable source of cobalt. At least one international consortium, International Hard Minerals Company, is moving into the study of the feasibility of crust mining. If the results of these investigations are positive, the first deep-sea mining operations may be for cobalt and other metals found in these crusts that lie within the EEZs of the United States or one of its allies. There are at least 100 of these seamounts in the EEZs of the Line Islands and Hawaiian Islands, and each could yield up to 4 million metric tons of ore (fig. 5–14B).

DISTRIBUTION OF OCEANIC SEDIMENTS

Oceanic sediments cover about 75% of the ocean bottom. Table 5–3 shows that calcareous oozes make up almost 48% of oceanic sediments, abyssal clay accounts for 38%, and siliceous oozes 14% of the total area. Calcareous oozes are the dominant oceanic sed-

iment in the Indian and Atlantic oceans, while abyssal clay is the principal oceanic sediment in the Pacific Ocean. This difference is undoubtedly related to the fact that the Pacific Ocean is deeper, so that more of its bottom lies beneath the calcium carbonate compensation depth. Siliceous oozes cover the smallest percentage of the ocean bottom in all of the oceans because regions of high productivity of diatoms and radiolarians, the major components of these deposits, are restricted.

Fecal Pellets

A major problem that once puzzled marine geologists studying oceanic sediment is that sediments on the deep-ocean floor very closely reflect the particle composition of the surface water directly above. This was difficult to understand because it would typically take these very tiny particles from 10 to 50 years to sink from the ocean surface to abyssal depths. During this interval a horizontal ocean current of only 1 cm/s (0.03 ft/s) (about 0.035 km/h, or 0.02 mi/h) would carry them from 3000 to 15,000 km (1860 to 9300 mi)

TABLE 5–3   Distribution of Oceanic Sediment.
This graph shows that the percentage of an ocean-basin floor covered by calcareous ooze decreases with increasing mean depth of the basin. This shift probably occurs because the deeper an ocean basin is the greater the percentage of its floor that lies beneath the calcium carbonate compensation depth. The mean depths calculated for this table exclude the shallow adjacent seas, where little oceanic sediment accumulates. Notice that the dominant oceanic sediment found in the deepest ocean basin, the Pacific Ocean, is abyssal clay, while calcareous ooze is the most widely deposited abyssal sediment in the shallower Atlantic and Indian oceans.

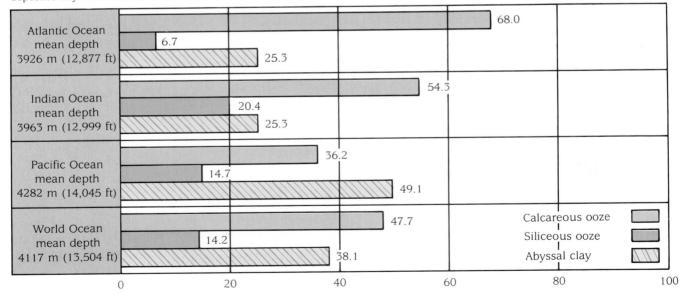

After Sverdrup, Johnson, and Fleming, 1942.

laterally before they reached the deep-ocean floor. To further complicate the picture, the particle assemblage at middepth is not similar to that found in the surface water and sediment below.

Such findings could be understood only if there was some mechanism by which the small surface particles could be aggregated into larger particles to sink faster. Continued study revealed such a mechanism—**fecal pellets**. Most particles in the surface water, organic and inorganic, are eaten at least once. Recovery of samples during the GEOSECS study of the ocean's chemical makeup and circulation confirmed the presence of fecal pellets among the flux of material falling from the ocean's surface to the ocean floor. The GEOSECS samples revealed an interesting pattern of concentration of suspended particulate matter in the ocean. They show that the highest concentrations of particulate matter in the water column at high latitudes are found at the surface and near the bottom. The surface concentration results from high biological productivity of microscopic plants and animals, and the deep increase results from the resuspension of bottom sediment into the benthic nepheloid layer.

Further study of these samples shows that 99% of particles that fall to the ocean floor do so as a part of fecal pellets produced by small animals and fish. These pellets (fig. 5–15), though still small (50 to 100 μm, or 0.02 to 0.04 in. in their smallest dimension), are large enough to allow the particulate matter to reach the abyssal ocean floor in 10 to 15 days. If this

process is important in transporting particles from surface water to abyssal depths, it could readily explain the similarity between particle composition of surface waters and bottom sediments immediately beneath them. Figure 5–16 shows a sediment trap set on the deep-ocean floor to intercept the flow of particles from the surface during particle flux studies.

Another process that may contribute significantly to increasing the settling rates of particles is the aggregation that results from particles being trapped in mucous networks produced by organisms such as salps and coccolithophores.

There also appears to be a seasonal pattern of flux of material from the surface zone to the deep sea. The highest levels of flux have been observed to follow periods of high biological productivity in the surface waters of tropical to temperate oceans.

Calcareous Ooze

Another look at table 5–3 will show that the percentage of calcareous ooze covering the deep-ocean floor of the three ocean basins decreases with increasing average depth of the basins. The shallowest ocean, the Atlantic Ocean, has 68% of its floor covered with calcareous ooze; in the slightly deeper Indian Ocean the coverage is 54.3%, and the deepest ocean, the Pacific Ocean, has a 47.7% coverage of calcareous ooze. Another factor at work here is the fact that the smaller surface areas of the Atlantic and Indian oceans increase the percentage of the surface waters

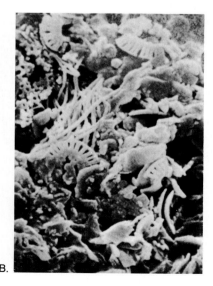

A.

B.

FIGURE 5–15 Fecal Pellets.
A, fecal pellet produced by zooplankton. The pellet is 200 μm (0.08 in.) long. *B*, view of the surface of the fecal pellet showing the remains of small phytoplankton and detritus. Photos courtesy of Susumu Honjo, Woods Hole Oceanographic Institution.

FIGURE 5–16   Sediment Traps.
Sediment traps that can be set on the ocean floor to catch particles falling from the surface. Photo courtesy of Susumu Honjo, Woods Hole Oceanographic Institution.

that have high levels of productivity. This results from a general pattern of decreasing biological productivity with increasing distance from shore.

Figure 5–17 shows that calcareous oozes are being deposited primarly on the oceanic ridges and rises

where depths are generally above the calcium carbonate compensation depth. The crests of these mountain ranges usually lie at depths between 2500 and 3000 m (8200 and 9800 ft). The North Pacific basin is free of calcareous ooze except for along the shallow Hawaiian Island Ridge. This is probably because this is the deepest of all ocean basins and the biological productivity of its surface waters is low throughout most of this expanse. The decreasing concentration of calcium carbonate in surface sediment with increasing depth shows up well in figure 5–18, which shows the distribution of calcium carbonate in the surface sediments of the oceans.

Calcareous oozes are composed primarily of planktonic (floating) foraminifers. However, up to 10% of the carbonate in some oozes is provided by benthic (bottom-dwelling) foraminifers. Also contributing to the calcareous oozes are the **Coccolithophoridae,** algae with minute calcareous plates that form a protective covering. The aragonite shells of **pteropods** are also found in calcareous oozes to a restricted degree. Aragonite is more readily dissolved than the calcite form of calcium carbonate, so pteropod remains are found only in restricted shallow, warm-water regions in the Atlantic Ocean (fig. 5–17).

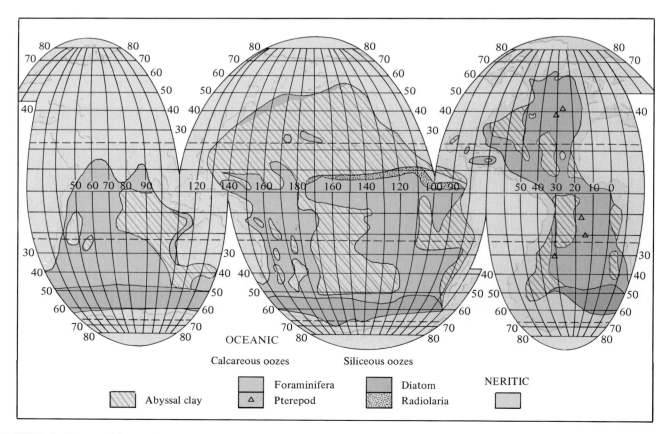

FIGURE 5–17   World Distribution of Neritic and Oceanic Sediments.
Base map courtesy of National Ocean Survey.

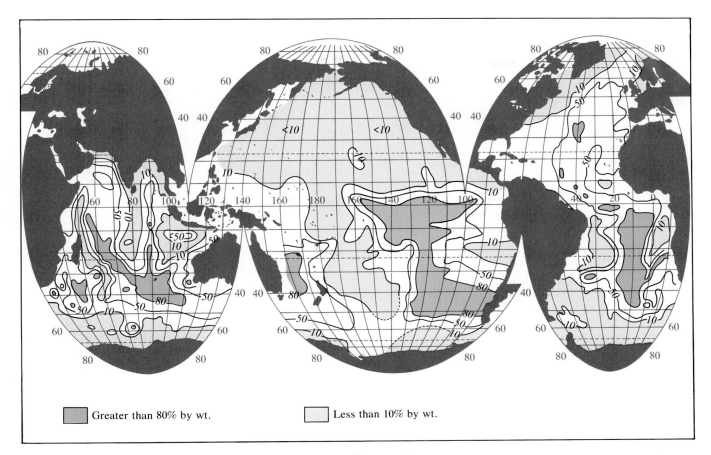

FIGURE 5–18   Calcium Carbonate in Surface Sediments of the World's Oceans.
Concentrations of calcium carbonate ($CaCO_3$) in ocean sediments is negatively correlated with ocean depth. The deeper the ocean floor, the less $CaCO_3$ found in the sediment. It is also low in sediments accumulating beneath cold, high-latitude waters. Extensive areas of concentrations of $CaCO_3$ in the sediment in excess of 80% are associated with relatively shallow ocean floor on the Carlsberg Ridge in the Indian Ocean, the East Pacific Rise, and the Mid-Atlantic Ridge. Note that the deepest ocean basin, the North Pacific, lies for the most part beneath the $CaCO_3$ compensation depth and has very little $CaCO_3$ in its accumulating sediment. After Biscaye et al., 1976; Berger et al., 1976; and Kolla et al., 1976.

Siliceous Ooze

Biogenous silica can be distinguished from lithogenous silica because biogenous silica incorporates water into its structure ($SiO_2 \cdot nH_2O$) and is called **opal.** Siliceous oozes are less abundant in the ocean basins because ocean water is never saturated with silica (SiO_2). Diatoms and radiolarians provide most of the skeletal material for siliceous oozes, but few of the diatom or radiolarian skeletons ever reach the bottom or become incorporated into the sediment before they are dissolved. Diatoms are plentiful in all areas of high productivity, but diatom oozes can be found only in areas below the calcium carbonate compensation depth. Above this depth, they are usually diluted to low concentrations by higher rates of deposition of $CaCO_3$ biogenous particles.

Although the sampling has been limited, there is presumed to be a significant belt of diatom ooze surrounding Antarctica. This accumulation would be made possible by the high diatom productivity of the region. Conditions for the formation of diatom ooze in the Antarctic are enhanced by the fact that calcium carbonate–secreting organisms do not occur in high concentrations where water temperatures are extremely low. The concentration of foraminifers decreases greatly south of the Antarctic Convergence at about 50°S latitude (fig. 5–18). Diatom ooze is also present in the extreme North Pacific because of the low-temperature waters that prohibit high productivity of carbonate-producing forms. The diatom ooze of the North Pacific does not have the high opal concentration of the Antarctic ooze because of dilution by the greater rate of deposition of lithogenous material in the North Pacific.

Only minor developments of radiolarian ooze are found in the equatorial Atlantic and Indian oceans, and a long band of radiolarian ooze exists in the Pa-

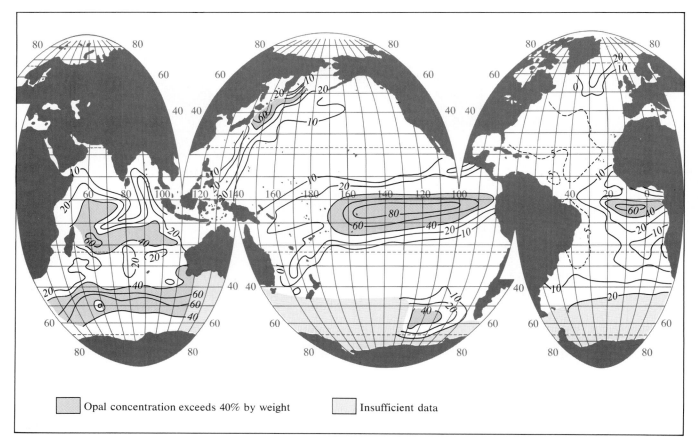

FIGURE 5–19   Biogenic Silica (SiO$_2 \cdot n$H$_2$O—Opal) in Surface Sediments of the World's Oceans.

The distribution pattern of opal in surface sediments of the ocean shows maximum concentrations in areas of highest biological productivity. These particles are produced by diatoms and radiolarians in the surface waters. In the equatorial and northwest Pacific, the opal is produced predominantly by radiolarians. Elsewhere, diatom remains are more abundant. The highly productive waters south of the Antarctic Convergence show up well in the south Indian Ocean and southeast Pacific Ocean. In the southern Atlantic Ocean and southwest Pacific Ocean there are insufficient data to determine the relationship between highly productive surface waters and opal concentrations in the surface sediment. After Leinen et al., 1986.

cific Ocean. The equatorial surface waters are quite productive in the eastern Pacific Ocean, and the production of calcium carbonate and opal particles is large. It is the deepening of the ocean floor to the north that makes this band of radiolarian ooze possible. The southern limit of the band of radiolarian ooze exists because of dilution by calcium carbonate particles that produce a calcium carbonate ooze to the south. The limit marks the intersection of the calcium carbonate compensation depth with the ocean floor. To the north of this line, calcium carbonate particles do not reach the bottom, so radiolarian particles can account for greater than 30% of the sediment mass—thus producing a radiolarian ooze. The northern limit of the radiolarian ooze band represents the

decreased production that results in too much dilution by lithogenous clays and a change of sediment type to abyssal clay (see fig. 5–17).

Figure 5–19 shows the distribution of biogenous silica (opal) in the surface sediment of the oceans and shows that large-percentage representations are restricted to areas of high productivity in equatorial and high-latitude regions.

Abyssal Clay

An interesting pattern of abyssal clay distribution is found in the Atlantic Ocean. The greatest portion of the ocean bottom covered by abyssal clays is on the west side of the Mid-Atlantic Ridge, which divides the ocean basin in half. This distribution pattern is a result of the different calcium carbonate compensation depths on the two sides of the ridge. Calcium carbonate is found in the sediments to depths in excess of 6000 m (19,680 ft) on the east side of the ridge, while deposition below 5000 m (16,400 ft) is rare in the basins to the west. This difference results from the fact that the Antarctic Bottom Water originating in the Weddell Sea to the south and the North Atlantic Deep Water originating to the north carry low-temperature water with high carbon dioxide content in the deep basins on the west side of the ridge. The ridge protects the basins to the east from the invasion of this cold water with its high capacity for dissolving calcium carbonate.

A far greater percentage of the Pacific Ocean basin is covered by abyssal clay (49.1%) than is true for the Atlantic and Indian oceans (25.3%). This is because the North Pacific Ocean is so deep and low biological productivity in the North Pacific has created a calcium carbonate compensation depth as shallow as 3000 m (9800 ft).

Transportation of land-derived clays from the continents into the Pacific Ocean is restricted by the system of trenches that separates all but a segment of North America from the basin. Prevailing westerly winds transport clay particles from the Asian mainland (fig. 5–20). Prevailing winds are an important means of transport of lithogenous clay-sized particles into all ocean basins.

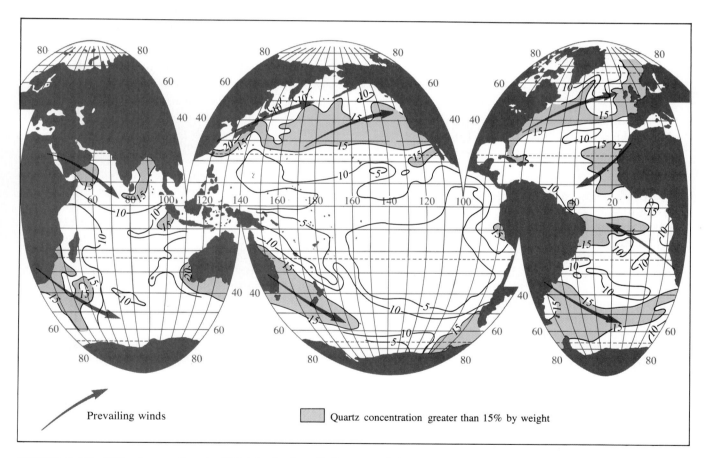

Prevailing winds

Quartz concentration greater than 15% by weight

FIGURE 5–20 Lithogenous Quartz (SiO$_2$) in Surface Sediments of the World's Oceans.

Occurring as chips and shards mostly within the 5–10 μm (0.002–0.004 in.) range, most quartz found in oceanic sediments is believed to have been transported by winds. Desert areas of Africa, Asia, and Australia are important sources of eolian (wind-blown) quartz. Turbidity flows have been significant sources of quartz in the equatorial, northwestern, and northern Atlantic Ocean and the Bay of Bengal, where deep-bottom currents have also modified sediment distribution. The distribution of lithogenous clays transported by winds may show a similar pattern. After Leinen et al., 1986.

HOW FAST ARE THE CONTINENTS GROWING?

After we consider all of the volcanic activity and subduction associated with global plate tectonics, the erosion of continents, and deposition of marine sediment, we may ask, "How fast are continents growing?"

A study conducted by the U.S. Geological Survey has created a possible budget for the growth and denudation of continents. By calculating the yearly flux of various components of marine sediment, seamounts, and island arc volcanoes, a somewhat tentative rate of growth for continents since the breakup of Pangaea has been determined.

It was determined that

1. Ocean crust is recycled every 110,000,000 years and its average age is 55,000,000 years.
2. Lithogenous particles accumulate in the deep ocean at a rate of 1.27 km^3/yr. Biogenous particles, the chemical components of which come ultimately from continents, accumulate at the rate of 0.38 km^3/yr. This totals to a **rate of continental denudation of 1.65 km^3/yr.**
3. Adding to this the 0.05 km^3/yr of volcanic sediment that accumulates gives a total yearly rate of sediment accumulation of **1.7 km^3/yr.**
4. Seamounts add 0.2 km^3/yr of mass to the ocean crust, and volcanic island arcs add 1.1 km^3/yr.
5. **Total material available to accrete to continents during subduction is 1.7 + 0.2 + 1.1, or 3 km^3/yr.**
6. Subtracting the rate of continental denudation (1.65 km^3/yr) from this total leaves **1.35 km^3/yr of possible continental growth.**
7. Whatever amount of deep-sea sediment or seamounts is subducted will have to be subtracted from this growth rate. The numbers for these variables are not yet known.

SUMMARY

Sediments that accumulate on the ocean floor are classified by origin as **lithogenous** (derived from rock), **biogenous** (derived from organisms), **hydrogenous** (derived from water), and **cosmogenous** (derived from beyond the earth's atmosphere). **Sediment texture**, determined in part by the size and sorting of the particles, is affected greatly by the type of transportation (water, wind, or ice) that brought it to the deposit and the energy conditions under which it was deposited.

Lithogenous sediment is composed of fragments of inorganic compounds called **minerals**, which make up rocks. **Ferromagnesian minerals**, rich in iron and magnesium, are abundant in the dark igneous rock **basalt**, characteristic of ocean basins. Minerals low in iron and magnesium, **nonferromagnesian minerals**, are characteristic of the igneous rock **granite**, which makes up the continents. **Biogenous** sediment is composed primarily of the compounds **calcium carbonate** (CaCO$_3$) from the remains of foraminifers, pteropods, and coccolithophores and **silica** (SiO$_2$) from the remains of diatoms and radiolarians. **Hydrogenous** sediment includes a wide variety of materials that precipitate from the water or form from interaction of substances dissolved in the water with materials on the ocean floor. **Manganese nodules, phosphorite, glauconite, carbonates,** and **zeolites** are examples. **Cosmogenous** sediment is composed of nickel-iron spherules and silicate chondrules rich in pyroxenes that probably represent sparks created as asteroids collided in the asteroid belt.

Neritic sediment accumulates rapidly along the margins of continents. It is dominated by particles of lithogenous origin. Owing to the recent rise in sea level from melting glaciers, many rivers throughout the world, including those flowing into the Atlantic along the east coast of the United States, are not now depositing sediment on the continental shelves, but rather in their estuaries. About 70% of the sediment covering the continental shelves is relict sediment from 3000 to 7000 years old. Of the 20 billion MT of sediment carried each year to the oceans by rivers, almost 80% is provided by the rivers of Asia.

Turbidity currents are thought to transport shelf sediment down submarine canyons and deposit it as **turbidite** on the deep-ocean floor. More than 75% of the sediment mass found on the ocean floor is part of the thick sediment wedge underlying the continental shelves, slopes, and rises. In high latitudes this accumulation includes poorly sorted **glacial deposits**.

Stromatolites are dome-shaped, CaCO$_3$-depositional structures that result from the entrapment of CaCO$_3$ particles in algal mats in environments that inhibit the development of algae-grazing gastropods.

The **Blake Plateau**, which extends from Cape Hatteras south off the Florida coast and ranges in depth from 700 to 1000 m (2296 to 3280 ft), does not receive sediment because of the high velocity of the northward-flowing Gulf Stream.

Oceanic sediment accumulates at low rates on the floor of the deep-ocean basins far from the continents. In the deeper basins, where most biogenous sediment is dissolved before reaching bottom, **abyssal clay** deposits predominate. **Oozes** composed of over 30% by weight of biogenous sediment are found in shallower basins and on ridges and rises. The rate of biological productivity measured against the rates of destruction and dilution of biogenous sediment determines whether abyssal clay or oozes will form on the ocean floor. Much of the destruction of $CaCO_3$ occurs because the CO_2 content of the ocean water increases with depth. At the **calcium carbonate compensation depth**, ocean water dissolves $CaCO_3$ at a rate equal to the rate it settles from above, thus preventing accumulation on the ocean floor below that depth.

Manganese nodules are found on the deep-ocean floor where other sediments are accumulating at rates of less than 7 mm (0.3 in.)/1000 yr. There is a relationship between the biological productivity of overlying waters and metallic content of the nodules.

Although it would take from 10 to 50 years for individual sediment particles in the ocean surface waters to settle to the bottom, most appear to be combined into larger aggregates as **fecal pellets** of small marine animals and reach the ocean floor in 10 to 15 days. This process of particle transport explains how the particle composition of the surface waters is similar to that of the sediment on the ocean floor immediately below.

Because the bottom water in the western Atlantic Ocean is colder, the calcium carbonate compensation depth is shallower on the west side of the Mid-Atlantic Ridge and abyssal clay in the western basins has a broader distribution. In the Pacific Ocean, abyssal clay covers more of the ocean floor than in other oceans because it is deeper (especially in the North Pacific) and lies beneath the calcium carbonate compensation depth. Siliceous oozes are well developed in the North Pacific Ocean and around the southern margins of all oceans because little lithogenous sediment reaches these areas and foraminifers that produce $CaCO_3$ do not thrive in these cold waters.

It is estimated that the rate of continental growth is 1.35 km^3/yr as a result of an annual rate of continental denudation of 1.65 km^3 and a rate of continental accretion of 3 km^3. The annual rate of sediment accumulation in the ocean basins from all sources is 1.7 km^3.

QUESTIONS AND EXERCISES

1. List the four basic sources of marine sediment.

2. What characteristics of marine sediment indicate increased maturity?

3. Why is a higher-velocity current required to erode clay-sized particles than larger, sand-sized particles?

4. List the four ferromagnesian and the three nonferromagnesian minerals designated as common rock-forming minerals. Why are the nonferromagnesian minerals of lower density?

5. What igneous rocks are most characteristic of the oceanic and continental crusts?

6. List the two major chemical compounds of which most biogenous sediment is composed and the organisms that produce them.

7. What is the chemical composition of hydrogenous deposits of manganese nodules, phosphorite, glauconite, and carbonate? In which environments do they form? What is the relationship between a deposit's formation and its enviroment?

8. What are the two most common forms in which calcium carbonate precipitates from ocean water? Describe the environment in which such precipitation is most likely to occur. Include a discussion of the pH, biological processes, and water temperature conditions that enhance precipitation.

9. Describe the most common types of cosmogenous sediment and give the probable source of these particles.

10. Describe the basic differences between neritic sediment and ocean sediment.

11. Discuss the processes by which sediments are carried to and distributed across the continental margin.

12. In the two locations of stromatolite occurrence discussed in the text, what processes inhibit the development of algae-grazing gastropods?

13. How do oozes differ from abyssal clay? Discuss how productivity, destruction, and dilution combine to determine whether an ooze or abyssal clay will form on the deep-ocean floor.

14. Refer to figure 5–13 and explain why abyssal clay deposits on the floor of deep-ocean basins are commonly underlain by calcareous ooze.

15. Describe the apparent relationships that exist between biological productivity of overlying waters and metallic content of manganese nodules.

16. How do fecal pellets help explain that the particles found in the ocean surface waters are closely reflected in the particle composition of the sediment directly beneath? Explain why you would not expect this.

17. Explain why the abyssal clay deposition on the east side of the Mid-Atlantic Ridge is more restricted than on the west side.

REFERENCES

Berger, W. H.; Adelseck, C. G., Jr.; and Mayer, L. A. 1976. Distribution of carbonate in surface sediments of the Pacific Ocean. *Journal of Geophysical Research* 81:15, 2617–29.

Billett, D. S.; Lampitt, R. S.; Rice, A. L.; and Mantoura, R. F. C. 1983. Seasonal sedimentation of phytoplankton to the deep-sea benthos. *Nature* 302:5908, 520–22.

Biscaye, P. E.; Kolla, V.; and Turekian, K. K. 1976. Distribution of calcium carbonate in surface sediments of the Atlantic Ocean. *Journal of Geophysical Research* 81:15, 2595–602.

Deuser, W. G.; Brewer, P. G.; Jickells, T. D.; and Commeau, R. F. 1982. Biological control of the removal of abiogenic particles from the surface ocean. *Science* 219:4583, 385–86.

Dravis, J. J. 1983. Hardened subtidal stromatolites, Bahamas. *Science* 219:4583, 385–86.

Heath, G. R. 1982. Manganese nodules: Unanswered questions. *Oceanus* 25:3, 37–41.

Hildebrand, A. R., and Boynton, W. V. 1990. Proximal Cretaceous-Tertiary boundary impact deposits in the Caribbean. *Science* 248:4957, 843–46.

Howell, D. G., and Murray, R. W. 1986. A budget for continental growth and denudation. *Science* 233:4762, 446–49.

Kolla, V., and Biscaye, P. E. 1976. Distribution of calcium carbonate in surface sediments of the Atlantic Ocean. *Journal of Geophysical Research* 81:15, 2602–16.

Leinen, M.; Cwienk, D.; Heath, G. R.; Biscaye, P. E.; Kolla, V.; Thiede, J.; and Dauphin, J. P. 1986. Distribution of biogenic silica and quartz in Recent deep-sea sediments. *Geology* 14:3, 199–203.

Shepard, F. P. 1973. *Submarine geology*. New York: Harper and Row.

Smith, R. V., and Kinsey, D. W. 1976. Calcium carbonate production, coral reef growth, and sea level change. *Science* 194:937–38.

Spencer, D. W.; Honjo, S.; and Brewer, P. G. 1978. Particles and particle fluxes in the ocean. *Oceanus* 21:1, 20–26.

Sverdrup, H. U.; Johnson, M. W.; and Fleming, R. H. 1942. Renewal 1970. *The oceans: Their physics, chemistry, and general biology*. Englewood Cliffs, N.J.: Prentice-Hall.

Udden, J. A. 1898. Mechanical composition of wind deposits. *Augustana Library Pub.* 1.

Weaver, P. P. E., and Thomson, J., eds. 1987. *Geology and geochemistry of abyssal plains*. Palo Alto: Blackwell Scientific.

Wentworth, C. K. 1922. A scale of grade and class terms for clastic sediments. *Journal of Geology* 30, 377–92.

SUGGESTED READING

Sea Frontiers

Dietz, R. S. 1978. IFOs (Identified Flying Objects). 24:6, 341–46.
The source of Australasian tektites and microtektites is discussed.

Dudley, W. 1982. The secret of the chalk. 28:6, 344–49.
An informative discussion of the formation of marine chalk deposits.

Dugolinsky, B. K. 1979. Mystery of manganese nodules. 25:6, 364–69.
The problems of origin, growth, and the environmental implications of manganese nodule mining.

Feazel, C. T. 1986. Asteroid impacts, seafloor sediments, and extinction of the dinosaurs. 32:3, 169–78.
A discussion of the possibility that high concentrations of iridium and osmium in a marine clay deposited at the time dinosaurs and many other species died out 65,000,000 years ago may have resulted from the collision of the earth with a meteor 6 mi in diameter.

Feazel, C. T. 1989. Inner space: Porosity of seafloor sediments. 35:1, 49–52.
A petroleum geologist discusses how porosity forms in marine sediments and the significance of the pores.

La Que, F. L. 1979. Nickel from nodules? 25:1, 15–21.
The economic and political problems related to nodule mining are discussed.

Shinn, E. A. 1987. Sand castles from the past: Bahamian stromatolites discovered. 33:5, 334–43.
The formation of stromatolites, sand domes trapped by algal growth, in the Bahamas is discussed.

Victory, J. J. 1973. Metals from the deep sea. 19:1, 28–33.
The formation and economic potential of mining manganese nodules are discussed.

Wood, J. 1987. Shark Bay. 33:5, 324–33.
The history of Shark Bay, Australia, since its discovery by Dirk Hartog in 1616 is summarized. The stromatolites and "tame" dolphins that attract tourists are also discussed.

Scientific American

Nelson H., and Johnson, K. R. 1987. Whales and walruses as tillers of the sea floor. 256:2, 112–18.
The 200,000 walruses and 16,000 gray whales that feed by scooping sediment from the Bering Sea continental shelf suspend large amounts of sediment that is transported by bottom currents.

6
PROPERTIES OF WATER

Before we can go further in our investigation of the ocean, we must establish a background that will allow us to understand the various properties of the substance that makes up 96.5% of the mass of the oceans—water. We at first may not consider water to be an unusual substance, inasmuch as it is probably the one we would mention if we were asked to name the most common or prevalent substance that exists on the earth. Water most certainly is abundant, but it does have unique properties; we find that it is a very uncommon substance.

It is the presence of water on the planet earth that makes life possible. Water controls the distribution of heat over the earth's surface as well as other conditions for life.

THE WATER MOLECULE

When two hydrogen atoms combine with one oxygen atom to form one water molecule, they join in such a manner that the hydrogen atoms are separated by an angle of 105° rather than being at opposite ends of a linear molecule and separated by 180° (see fig. 6–1A). As the electrons move around within the structure, this arrangement of atoms within a molecule produces a greater concentration of electrons around the nucleus of the oxygen atom than around the hydrogen nuclei. As a result, a positive charge associated with the unshielded proton in the nucleus of each hydrogen atom is concentrated at the location of the hydrogen atoms. This situation produces polarity of electrical charge, in which the end of the molecule represented by the oxygen atom is slightly more negatively charged and the end with the hydrogen atoms is slightly more positively charged: the water molecule is **dipolar**, and water is known as a polar substance.

SOLVENT PROPERTIES

The polarity in the distribution of electrical charge does not make the water molecule behave as an ion. Pure water is actually a very poor conductor of electricity, since the molecule will not move toward either the negatively charged pole or the positively charged pole in an electrical system. Instead, the water molecule simply becomes oriented in an electrical field

Preceding page: Surf as seen from lighthouse, Point Reyes, California. Photo by Stephen Trimble.

with its positively charged hydrogen nuclei toward the negative plate and its negatively charged oxygen end toward the positively charged plate. This orientation of water molecules in an electrical field tends to neutralize the field.

Owing to their polar nature, water molecules form bonds between one another. These **intermolecular bonds** are referred to as **hydrogen bonds**. The positively charged hydrogen areas of the water molecule attract the negatively charged oxygen portion of the neighboring water molecules and bond together by electrostatic forces. When polar molecules are bound together in complexes, as are the water molecules, their ability to reduce the intensity of an electrical field acting on the water and, therefore, the electrostatic attraction between ions of opposite charges introduced into water, is greatly increased. In fact, these forces can be reduced to 1/80 their value out of water.

If we consider sodium chloride, a compound containing **ionic bonds**, we can see that simply by placing that substance in water, we have reduced the electrostatic attraction between the sodium and chloride ions by 80 times. This reduced attraction makes it much easier for the ions to be dissociated. Once they do become dissociated, the sodium ions are attracted to the negative pole of the water molecule and the chloride ions are attracted to the positive pole of the water molecule (fig. 6–1).

As more and more ions of sodium and chlorine are freed by the weakening of the electrostatic attraction that is holding them together, they become surrounded by the polar molecules of water, or **hydrated**. Sodium ions are surrounded by water molecules with their negative poles toward the ion, and chloride ions are surrounded by water molecules oriented so that their positively charged poles are directed toward the ion.

THERMAL PROPERTIES

Freezing and Boiling Points of Water

The Celsius (centigrade) temperature scale was constructed on the basis of the characteristics of water at one atmosphere of pressure, with the **freezing point** of water, the temperature at which it changes from a liquid to a solid, representing 0° and the **boiling point** of water, the temperature at which it changes from a liquid to a gas, representing 100°. If we compare these values to those of some compounds similar to water in makeup, we will see that the freezing and boiling points of water are uniquely high. The other

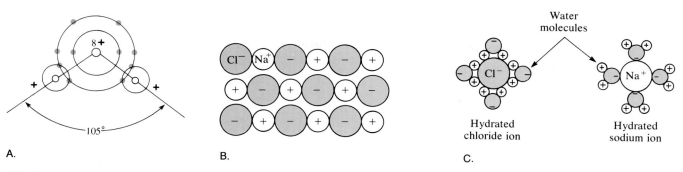

A.

B.

C.

FIGURE 6–1 Water as a Solvent.

A, water molecule. With the angle of 105° between the two hydrogen atoms in the water molecule, a dipolar molecule results. The oxygen end of the molecule is negatively charged, and the hydrogen regions exhibit a positive charge. *B,* sodium chloride solid state. Ionic bonds hold sodium ions (Na^+) and chloride ions (Cl^-) together in the compound sodium chloride ($NaCl$). *C,* sodium chloride in solution. The positively charged hydrogen end of the dipolar water molecule is attracted to the negatively charged Cl^- ion; the negatively charged oxygen end is attracted to the positively charged Na^+ ion. The ions become hydrated when surrounded by water molecules that are attracted to them. In this state they have become solute ions in a water solution.

compounds occur in nature as gases, while water occurs in the gaseous, liquid, and solid states of matter within the narrow range of atmospheric conditions.

We should now consider the nature of the forces that must be dealt with to change the state of a compound. There exists between all molecules of any compound a relatively weak attraction that results from the electrostatic attraction between the nuclear parts of one molecule and the electrons of another. This attraction slightly exceeds the electrostatic repulsion that exists between the nuclear parts of the same molecules and the electrons of the same molecules.

This intermolecular force is known as **van der Waals force,** and it becomes significant only when molecules are very close together, such as in the solid or liquid states. Generally, the heavier the molecule, the greater the van der Waals attraction between individual molecules of the compound. Therefore, with increasing molecular weight, a greater amount of energy is needed to overcome that attraction and allow a change of state, say from a solid to a liquid or a liquid to a gas, to occur. Consequently, the melting and boiling points of compounds generally increase as the molecular weight increases.

What form of energy is used to change the state of matter? We should first define a few terms before we go further in our discussion. **Heat,** a form of energy associated with and proportional to the energy level of molecules, is the total **kinetic energy** of a sample. It may be generated by combustion, chemical reaction, friction, or radioactivity. So that we can measure the amount of energy that is being added to or removed from the molecules, we must introduce the

term **calorie** as a unit for measuring quantity of heat. A calorie is defined as the amount of heat required to raise the temperature of one gram of water one degree Celsius. So that we can measure the amount of energy that the molecules within the substance have, we will need to understand the meaning of the word **temperature.** Temperature is defined as the direct measure of the average kinetic energy of the molecules that make up a substance. Kinetic energy is energy of motion, so the higher the temperature, the greater the velocity of the molecules of the substance for which the temperature is being measured.

Let us return now to the van der Waals force that causes molecules of a given substance to be attracted to one another. We can see that if this attraction is to be broken, energy must be given to the molecules so that they can move more rapidly, in the ways described in figure 6–2, to overcome this force. It is this attraction that must be broken if we are going to produce a change of state in any substance from solid to liquid or liquid to gas.

We find that in a crystalline **solid state,** although bonds are constantly being broken and reformed, the predominant relationship between molecules is that of a rather firm attachment produced by the nearness of the molecules and the great effect the van der Waals force has upon molecules that are close together. **Vibration** is the dominant type of molecular movement in the solid state as the molecules vibrate but remain in relatively fixed positions.

In the **liquid state** the molecules have gained enough energy to overcome many of the van der Waals forces that bound them together in the solid state. The molecules have enough freedom to move relative to one another. In this state we see all forms

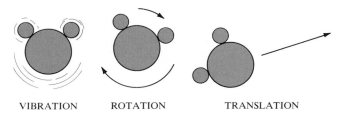

VIBRATION   ROTATION   TRANSLATION

FIGURE 6–2   Motion of Molecules.
Vibration is the typical movement of molecules in a crystalline solid; *rotation* is the movement in liquids and gases; and all gas molecules move randomly by *translation*.

of molecular movement—vibration, **rotation**, and **translation**. The molecules are free to move relative to one another but are still attracted by one another. Bonds are being formed and broken at a much greater rate than in the solid state.

In the **gaseous state**, translation has become the dominant type of motion. Molecules are now moving at random, and there exists no attraction between individual molecules. The only effect that one molecule will have on another is that which is produced by collision during random movement.

Returning to our comparison of the properties of water and compounds of similar composition containing two hydrogen atoms and an atom of another element, we will look at H_2S, H_2Se, and H_2Te. The molecular weights of the four compounds are H_2O (18), H_2S (34), H_2Se (80), and H_2Te (129). Figure 6–3 shows that, as was predicted in the discussion of the van der Waals force, the freezing points and boiling points for H_2S, H_2Se, and H_2Te increase with increased molecular weight. When we insert H_2O into the scale, the freezing point and boiling point based on its molecular weight should be about −90°C (−130°F) and −68°C (−108.4°F), respectively. The fact that water freezes at 0°C (32°F) and boils at 100°C (212°F), seems to be a violation of natural order. Here we see again the great significance of the polarity of the water molecule and the hydrogen bond that this structure produces. The high freezing and boiling points of water are the manifestations of the additional kinetic energy required to overcome not only van der Waals bonds but also the hydrogen bonds to achieve a change of state.

Heat Capacity

Another direct result of the hydrogen bond is the high heat capacity of water. As noted, a calorie is the amount of heat required to raise the temperature of 1 g (0.035 oz) of water 1°C (1.8°F). The heat capacity of water compared to that of most other substances

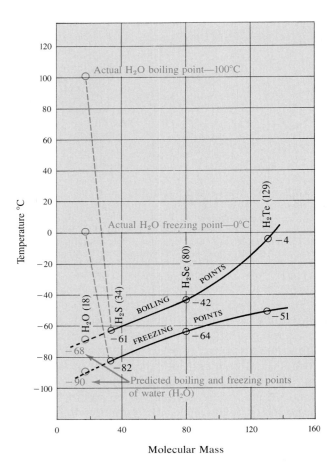

FIGURE 6–3   Molecular Mass—Freezing and Boiling Points.
Because of the hydrogen bonds that must be broken in addition to overcoming the van der Waals force in achieving changes of state, the freezing and boiling points of water are higher than would be expected for a compound of its molecular makeup.

is great, and we use the capacity of water to absorb heat as a standard against which we compare the heat capacities of other substances. The amount of heat that is required to raise the temperature of 1 g of any substance 1°C is the **heat capacity** of the substance. It is 1 cal for water and less than 1 cal for most other substances. It is this property that produces the **mild climates** in coastal regions. Great amounts of heat can be gained or lost by the coastal water without causing extreme changes in temperature.

Latent Heats of Melting and Vaporization

Closely related to water's unusually high heat capacity are its high latent heat of melting and latent heat of vaporization. The continuous addition of heat to a substance in the solid or liquid state will bring about a change of state in the substance. A solid converts to a liquid at a temperature called its **freezing** (melting) **point** and a liquid is changed to a gas at a temperature defined as its **boiling** (condensation) **point.**

When changing the state of any substance, there may be no increase of temperature at that point where a change of state occurs even though heat is continuously being added. The heat energy is being used entirely to break all the intermolecular bonds required to complete the change of state. While the bonds are being broken, a mixture of the substance in both states is in equilibrium. Only upon completion of the change of state will the temperature again rise. The heat that is added to 1 g of a substance at the melting point to break the required bonds to complete the change of state from solid to liquid is the **latent heat of melting.** The heat applied to effect a change of state at the boiling point is the **latent heat of vaporization.**

The use of the word *latent* in describing the heats of vaporization and melting is necessary because the heat that must be added to a given mass of ice or water to convert it to a higher energy state, water or water vapor, is held in reserve or "hidden" by that mass of water or water vapor. When water vapor returns to a liquid, it **condenses** and heat is released into the surrounding air. Release of heat also occurs with the freezing of water, which is the change of state from liquid to solid. During condensation and freezing, the same amount of heat that is added to change the state of that water from a liquid to a gas or from a solid to a liquid is released.

A practical application of these principles of heat transfer can be seen in the use of ice for the purpose of refrigeration. A block of ice set in a closed container among food articles will lower their temperature because heat energy is drawn from the food articles and added to the molecules of ice to change them to the liquid state as the ice melts. The principle of cooling air with water is similar: in arid climates, the hot dry air that passes through a surface coated with liquid water will lose heat to the water. The water is converted to a vapor. Thus, after passing through or across the water-covered surface, the air is considerably cooler.

To help clarify this phenomenon, we will discuss the heat transfer and the change of state of H_2O from a solid at $-40°C$ ($-40°F$). Observe the heat-temperature graph, which has a vertical temperature scale ranging from $-40°$ to $120°C$ ($248°F$) and a horizontal scale which begins at 0 and continues through 800 cal (fig. 6–4).

For 1 g of ice, we find that the addition of 20 cal of heat raises the temperature $40°$, from $-40°C$ to $0°C$. Thus, the heat capacity of ice is 0.5 cal/g, half that of liquid water.

Once the temperature has been raised to $0°C$ ($32°F$), the continuous addition of heat does not result in an increase in temperature until 80 cal have been added. We do not observe an increase in temperature during the addition of these 80 cal of heat because all the heat energy added was used to break the intermolecular bonds that were holding the water molecules in the solid state. The temperature will remain unchanged until all of the necessary bonds are broken and the mixture of ice and water has been changed to 1 g of water. The amount of heat required to convert 1 g of ice to 1 g of water, 80 cal, is termed the latent heat of melting and it is

FIGURE 6–4   Change of State Graph—Water.

The latent heat of melting (80 cal) is much less than the latent heat of vaporization (540 cal) because only a few hydrogen bonds must be broken to convert 1 g of ice to a liquid, while all remaining hydrogen bonds must be broken to convert 1 g of liquid water to a gas.

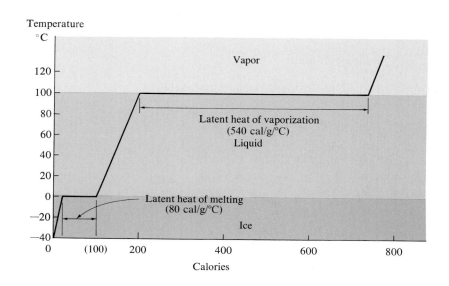

higher for water than for any other commonly occurring substance.

The bonds that are broken in converting most substances from a solid to a liquid are the van der Waals bonds. In water, however, not only the van der Waals bonds but some of the hydrogen bonds must also be overcome. It is necessary only to break the ice structure down into numerous small clusters surrounded by individual water molecules to the extent that these remaining ice clusters can move relative to one another and allow the mass to assume a liquid state. Liquid water, particularly at low temperature near the freezing point, could well be described as a pseudo-crystalline liquid, as there are still many small clusters of ice crystals contained within it. Once enough hydrogen bonds have been broken to allow freedom of movement among the clusters and individual water molecules, the temperature of the water will again rise with the addition of heat.

After the phase change from ice to liquid water has occurred at 0°C, the addition of heat to the water causes the temperature to rise again. As it does, note that it requires 1 cal of heat to raise the temperature of water 1°C (1.8°F). We must therefore add another 100 cal or a total of 200 cal, before one gram of water reaches the boiling point of 100°C (212°F). At this temperature we note the development of another plateau, far more prominent on the graph than that which represents the latent heat of melting, 80 cal/g. This plateau at 100°C represents an addition of 540 cal, the latent heat of vaporization, to the gram of water before complete conversion to the vapor state.

Why is so much more heat energy required to convert 1 g of water to water vapor than was required to convert 1 g of ice to water? We defined a gas as a substance in which the molecules were moving at random, free from the influence of other molecules, except when they collide. To make the conversion from ice to water, not all the hydrogen bonds had to be broken, but only enough to allow freedom of movement along the various ice clusters that remained and the individual molecules that were also present in the system. To convert water to water vapor, every molecule must be freed from the attraction of other water molecules. Therefore, every hydrogen bond must be broken (fig. 6–5). To do this requires a great amount of heat energy.

We can now see that large amounts of heat energy are absorbed or released by water as it changes state, particularly if the change of state is from a liquid to a vapor or vice versa. Of course, we do not have temperatures of 100°C at the surface of the ocean where this phenomenon of conversion of water to vapor occurs in nature. Sea-surface temperatures average about 20°C (68°F) or less.

The conversion of a liquid to a gas below the boiling point is called **evaporation**. At ocean-surface temperatures, individual molecules that are being converted from the liquid to the gaseous state have a lower amount of energy than do the molecules of water at 100°C. Therefore, to gain the additional energy necessary to break free of the surrounding water molecules, an individual molecule must capture energy from its neighbors. This phenomenon explains the cooling effect of evaporation. The molecules left behind have lost heat energy to those that escape. To produce 1 g of water vapor from the ocean surface at temperatures less than 100°C requires more than the 540 cal of heat that are required to make this conversion at the boiling point. For instance, the **latent heat of evaporation** is 585 cal/g at 20°C. This higher value is due to the fact that more hydrogen bonds must be broken at this lower temperature for a gram of water molecules to enter the gaseous state.

FIGURE 6–5 Hydrogen Bonds in H₂O.

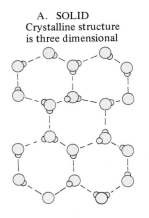

A. SOLID
Crystalline structure
is three dimensional

B. LIQUID

C. GAS

We can readily see the significance of the evaporation-condensation cycle in the earth's surface temperatures. Evaporation removes heat energy provided by the sun and stored in the oceans. This energy is carried high into the atmosphere as the water vapor rises and is released there when the vapor condenses and falls as precipitation. It is water's latent heat of evaporation that accounts for the removal of great quantities of heat from the low-latitude oceans by evaporation. It is later released in the heat-deficient higher latitudes after the vapor is transported through the atmosphere and condenses there as rain and snow (see *Hurricanes—Nature's Safety Valves* on p. 144).

SURFACE TENSION

Next to mercury, water has the highest **surface tension** of all commonly occurring liquids. We observe a surface tension phenomenon in filling a container with water to the brim and even beyond (fig. 6–6A). You will note that the water can be piled up above the rim of the container, forming a **convex** surface that is the water interface with the atmosphere. Water drops are also a very common manifestation of surface tension. These phenomena result from the tendency for water molecules to attract one another or to cohere at the surface of any accumulation of water. Because of this cohesive tendency, it is possible to float on water objects that are much heavier than

A. A container can be filled above its rim due to the strength of the surface layer.

C. Capillarity-water climbs the walls of a tubular container. The smaller the diameter, the higher it will climb.

B.   Hydrogen bonds

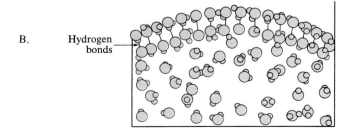

FIGURE 6–6   Surface Tension Phenomena.
A surface tension layer forms as hydrogen bonds form a strong attraction between the top layer of water molecules and the underlying molecules.

water. A razor blade carefully laid on the surface will float, although it is normally five times as dense as water. Many insects use the surface of water as if it were a solid surface, moving across it at will.

Surface tension is a manifestation of the presence of the hydrogen bond. Those molecules of water that are at the surface are strongly attracted to the molecules of water below them by their hydrogen bonds. The air above the surface has a very low density of molecules as compared to the water itself, and although water molecules are attracted to molecules of other substances, the attraction of the hydrogen bond holds the water molecules in the surface layer to those below (fig. 6–6B).

Water clings to the surface of many substances; we refer to this phenomenon as **wetting**. Water will adhere strongly to glass, organic substances, and inorganic material such as rocks and soil. When water is poured into a container made of a substance to which it strongly adheres, the **adhesive force**, or the force of attaction between water molecules and molecules of another substance, will cause the surface tension layer to take on a form unlike the convex one described previously. In this case the surface tension layer is **concave**, being drawn up on the container's sides (*see* fig. 6–6C).

If the container is glass, the positively charged portions of the water molecules are attracted to unbonded electrons in the oxygen atoms that are part of the makeup of the glass container. Being strongly attracted by these oxygen atoms, the water molecules will "climb" up the side of the container and will be held back only by the hydrogen bond attraction that exists between individual water molecules. In fact, if the diameter of the container is decreased to a very fine bore, the combination of cohesion, which holds the water molecules together, and the adhesive attraction between the water molecules and the glass container will pull the column of water to great heights. This phenomenon is known as **capillarity**.

SALINITY OF OCEAN WATER

Though the proportion of water to dissolved salts may vary within the ocean, the major component ions are distributed in ocean water in relatively constant proportions. This **constancy of composition** was first established by Dittmar in his analysis of the *Challenger* expedition water samples discussed in chapter 1. Water makes up, on the average, 96.53% of the ocean's mass, so it determines most of the physical properties that we observe in ocean water.

HURRICANES—NATURE'S SAFETY VALVES

Tropical storms usually develop near the end of the summer season in both hemispheres (fig. 6A). These storms begin as low-pressure cells that break away from the equatorial low-pressure belt and grow as they pick up heat energy from the warm ocean. By means of the large heat capacity and latent heat of evaporation of the ocean's water, they transport great quantities of heat energy into heat-deficient higher-latitude regions by moving in curved paths around the western sides of the oceanic subtropical high-pressure cells (fig. 6B). When wind velocity within the storms reaches 120 km/h (74 mi/h), tropical storms are classified as hurricanes in the Western Hemisphere and typhoons in the Eastern Hemisphere.

Hurricanes can have diameters exceeding 800 km (496 mi). More typically, they will have a diameter measuring less than 200 km (124 mi). As air moves across the ocean surface toward the low-pressure eye of the hurricane and is drawn up to the outer reaches of the atmosphere, its sustained velocity increases to as much as 300 km/h (186 mi/h). Once near the eye, the air begins to spiral upward and horizontal wind velocities may be negligible. Because of the Coriolis effect, the air flowing toward the eye of a storm veers to the right in the Northern Hemisphere and to the left in the Southern Hemisphere, producing counterclockwise and clockwise rotation, respectively (fig. 6B).

The energy contained in one hurricane is greater than that generated by all energy sources over the last twenty years in the United States. The energy released by an atomic bomb detonation in the center of a hurricane would have no effect on the storm. It would quickly be sucked up through the eye and distributed throughout the system.

The worst natural disaster in U.S. history occurred on September 8, 1900, when a storm surge associated with a 135-km/h (84-mi/h) hurricane inundated the barrier island city of Galveston, Texas, killing 6000 people.

FIGURE 6A Hurricanes and Typhoons.

Hurricanes begin as low-pressure cells moving out of the equatorial low-pressure belt toward the higher latitudes in curved paths around the oceanic subtropical highs. They are called typhoons in the western Pacific and Indian oceans and hurricanes in the western Atlantic and eastern Pacific oceans.

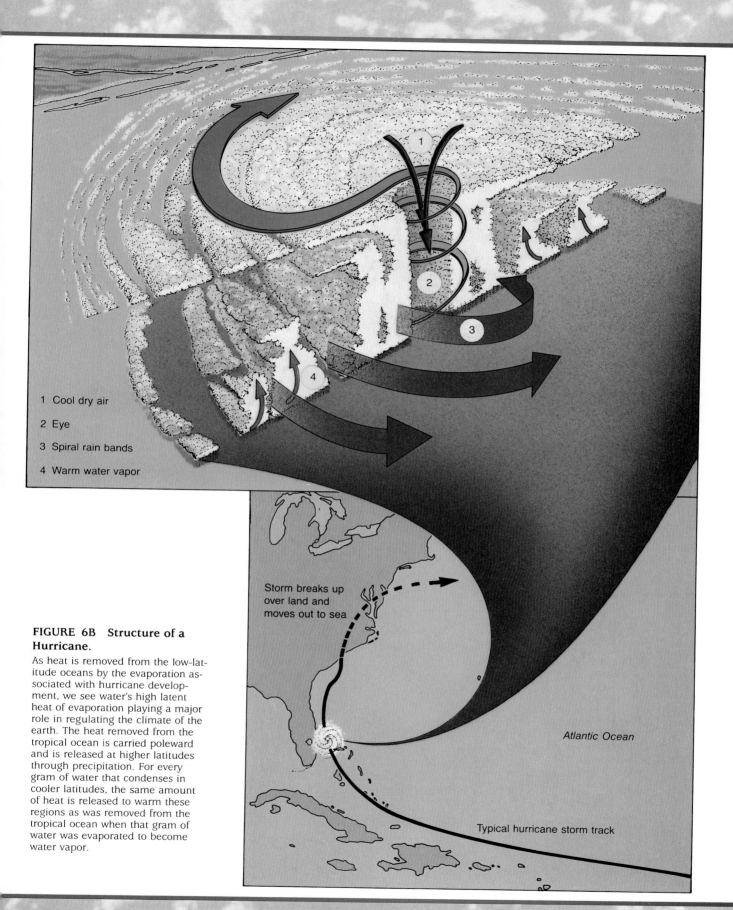

1 Cool dry air

2 Eye

3 Spiral rain bands

4 Warm water vapor

Storm breaks up over land and moves out to sea

Atlantic Ocean

Typical hurricane storm track

FIGURE 6B Structure of a Hurricane.

As heat is removed from the low-latitude oceans by the evaporation associated with hurricane development, we see water's high latent heat of evaporation playing a major role in regulating the climate of the earth. The heat removed from the tropical ocean is carried poleward and is released at higher latitudes through precipitation. For every gram of water that condenses in cooler latitudes, the same amount of heat is released to warm these regions as was removed from the tropical ocean when that gram of water was evaporated to become water vapor.

An illustration of the cooling effect of hurricanes on sea-surface temperature (SST) is change in SST that resulted from the passage of hurricane Gloria in September 1985. Its path out of the warm Sargasso Sea east of the Bahamas, along the east coast, and across Long Island onto the mainland had the cooling effect shown in figure 6C. The cooling of the ocean surface by as much as 5°C (9°F) resulted in part from evaporation and conduction, but a significant amount of the cooling resulted from cold water from beneath the surface mixing with the warmer surface water.

Hurricane Gloria was a strong hurricane, with a minimum central pressure of 919 millibars (mbar) compared to standard sea-level pressure of 1013 mbar. The lowest atmospheric pressure ever recorded was 865 mbar in a typhoon off Okinawa in September 1945. The lowest pressure recorded in the Western Hemisphere was measured September 12, 1988, in hurricane Gilbert between Jamaica and the Cayman Islands. Gilbert at that point had sustained winds of 296 km/hr (184 mi/hr) and gusts of up to 320 km/hr (200 mi/hr). Figure 6D shows Gilbert two days later moving onto the Yucatán peninsula of Mexico, where it produced a 4-m (13-ft) storm surge.

FIGURE 6C Sea-surface Temperature (SST) Cooling—Hurricane Gloria, 1985.

The path of hurricane Gloria is indicated by the black-and-white lines. The images to the left and right show the SST measured by the Advanced Very High Resolution Radiometer (AVHRR) aboard the *NOAA 9* satellite. The September 19–23 image reflects conditions just before the hurricane entered the area, and the September 27–28 image represents conditions just after the storm made landfall at Long Island. Note that the Gulf Stream, represented by the dark orange band just north of the center of the left-hand image, is broken up in the right-hand image. The temperatures represented by the colors on the end images are shown in the color band at the top. The center image shows the amount of cooling caused by Gloria, and the values in °C are shown on the short color band immediately above the center image. The greater cooling along the north end of the track resulted from the fact that cold subsurface water was closer to the surface there than to the south. The stippled areas represent cloud cover.

In all, 316 deaths were attributed to hurricane Gilbert. The island nations of Haiti, the Dominican Republic, and Jamaica suffered losses, as did Central American nations from Costa Rica to Guatemala. Mexico was hardest hit, with 202 deaths, most of which resulted from flooding caused by torrential rains as Gilbert came ashore near Monterrey.

Property damage from Gilbert exceeded 10 billion dollars, with Jamaica sustaining 6.5 billion of that total. Only 33 million dollars in damage occurred in the United States. Most of the U.S. damage and the only U.S. death was from 29 confirmed tornadoes spawned by Gilbert in south Texas on September 16 and 17.

Even islands near the centers of ocean basins may be struck by hurricanes. The Hawaiian Islands were hit hard by Hurricane Dot in August 1959 and Hurricane Iwa on November 23, 1982. Hurricane Iwa hit very late in the season with winds of up to 130 km/h (81 mi/h) (fig. 6E). Well in excess of $100 million in damage was done to the islands of Kauai and Oahu. Niihau, a small island occupied by 230 native Hawaiians, was directly in the path of the storm and suffered severe damage, but none of the population received serious injuries.

Hurricanes such as Gloria, Gilbert, and Iwa will always be a threat to life and property. There is little hope that they will ever be controlled. Inhabitants of areas subject to a hurricane's destructive force need to be cautious of this important natural phenomenon. With present-day warning capabilities, prudent responses to hurricane warnings should help prevent much property damage and eliminate any great loss of life. By ignoring safety precautions, however, many lives will needlessly be lost in future hurricanes.

FIGURE 6D Hurricane Gilbert.

The strongest hurricane ever recorded in the Western Hemisphere, Hurricane Gilbert, moves onto the Yucatán Peninsula of Mexico after having reached its fullest development. Photo courtesy of Dave Clark, NOAA.

FIGURE 6E *GOES* Satellite Infrared Image of Hurricane Iwa as Observed at 2:15 A.M., November 24, 1982.

The three most northerly inhabited islands, Niihau, Kauai, and Oahu, suffered the most damage.

Hurricanes may well be thought of as natural safety valves that help to release the heat energy stored up during summer months in the low latitudes. For every gram of water evaporated from the ocean's surface, the hurricane gains 585 cal of heat energy. As destructive as they are, they play a major role in transferring the excess heat energy accumulated in the lower latitudes to the heat-deficient higher latitudes. Hurricanes are an integral part of the earth's "heat engine" system.

It seems probable that when the techniques are developed for making such measurements, all of the known elements will be found to be dissolved in ocean water. Merely six elements, however, account for over 99% of the dissolved solids in ocean water. These major components are chlorine, sodium, magnesium, sulfur (as SO_4), calcium, and potassium (table 6–1 and fig. 6–7).

Salinity is defined as the total amount of solid material dissolved in a kilogram of seawater when all the carbonate has been converted to oxide, all bromine and iodine replaced by chlorine, and all organic matter completely oxidized. Given this definition, it may appear that salinity is a very complex property and very difficult to measure. However, since the oceans are so very well mixed and the relative abundance of the major constituents is essentially constant, we have a condition that makes the chemical measurement of salinity relatively simple.

Because of constancy of composition, it is necessary to measure the concentration of only one of the major constituents in order to determine the salinity of a given water sample. The constituent that occurs in the greatest abundance and is the easiest to measure accurately is the chloride ion, Cl^-. The portion of the weight of a given sample of water that is the direct result of the presence of this ion is called **chlorinity** and is usually expressed in the terms grams per kilogram of ocean water (g/kg) or parts per thousand (‰). The chloride ion always accounts for 55.04% of the dissolved solids in any sample of ocean water. Therefore, by measuring its concentration, we can determine the total salinity in parts per thousand by the following relationship:

$$\text{Salinity (‰)} = 1.80655 \times \text{Chlorinity (‰)}$$

For example, given the average chlorinity of the ocean, 19.2‰, the ocean's average salinity can be calculated as 1.80655×19.2‰, or 34.7‰. Although 1/0.5504 equals 1.81686 (not 1.80655), oceanographers have agreed on 1.80655 because the constancy of composition has been found to be more of an approximation than an absolute condition.

From 1902 to 1975, the Hydrographic Laboratory at Copenhagen, Denmark, provided Standard Seawater samples to assure that salinity determinations made throughout the world would be based on the same reference. In 1975, this duty was taken over by the Institute of Oceanographic Services located in Wormly, England. **Standard Seawater** consists of ocean water analyzed for chloride ion content to the

TABLE 6–1 Ocean Salinity.

Considering the average salinity of the world ocean, 34.7 by weight, a 1000 g sample of this water would contain 34.7 g of dissolved solids. Six ions account for 99.28% of the dissolved solids.

Major Constituents (each is over 100 parts per million)	
Ion	Percentage
Chloride, Cl^-	55.04
Sodium, Na^+	30.61
Sulfate SO_4^{-2}	7.68
Magnesium, Mg^{+2}	3.69
Calcium, Ca^{+2}	1.16
Potassium, K^+	1.10
	99.28

Minor Constituents (each is 1–100 parts per million)	
	p.p.m.
Bromine	65.0
Carbon	28.0
Strontium	8.0
Boron	4.6
Silicon	3.0
Fluorine	1.0

Trace Elements (each is less than 1 part per million)	
Nitrogen	Iodine
Lithium	Iron
Rubidium	Zinc
Phosphorus	Molybdenum

nearest ten-thousandth of a part per thousand. It is then sealed in ampules and labeled to be sent to laboratories throughout the world (fig. 6–8). The chlorinity of water samples taken and measured in other parts of the world can then be compared if the titration solution or conductivity instrumentation is calibrated with the standard water sample.

The chemical determination of the chloride content in seawater can be made with a high degree of accuracy but requires a great deal of time and care. It is tedious to carry out the chemical titration with silver nitrate aboard ship. With the advancements in instrumentation in the field of oceanography, this task has been greatly simplified. Since electrical conductivity of ocean water is known to increase with increased temperature and salinity, salinity is now most commonly determined by measuring the electrical conductivity of ocean water (at a constant temperature of 15°C, or 59°F) with a modern conductance-measuring instrument, the **salinometer**. Salinity can be determined to better than 0.003‰ by this method.

FIGURE 6–7 Constituents of
Ocean Salinity.

FIGURE 6–8 An Ampule of
Standard Seawater.

WATER DENSITY

The density of water in its various states and at different temperatures is of great importance in considering the movement of water in the ocean. Anything that is more dense than water will sink into it, and a substance that is less dense will float on the surface. We define **density** as mass per unit of volume, usually grams per cubic centimeter (g/cm^3).

Density is affected by temperature, salinity, and pressure. With most substances we observe that a decrease in temperature produces an increase in the density of the substance. The increase in density is the result of the same number of molecules occupying less space as they lose energy. This condition of thermal contraction is also found in water. As the temperature of water decreases, the density increases as long as this temperature decrease occurs above 4°C (39.2°F). If we consider the change in the density of water below 4°C, we encounter considerations that must include the hydrogen bond. As the temperature of water is lowered from 4° to 0°C (32°F), we observe that its density decreases owing to expansion of the water.

This anomalous behavior can be explained only by considering the molecular structure of water and the hydrogen bond. As we lower the temperature of water from 20°C (68°F) and reduce the amount of thermal agitation of the water molecules, the unbonded water molecules occupy less volume because of their decreased energy. But, as we approach the freezing point below 4°C, this reduction in volume owing to

the decreased energy of unbonded molecules is not sufficient to compensate for another phenomenon that is occurring. Ice crystals, open six-sided structures in which water molecules are widely spaced, are becoming more abundant. The greater rate of increase of ice crystals as the temperature approaches the freezing point accounts for the decreased density of water below 4°C (fig. 6–9).

The **temperature of maximum density** for fresh water, 4°C, is lowered by increasing pressure or adding solid particles such as salt, both of which will inhibit the formation of the ice clusters. Thus, more energy must be removed to produce crystals equal in volume to those that could be produced at 4°C in fresh water, causing a reduction in temperature of maximum density. This interference with the formation of ice crystals also produces a progressively lower freezing point for water as more solids are added. The freezing point and temperature of maximum density converge as they drop and coincide when the salinity reaches 24.7 parts per thousand (‰) at a temperature of −1.33°C (29.61° F)(fig. 6–10). When salinity is higher than 24.7‰, the density of water increases with decreasing temperature until the water freezes. Thus, average seawater (34.7‰) has no maximum density anomaly.

Because the components of salt have a density greater than the density of water, density of ocean water increases with increasing salinity. Physical oceanographers are greatly concerned with the relationships among water density, salinity, and temperature since density distribution is an important determinant of ocean circulation. Density of ocean water

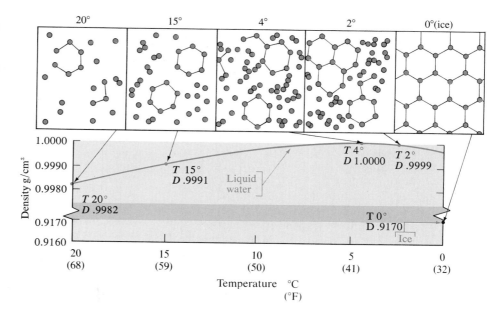

FIGURE 6–9 Formation of Ice Clusters in Water.

Note that the density ranges from 0.9982 g/cm^3 at 20°C to a maximum of 1.0000 g/cm^3 at 4°C. The density of ice (solid water) is only 0.9170 g/cm^3, so it floats on water.

FIGURE 6–10 Salinity, Freezing Point, and Temperature of Maximum Density.

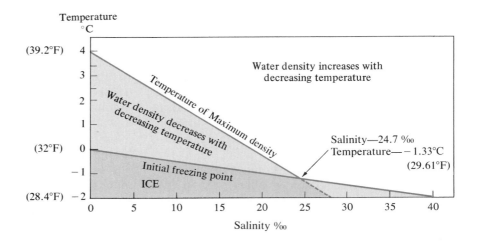

cannot easily be measured directly, so to determine density distribution within a region of study, oceanographers measure temperature and salinity and then use these values to calculate density.

RESIDENCE TIME

The **residence time**, or the average length of time an atom of an element will spend in the ocean, can be calculated by the following equation:

$$\text{Residence Time} = \frac{\text{Amount of element in the oceans}}{\text{Rate at which element is added to or removed from the oceans}}$$

The oceans are considered to be well mixed and in a steady state; that is, the rate at which an element is removed from the oceans must be equal to the rate at which it is being added. For that reason, we could use either of these values as the denominator in the equation.

The residence time of an element in the ocean depends on how reactive it is with the marine environment. Those elements that are more reactive have a shorter residence time. The reactive elements aluminum and iron have residence times of 100 and 140 years, respectively, while less reactive sodium has a residence time of 260 million years (*see* table 6–2).

The short residence time for aluminum shown in table 6–2 (100 years) is the average for particulate and dissolved aluminum. Recent investigation has shown that dissolved aluminum has a residence time of 1400 years, indicating that particulate aluminum must be removed from ocean water at a very rapid

rate. It is believed that comparisons of the residence times of soluble fractions of elements in the ocean must be known to understand their chemical reactivity in the oceans since the residence time of particles is affected primarily by physical mixing processes.

Elements are removed from ocean water through biological reactions, or they are incorporated into sediment. Each year a 1-m (3.28-ft) layer of the ocean's surface is evaporated. This water is returned directly by precipitation or indirectly by runoff from the continents. The ocean, which has an average depth of 4000 m (13,120 ft), would completely evaporate in 4000 years, but because the water is returned, this is the time required to completely recirculate the ocean. Thus, the residence time of water in the oceans is 4000 years.

The process by which water is recycled among the ocean, atmosphere, and continents is called the **hydrologic cycle** (fig. 6–11). Of the total yearly evaporation of water from the earth's surface, 83% is re-

TABLE 6–2 Residence Time of Some Elements in the Oceans.

Element	Amount in Ocean (g)	Residence Time in Years
Sodium, Na	147×10^{20}	2.6×10^{8}
Potassium, K	5.3×10^{20}	1.1×10^{7}
Calcium, Ca	5.6×10^{20}	8.0×10^{6}
Silicon, Si	5.2×10^{18}	1.0×10^{4}
Manganese, Mn	1.4×10^{15}	7.0×10^{3}
Iron, Fe	1.4×10^{16}	1.4×10^{2}
Aluminum, Al	1.4×10^{16}	1.0×10^{2}

Data from Goldberg and Arrhenius, 1958

FIGURE 6–11 Hydrologic Cycle.
Ninety-seven percent of the earth's water is in the world ocean, 2% is in glaciers, 0.6% is groundwater, 0.3% is in the atmosphere, and about 0.1% is in rivers and lakes. Although there is some evidence that indicates the glaciers are entering a stage of melting, we will consider them to be in a steady state. Water is removed from the liquid reservoirs by evaporation and returned to them by precipitation. There is a net atmospheric transfer of water vapor from the oceans to the continents, where it is deposited as precipitation and returns to the oceans by stream flow or as groundwater. Reprinted by permission from Tarbuck, E. J., and Lutgens, F. K., *The earth: An introduction to physical geology,* 3d ed. (Columbus: Merrill, 1990), fig. 10.1.

moved from the oceans and 17% is removed from the surface of the continents. Seventy-six percent of this water is removed from the atmosphere by precipitation directly back into the oceans. Twenty-four percent falls as precipitation onto the continents. During a year, the atmosphere transports water vapor equivalent to 8.5% of that which is being evaporated from the oceans to the continents. This water is returned to the oceans by stream runoff and as groundwater. At any given time, 97% of the earth's water is contained in the ocean basins, 2% is in glaciers, 0.6% is groundwater, 0.3% is in the atmosphere, and about 0.1% is in rivers and lakes.

The **World Climate Research Program** (WCRP) has proposed a study of the world hydrologic cycle called the **Global Energy and Water Cycle Experiment** (GEWEX). Planned for the 1990s, it will use satellite data to aid in improving long-range weather forecasts and in increasing our understanding of hydrology and climate trends.

GEOSECS (GEOCHEMICAL OCEAN SECTIONS)

The GEOSECS project mentioned in chapter 1 is a huge systematic program designed to increase understanding of circulation patterns and mixing processes in the oceans. All of the major ocean basins have been sampled throughout the depth of the water column. The sampling system includes electronic sensors that telemeter temperature, salinity, dissolved oxygen, pressure, and particulate content to a console aboard ship. As rosettes of 30-liter (7.9-gal) sampling bottles are lowered into the ocean, oceanographers have the option of deciding to sample at any depth where an interesting value for any of the properties shows on the console (fig. 6–12).

The collected water is analyzed for some 23 chemical species, 15 isotope species, and the amount and types of particulate matter. From the results thus far obtained, oceanographers think the GEOSECS program will not only provide a much improved understanding of physical processes but also will supply much information about biological cycles in the ocean (fig. 6–13).

FIGURE 6–12 GEOSECS Sampling Bottles.
One of the two rosettes carried aboard Scripps Institution of Oceanography's research vessel *Melville* during GEOSECS Pacific operations shows nonmetallic bottles used in collecting sea water samples. The ends of each bottle can be triggered to close by the scientist operating the shipboard console. Photo from Scripps Institution of Oceanography, University of California, San Diego.

SUMMARY

The **hydrogen bond**, the bond between water molecules resulting from the dipolar nature of the molecule, plays the major role in giving water its many unusual properties.

Water is a great **solvent** because its dipolar molecules can attach themselves to charged particles, **ions**, that make up many substances such as sodium chloride, hydrate them, and put them into solution.

The presence of the hydrogen bond also accounts for the unusual **thermal properties** of water such as its high **freezing point** and **boiling point**, high **heat capacity**, and high **latent heat of melting** and **latent heat of vaporization**.

The **surface tension** phenomenon that makes water form drops and allows it to be poured into a container until it stands well above the sides of the container results from the water molecules being strongly attracted by the hydrogen bond to those water molecules beside and beneath them. This **cohesive attraction** combined with the **adhesive attraction** of water to glass, organic substances, rock, and soil produces the **capillarity** that pulls water to great heights in small tubular openings.

Salinity is the amount of dissolved solids in ocean water, averaging about 34.7 g of dissolved solids per kg of ocean water. Salinity is usually expressed in **parts per thousand (‰)**. Over 99% of the dissolved solids in ocean water are accounted for by six ions, in order of decreasing abundance: **chloride, sodium, sulfate, magnesium, calcium,** and **potassium**. In any sample of ocean water these ions will be in the same relative proportions, making it possible to determine salinity by measuring the concentration of only one of them, usually the chloride ion.

A.

B.

FIGURE 6–13 GEOSECS Data.
A, the map shows the location of GEOSECS Station 202 located at latitude 26°N, longitude 139°W. *B,* the four curves show the plot of temperature, salinity, dissolved oxygen, and nitrate measured from the surface to a depth of 5000 m (16,400 ft) at Station 202. They show that ocean water properties may change significantly from the surface to a depth of 1000 m (3280 ft). The rate of change below that depth is much less. Plots courtesy of Woods Hole Oceanographic Institution.

The **density** of pure water increases with decreased temperature, as does the density of most substances, to a temperature of 4°C (39.2°F). Below 4°C, its density decreases with decreased temperature. This change is due to the fact that the rate of formation of open ice crystals increases dramatically below 4°C, the **temperature of maximum density** for fresh water. By the time water freezes, the density of the ice is only about 0.9 that of water at 4°C. Density increases with increased salinity and pressure. Due to the fact that the components of salinity have a greater density than water, increased salinity increases the density of the ocean water. The **residence time** of elements in the ocean water ranges from 100 yr for aluminum to 260,000,000 yr for sodium. The time required to totally recirculate the water in the oceans, its residence time, is 4000 yr.

The ocean contains 97% of the earth's water. Water is recycled among the ocean, atmosphere, and continents by the **hydrologic cycle**, which involves evaporation, precipitation, and surface and groundwater runoff from the continents.

The GEOSECS program has helped give oceanographers an increased understanding of physical and biological processes of the oceans. It features an advanced system of sampling and chemically analyzing the properties and content of ocean water.

QUESTIONS AND EXERCISES

1. Describe the condition that exists in the water molecule to make it dipolar.

2. Discuss how the dipolar nature of the water molecule makes it such an effective solvent for ionic compounds.

3. Define freezing point and boiling point.

4. There is a fundamental difference between the intermolecular bonds that result from the dipolar nature of the water molecule (hydrogen bond) and the van der Waals force as compared to the chemical bonds (covalent and ionic). What is it?

5. Why are the freezing and boiling points of water higher than would be expected for a compound of its molecular makeup?

6. How does the heat capacity of water compare with that of other substances? Describe the effect this characteristic of water produces in climate.

7. Why does the heat energy added as latent heats of melting and vaporization for water not produce an increase in temperature? Why is the latent heat of vaporization so much greater than the latent heat of melting?

8. Describe how excess heat energy absorbed by the earth's low-latitude regions could be transferred to the heat-deficient higher latitudes through a process that makes use of water's latent heat of evaporation.

9. How does hydrogen bonding produce the surface tension phenomenon of water?

10. As water cools, two distinct changes in the behavior of molecules take place. One tends to increase density while the other decreases density. Describe how the relative rates of their occurrence cause pure water to have a temperature of maximum density at 4°C (39.2°F) and makes ice less dense than liquid water.

11. As water becomes more saline, the temperature of maximum density and the freezing temperature of water decrease and converge. At what salinity does water cease to have a temperature of maximum density above its freezing temperature?

12. What condition of salinity makes it possible to chemically determine the total salinity of ocean water by measuring the concentration of only one constituent, Cl^-?

13. If there is an estimated 18×10^{20} g of magnesium (Mg) in the ocean, and it is being added (or removed) at the rate of 56.5×10^{12} g per year, what is the residence time of magnesium in the ocean? (*See* appendix I.)

14. List the reservoirs of water on the earth and the percentage of the earth's water each holds. Explain how the hydrologic cycle moves water among these reservoirs.

REFERENCES

Davis, K. S., and Day, J. S. 1961. *Water: The mirror of science*. Garden City, N. Y.: Doubleday.

Hammond, A. L. 1977. Oceanography: Geochemical tracers offer new insight. *Science* 195:164–66.

Harvey, H. W. 1960. *The chemistry and fertility of sea waters*. New York: Cambridge University Press.

Kuenen, P. H. 1963. *Realms of water*. New York: Science Editions.

MacIntyre, F. 1970. Why the sea is salt. *Scientific American* 223:104–15.

Pickard, G. L. 1975. Descriptive physical oceanography. 2nd ed. New York: Pergamon Press.

Revelle, R. 1963. Water. *Scientific American* 209:93–108.

Yokoyama, Y.; Guichard, F.; Reyss, J-L; and Van, N. H. 1978. Oceanic residence times of dissolved beryllium and aluminum deduced from cosmogenic tracers ^{10}Be and ^{26}Al. *Science* 201:1016–17.

SUGGESTED READING

Sea Frontiers

Gabianelli, V. J. 1970. Water—the fluid of life. 16:5, 258–70.
The unique properties of water are lucidly described and explained. Topics covered include hydrogen bond, capillarity, heat capacity, ice, and solvent properties.

Smith, F., and Charlier, R. 1981. Saltwater fuel. 27:6, 342–49.
The potential for using the salinity difference between river water and coastal marine water to generate electricity is considered.

Scientific American

Baker, J. A., and Henderson, D. 1981. The fluid phase of matter. 245:5, 130–39.
The structure of gases and liquids is modeled using hard spheres.

MacIntyre, F. 1970. Why the sea is salt. 223:5, 104–15.
A summary of what is known of the processes that add and remove the elements dissolved in the ocean.

7
AIR-SEA INTERACTION

In this first chapter on physical oceanography we will consider some properties of ocean water, the Coriolis effect, the heat budget of the ocean, and the process by which solar energy is converted into kinetic and heat energy of moving masses of ocean water. All will be discussed in terms of their relationship to the interaction between the atmosphere and oceans.

This interaction has received increasing amounts of attention over the years. One of the first major phenomena to spur the quest for an increased understanding of how the oceans and atmosphere interact was the recognition of the El Niño–Southern Oscillation as an event that was tied to worldwide changes in climate. This phenomenon is discussed in some detail in the special feature, "El Niño–Southern Oscillation Events," on page 160. Then, in 1988, studies that confirmed that the world's average temperature had increased over the past 100 years led to much investigation into the possibility that the "greenhouse effect" of the earth's atmosphere was increasing because of the increasing amounts of carbon dioxide, methane, chlorofluorocarbons, and other gases that increase the atmosphere's ability to absorb heat.

A key factor in all of this investigation is the role of the oceans. Do changes in the ocean produce changes in the atmosphere that lead to the El Niño phenomenon—or vice versa? Or does there exist a complex feedback system that involves both to a more or less equal degree? Because the atmospheric carbon dioxide has increased only half as much as would be predicted from the activities of humans during the past century, where did the rest of the carbon dioxide go? Did it go into the oceans or maybe the biosphere? Many questions are being asked, and answers are probably going to be a long time in coming. Increased understanding of some of the phenomena we will discuss in this chapter will surely be required before the answers are found.

Preceding page: Hurricane Diana developed winds of up to 217 km/h (135 mph) before coming ashore at Cape Fear, South Carolina, in September 1984. The precautions taken by local authorities prevented any deaths occurring as a direct result of the storm, although more than $60 million in property damage occurred. Courtesy of NASA, Goddard Space Flight Center.

PHYSICAL PROPERTIES OF OCEAN WATER

Light Transmission

The color of ocean water ranges from a deep indigo blue in tropical and equatorial regions of little biological productivity to a yellow green in the coastal waters of high-latitude areas where biological productivity is carried on seasonally at a very high rate. The blue of the tropical waters where there is little particulate matter is due to greater scattering by water molecules of short wavelengths of solar radiation, which we perceive as the color blue. This process is also responsible for the blue color of the sky. Higher concentrations of dissolved organic matter, especially where microscopic plants are abundant, increase the average particle size and result in a greater scattering of longer wavelengths. This, along with the pigment chlorophyll, produces the greenish color characteristic of coastal waters.

The many forms of electromagnetic energy radiated by the sun may be seen in figure 7–1—the **electromagnetic spectrum**. The very narrow segment of this spectrum designated as **visible light** may be broken down by wavelength into violet, blue, green, yellow, orange, and red. Combined, these different wavelengths of light produce white light. The shorter-wavelength forms of energy to the left of visible light represent highly dangerous forms of electromagnetic energy, including X rays and gamma rays. To the right of the visible segment of the electromagnetic spectrum, we find longer-wavelength forms of energy that are the basis for the technological fields dealing with heat transfer and communication. We will consider further the electromagnetic spectrum when we consider the role of solar energy in setting the ocean masses in motion. Now only that portion of the spectrum that is manifested as visible light will be discussed.

The reason we see things in color is because objects reflect wavelengths of light that represent the colors of the visible spectrum. If those wavelengths of light are not present in the light that falls upon an object, the object will be black. In the ocean the absorption of visible light is greater for the longer-wavelength colors; thus the shorter-wavelength portion of the visible spectrum is transmitted to greater depths. As a result of this condition, the red wavelengths are absorbed within the uppermost 15 to 20 m (49 to 66

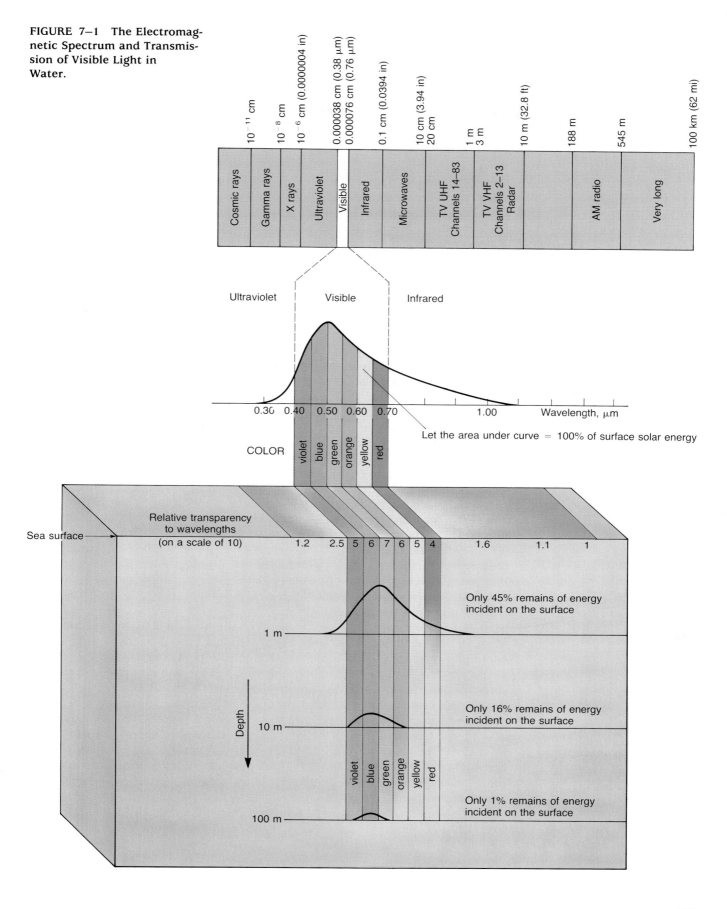

FIGURE 7–1 The Electromagnetic Spectrum and Transmission of Visible Light in Water.

El NIÑO–SOUTHERN OSCILLATION EVENTS

During the 1920s, G. T. Walker identified what he called the Southern Oscillation (SO), associated with a condition in which the summer high-pressure system in the southeastern Pacific occurs in conjunction with a low-pressure system over the Indo-Australian region (*see* figs. 17A and 7–12). This pressure difference lessens when trade winds diminish, the temperature of surface waters in the eastern Pacific increases, and the Equatorial Countercurrent flow increases. The average period of this oscillation is three years, but it ranges from two to ten years. When the oscillations are extreme—producing very widespread warm surface-water conditions and minimal pressure differences across the tropical Pacific—they are called El Niño–Southern Oscillations (ENSO). There were seven ENSO events between 1950 and 1983.

Normally the circulation cell called the Walker Circulation affects the southeast trade winds, which converge on the Indo-Australian low-pressure cell, rise, and produce high rates of precipitation in the low-pressure area. Dry air descends within the southeastern Pacific high-pressure cell off the west coast of South America. This coast is characterized by high rates of evaporation.

One precursor of an ENSO event is the movement of the Indo-Australian low-pressure cell to the east beginning in October or November. In extreme cases such as the 1982–83 event, severe droughts can occur in Australia because the low-pressure cell moves so far east.

Concurrent with the eastward shift of the Indo-Australian low-pressure cell, the meteorological equator or Intertropical Convergence Zone (ITCZ), where the northeast trade winds and southeast trade winds meet and rise, moves south. Its normal seasonal migration is from 10°N latitude in August to 3°N in February, but during ENSO events it may move south of the equator in the eastern Pacific. Associated with this shift are weak trade winds, a decrease in coastal upwelling, and an unusually thick column of abnormally warm surface water in the eastern Pacific (fig. 17B*A*). These initial events are amplifications of normal seasonal fluctuations.

As the ENSO develops, the weakened trade winds and anomalous warmth of surface waters observed in the eastern Pacific spread toward the west. The coming of unusually warm surface waters to Kiritimati (Christmas Island; 2°N, 157°W) can be predicted by earlier observation of increased surface temperatures off the coast of Peru. The event is fully developed by January (fig. 17B*B*). Heavy rainfall from the southward shift of the ITCZ strikes the coast of Ecuador and Peru, which is usually arid, and spreads west across the tropical Pacific. The intense eastward flow of the Equatorial Countercurrent causes a rise in sea level along the western coast of the Americas that progresses poleward in both hemispheres. The event ends 12 to 18 months after it starts, with a gradual return to normal conditions that begins in the southeastern tropical Pacific and spreads to the west (fig. 7B*C*).

Although the below-average temperatures in the eastern Pacific at the end of an El Niño, shown in figure 7B*C*, are usual, such an event had not occurred

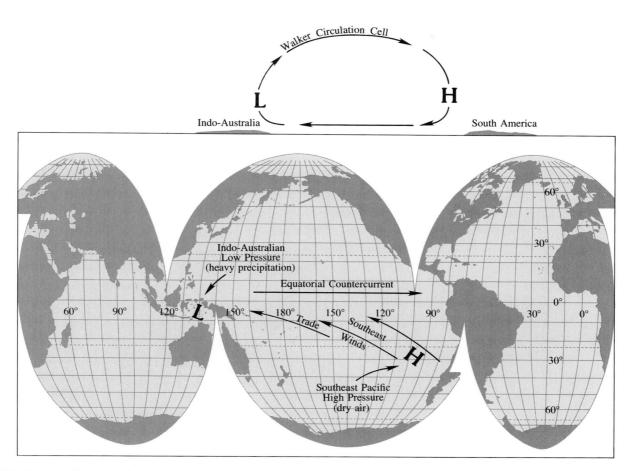

FIGURE 7A Walker Circulation.

Normal oceanic and atmospheric conditions that are altered by El Niño–Southern Oscillation events.

between 1975 and 1988. This cooling has been given the name La Niña and seems to be associated with weather phenomena opposite to those of El Niño at other locations on the earth. For instance, Indian Ocean monsoons are drier than usual in El Niño years and wetter than usual in La Niña years.

A very intense 1982–83 ENSO event that caused a severe drought in Australia and Indonesia was anomalous in that it was initially confined to the central and western tropical Pacific and spread to the east late in its development.

Well before the overall pattern of development of the ENSO was recognized, a periodic movement of warm waters over the rich, cold fishing waters off the coast of Peru was recognized. This event, called El Niño, causes severe coastal rains and drives away the anchovy that are the basis of fishing there

A.

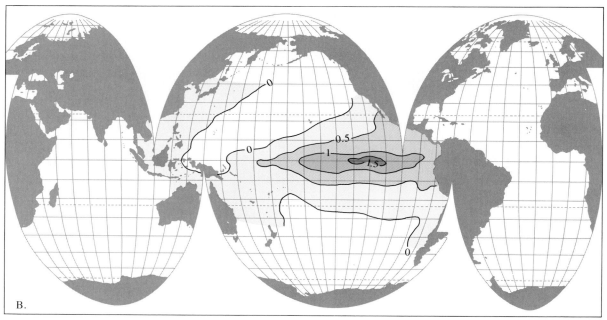

B.

and serve as a food supply for a large bird population. El Niños have especially disastrous effects on the economy of Peru because of its heavy dependence on this fishery and the guano industry.

In November 1982, it was observed that virtually all of the 17 million adult birds that normally inhabited Kiritimati had abandoned their nestlings. This event indicated the severity of the ENSO in the central Pacific, as such an abandonment is not known to have occurred before. It is assumed that the spread of a thick layer of warm water over the surrounding ocean prevented

C.

0 to −0.5°C below the average surface-water temperature for non-El Níno years.

0 to 0.5°C (0.9°F) above the average surface-water temperature for non-ENSO years

0.5° to 1.0°C (0.9° to 1.8°F) above average

1.0° to 1.5°C (1.8° to 2.7°F) above average

More than 1.5°C (2.7°F) above average

FIGURE 7B   Sea Temperature Anomaly (°C) Resulting from the Averaging of ENSO Event Temperature Anomalies from 1950 through 1973.

A, after onset: average of March, April, and May temperature anomalies. *B,* maximum development: average of December, January, and February anomalies. *C,* ending of event: average of May, June, and July anomalies. The typical progress of an ENSO event shows that the abnormally high surface-water temperatures first appear in the eastern Pacific off the coast of Ecuador and Peru *(A)*. This condition begins to be observable by December or January along the coast of South America and is well developed during the March–May period shown. In *B* it can be seen to develop in a westerly direction along the equator and reaches maximum development in the Central Pacific by the following February. By July *(C)*, conditions return to near normal, with the exception of a significant negative temperature anomaly in the eastern Pacific. Courtesy of NOAA.

the rise of nutrients into the surface waters, causing the fish to leave in search of better feeding grounds. The birds, in turn, were forced to leave in search of the fish. If the birds did not find their necessary supply of fish, a large percentage of the adults may have died in addition to the nestlings.

Figure 7C (top) shows the warming that occurred off southern California from January 1982 to January 1983 as a result of the 1982–83 ENSO. This warming was accompanied by major reduction in biological productivity, as is shown by the Coastal Zone Color Scanner phytoplankton pigment images for April 1982 compared to those of March 1983 (fig. 7C, bottom).

Reference: Philander, S. G. H. 1983. El Niño Southern Oscillation phenomena. *Nature* 302, 5906:295–301.

FIGURE 7C Water Temperatures off Southern California.

The Advanced Very High Resolution Radiometer sea-surface temperature image (top) shows that waters off southern California in January 1983 are an average of 1.72°C warmer than a year before. The Coastal Zone Color Scanner phytoplankton pigment images (bottom) show that there was a reduction in plant productivity over the same time span. The strongest El Niño ever recorded occurred in 1982–1983 and pushed a layer of warm water north from the equator. Coinciding with this event was a reduction of upwelling of nutrient-rich, cold water along the coast. This reduction in the nutrient level of the surface water reduced the level of biological productivity. Spectacular physical phenomena such as this and their biological consequences help highlight the close relationship between the physical and biological phenomena of the oceans. Courtesy Paul C. Fiedler/National Marine Fisheries Service.

ft), the yellow will disappear before a depth of 100 m (328 ft) has been reached, but green light can still be perceived down to 250 m (820 ft). Only the blue and some green wavelengths extend beyond these depths, and their intensity becomes low. It is because of this pattern of absorption that objects in the ocean usually appear to be blue green. It is only in the surface waters that the true colors of objects can be observed in natural light, since it is only in the surface waters that all of the wavelengths of the visible spectrum can be found.

A **photometer** measures the transmission of visible light in the ocean. Increased scattering of light indicates increased **turbidity**—the amount of suspended material in the water. Increased turbidity increases the rate of absorption, thus decreasing the transmission of visible light.

Density

Density ranges from 1.02200 to 1.03000 g/cm^3 in the open ocean. It is an important property of ocean water because it determines the positions and motions of water masses in the ocean. In a stable system, low-density water rests at the top of a column of water in which the density increases with increasing depth. The **in situ density (σ or sigma)** of ocean water is its density *in place,* that is, at whatever depth it is in the ocean. In situ density in the deeper trenches is about 5% greater than at the ocean surface because of the effect of pressure.

For many oceanographic applications, it is useful to remove the **adiabatic** effects of pressure on water temperature. Adiabatic refers to the increase in temperature that results from compression of water at increasing depths because of increasing pressure. The density value that has the adiabatic temperature effect removed is called the **potential density (σ_θ, or sigma theta)**. Potential density is always less than in situ density except at the surface, where the adiabatic heating of water is zero. Potential density is a useful property of ocean water for oceanographers to know because subsurface currents flow along lines of equal potential density.

Oceanographers usually use only the last four digits in a water density expession and eliminate the units, g/cm^3. To derive this simplified form, water density is first converted to specific gravity by dividing it by the density of pure water at 4°C (39.2°F). Suppose we have a water sample with an in situ density (σ) of 1.02567 g/cm^3. We would convert this value to the simplified form by first converting it to specific gravity.

$$\text{Specific gravity} = \frac{\text{density of ocean water}}{\text{density of pure water}}$$
$$= \frac{1.02567 \text{ g/cm}^3}{1.00000 \text{ g/cm}^3}$$
$$= 1.02567$$

This eliminates the density units, g/cm^3. The specific gravity of ocean water is then inserted into the following equation to convert in situ density to the simpler form.

$$\text{In situ density } (\sigma) = (\text{Specific gravity} - 1) \times 1000$$
$$25.67 = (1.02567 - 1) \times 1000$$

It can be determined by observing the temperature-salinity-density relationships in figures 7–2 and 7–3 that temperature change can be expected to have a much greater effect on density in the high-temperature, low-latitude areas than in polar regions. It can be noted that the σ *isopleths* (lines of constant density) are more nearly parallel to the temperature axis for low temperatures than for high temperatures, showing a greater density change per unit of temperature change in the high-temperature range.

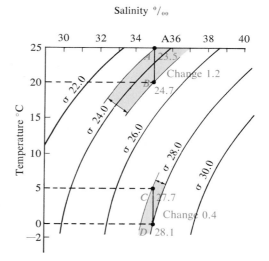

FIGURE 7–2 Temperature-Salinity-Density Relationship.

In high-latitude areas characterized by low-temperature water, temperature has less effect on density (sigma, σ) than in the high-temperature, low-latitude areas. Measured at a salinity of 35‰, σ values determined for points A, B, C, and D show the density change is greater over a 5° temperature span at a higher temperature range. The change in σ across a temperature range of 20–25°C (68°–77°F) is 1.2 compared to a change at only 0.4 across an equal range at the lower temperatures of 0–5°C (32–41°F). Thus, a change in the temperature of warm, low-latitude water has about three times the effect on density that an equal change in temperature occurring in colder, high-latitude waters has. After G. L. Pickard, *Descriptive physical oceanography* © 1963. By permission of Pergamon Ltd., Oxford, England.

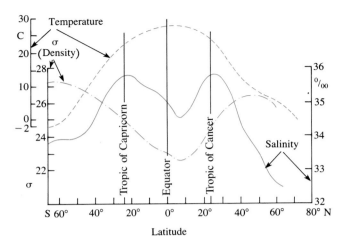

FIGURE 7–3   Average Surface Temperature, Salinity, and Density of World Ocean by Latitude.

Density (sigma) increases from about 22 near the equator to maxima of 26–27 near 50°–60° latitude in the surface water of the ocean. At higher latitudes it decreases slightly. Salinity maxima in the tropics do not seem to affect density, pointing out the importance of temperature in controlling density in low latitudes. After G. L. Pickard, *Descriptive physical oceanography* © 1963. By permission of Pergamon Press Ltd., Oxford, England.

Sigma and the variables that most affect it, temperature and salinity, are important water characteristics, defined as **conservative properties**, used in the identification of water masses. Conservative properties of water are those that are not significantly altered by any processes other than mixing and diffusion once the water sinks beneath the surface.

A **nonconservative property** of water is changed by some process other than mixing and diffusion after ocean water sinks beneath the surface. An example is dissolved oxygen, which will be changed by biological processes.

The density of surface water is affected primarily by temperature changes in the open ocean. In the extreme high-latitude areas of the ocean, however, where temperatures remain relatively constant, salinity changes can have a significant effect on density. Figure 7–4 shows the vertical distribution of density in various regions of the ocean. In the equatorial region, there is a shallow zone of low-density water near the surface, and below this lies a zone where the increase in density is very rapid. This zone of rapidly increasing density is called the **pycnocline** (density slope) and represents a very stable barrier to the mixing of the low-density water above and the high-density water below. The pycnocline is considered to have high **gravitational stability**. It would require a greater amount of energy to move a given mass of water from some point in the pycnocline either up or

down than would be required to move an equal mass an equivalent distance at depths where density change occurred very slowly with increasing depth.

The pycnocline is developed as a result of the combined effects of zones of rapid vertical changes in temperature, the **thermocline** (temperature slope), and salinity, the **halocline** (salinity slope). The interrelatedness of these three zones, which determines the degree of separation between the **upper-water** and **deep-water** masses, is shown in figure 7–4C.

The pycnocline is lacking in the high latitudes. Because there is not a great difference in the density of the surface waters as compared to the bottom waters in high-latitude areas, the high-latitude water columns are less stable than low-latitude water columns possessing a pycnocline.

Sound Transmission

Sound can be transmitted much more efficiently through water than through air. This characteristic has made it possible to develop systems for determining the positions of objects in the ocean as well as determining the distance to the bottom of the ocean by **sonar** (sound navigation and ranging). The average velocity of sound in the ocean is 1450 m/s (4756 ft/s), compared to 334 m/s (1096 ft/s) in a dry atmosphere at 20°C (68°F). Velocity of sound increases with increases in temperature, salinity, and pressure. Therefore, the velocity with which sound travels through water in a particular area will vary by season. To determine the exact velocity of sound for any point in the ocean, these three variables—temperature, salinity, and pressure—must first be determined, or a **velocimeter** can measure it directly.

At a depth of around 1000 m (3280 ft) there exists a layer of ocean water where these three variables have values that result in relatively low velocity for the transmission of sound. Sound originating above and below this layer will be refracted, or bent, into the low-velocity layer and be trapped there. Thus by following this channel, sound can be transmitted for unusually great distances (fig. 7–5). This channel is called the **sofar** (sound fixing and ranging) channel because of its practical application in distance determination.

Such determinations are made by emitting a sound signal that will travel through the water, bounce off an object, and return to be picked up by a receiver. If the velocity at which the sound will travel in water has been determined, the distance to the object is equal to the velocity times one-half the time required for the sound signal to make the round trip.

A

B

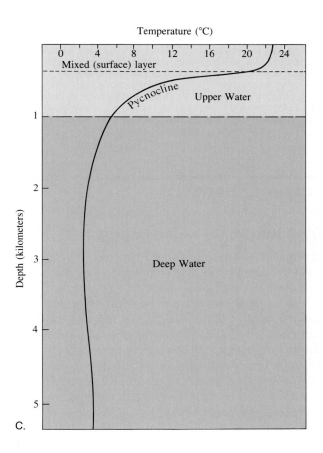

FIGURE 7–4 Vertical Density Profiles for Various Latitudinal Regions.

A, because of natural tendencies toward stability, density can normally be expected to increase with depth. In low-latitude equatorial and tropical regions a thin layer of low-density surface water is separated from high-density deep water by a zone of rapid density change, the pycnocline. The pycnocline is poorly developed or missing in high-latitude waters. It is the absence of the stable pycnocline that facilitates vertical mixing in some high-latitude areas. B, the gravitationally stable pycnocline described above is primarily the result of and coincides with a zone of rapid vertical decrease in temperature, the thermocline. The thermocline is also more highly developed in low latitudes. A zone of rapid vertical change in salinity, the halocline, may also affect density. The halocline usually represents a decrease in salinity with increasing depth in low latitudes but may represent the opposite condition in high-latitude areas (note that in B the temperature increases to the left. Although this is unconventional, it was done to make the curves easier to compare). C, extending from the surface down through the mixed, or surface, layer and the pycnocline is the relatively light *upper water*. It is well developed throughout the low- and midlatitudes. It is underlain by the denser, cold *deep-water* mass that extends to the deep-ocean floor. The mixed layer represents water of uniform temperature resulting from wave mixing. A and B after G. L. Pickard, *Descriptive physical oceanography* © 1963. By permission of Pergamon Press Ltd., Oxford, England.

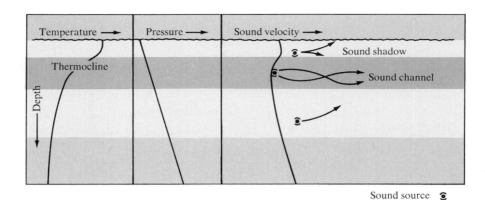

FIGURE 7–5   Transmission of Sound in Ocean.

Velocity of sound in the ocean increases with increases in temperature, salinity, and pressure. Generally salinity changes are not as important in affecting the velocity of sound transmission as are the changes in temperature and pressure. Pressure increases steadily with depth, which increases the velocity of sound. However, the rapid decrease in temperature represented by the thermocline offsets the pressure increase and produces a low-velocity sound channel. Wave phenomena, such as sound, bend (refract) into low-velocity areas. This refraction produces shadow zones where velocity maxima exist as well as a channel that traps sound energy in the low-velocity zone, the *sofar* channel. Because the thermocline is primarily responsible for producing the sound channel, the sound channel gets shallower and weaker with increased latitude.

SOLAR ENERGY

Distribution of Solar Energy

Essentially all of the energy available to the earth comes from the sun. Solar energy strikes the earth at an average rate of 2 cal/cm^2/min. Although energy radiated by the sun covers the full electromagnetic spectrum, most of the energy that reaches the earth's surface is contained in wavelengths close to and including the visible portion of the spectrum.

In the upper atmosphere, molecular oxygen (O$_2$) is broken into individual atoms of oxygen (O) that recombine with molecular oxygen to produce ozone (O$_3$). This process absorbs most of the solar radiation with wavelengths shorter than 0.29 μm. Water vapor absorbs a high percentage of solar radiation with wavelengths greater than 0.8 μm leaving most of the solar radiation within the visible spectrum 0.38–0.76 μm to penetrate the atmosphere and reach the earth's surface. After scattering by atmospheric molecules and reflection by clouds in the atmosphere, about 47% of the solar radiation that is directed toward the earth is absorbed by the oceans and continents. Nineteen percent is absorbed by the atmosphere and 34% is lost to space.

If it is true that the earth has maintained a constant average temperature over the years, the earth must be radiating energy back to space at the same rate it is absorbing solar energy. The energy radiating from the earth falls within the infrared range (0.76 μm–0.1 cm). Figure 7–6 shows the intensity of energy radiated by the sun peaks at 0.48 μm in the visible spectrum. It shows also that the radiation from the earth peaks at 10 μm in the infrared range. While the atmosphere absorbs only 19% of the short-wavelength solar radiation, the water vapor and carbon dioxide, as well as other gases, in the atmosphere absorb a large amount of the longer-wavelength infrared radiation emitted by the earth. It is this change of wavelength that is the essence of the **greenhouse effect**. Some of the infrared energy absorbed in the atmosphere will be reabsorbed by the earth (the rest being lost to space) to continue the process. Therefore, the solar radiation received is retained for a time within our atmosphere to moderate the temperature fluctuations between night and day and between seasons. The amount of atmospheric heating varies with latitude because of the unequal distribution of solar radiation on the earth's surface and water vapor in the atmosphere.

Considering the earth as a sphere, figure 7–7 shows that radiation reaching its surface will strike at an angle of 90° near the center of the sphere that is lit by this radiation, while it will be striking at 0° at the edge of this circle (solar rays will be tangent to the earth's surface). If we consider a 1 km^2 (0.4 mi^2) cross-sectional area of solar radiation that falls near the center of the illuminated portion of the sphere,

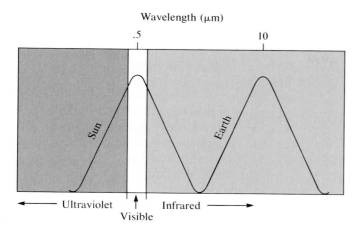

Wavelength (μm)

Ultraviolet / Visible / Infrared

FIGURE 7–6   Comparison of Spectrums of Energy Radiated by the Sun and the Earth.
Most of the energy coming to the earth from the sun is within the visible spectrum and peaks at 0.48 μm (0.0002 in.). The atmosphere is transparent to most of this radiation, and it is absorbed at the earth's surface. When the earth reradiates this energy back toward space, it does so as infrared radiation with a peak at a wavelength of 10 μm (0.004 in.). The CO_2 and H_2O in the atmosphere absorb this infrared radiation to produce the greenhouse effect, which keeps the atmosphere's temperature higher than it otherwise would be.

this ray of light will cover exactly 1 km² of the earth's surface. By contrast, if we consider the same cross-sectional area falling on a high-latitude portion of the earth's surface, where it does not strike the earth at a right angle, this square-kilometer beam will be spread out over considerably more than 1 km² of the earth's surface. The intensity of radiation available

FIGURE 7–7   Solar Radiation.

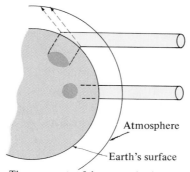

The amount of heat received at higher latitudes is less than that at lower latitudes for three reasons: (1) a ray of solar radiation that strikes the earth at a high latitude is spread over more area than an equal ray that is perpendicular to the earth's surface at a lower latitude, (2) the high latitude ray also passes through a greater thickness of atmosphere, and (3) more of its energy is reflected due to the low angle at which it strikes the earth's surface.

per unit of surface area will be greatly decreased compared to that available in the lower latitudes.

As the earth rotates on its axis, which is inclined 23.5° from being perpendicular to the **ecliptic** (plane of earth's orbit), there will be significant portions of its surface above 66.5°N latitude, the **Arctic Circle**, and 66.5°S latitude, the **Antarctic Circle**, that will spend up to six months in darkness. The direct rays of the sun will migrate back and forth across the equator between the **Tropic of Cancer** and the **Tropic of Capricorn** 23.5° north and south of the equator. The belt between the two tropics will receive a much greater amount of radiation per unit of surface area during a year than will the portions of the earth's surface north of the Arctic Circle and south of the Antarctic Circle.

Oceanic Heat Flow and Atmospheric Circulation

Because of the different angles at which solar radiation strikes the earth's surface and the very highly reflective characteristics of the ice cover found in high-latitude areas of the earth, there is more energy absorbed than is radiated back into space in latitudes between 35°N and 40°S, while there is less energy absorbed than is lost to space in latitudes higher than 35°N and 40°S. Figure 7–8 shows how this phenomenon is manifested in the average daily heat flow of the oceans of the Northern Hemisphere.

Arctic Circle

Tropic of Cancer

During the Northern Hemisphere winter, areas north of the Arctic Circle receive no direct solar radiation, while areas south of the Antarctic Circle receive continuous radiation for months.

Tropic of Capricorn

Antarctic Circle

The reverse of the above situation is true during the Northern Hemisphere summer. Throughout the year, the sun is directly over some latitude between the tropics.

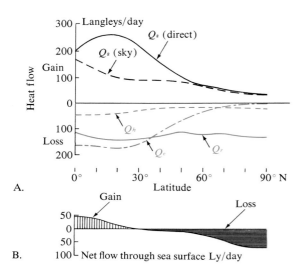

A.

B.

FIGURE 7–8 Heat Flow: Ocean-Atmosphere.

A, heat flow through the ocean surface is represented as a function of latitude. At all latitudes there is a gain from solar radiation *(Qs)* and a loss from back radiation *(Qr)*, conduction *(Qh)*, and evaporation *(Qe)*. The decrease in *Qs* (direct) at the equator results from cloud cover associated with the equatorial low-pressure belt. *B,* the amount of heat lost from the ocean by back radiation, conduction, and evaporation is less than that gained from solar radiation below 30° latitude. The heat loss exceeds heat gain above 30° latitude. The langley (ly) is a calorie of heat flow through 1 cm² of the ocean surface. After G. L. Pickard, *Descriptive physical oceanography* © 1963. By permission of Permagon Press Ltd., Oxford, England.

FIGURE 7–9 Air Masses.

Low-pressure belts develop along the equator and the 60° latitude regions because of rising columns of warm air. Descending cool dry air produces high-pressure belts in the 30° latitude regions (subtropics). Air movements are primarily vertical in these belts. Strong lateral movements of air between the belts produce the westerlies and trade winds. Because of cold air masses overlying the polar regions, high-pressure conditions exist there also. Reprinted by permission from Tarbuck, E. J., and Lutgens, F. K., *Earth science,* 5th ed. (Columbus: Merrill, 1988), fig. 14.11.

As a direct result of this condition, one might expect that the equatorial zone will continually get warmer as the years pass and the polar regions will become progressively cooler. Such is not the case. Although the polar regions are considerably colder than the equatorial zone, the temperature difference does not appear to be increasing with time. In order to explain this condition, we must conclude that the excess heat that is absorbed in the low latitudes is transferred by some mechanism into the higher latitudes. Thus a more or less stable climatic difference is maintained between the two regions. The mechanism by which such transfer is achieved involves both the oceans and atmosphere.

Figure 7–9 shows graphically that the greater heating of the atmosphere over the equator causes air to decrease in density and rise. As it rises, it cools by expansion, and the water vapor contained in the rising air mass condenses and falls as precipitation in the equatorial zone. After losing its moisture, this dry air mass descends. The descents occur in the subtropical regions (30° latitude) of the Northern and Southern hemispheres. As the descending air approaches the earth's surface, it is warmed by compression. Upon reaching the surface it moves away from the tropics, toward either the equator or higher latitudes. A high-pressure belt develops in the

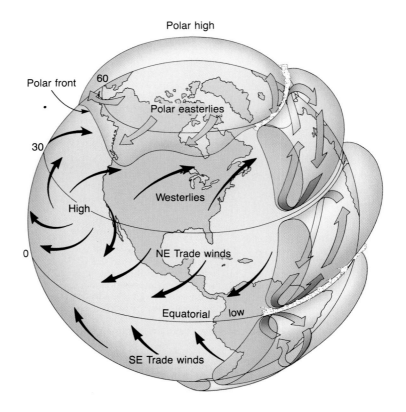

subtropics owing to the dense air that descends in these regions. Between the **subtropical high-pressure belts,** the horse latitudes, lies an **equatorial low-pressure belt** resulting from the low density of the rising air column above the equator.

The masses of air that move across the earth's surface from the subtropical high-pressure belts toward the equatorial low-pressure belt constitute the **trade winds.** Some of the air that descends in the subtropical regions moves along the earth's surface to higher latitudes as the **westerly wind belts.** These masses rise over the dense, cold air moving away from the **polar high-pressure caps** at the **subpolar low-pressure belts** located near 60°N and 60°S latitude. The air moving away from the poles produces the **polar easterly wind belts.** The air that rises at the 60° latitudes cools, releases precipitation in these regions, and ultimately descends in the polar regions or in the subtropics. These idealized latitudinal patterns are significantly altered by the uneven distribution of land and ocean over the earth's surface. The general effects of this idealized system are, however, clearly visible on a very broad scale. In the following paragraphs, we will examine how these air masses are affected by the earth's rotation as they move across the earth's surface.

CORIOLIS EFFECT

Any freely moving object on the earth's surface moving horizontally through a long distance for a relatively long period of time will veer to the right in the Northern Hemisphere and to the left in the Southern Hemisphere. The magnitude of this effect will increase with the velocity and the latitude of the object. This phenomenon is called the **Coriolis effect.** It is zero at the equator and maximum at the poles.

We will, for the sake of our following discussion, consider only the Northern Hemisphere. The air masses moving away from the subtropics toward the equator and high latitudes are moving basically south in the first case and north in the second. In the case of the trade winds, they do not appear to blow directly out of the north but out of the northeast, while the westerlies, as their name implies, blow out of the southwest. Why do we observe this east-west component of motion in air masses that are moving north and south?

Figure 7–10A shows that as the earth rotates on its axis, points at different latitudes rotate at different velocities. The rotational velocity is proportional to the

distance from the earth's surface to the axis of rotation and ranges from 0 km/h at the poles to over 1600 km/h (994 mi/h) at the equator.

The trade winds blow out of the northeast because as the air mass moves south from the subtropical region near 30°N latitude, it is also moving with the rotating earth in an easterly direction at about 1400 km/h (870 mi/h). The air mass starts moving toward a point on the equator at the same longitude that is moving east at 1600 km/h (994 mi/h). The distance that the air mass must cover to reach the equator is about 3200 km (2000 mi). If it moves to the south at 32 km/h (20 mi/h), it will require 100 h to arrive at the equator (fig. 7–10B).

The air mass is moving south at 32 km/h and east at 1400 km/h. For every hour that the air mass moves in the southerly direction, the point on the equator

A.

B.

FIGURE 7–10 Coriolis Effect.
A, points at different latitudes on the earth's surface rotate at different velocities, ranging from 0 km/h at the poles to 1600 km/h (994 mi/h) at the equator. *B,* an air mass that leaves the 30° latitude where it is rotating with the earth at 1400 km/h (870 mi/h) and heads south toward the equator at 32 km/h will not reach the equator until 100 hours later. Since the equator is rotating east at 200 km/h (124 mi/h) faster than the air mass, it will meet the equator at a point 20,000 km (12,420 mi) west of the point on the equator that was directly south of the air mass when it left its position on 30° latitude. This air mass could represent the northeast trade winds, which follow a similar path.

toward which it started moves east 200 km (124 mi) farther than does the air mass (1600 km/h − 1400 km/h). In the period of 100 h it takes the air mass to make the trip, the point on the equator toward which it started will be 20,000 km (100 h × 200 km/h) (12,420 mi) east of the air mass when it reaches the equator. While the air mass was moving the 3200 km in a southerly direction, it appears, as a result of the earth's rotation, to have moved 20,000 km in a westerly direction. Stated another way, as the air mass covered 30° of latitude from north to south, it appeared to move across 180° of longitude in a westerly direction. Certainly, if we were to encounter this air mass aboard ship at some point between the equator and 30°N latitude, it would be coming out of the east-northeast.

A westerly wind air mass moves from 30° latitude in a northerly direction to 60° latitude. This mass also is moving in an easterly direction at 1400 km/h. The point at 60° latitude toward which the air mass is headed is moving in an easterly direction at only 800 km/h (497 mi/h). Unlike the situation discussed in regard to the trade winds, the westerly air mass is moving in an easterly direction at a velocity greater than that of the point at 60°N latitude toward which it started. For every hour that this mass moves in a northerly direction, it moves 600 km (373 mi) farther east than the point at 60° latitude toward which it started. If we were to encounter this air mass at some point between 30° and 60°N latitude, it would appear to come out of the west-southwest.

If the air mass moves north at the same velocity as a trade wind mass moves south, the amount of deviation to the right in the case of the westerlies will be greater than for the trade winds. This results from the fact that the condition responsible for the intensity of the Coriolis effect is the different rates at which points at different degrees of latitude rotate about the earth's axis. The rotational velocity of these points ranges from 0 km/h at the poles to more than 1600 km/h at the equator, but the rate of change per degree latitude is not constant. As we approach the pole from the equator, the rate of change of rotational velocity per degree of latitude change increases. As a comparison, we can note that there is a difference of 200 km/h in velocity across 30° of latitude from the equator to 30°N latitude, while there is a difference of 600 km/h in the velocity of rotation from 30°N to 60°N latitude. If we carry this consideration over another 30° of latitude from 60°N latitude to the pole, where the velocity is zero, the difference is over 800 km/h. This explains the fact that the Coriolis effect increases with increased latitude.

The primary factor affecting the amount of deflection resulting from the Coriolis effect is not, however, the magnitude of the effect since it is very small, even at high latitudes. It is the length of time a particle is in motion that will have the greatest influence on how much it is deflected. Thus, even at low latitudes, a large Coriolis deflection is possible if an object is in motion for a long time.

Although we will not attempt to discuss it here, a complete mathematical treatment of the rotating earth shows that the effects described for these north–south-moving masses are valid for movements in any direction. Freely moving objects veer to the right of their intended path in the Northern Hemisphere and to the left in the Southern Hemisphere.

HEAT BUDGET OF THE WORLD OCEAN

As a result of variations in the absorption of solar energy, ocean temperatures will vary from place to place and from time to time. We previously discussed the fact that the temperature difference between polar and equatorial regions remains constant as a result of a transfer of heat energy from the equator to the higher latitudes by moving air and ocean masses. Driven by the moving air masses, large surface current systems are set up in the world's oceans and play an important role in transfer of heat energy. The temperatures at various places in the ocean will depend upon the rate at which heat flows in and out of these areas. Quantitative considerations of heat transfer in ocean-water masses constitute the **heat budget** of these masses (fig. 7–11).

The heat budget for a given *locality* in the ocean is expressed by the following equation:

Rate of heat gain − Rate of heat loss
= Net rate of heat loss or gain
$$(Q_s + Q_c) - (Q_r + Q_e + Q_h) = Q_t$$

in which

Q_s = the rate of heat gain from solar radiation

Q_c = the rate of heat gain (or loss) through ocean current

Q_r = rate of heat loss through radiation

Q_e = rate of heat loss through evaporation

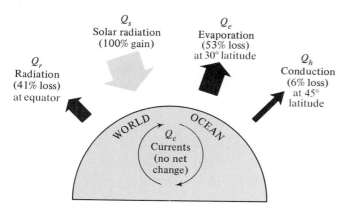

FIGURE 7-11   Avenues of Heat Flow between the Ocean and the Atmosphere.

Solar radiation (Q_s) provides heat to the oceans. It is circulated within the ocean by currents (Q_c). Heat is lost from the ocean through evaporation (Q_e), radiation (Q_r), and conduction (Q_h). Heat loss equals heat gain.

Q_h = rate of heat loss through conduction into the atmosphere

Q_t = net rate of heat loss or gain in the locality

If Q_t is a positive value, it means the temperature of the ocean in that locality is rising. If it is a negative value, the temperature of the mass of water is decreasing. If it is zero, the temperature is not changing.

Theoretically, Q_c and Q_t should be zero when we consider the **world ocean** over a long period of time. Current flow is internal and seasonal gains and losses of heat should average out so that Q_t becomes zero. Therefore, the rate at which heat is gained in the world ocean through solar radiation should equal the rate at which heat is lost through radiation, evaporation, and conduction into the atmosphere (fig. 7-11). Symbolically the heat budget of the world ocean is expressed using average values as

$$Q_s = Q_r + Q_e + Q_h$$

$$100\% = 41\% + 53\% + 6\%$$

The amount of heat lost by radiation, primarily of infrared wavelengths, is greatest near the equator, where the greatest amount of solar energy is absorbed. The greatest heat loss by evaporation is in dry subtropics, whereas the heat lost through conduction reaches a maximum along the western margins of ocean basins, where warm-water currents carry water into regions where the atmosphere is much colder than the water.

THE OCEAN'S WEATHER AND CLIMATE

The idealized pressure belts and consequent wind systems previously discussed are significantly modified as a result of two factors. First, the tilt of the earth's axis of rotation produces seasons, and second, the air over continents gets colder in winter and warmer in summer than the air over adjacent oceans. As a result, continents usually develop atmospheric high-pressure cells that reflect the weight of the cold air centered over them during winter and low-pressure cells during the summer (fig. 7-12). As air moves away from the high-pressure cells and toward the low-pressure cells, the Coriolis effect produces a cyclonic (counterclockwise) flow of air around low-pressure cells and an anticyclonic (clockwise) flow of air around high-pressure cells in the Northern Hemisphere (fig. 7-13). Thus, wind patterns associated with continents may reverse themselves on a seasonal basis as winter high-pressure cells are replaced by summer low-pressure cells.

Day-to-day weather may change little at high and low latitudes. Polar regions are usually cold and dry regardless of the season. Near the equator, the air is warm, damp, and still because the dominant direction of air movement in the doldrums belt is up. Midday rains are common. It is at the midlatitudes that weather gets interesting. Owing to the seasonal change of pressure systems over continents, air masses from the high and low latitudes may move into the midlatitudes, meet, and produce severe storms. In the United States, we may be invaded by four major polar air masses and three tropical air masses (fig. 7-14).

As polar and tropical air masses move into the midlatitudes, they are also gradually moving in an easterly direction. A **warm front** is the contact between a warm air mass moving east into an area occupied by cold air. A **cold front** is the contact between a cold air mass moving east into an area occupied by warm air (fig. 7-15). These eastward movements are brought about by the movement of the **jet stream**, an easterly moving air mass centered at about 10 km (6.2 mi) elevation above the midlatitudes. It usually follows a wavy path and may cause unusual weather by pulling a polar air mass far south or a tropical air mass far to the north. Regardless of whether the colliding air masses create a cold front or a warm front, the warm air rises above the denser cold air, cools, and the moisture in it condenses as precipitation. A cold front will usually be steeper, and the tempera-

FIGURE 7–12   Sea-Level Atmospheric Pressures and Winds.

The idealized high- and low-pressure belts shown in figure 7–9 are modified by the seasons and distribution of continents. Pressure patterns shown in this figure are the averages for the month of January. Because continental rocks have a lower heat capacity than the water of the oceans, continents get hotter in the summer and colder in the winter than the adjacent oceans. The cold air over the continents of the Northern Hemisphere and warm air over the continents in the Southern Hemisphere produce the high- and low-pressure cells centered over the continents. Courtesy of NOAA.

FIGURE 7–13   High- and Low-Pressure Cells and Air Flow.

As air moves away from high-pressure cells *(A)* and toward low-pressure cells *(B)*, the Coriolis effect causes the air masses (winds) to veer to the right in the Northern Hemisphere. This results in anticyclonic clockwise winds around high-pressure cells and cyclonic counterclockwise winds around low-pressure cells.

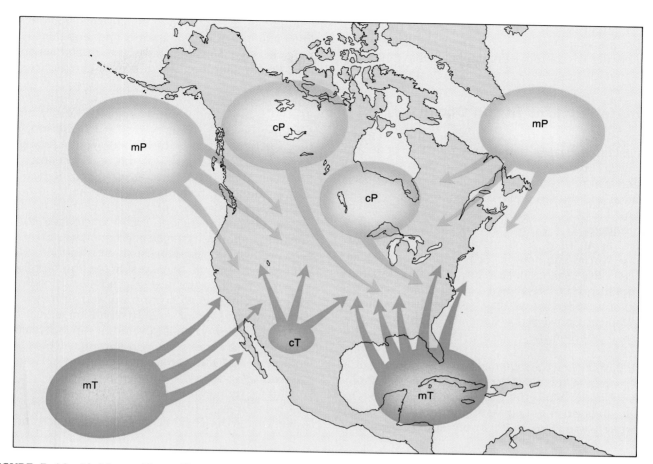

FIGURE 7–14   Air Masses That Affect Weather in the United States.

The polar air masses are more likely to invade the United States during winter, and the tropical air masses tend to move in from the south during summer. Air masses are classified on the basis of their source region. The designation continental (c) or maritime (m) gives an indication of moisture content, whereas polar (P) and tropical (T) indicate temperature conditions. Reprinted by permission from Tarbuck, E. J., and Lutgens, F. K., *Earth science,* 5th ed. (Columbus: Merrill, 1988), fig. 15.1.

A.

FIGURE 7–15   Warm and Cold Fronts.

Cross sections through a warm front *(A)* and cold front *(B).*

ture differences across it will be greater. Therefore, rainfall associated with a cold front will usually be heavier and of shorter duration than that resulting from a warm front.

Climate Patterns in the Ocean

The open ocean can be divided into climatic regions with relatively stable boundaries that run generally east–west (fig. 7–16). Temperature and salinity of surface waters are determined by the amount of solar radiation and the amounts of evaporation and precipitation.

In **equatorial** regions the major air movement is vertical as air rises, so winds are weak. Surface waters are warm, and heavy precipitation keeps salinity relatively low. Sailors once referred to this region as the **doldrums** because their sailing ships were becalmed by the lack of winds. Meteorologists call it the **Intertropical Convergence Zone** (ITCZ) because it is the zone between the tropics where the trade winds converge.

Tropical regions are characterized by northeasterly trade winds in the Northern Hemisphere and southeasterly trade winds in the Southern Hemisphere. These winds push the equatorial currents and create moderately rough seas. Relatively little precipitation

falls at higher latitudes within tropical regions, but precipitation increases toward the equator. Hurricanes and typhoons that carry large quantities of heat into higher latitudes are initiated here as tropical storms.

The belts of high pressure previously described are centered in the **subtropical** regions. The dry air descending on the subtropics results in little precipitation and a high rate of evaporation, producing the highest surface salinities in the open ocean. Winds are weak in the open ocean as are currents. However, strong boundary currents flow north and south, particularly along the western margins of the subtropics.

The **temperate** regions are characterized by seasonal temperature changes and strong westerly winds blowing from the southwest in the Northern Hemisphere and from the northwest in the Southern Hemisphere. Severe storms are common when cold polar air masses interact with warm tropical air masses, especially during winter, and precipitation is heavy.

The **subpolar** ocean is covered in winter by sea ice, some of which melts away in summer. Icebergs are common, and the surface temperature seldom exceeds 5°C (41°F) in the summer months.

Surface temperatures remain at or near freezing in the **polar** areas, which are covered with ice throughout most of the year. In these areas, which include

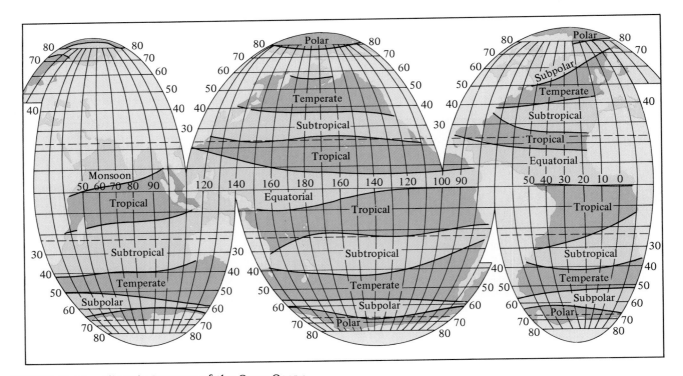

FIGURE 7–16 Climatic Patterns of the Open Ocean.
Base map courtesy of National Ocean Survey.

FIGURE 7–17   Sea Ice Distribution in the Arctic and Antarctic.
A, pancake ice. Photo from Oceanographer of the Navy, Public Affairs Office. *B,* Banks Island, Canada, to the right, is bounded by fast ice (black). A strip of open water (light) separates the fast ice from pack ice along the left margin. Photo courtesy of NASA. *C,* rafted ice.

the Arctic Ocean and the ocean adjacent to Antarctica, there is no sunlight during the winter and no night during the summer.

Sea Ice

Resulting directly from the low-temperature conditions that are characteristic of high-latitude areas is the development of a permanent to nearly permanent ice cover at the sea surface. The term **sea ice** is used to distinguish these masses of ice from **icebergs**, which may also be found at sea but originate by breaking away from glacial masses of ice that form on land.

The freezing point of water with a salinity of 35‰ is −1.91°C (28.6°F). As water freezes at the ocean surface, the dissolved solids do not fit into the crystalline structure of the ice and are left behind so that the salinity of the surrounding water increases. This greater salinity tends to lower the freezing point of the remaining water. However, the low-temperature, high-density water that is excluded from the ice tends to sink and be replaced by warmer, less dense water at the surface. This circulation enhances the formation of sea ice, since freezing is aided by low salinity and calm water conditions. Sea ice is found through-

out the year around the margin of Antarctica, within the Arctic Sea, and in the extreme high-latitude region of the North Atlantic Ocean.

Sea ice begins to form as small needlelike crystals of hexagonal shape that eventually become so numerous that a **slush** develops. As the slush begins to form into a thin sheet, it is broken up by wind stress and wave action into small disc-shaped **pancakes** (fig. 7–17A). As further freezing occurs, the pancakes coalesce to form **floes**. The rate at which sea ice forms is closely tied to temperature conditions, and large quantities of ice form in relatively short periods of time when the temperature is very low; for example, −30°C (−22°F). Even at low temperatures, the rate of new ice formation will decrease as ice thickness increases, owing to the poor heat conduction of ice. Newly formed sea ice contains a significant quantity of brine that is trapped during the freezing process.

Depending upon the rate of freezing, newly formed ice, **new ice**, may have a total salinity that ranges from 4‰ to 15‰. The more rapidly it forms, the more brine will be captured and the higher the salinity. After a period of time, the brine will trickle down through the coarse structure of the sea ice, and the salinity of the sea ice will decrease. Normally by the time it is a year old, **old ice**, sea ice has become relatively pure.

In the Arctic Sea, ice that forms at the sea surface can be classified into one of three categories—pack ice, polar ice, or fast ice. **Pack ice** forms around the margin of the Arctic Sea, extending through the Bering Strait into the Bering Sea and as far south as Newfoundland and Nova Scotia in the North Atlantic. Pack ice reaches its maximum extent during the month of May and breaks up to cover its least area in September (fig. 7–18). This ice, which can be penetrated by icebreakers, reaches a maximum thickness of about 2 m (6.6 ft) in the winter period. The pack ice is driven primarily by the winds, although it also responds to surface currents, which produce stresses that continually break and reform its structure. Its formation is achieved by the expansion of floes that begin to raft onto one another as they expand to cover the sea's surface (figs. 7–17B and 7–17C).

The **polar ice** that covers the greatest portion of the Arctic Sea, including the polar region, attains a maximum thickness in excess of 50 m (164 ft). During the summer months, melting may produce enclosed bodies of water called **polynyas**, although the polar ice never totally disappears. Its average thickness during the summer months is over 2 m (6.6 ft). The polar ice is constantly being exchanged as floes from the pack ice are carried into the polar region during the winter season, and floes break out of the polar ice and reenter the pack ice during the summer. Circling in a clockwise direction around the Arctic Ocean, about one-third of the pack and polar ice is carried into the North Atlantic by the East Greenland Current each year.

Developing in the winter from the shore out to the pack ice is the **fast ice**, which completely melts during the summer. The fast ice, which is firmly attached to the shore, attains a winter thickness in excess of 2 m (6.6 ft).

In the Southern Hemisphere, where the polar region is covered by a continental mass, we might characterize all of the sea ice that forms around the margin of the Antarctic continent as pack ice and fast ice that have a rather temporary existence. The pack ice rarely extends north of 55°S latitude and breaks up rather completely from October to January except in the very quiet bays. Winds that are frequently very strong help prevent the formation of a greater pack ice accumulation around the Antarctic continent.

Icebergs

Icebergs are derived from land ice in the form of glaciers that cover the continent of Antarctica and most of the island of Greenland and part of Ellesmere Island in the Arctic (fig. 7–19). These vast sheets of ice that grow from the snow accumulation on the landmasses spread outward toward the margins until they push their edges out into the marginal sea. There the ice sheets are buoyed up by the water and break up under the stress of current, wind, and wave action. This breakup that produces icebergs is called **calving**.

In the Arctic, the icebergs that are produced originate primarily from the ice that follows narrow valleys into the sea along the western coast of Greenland. Icebergs are also produced along the east coast of Greenland and the east coast of Ellesmere Island. The East Greenland Current and the West Greenland Current carry the icebergs at rates of up to 20 km (12.4 mi) per day into the North Atlantic, where the Labrador Current may move them into North Atlantic shipping channels. They are seldom carried south of 45°N latitude, but during some seasons icebergs will move as far south as 40°N latitude and become shipping hazards.

Such an accumulation of icebergs developed when the *Titanic* was sunk in April 1912. After receiving repeated warning of the existence of such a hazard, the 46,000-ton ship containing 2224 passengers proceeded at an excessive speed of 41 km/h (25.5 mi/n) until it came to a grinding halt after hitting the iceberg with its starboard bow. The sinking of the *Titanic* at 41°46'N latitude, 50°14'W longitude near the Grand Banks cost the lives of 1517 people and brought about the formation of the ice patrol that has prevented further loss of life from such accidents. The U.S. Navy began this patrol immediately following the *Titanic* disaster; it became international in 1914. The patrol, now maintained by the U.S. Coast Guard, concentrates its efforts between 40°30' and 48°N latitude and 43° and 54°W longitude.

The most recent major iceberg ramming occurred in the early hours of June 20, 1989. The Soviet cruise liner *Max Gorky* rammed an iceberg well north of the Arctic Circle between Greenland and Spitzbergen. Quick action by the crew and the Norwegian Coast Guard prevented any loss of life among the 953 passengers and 378 crew members.

The **shelf ice** that represents the edges of glaciers pushing into the marginal seas of Antarctica produces icebergs that have received less attention than those of the North Atlantic, owing to the fact that they interfere less with shipping. Vast tabular bergs that break from the edges of the shelf ice have lengths over 100 km (62 mi) (fig. 7–19C). They may stand as much as 200 m (656 ft) above the ocean surface, though most are probably less than 100 m (328 ft) above sea level. Most of the calving occurs, as it does in the Arctic region, during the summer months. When the sea ice breaks up and allows the swells driven by the strong westerly winds to reach the edge

FIGURE 7–18 Extent of Ice in Arctic Sea.

These 1974 images showing percent of sea-ice concentration are accurate to within 15% for first-year ice (pack ice) and 25% for multiyear-ice (polar ice cap). The scales of ice concentration along the right margins of each image are for first-year ice. The scale above can be used for first-year ice (use left side of scale) and multiyear-ice (use right side of scale). The images were developed from the ESMR (Electrically Scanning Microwave Radiometer) aboard the *Nimbus 5* satellite. Courtesy of NOAA.

A.

B.

C.

D.

FIGURE 7–19 Icebergs.

A, North Atlantic icebergs off Cape York, Greenland. They are generally smaller and more irregular in shape than Antarctic icebergs. *B,* the USS *Wyandotte* moves past a tabular iceberg in the Weddell Sea, Antarctica. Many of the icebergs are kilometers in length and have been mistaken for land. Photographs *A* and *B* from Oceanographer of the Navy, Public Affairs Office. *C,* B-1 Iceberg. This infrared image shows the huge iceberg that broke from the Ross Ice Shelf in October 1987. The longitude and latitude lines intersecting near the east end of B-1 are 160°W and 78°S. This monster is 136 km (85 mi) long, but two larger icebergs formed the year before in the Weddell Sea. Infrared image in *C* courtesy of Robert Whritner, Manager, Antarctic Research Center, Scripps Institution of Oceanography, University of California, San Diego. *D,* a map showing North Atlantic current and typical iceberg distribution (Δ).

of the shelf ice, large icebergs are calved and move to the north into the warmer and rougher water, in which they disintegrate. Carried by the strong West Wind Drift, these icebergs move in an easterly direction around Antarctica, rarely moving farther north than 40°S latitude.

RENEWABLE SOURCE OF ENERGY

The potential of extracting energy from the motions and heat distribution patterns of the atmosphere and ocean, which are maintained by the sun, is attractive for the following reasons:

1. Work can be achieved without significant pollution.
2. The amount of energy available at any time is far greater than that in fossil and nuclear fuels.
3. Such forms of energy are renewable and will not be depleted.

Considered in order of decreasing energy potential, the sources of renewable energy are (1) heat stored in the oceans, (2) kinetic energy of the winds, (3) potential and kinetic energy of waves, and (4) potential and kinetic energy of tides and currents. In later chapters we will consider the use or potential use of winds, waves, tides, and currents. Here we will consider only the renewable source with the greatest store of potential energy: the surface layers of the tropical oceans.

Ninety percent of the earth's surface between the Tropic of Cancer and Tropic of Capricorn is ocean. What makes this warm tropical surface water such an important source of energy is the presence of much colder water beneath the thermocline. With a temperature difference as small as 17°C (30.6°F), useful work can be done by Ocean Thermal Energy Conversion (OTEC) systems (fig. 7–20).

The system works in the opposite way from a typical refrigeration system and on a much larger scale.

A.

B.

FIGURE 7–20 Proposed Uses of Ocean Thermal Power.

A, Ocean Thermal Energy Conversion (OTEC) system with crew quarters and maintenance facilities. Attached around the outside are turbine-generators and pumps. It is over 75 m (246 ft) in diameter, 485 m (1595 ft) long, and weighs about 300,000 tn. This unit is designed to generate 160 million watts of power, which is enough to meet the needs of a city with a population of 100,000. Photo courtesy of Lockheed Missiles and Space Co., Inc. *B,* OTEC plant for production of ammonia. Moving slowly through tropical waters, this plant could produce 1.4% of the ammonia requirements for the United States each year and save 22.6 billion cubic feet of natural gas. Courtesy of U.S. Department of Energy.

The warm surface water heats a fluid such as propane or ammonia that is under pressure in evaporating tubes. The fluid is vaporized and passes through a turbine that drives an electrical generator. After passing through the turbine, the fluid is condensed by cold water that has been pumped up from the deep ocean. It is again ready for heating by warm surface water that will cause it to vaporize and pass through the turbine.

The only region with such potential along the conterminous United States coast is a strip about 30 km (18.6 mi) wide and 1000 km (621 mi) long. It extends north from southern Florida along the western margin of the Gulf Stream. The National Science Foundation is supporting an experimental OTEC system to be established 25 km (16 mi) east of Miami, Florida.

Mini-OTEC, a unit mounted on a U.S. Navy barge, was tested off Ke-Ahole Point, Hawaii, in 1979. Designed primarily to test equipment under ocean conditions, it generated only 15 net kilowatts of electricity. A much larger unit, OTEC 1, designed to generate 1 megawatt, tested heat exchanger designs and biofouling systems under ocean conditions for 3 months in 1981 in Hawaiian waters.

Testing is presently under way on the Seacoast Test Facility, a shore-based OTEC installation, on the island of Hawaii. Although it will require a large initial investment, Hawaii has high hopes that OTEC power generation will be a commercially successful means of power generation by the mid-1990s.

Another proposed use of the OTEC concept is producing ammonia. A floating factory that could produce 586,000 tn of ammonia per year has been described in a feasibility report from the Johns Hopkins Applied Physics Laboratory to the U.S. Maritime Administration. The proposed demonstration ship would weigh 68,000 tn, have a width of almost 60 m (197 ft) and have a length of over 144 m (472 ft). It would move at less than 2 km/hr through tropical waters and use energy derived from the temperature difference between the warm surface waters and cold deep waters to produce ammonia. Each such plant could produce about 1.4% of our nation's ammonia requirements—75% of which is used for fertilizer—and save 22.6 billion cubic feet of natural gas each year (fig. 7–20).

SUMMARY

In ocean water of low biological productivity, the molecular-sized particles scatter the short wavelength of visible light, producing a blue ocean. Greater amounts of dissolved organic matter in more-productive ocean water scatter more green light which, along with the chlorophyll pigmentation of the plants, produces a green ocean. With increasing depth in the ocean, the colors of the visible spectrum are absorbed by ocean water in a way that removes red, yellow, and violet at relatively shallow depths. The blue and green wavelengths of light are the last to be removed. High **turbidity**, the measure of the amount of suspended material in water, greatly reduces the depth to which light will penetrate.

Density of ocean water is increased by increased salinity and decreased temperature. Density change per unit of temperature change is greater in warmer water. Density, temperature, and salinity are **conservative properties** of water, properties that are affected only by mixing and diffusion once a surface water mass sinks. Zones of rapid change in the density, temperature, or salinity in the water column are called a **pycnocline, thermocline,** and **halocline,** respectively.

Velocity of sound transmission in the ocean increases with increases in temperature, salinity, and pressure. A low-velocity sound channel is caused by a thermocline, in which temperature decreases with increased depth.

Radiant energy reaching earth from the sun is mostly in the **ultraviolet** and **visible light** range, while that radiated back to space from earth is primarily in the **infrared spectrum.** The atmosphere is unevenly heated and set in motion because more energy is received and radiated back into space at low latitudes and water vapor, which absorbs infrared radiation well, is unevenly distributed in the atmosphere. **Belts of low pressure,** where air rises, are generally found at the equator and at about 60° latitude. **High-pressure regions,** where dense air descends, are located at the poles and at about 30° latitude. The air at the earth's surface that is moving away from the subtropical highs produces the **trade winds** moving toward the equator and **westerlies** moving toward higher latitudes.

Because the earth's surface rotates at different velocities at different latitudes, increasing from 0 km/h at the poles to over 1600 km/h (994 mi/h) at the equator, objects in motion tend to veer to the right in the Northern Hemisphere and to the left in the Southern Hemisphere. This is called the **Coriolis effect.**

In the **heat budget** of the world ocean, the only source of heat gain is the sun, while heat may be lost by **radiation, evaporation,** and **conduction** into the atmosphere. The amount of heat lost by *radiation,* primarily of infrared wavelengths, is greatest near the equator, where the greatest amount of solar energy is absorbed. The greatest heat loss by *evaporation* is in dry subtropics, while the heat lost through *conduction* reaches a maximum along the western margins of oceans in temperate latitudes where warm water currents carry water into regions where the atmosphere is much colder than the water.

The tilt of the earth's axis of rotation and the distribution of continents modify the idealized pressure belts discussed above. High-pressure cells form over continents in winter and are replaced by low-pressure cells in summer. There is a **cyclonic movement** of air around the low-pressure cells and an **anticyclonic movement** around high-pressure cells. Cold air masses of high latitudes meet warm air masses of lower latitudes and create **cold and warm fronts** that move from west to east across the earth's surface at midlatitudes.

Nonetheless, ocean climate patterns are closely related to the idealized pressure belts previously discussed.

Sea ice forms as sea water is frozen in high latitudes. The process usually involves the formation of a **slush**, which breaks into **pancakes** that ultimately grow into **floes**. Sea ice develops as **pack ice** that forms each winter and melts almost entirely each summer, **polar ice** that is a permanent accumulation in polar regions of the Arctic Ocean, and **fast ice** that forms frozen to shore during the winter. **Icebergs** form as large chunks of ice break away from the large continental glaciers that form on Ellesmere Island and Greenland in the Northern Hemisphere and Antarctica in the Southern Hemisphere.

From the motions and patterns of heat distribution in the atmosphere and oceans, **renewable, nonpolluting sources of energy** can be exploited. These include *heat stored in the ocean* and potential and kinetic energy stored in the *winds, waves, tides,* and *currents. Ocean Thermal Energy Conversion* (OTEC) is a process developed to use the difference in temperature between warm tropical surface water and the cold water below the thermocline to produce electricity and ammonia.

QUESTIONS AND EXERCISES

1. How is the color of the ocean surface water related to biological productivity? Why does everything in the ocean at depths below the shallowest surface water take on a blue-green appearance?

2. Describe the relative effect of temperature change on water density at high- versus low-temperature ranges.

3. The position of the pycnocline in the water column is determined by the combined effects of the thermocline and halocline. Describe these relationships in tropical waters (*see* figs. 7–3 and 7–4).

4. How does the development of a thermocline produce a *sofar,* or sound channel, below the ocean's surface?

5. Explain why the atmosphere is heated primarily by back radiation from the earth rather than by direct radiation from the sun.

6. Describe the effect on the earth as a result of the earth's axis of rotation being angled 23.5° to the ecliptic.

7. Since there is a net annual heat loss at high latitudes and a net annual heat gain at low latitudes, why does the temperature difference between these regions not increase?

8. Why are there high-pressure caps at each pole and high-pressure belts at 30° latitudes compared to low-pressure belts in the equatorial region and at 60° latitudes?

9. Describe the Coriolis effect in the Northern and Southern hemispheres and include a discussion of why the effect increases with increased latitude.

10. In considering the heat budget of the world ocean over a long period, why should Q_c (rate of heat gain or loss through ocean currents) and Q_t (net rate of heat gain or loss) be considered to be zero?

11. Describe the ocean regions where the maximum rates of heat loss by radiation, evaporation, and conduction occur, and explain why.

12. Discuss why the idealized belts of high- and low-atmospheric pressure shown in figure 7–9 are modified (*see* fig. 7–12).

13. Name the polar and tropical air masses that affect U.S. weather. Describe the pattern of movement across the continent and patterns of precipitation associated with warm and cold fronts.

14. How are the ocean's climatic belts (fig. 7–16) related to the broad patterns of air circulation described in figure 7–9?

15. Describe the formation of sea ice from the initial freezing of the ocean surface water through the development of polar ice in the Arctic Ocean.

16. What is the difference in the average size and shape of icebergs in the Arctic and Antarctic? Why do these differences exist?

17. Construct your own diagram of how an Ocean Thermal Energy Conversion unit might generate electric-

ity, or make a flow diagram presenting the steps of the process.

REFERENCES

Changing climate and the oceans. 1987. *Oceanus* 29:4, 1–93.

Miller, A. 1971. *Meteorology*. Columbus, Ohio: Charles E. Merrill.

Ocean energy. 1979. *Oceanus* 22:4, 1–68.

Oceans and climate. 1978. *Oceanus*. 21:4, 1–70.

The oceans and global warming 1989. *Oceanus*. 32:2, 1–75.

Pickard, G. L. 1975. *Descriptive physical oceanography*. 2nd ed. New York: Pergamon Press.

SUGGESTED READING

Sea Frontiers

Boling, G. R. 1971. Ice and the breakers. 17:6, 363–71.
An interesting history of people in icy waters and the development and improvement of ice breakers.

Charlier, R. 1981. Ocean-fired power plants. 27:1, 36–43.
The potential of ocean thermal-energy conversion is discussed.

Houghton, R. A., and Woodwell, G. M. 1989. *Global climate change*. 260:1, 36–47.
The history of global climate change is discussed along with the potential for future climatic change.

Land, T. 1976. Europe to harness the power of the sea. 22:6, 346–49.
An article emphasizing the clean, safe, permanent nature of ocean tides and waves as a source of power.

Mayor, A. 1988. Marine mirages. 34:1, 8–15.
The nature of marine mirages and the research efforts that lead to our understanding them are considered.

Rush, B., and Lebelson, H. 1984. Hurricane! The enigma of a meteorological monster. 30:4, 233–39.
An overview of the nature of hurricanes and the problem of predicting where they will go.

Scheina, R. L. 1987. The Titanic's legacy to safety. 33:3, 200–209.
A brief summary of the *Titanic* sinking and an overview of the ice patrol and iceberg collision history subsequent to the sinking of the "unsinkable" luxury liner.

Smith, F. G. W. 1974. Planet's powerhouse. 20:4, 195–203.
A description of how the earth's "heat engine" works, with an emphasis on the nature of tropical cyclonic storms.

_____. 1974. Power from the oceans. 20:2, 87–99.
A survey of the many tried and untried proposals for extracting energy from the oceans.

Sobey, E. 1979. Ocean ice. 25:2, 66–73.
The formation of sea ice and icebergs as well as their climatic and economic effects are included.

_____. 1980. The ocean-climate connection. 26:1, 25–30.
The increasing amount of knowledge of the effect of the oceans on the earth's climate may be used to predict climatic trends of the future.

Scientific American

Gregg, M. 1973. Microstructure of the ocean. 228:2, 64–77.
A discussion of the methods of studying the detailed movements of ocean water by observing temperature and salinity changes over distances of one centimeter and the motions they reveal.

MacIntyre, F. 1974. The top millimeter of the ocean. 230:5, 62–77.
Processes that are confined to a thin film at the surface of ocean water and their role in the overall nature of the oceans are discussed.

Penney, T. R., and Bharathan, D. 1987. Power from the sea. 256:1, 86–93.
A prediction that generating electricity by ocean thermal energy conversion will be competitive with fossil fuel plants as the price of oil rises.

Revelle, R. 1982. Carbon dioxide and world climate. 247:2, 35–43.
Some of the possible effects of increasing atmospheric temperature due to CO_2 accumulation are considered.

Stanley, S. M. 1984. Mass extinctions in the oceans. 250:6, 64–83.
Geological evidence suggests most major periods of species extinction over the last 700 million years occurred during brief intervals of ocean cooling.

8
OCEAN CIRCULATION

Currents, or water masses in motion, are driven ultimately by energy from the sun. The circulation of these masses can be categorized as being either **wind-driven** or **thermohaline**. The wind-driven currents are set in motion by moving air masses, and this motion is confined primarily to horizontal movement in the upper waters of the world ocean. The thermohaline circulation has a significant vertical component and accounts for the thorough mixing of the deep masses of ocean water. Thermohaline circulation is initiated at the ocean surface by temperature and salinity conditions that produce a high-density mass, which sinks and spreads slowly beneath the surface waters.

Even though we think of the ocean as having great depth, ocean basins are very shallow in comparison with their widths. If the ocean basins were scaled down so their widths equaled the width of this page, they would be no deeper than the page is thick. Yet within this thin shell of water, there is a rich dynamic structure in which the wind-driven surface current system is separated by the pycnocline from the deep thermohaline circulation. Nevertheless, these separate systems do communicate across the pycnocline in a way that is only now being recognized.

HORIZONTAL CIRCULATION

Wind-driven horizontal circulation in the surface waters develops from stress at the interface between the ocean and the wind. The trade winds blowing out of the southeast in the Southern Hemisphere and out of the northeast in the Northern Hemisphere provide the backbone of the system of ocean surface currents. Setting the water masses between the tropics into motion, the trade winds develop the equatorial currents that can be found in all of the world's oceans.

These currents move west parallel to the equator, and owing to the Coriolis effect and the deflection along continental margins, they move away from the equator as **warm currents** or **western boundary currents**. This motion is the start of the dominant surface feature, the **subtropical gyres**, of each of the major ocean basins. The gyres rotate clockwise in the Northern Hemisphere and counterclockwise in the Southern Hemisphere. They are so named because their centers are in the subtropics (near 30° latitude); they

Preceding page: Satellite image of the Gulf Stream moving out into the Atlantic off Chesapeake Bay. Photo courtesy of NOAA.

dominate the upper ocean motion to at least 45° latitude. As the warm western boundary currents move into higher latitudes along the western margins of their respective ocean basins, the westerly wind systems blowing out of the northwest in the Southern Hemisphere and out of the southwest in the Northern Hemisphere add more energy and push the circulating masses on in an easterly direction. Deflection of **cold boundary currents**, or **eastern boundary currents**, toward the equator along the eastern margins of their respective basins completes the path of circulation (fig. 8–1).

Ekman Spiral

To explain why water masses move as they do relative to wind direction, we will recall the observations made by Fridtjof Nansen during the voyage of the *Fram.* Nansen determined that the Arctic sea ice moved 20° to 40° to the right of the wind blowing across its surface. He passed this information on to V. Walfrid Ekman, a physicist who developed the mathematical relationships that explain the observations.

Ekman developed a circulation model that has been called the **Ekman spiral** (fig. 8–2). The model assumes that a homogeneous water column is being set in motion by wind blowing across its surface. Owing to the Coriolis effect, the surface current moves in a direction 45° to the right of the wind in the Northern Hemisphere. This surface mass of water, moving as a thin **lamina**, or sheet, sets another layer beneath it in motion. The surface layer moves with a velocity no more than 3% of the wind speed. The energy of the wind is passed through the water column from the surface down, with each successive layer of water being set in motion with a lower velocity than and in a direction to the right of the one that set it in motion. At some depth the momentum imparted by the wind to the moving water laminae will be lost and there will be no motion as a result of wind stress at the surface. The depth at which motion ceases is called the **depth of frictional influence**. Though it depends on wind speed and latitude, this stillness normally occurs at a depth of about 100 m (328 ft). Figure 8–2 shows the spiral nature of movement with increasing depth from the ocean's surface. The length of each arrow in the figure is proportional to the velocity of the individual lamina, and the direction of each arrow indicates its direction of movement.

Under these theoretical conditions, the surface current should flow at an angle of 45° to the direction of the wind. From the surface to the depth of frictional influence, the net water movement, the **Ekman**

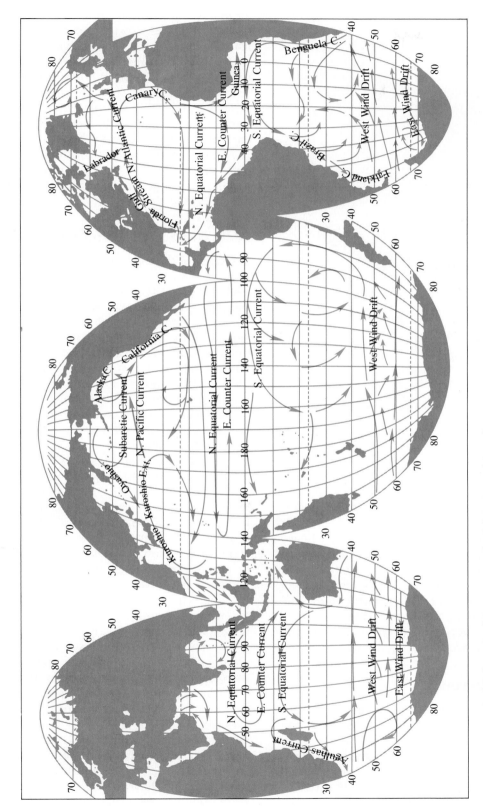

FIGURE 8–1 Wind-Driven Surface Currents in February and March.
After Sverdrup et al., 1942.

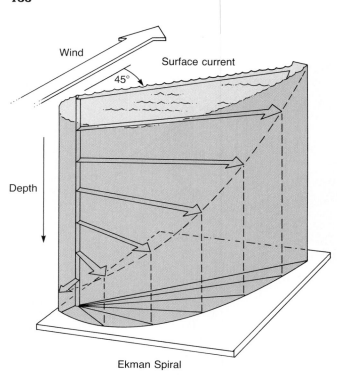

FIGURE 8–2 Ekman Spiral.
Wind drives surface water in a direction 45° to the right of the wind in the Northern Hemisphere. Deeper water continues to deflect to the right and move at a slower speed with increased depth. Ekman transport, the net water movement, is at right angles to the wind direction.

transport, will be at right angles to the direction of the wind.

There are, however, no ideal conditions in existence in the ocean, and movements actually occurring as a result of wind stress on the ocean surface will deviate from this idealized picture. Generally, we can say that the surface current will move at an angle of less than 45° to the direction of the wind and that the Ekman transport will be at an angle less than 90° to the direction of the wind. This is particularly true in shallow coastal waters, where all of the movement may be in a direction very nearly that of the wind and the turning with increased depth is minor.

Geostrophic Currents

If we consider a gyre in the North Atlantic Ocean, for example, and remember the Ekman transport, it can be seen that a clockwise rotation in the Northern Hemisphere will tend to produce a convergence and a piling up of water in the center of that gyre. We find within all such ocean gyres hills of water that rise as much as 2 m (6.6 ft) above the water level at the margins of the gyres. As Ekman transport pushes water into the "hill" structure, the gravitational force acts to move particles down the surface slope. The Coriolis force deflects the water flowing down the slope to the right in a curved path (fig. 8–3). The water piles up on these hills until the down-slope component of gravitational force acting on individual particles of water balances the Coriolis force. When these two forces finally balance, the net effect is a **geostrophic current** moving around the hill.

Again, this idealized flow would exist if there were no friction. Owing to the friction between water molecules, the water follows a path that moves it gradually down the slope of the hill.

Westward Intensification

The apex of the hills formed within the rotating gyres is not in the center of the gyre. The highest part of each hill is located closer to the western boundary of the gyre. The causes of **westward intensification** are complex, but a good part of the phenomenon can be explained in terms of the Coriolis effect. Remember that this effect increases in strength toward higher latitudes. Thus, the water that is driven eastward by the westerly winds tends more strongly to veer right toward the equator than the water that is driven westward along the equator tends to veer toward higher latitudes. This causes a broad equatorward flow of water across most of the subtropical gyres.

Assuming we have a steady state with the same volume of water rotating around the apex of the hill, it can be seen in figure 8–3B that the velocity with which the water moves along the western margin will be much greater than that with which it will move around the eastern side of the hill. These relationships can be seen particularly well in the North Atlantic and North Pacific gyres, where the western boundary currents move with speeds in excess of 5 km/h (3.1 mi/h) in a northerly direction, while a flow along the eastern margin of each basin is better characterized as a drift moving at velocities well below 0.9 km/h (0.6 mi/h). Western boundary currents commonly flow 10 times faster and to greater depths than eastern boundary currents. The warm western boundary currents are usually less than one-twentieth the width of the broad, cool drifts flowing toward the equator on the east side of the subtropical gyres.

Directly related to this difference in speed is the steepness of the hill's slope. The slope of the hill on the side of the slow-moving eastern boundary current is quite gentle; the western margin has a steep slope

FIGURE 8-3 Geostrophic Current.

A, as the earth's wind systems set ocean water in motion, circular gyres are produced. Water piles up inside the subtropical gyre with the apex of the "hill" closer to its west margin because of the earth's rotation to the east. The theoretical geostrophic current flows parallel to the contour of the hill and represents an equilibrium between the Coriolis force pushing water toward the apex through Ekman transport and the downslope component of gravity acting to move the water down the slope. Because of friction between water molecules, the path of the current is gradually down the slope of the hill. *B,* because the Coriolis effect is stronger on water farther from the equator, the eastward-flowing high-latitude water tends to turn equatorward more strongly than the westward-flowing equatorial water tends to turn toward higher latitudes. This causes a broad, slow, equatorward flow of water across most of the subtropical gyre and forces the apex of the geostrophic hill to the west. This phenomenon is referred to as westward intensification, the main manifestation of which is a high-speed, warm, western boundary current flowing along a westward slope of the hill that is much steeper than that of the eastern slope associated with the slow drift of cold water toward the equator.

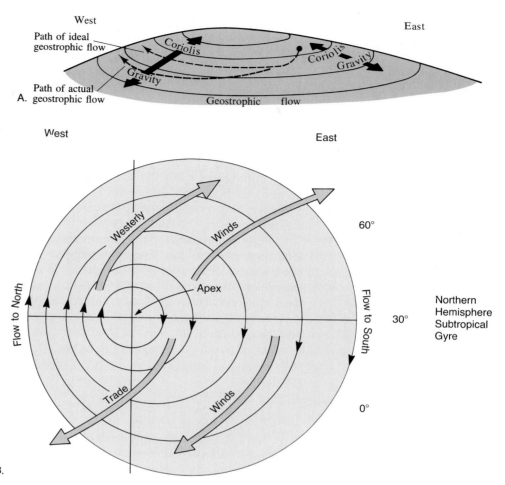

corresponding to the high velocity of the western boundary current.

An indirect method of determining current velocity in the upper-water mass is measuring the internal distribution of density and the pressure field it creates. This pattern of distribution is determined by collecting water samples along a number of vertical columns distributed throughout the area to be studied. At each station, temperature and salinity are measured. Average density of the water column is computed from the temperature-salinity characteristics measured above an arbitrary depth at which the surface current is assumed to have died out (possibly 100 m, or 328 ft). No current or horizontal pressure gradient is thought to exist at this depth. The pressures of the overlying water columns are assumed to be equal.

If our assumption of no horizontal pressure gradient existing at our reference level is correct, all measured water columns above this datum must contain equal mass. However, a water column with a low average density will, because of its greater volume, stand higher above the datum than a water column of high density. By computing the heights of columns necessary at all stations to produce the zero horizontal pressure gradient at the datum depth, a topography for the ocean surface can be determined (fig. 8–4).

Maps showing this type of water motion can be prepared for any depth between the datum and the ocean surface; they represent the **dynamic topography** that is used to compute current velocity. The steeper the slope, the higher the velocity of the geostrophic current flowing in a direction generally parallel to the topography **contours** (lines of equal height above the equal pressure datum). Rapid progress is being made in the use of radar altimeters mounted on satellites to map the dynamic topography of the oceans. Another technique that may replace the laborious procedures involved in the classic method of determining dynamic topography is described in "Ocean Acoustic Tomography," on page 190.

OCEAN ACOUSTIC TOMOGRAPHY

When we observe the motions of ocean water, we find they range across broad spectrums of size and time. The identifiable patterns of ocean circulation range in size from eddies a few tens of kilometers across, created by water flowing over and around seamounts, to the slowly rotating subtropical gyres the size of ocean basins. Between these extremes are features such as the Gulf Stream rings, which are a few hundred kilometers in diameter, and coastal upwelling such as that responsible for the highly productive waters along the coast of Peru.

Fully understanding these motions and their effects on climate and biological productivity has long been the goal of oceanographers. However, it has not been possible to continuously gather data over sufficient area for the necessary length of time to achieve this goal because of the high cost of sending the required number of ships on long sea voyages.

Recently, Walter Munk of Scripps Institution of Oceanography (fig. 8A*A*) and Carl Wunsch (fig. 8A*B*) of Massachusetts Institute of Technology have led the way in developing a system that may make such data gathering possible— ocean acoustic tomography. "Tomo" is derived from the Greek word *tomos,* meaning a section or slice. A *tomogram* is a picture of a slice. Tomography has already found use in medicine through the use of x-rays. The CAT (computerized axial tomography) scan uses the x-rays to provide tomographic images of parts of the body, usually the brain.

The use of tomography to map the features of ocean circulation involves the transmission of low-frequency sound. Increases in water temperature, salinity, and pressure increase the speed of sound travel in ocean water. As sound travels through features such as cold or warm rings, the sound speed

FIGURE 8A Munk and Wunsch.

Walter H. Munk of Scripps Institution of Oceanography (*A*) and Carl Wunsch of Massachusetts Institute of Technology (*B*) are leaders in the development of ocean acoustic tomography techniques for the syntopic study of mid- to large-scale oceanic circulation patterns. Walter Munk's photo courtesy of University of California at San Diego, Office of Learning Resources S.I.O. Photo Lab.

A.

B.

will be decreased or increased. Theoretically, if sound is transmitted across a slice of ocean from a series of transmitters on one side to a series of receivers on the other, features that cause a change in the speed of sound can be mapped.

A 1981 demonstration covering a 300-km (186-mi) square of ocean southwest of Bermuda involved the use of four transmitters and five receivers over a period of six months (fig. 8B). Two sensors of temperature and current flow were included and survey ships periodically entered the area to record data against which the tomography could be checked.

During the demonstration, a large cold eddy moved across the area and a frontal system separating cold and warm water moved into the area. The data gathered conventionally confirmed the existence of the features and proved the feasibility of such a system to provide instantaneous pictures of large slices of the world's oceans. Subsequent tests (1983) proved the technique could be used to measure current flow and deep-sea circulation at a depth of 760 m (2500 ft). The latter experiment, conducted in the North Pacific Ocean, involved three transceivers arranged in a triangle with sides 1000 km (620 mi) in length. A seven-transceiver system was set up off New England in 1987 to study the Gulf Stream, and a large, five-year study is under way to investigate the contribution of the Greenland Sea to the formation of North Atlantic Deep Water. In the near future it may well be possible to map such features within large ocean basins—something that was only a dream a few years ago.

Reference: Behringer, D.; Birdsall, T.; Brown, M.; Cornuelle, B.; Heinmiller, R.; Knox, R.; Metzger, K.; Munk, W.; Spiesberger, J.; Spindel, R.; Webb, D.; Worcester, P.; and Wunsch, C. 1982. A demonstration of ocean acoustic tomography. *Nature* 299, 121–25.

FIGURE 8B   Ocean Acoustic Tomography.

A, during the 1981 demonstration of ocean acoustic tomography, transmitters (T1 to T4) emitted low-frequency sound signals that were received by the receivers (R1 to R5). Temperature and currents were measured by conventional means at moorings E1 and E2. *B,* results of the 1981 ocean acoustic tomography demonstration. Source of data: 1 and 3, ship surveys; 2, tomography. During the demonstration, data from a depth of 700 m showed a large cold eddy in the center of the survey area at the start of the survey. It was separated from warm water to the northeast by a well-defined temperature front. The cold water eddies observed moved off to the northwest, while the front progressed to the southwest during the demonstration. Cold eddies were indicated by the slowing of sound signals that passed through them. Data courtesy of Woods Hole Oceanographic Institution.

A.

1.

2.

3.

B.

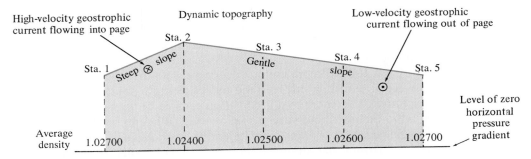

FIGURE 8–4 Dynamic Topography.
Comparison of average densities for the water columns above the arbitrary no-current datum tells us the apex of the hill is near station 2 because the lower the density of the water, the greater the volume a given mass will occupy. Northern Hemisphere geostrophic flow is indicated by circles beneath slopes.

Separating gyres in the northern and southern portions of the Atlantic and Pacific oceans are equatorial countercurrents that flow in an easterly direction. Although such a countercurrent exists throughout the breadth of the Pacific Ocean, it is restricted to the eastern Atlantic. In both of these oceans the current is centered about 5° north of the equator as a result of the northward shift of the ITCZ (Intertropical Convergence Zone) caused by the uneven distribution of land and water in the Northern and Southern hemispheres. In the Indian Ocean, such a current develops during the winter monsoon only. There it is shifted about 7° south of the equator, owing to the fact that the Indian Ocean extends only to about 20°N latitude (*see* fig. 8–1).

VERTICAL CIRCULATION

Lateral movements of water masses may bring about vertical circulation within the upper-water mass. We refer to this shallow vertical circulation system as **wind-induced circulation.** Of greater oceanwide importance in producing the thorough mixing within the ocean is the vertical circulation resulting from density changes at the ocean surface, which cause the sinking of water masses. Since the changes that increase the density are normally changes in temperature and salinity, this circulation is referred to as **thermohaline circulation.**

Wind-Induced Circulation

There are regions throughout the world ocean where water moves vertically to the surface or down away from the surface as a result of wind-driven surface currents carrying water away from or toward these regions. **Upwelling** occurs in areas where the surface flow of water is away from the area. If volume is to be conserved and horizontal surface flows bring insufficient water into the area, water must come from beneath the surface to replace that which has been displaced. This condition may occur in the open ocean or along the margins of continents.

As the North Equatorial Current of the Indian Ocean and South Equatorial currents in the Atlantic and Pacific oceans are driven by easterly winds in a westerly direction on either side of the equator, the water on the north side of the equator will tend to veer to the right, moving to a higher latitude, while that in the Southern Hemisphere will veer to the left to a higher latitude (fig. 8–1). The net effect is a water deficiency at the surface between the two currents. Water from deeper within the upper-water mass comes to the surface to fill the void resulting from the deflection of equatorial water to the north and south (fig. 8–5). This phenomenon is called **equatorial upwelling.**

FIGURE 8–5 Equatorial Upwelling.
The Coriolis effect acting on trade-wind-driven westward-flowing equatorial currents pulls surface water away from the equatorial region. This water is replaced by subsurface water.

Coastal upwelling is common along the margins of continents where the wind conditions are such that the surface waters adjacent to the continents are carried out to the open ocean via Ekman transport (fig. 8–6). The replacement of this water comes from the lower portions of the upper water, as is the case with the Peru Current. Such areas are characterized by low surface temperatures and high concentrations of nutrients that make them areas of high biological productivity. The reversal of the direction of coastal winds that are causing upwelling tends to push water toward shore, where it piles up and causes **coastal downwelling**, a sinking of surface water.

Near the center of the rotating gyres of the major ocean current systems the winds are relatively weak, and the water rotates at a very low rate. The winds that do blow in these regions may be steady in direction and are known to set up convection cells in the upper-water mass. This phenomenon was first recognized by Irving Langmuir while crossing the Sargasso Sea in 1938. He observed **streaks**, on straight rows, of seaweed parallel to the direction of the wind and concluded that the plants were trapped in zones of convergence between cells. Not only is macroscopic plant material concentrated in these regions, but also microscopic plants and dissolved organic material. By contrast, in the regions of divergence, where water is surfacing as a result of convection, the concentrations of organic material are relatively low (fig. 8–7). This phenomenon is called **Langmuir circulation**.

Thermohaline Circulation

The large-scale vertical circulation in the ocean results primarily from surface-density changes in oceanic water. Vertical mixing of ocean water is achieved primarily through the sinking and rising of water masses in high latitudes. It is only in the high-latitude areas that the water column has a gravitational stability sufficiently neutral to allow vertical movements of large masses of water. Surface masses may sink to the ocean bottom and deep-water masses may rise to the surface. The relatively strong density stratification, or pycnocline, that separates the upper- and deep-water masses throughout the lower-latitude regions does not exist in the high-latitude oceans.

Because the intensity of solar radiation is greater in the equatorial region, we might expect that heating of surface water there may cause the water to expand and move away from the equator toward the poles, spreading over the colder, more dense high-latitude waters (fig. 8–8). We may further expect that at some depth a return flow from the high latitudes toward

FIGURE 8–6   Coastal Upwelling and Downwelling.
A, in areas where wind-driven coastal currents flow along the western margins of continents and toward the equator, the Ekman transport carries surface water away from the continent. An upwelling of deeper water replaces the surface water that has moved away from the coast. *B,* a reversal of the direction of the winds that cause upwelling will cause water to pile up against the shore and force downwelling.

FIGURE 8–7 Langmuir Circulation.

Steady winds blowing across the ocean surface create convection cells with alternate right- and left-hand circulation. The axes of these cells run parallel to wind direction. Organic debris accumulates in downwelling zones of convergence and produces windrows that run parallel to the wind direction for great distances.

FIGURE 8–8   Average Sea-Surface Temperature (SST) Distribution Pattern for July (in °C).

the equator would be set in motion to replace the surface water moving away from the equatorial region. This exchange does not seem to occur in the oceans, although it is theoretically proper to assume that such a pattern might develop. The energy imparted to the surface waters by winds greatly exceeds the energy that develops as a result of density changes due to temperature differences. Thus the effect of the wind overcomes any tendency that may exist for such a pattern of circulation to develop.

Another variable that affects the density of surface water is salinity (fig. 8–9). Salinity appears to have a very minimal effect on the movement of water masses in the lower latitudes, and density changes resulting from salinity changes are of importance only in the very high latitudes, where low water temperature remains relatively constant. For example, the highest-salinity water in the open ocean is found in the subtropical regions, but there is no sinking water in these areas because water temperatures are high enough to maintain a low density for the surface-water mass and prevent it from sinking. In such areas, a strong halocline, or salinity gradient, may develop with a relatively thin surface layer of water having salinities in excess of 37‰. Salinity decreases

rapidly with increasing depth to typical ocean-water salinities below 35‰.

An important sinking of cold surface waters that become deep-water masses occurs in the subpolar regions of the Atlantic Ocean. In the North Atlantic, the major sinking of surface water is thought to occur in the Norwegian Sea. From there it flows as a subsurface current into the North Atlantic. This flow becomes part of the **North Atlantic Deep Water**. Additional surface water may sink at the margins of the Irminger Sea off southeastern Greenland and the Labrador Sea. In the southern subpolar latitudes, the most significant area of deep-water-mass formation is the Weddell Sea, where rapid winter freezing produces high-density water that sinks down the continental slope of Antarctica and becomes **Antarctic Bottom Water**, the densest water in the open ocean (fig. 8–10).

On a broad scale, there are latitudinal regions where surface-water masses converge and may cause sinking (fig. 8–10). These regions of convergence are found within the subtropical gyres and in the Arctic and Antarctic. The **subtropical convergences** do not produce sinking because surface waters have relatively low densities as a result of high temperature.

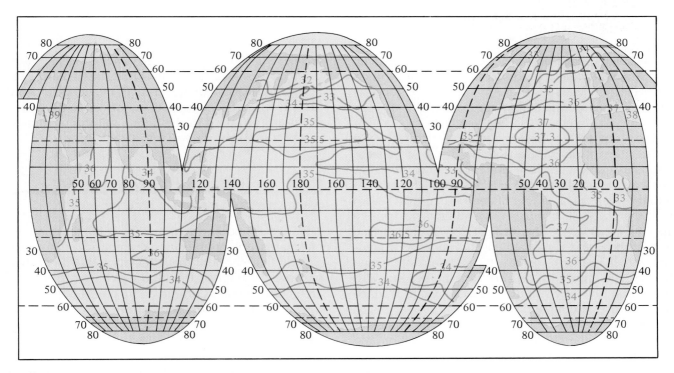

FIGURE 8–9   Surface Salinity of the Oceans in August (‰).
After Sverdrup et al., 1942. Base map courtesy of National Ocean Survey.

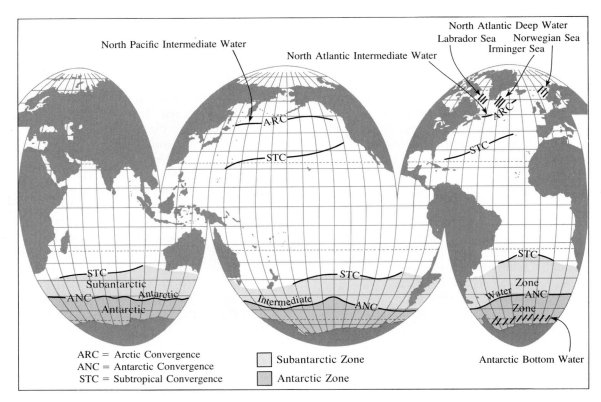

FIGURE 8–10   Regions Where Intermediate and Deep-Water Masses of the World Ocean Sink.
Base map courtesy of National Ocean Survey.

Major sinking does occur, however, along the **Arctic** and **Antarctic convergences.** The major water mass formed from sinking at the Antarctic convergence is the **Antarctic Intermediate** water mass (fig. 8–10).

Figure 8–11 shows what is thought to be the general pattern of deep-water circulation in the world ocean. The most intense flow occurs as a western boundary current because of the rotation of the earth. For every unit of water that sinks from the surface into the deep ocean, a unit of deep water must return to the surface. It is much more difficult to identify specific areas where this vertical flow to the surface is occurring, and it probably occurs more as a gradual rather uniform return throughout the ocean basins. This return to the surface may be somewhat greater in the low-latitude regions, where surface temperatures are higher.

Figure 8–12 shows a probable pattern of integration between the deep thermohaline circulation and surface currents in which the deep ocean is provided oxygen-rich water by the sinking of dense surface water primarily in the subpolar Atlantic Ocean. The return limb of this system begins with surface water

flowing out of the northeastern Pacific Ocean. Water is added as the surface flow crosses the Indian Ocean and Atlantic Ocean on its return to the subpolar North Atlantic Ocean. The unit of ocean current volume, **sverdrup (sv)**, used in the figure and throughout the rest of this chapter is equal to 1 million m³/s.

ANTARCTIC CIRCULATION

Although the oceanic mass surrounding the continent of Antarctica is not officially recognized as an ocean, we will first consider this circulation. We begin here because the mightiest of all ocean currents, the **Antarctic Circumpolar Current,** dominates the movement of water masses in the southern Atlantic, Indian, and Pacific oceans south of 50°S latitude. This latitude can be considered the northern boundary of the "Antarctic Ocean."

A surface flow called the **East Wind Drift** moves in a westerly direction around the margin of the Antarctic continent (fig. 8–13). This flow is driven by the

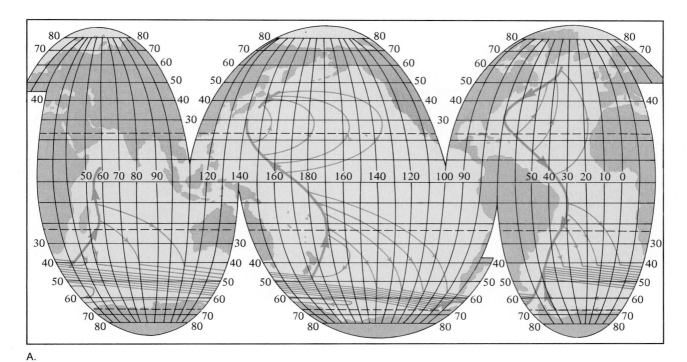

A.

FIGURE 8–11   Stommel's Model of the Deep-Water Circulation of the World Ocean.

A, based on information obtained to date, this highly schematic model of deep-water circulation developed by Henry Stommel in 1958 appears to be reasonably correct. Heavy lines mark the major western boundary currents. They result from the same forces that produce the more intense western boundary currents in the surface circulation. After H. Stommel, 1958. Base map courtesy of National Ocean Survey. *B,* Dr. Henry M. Stommel entered the field of physical oceanography as a research assistant at Woods Hole Oceanographic Institution in 1944. Although his formal education consisted of no more than a B.S. in physics, he soon became prominent in the field. Many of his colleagues consider him to be the world's preeminent physical oceanographer. Photo courtesy of Woods Hole Oceanographic Institution.

B.

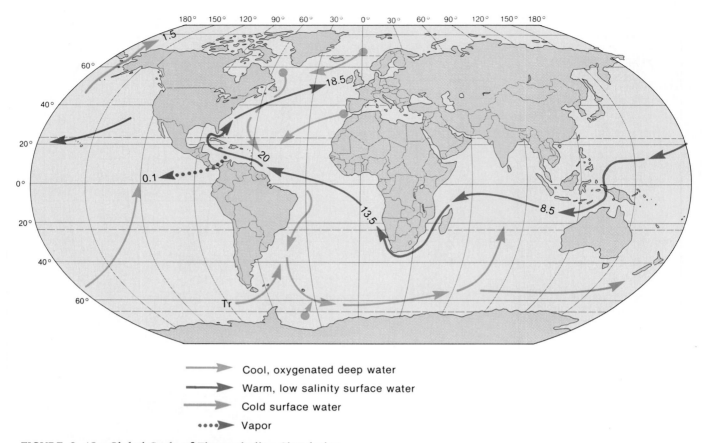

Cool, oxygenated deep water

Warm, low salinity surface water

Cold surface water

Vapor

FIGURE 8–12 Global Cycle of Thermohaline Circulation.

The deep ocean is replenished with oxygen by the sinking of the North Atlantic Deep Water (NADW). NADW is composed of water that sinks in the Norwegian Sea and is joined by water from Baffin Bay and the Mediterranean Sea. The total volume of this flow may be about 20 sv as it enters the South Atlantic. This deep flow is recooled in the Antarctic and flows on into the deep Indian Ocean and Pacific Ocean. A broad pattern of upwelling returns the bottom water to the surface in all oceans. A surface flow of warmer, low-salinity water of the North Pacific starts the return journey to the North Atlantic. A flow of 8.5 sv from the North Pacific increases to 13.5 sv after crossing the Indian Ocean and reaches a total of 18.5 sv by the time it is carried into the North Atlantic. A small amount of cold surface water enters the Atlantic from the Pacific through the Drake Passage and 1.5 sv flows into the Atlantic through the Bering Strait. At least 0.1 sv of water is transferred from the Atlantic Ocean to the Pacific Ocean as water vapor across the Isthmus of Panama. (After Gordon, 1986)

FIGURE 8–13 Antarctic Surface Circulation.

Near the Antarctic continent a westward-flowing current—the East Wind Drift—is driven by the polar easterly wind system. Ekman transport produces a zone of divergence—Antarctic Divergence—between the East Wind Drift and the West Wind Drift that circles Antarctica with an easterly flow. The cold West Wind Drift water converges with warmer water in each of the major ocean basins to the north at the Antarctic Convergence.

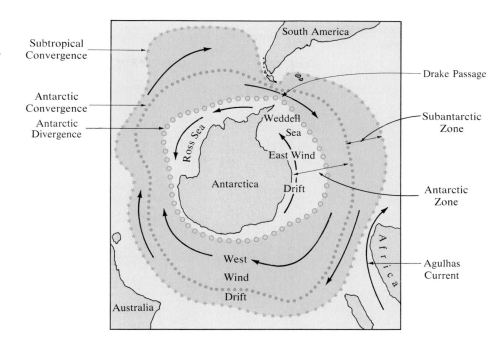

easterly winds that are manifestations of the polar air masses moving out to sea from Antarctica. The East Wind Drift is most extensively developed to the east of the Antarctic Peninsula in the Weddell Sea region and in the area of the Ross Sea. If this terminology seems confusing, it is because these currents are named after the winds that drive them, and winds are named on the basis of the direction they are coming from. For example, east winds blow out of the east.

The main circulation system in Antarctic waters is the **Antarctic Circumpolar Current**, extending northward to a position of approximately 40°S latitude. This mass is driven by the westerly winds that maintain very great strength throughout much of the year (*see* fig. 9D on page 224). The surface portion of this flow is named the **West Wind Drift**. As a result of the Coriolis effect deflecting moving masses to the left in the Southern Hemisphere, there is a zone of divergence between the East Wind Drift and the West Wind Drift.

The Antarctic Circumpolar Current meets its greatest restriction as it passes through the 1000-km (621-mi) Drake Passage between the Antarctic Peninsula and the southern islands of South America. Although the current does not move at high velocity, reaching a maximum velocity of about 0.5 km/h (0.3 mi/h) just north of the Antarctic Convergence, it does transport more water than any other ocean current, an average of about 130 sv.

All around the Antarctic continent the Antarctic Circumpolar Water has similar characteristics. Especially during winter, a mixture of circumpolar water

and shelf water from the **Weddell Sea** produces the densest water in the world ocean, the Antarctic Bottom Water. This mixture of shelf water with circumpolar water produces a water mass with σ value of 27.9 (density of 1.02790 g/cm³). Owing to its high density, the Antarctic Bottom Water flows down the Antarctic continental shelf toward the Atlantic Ocean. As it moves northward along the bottom, it is carried to the east by the Circumpolar Current into the Indian and Pacific portions of the Antarctic Ocean.

The surface water in the **Antarctic Zone** has a northward component of flow that produces a convergence between this cold-water mass and the warmer water of the **Subantarctic Zone**. At this convergence, Antarctic surface water sinks and becomes the subsurface mass called Antarctic Intermediate Water, which has a temperature of about 2°C (32.4°F) and salinity of 33.8‰. This combination results in a σ of 27.0. The water mass can be identified with its core at about 900 m (2952 ft) depth at 40°S latitude by its low salinity and relatively high dissolved oxygen content (fig. 8–14).

Sandwiched between the Antarctic Intermediate and the Antarctic Bottom water masses at 40°S latitude is the North Atlantic Deep Water, which moves southward with its core at about 2200 m (7216 ft) depth. It is characterized by a temperature between 2° and 3°C (35.6° and 37.4°F) and salinity that averages 34.7‰. The North Atlantic Deep Water can be identified on the basis of its temperature, which is slightly above that of the Intermediate Water and the Antarctic Bottom Water. This deep water of North At-

FIGURE 8-14 Antarctic Sub-surface Water Masses.

A.

B. Major subsurface water masses may be identified at 40° S latitude by observing vertical temperature and salinity profiles as follows:

   1. Antarctic Intermediate Water—Salinity minimum identifies core of mass at about 900 m (2952 ft).

   2. North Atlantic Deep Water—Temperature and salinity maxima identify NADW. Core at 2200 m (7216 ft) approaches 34.9 ‰.

lantic origin has been away from the surface for at least 300 yr and is sandwiched between two water masses that have more recently left the surface, so it is also possible to identify it on the basis of a dissolved oxygen minimum. Although it is low in dissolved oxygen content, it still has a very positive effect on the biological productivity of the Antarctic Zone. The major contribution made by this deep water to biological productivity as it surfaces within the Antarctic Zone is its high concentration of nutrients that have been accumulating during the hundreds of years it was beneath the photosynthetic zone.

ATLANTIC OCEAN CIRCULATION

Surface Circulation

In the Atlantic Ocean, the basic surface circulation pattern is that of two large subtropical gyres. The North Atlantic gyre rotates in a clockwise direction, while the South Atlantic gyre rotates in a counter-clockwise pattern. The driving force behind these rotations, which are separated by the **Atlantic Equatorial Countercurrent**, is that of the northeast and southeast trade winds.

The South Atlantic gyre is composed of the **South Equatorial Current**, which reaches its greatest strength just below the equator and is split in two by the topographic interference of the eastern extension of Brazil. Part of the South Equatorial Current moves off along the northeastern coast of South America toward the Caribbean Sea and the North Atlantic. The rest is turned southward as the **Brazil Current**, which ultimately merges with the West Wind Drift and moves across the South Atlantic. The gyre is completed by a slow-drifting movement of cold water, the **Benguela Current**, that flows up the western coast of Africa. There is also a significant flow of cold water moving in a northerly direction along the western margin of the South Atlantic. The **Falkland Current**, an important cold current, moves up the coast of Argentina as far north as 25° to 30°S latitude, wedging its way between the continent and the Brazil Current. The Brazil Current has a much smaller volume than its Northern Hemisphere counterpart, the Gulf Stream. This smaller size results partly from the splitting of the South Equatorial Current by the configuration of South America (fig. 8–15).

The **North Equatorial Current** moves parallel to the equator in the Northern Hemisphere, where it is joined by that portion of the South Equatorial Current that is shunted toward the north along the South American coast. This flow splits into two masses, the **Antilles Current**, which flows along the Atlantic side of the West Indies, and the **Caribbean Current**, which passes through the Yucatan Channel into the Gulf of Mexico. These masses reconverge as the water that entered the Gulf of Mexico exits between Florida and Cuba as the **Florida Current**. The Florida Current flows close to shore over the continental shelf, carrying a volume that at times exceeds 35 sv. As it moves off Cape Hatteras and flows across the open ocean in a northeasterly direction it becomes the **Gulf Stream**, flowing at velocities up to 9 km/h (5.6 mi/h). The western margin of the Gulf Stream can frequently be defined as a rather abrupt boundary that moves periodically closer to and farther away from the shore. The eastern boundary becomes very difficult to identify, as it is usually masked by filamentous meandering masses that continuously change their positions. Its character gradually merges with that of the water of the Sargasso Sea to the south and east of the Gulf Stream. A volume transport of over 90 sv off Chesapeake Bay indicates a large volume of Sargasso Sea water has been added to the flow provided by the Florida Current. However, south of Newfoundland, the volume carried by the Gulf Stream has been reduced to 40 sv. This lower volume would indicate that most of the Sargasso Sea water that joined the Florida Current and made up the Gulf Stream has returned to the diffuse flow of the Sargasso Sea.

CURRENTS

A—Antillean	G—Guinea
Bg—Benguela	GS—Gulf Stream
Br—Brazil	I—Irminger
C—Canary	L—Labrador
CC—Caribbean	NA—North Atlantic
EG—East Greenland	NE—North Equatorial
EW—East Wind Drift	N—Norwegian
EC—Equatorial Counter	SE—South Equatorial
Fa—Falkland	WW—West Wind Drift
F—Florida	

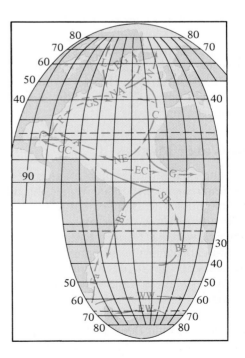

FIGURE 8–15 Atlantic Ocean Surface Currents.
Base map courtesy of National Ocean Survey.

Just how this loss of water occurs is yet to be determined, but much of it may be achieved by meanders (sinuous curves in the course of a current) pinching off to form large eddies. As is shown in figure 8–16, meanders north of the Gulf Stream pinch off and trap warm Sargasso Sea water in eddies rotating in a clockwise direction. These eddies move at speeds of 3 to 7 km/day (1.9 to 4.3 mi/day) southwest toward Cape Hatteras, where they rejoin the Gulf Stream. South of the Gulf Stream and west of 50°W longitude, eddies with a core of cold inshore water rotating counterclockwise move south and west. (See "Gulf Stream Rings and Marine Life," p. 336.) These large eddies, ranging in diameter from 100 to 500 km (62 to 310 mi) and extending to depths of 1 km (0.62 mi) in the warm rings and as much as 3.5 km (2.2 mi) in the cold rings, could remove large volumes of water from the Gulf Stream.

At the Tail of the Banks, about 40°N latitude and 45°W longitude, the Gulf Stream becomes the **North Atlantic Current** and continues in a more easterly direction across the North Atlantic. The North Atlantic Current is composed of numerous branches into which the Gulf Stream flow is ultimately broken.

Two branches that are composed of water produced through the mixing of the **Labrador Current** and the Gulf Stream are the **Irminger Current**, flowing up along the west coast of Iceland, and the **Norwegian Current**, which moves north along the coast of Norway. The other major branch of the North Atlantic Current follows 45°N latitude across the North Atlantic and turns south as the **Canary Current**, passing between the Azores and Spain. This southward flow is very diffuse and spreads over a broad area as it moves southward and eventually joins the North Equatorial Current.

Deep-Water Masses

Surface and near-surface water masses are underlain by the Antarctic Intermediate Water throughout the South Atlantic and to some extent in the North Atlantic. This intermediate water mass begins to lose its characteristics in the North Atlantic, and we find that **Mediterranean Water** that has passed into the Atlantic Ocean over the Gibralter Sill underlies much of the Central Water Mass in the North Atlantic. Figure 8–17 shows the temperature, salinity, and dissolved oxygen characteristics of the Atlantic water masses.

The North Atlantic Deep Water, transporting 14 sv, can still be identified as far south as 40°S latitude on the basis of temperature and salinity maxima ob-

served at depths around 3000 m (9840 ft). Antarctic Bottom Water, transporting 7 sv, can be identified as far north as 40°N latitude by its low temperature and salinity and high oxygen content. The Mediterranean Water is most easily identified by its salinity maximum at a depth between 1000 and 2000 m (3280 and 6560 ft). The Antarctic Bottom Water moves northward along the western margin of the Atlantic Ocean in the West Atlantic Basin, which is separated from the East Atlantic Basin by the Mid-Atlantic Ridge. As a result of this movement, the bottom temperatures in the West Atlantic Basin are maintained well below 2°C (35.6°F), while they are around 2.5°C (36.5°F) in the bottom of the East Atlantic Basin.

Although the exact volumes involved are difficult to determine, it is evident that there is a large transfer of water at depth across the equator between the North and South Atlantic. The residence time of deep water in the Atlantic Ocean is approximately 275 years. It appears that all of the deep and bottom-water masses throughout the world ocean originate in the Atlantic Ocean.

Changes in the Deep Ocean

During a study of deep-ocean circulation in 1981, observations in the North Atlantic showed that below a depth of a few hundred meters water between 50° and 60°N latitude was 0.02‰ less salty and 0.15°C (0.27°F) colder than it was in 1972. This finding flew in the face of a long-held conception of a deep ocean that changed significantly only on a scale of thousands of years. Observations conducted as part of the Transit Tracers in the Ocean Program (TTO) were published in 1982 and showed changes at 24° and 36°N latitude in the Atlantic Ocean compared to 1958. The water from 500 to 3000 m (1640 to 9840 ft) is 0.2°C (0.36°F) warmer. Above and below this warmer water, the water is now colder than in 1958.

The cooling and freshening of the 50°–60°N latitude waters may result from an increased rate of sinking of North Atlantic Deep Water during the 1970s. The temperature changes observed at 24° and 36°N latitude may be more closely related to changing rates of formation of intermediate- and bottom-water masses in the Antarctic.

These changes could result from atmospheric changes in the 1970s, a north-south sloshing of water back and forth in the Atlantic basin, or a number of other possible causes. Although the changes seem small, they are significant and need to be understood. Only a full understanding of ocean circulation may

FIGURE 8–16 Sea-Surface Temperatures in Northwest Atlantic Ocean.

A, sea-surface temperature data gathered by a NOAA satellite were processed to produce this false-color image of the northwest Atlantic Ocean. The warm Gulf Stream waters are shown as orange and red. The colder nearshore waters are blue and purple. Warm water from south of the Gulf Stream is transferred to the north as warm core rings (yellow) surrounded by cooler (blue and green) water. Cold nearshore water spins off to the south of the Gulf Stream as cold core rings (green) surrounded by warmer (yellow and red) water.

The rings form when meanders close to trap warm or cold water within them. The warm rings contain shallow, bowl-shaped masses of warm water about 1 km (0.62 mi) deep with diameters of about 100 km (62 mi). The cold rings have cones of cold water that extend to the ocean floor. They may be more than 500 km (310 mi) across at the surface. The diameter of the cone increases with depth. Image courtesy of Dr. Charles McLain/Rosensteil School of Marine and Atmospheric Science, University of Miami. *B,* this diagram is an overlay of the satellite image in *A.* It attempts to clarify the process of ring formation by showing the core of the Gulf Stream flow. It shows the warm ring and three cold rings as having formed by the pinching off of meanders. The cold rings have separated from the main Gulf Stream flow, but the warm ring has just formed as the meander that flowed around it pinches off from the main flow.

A.

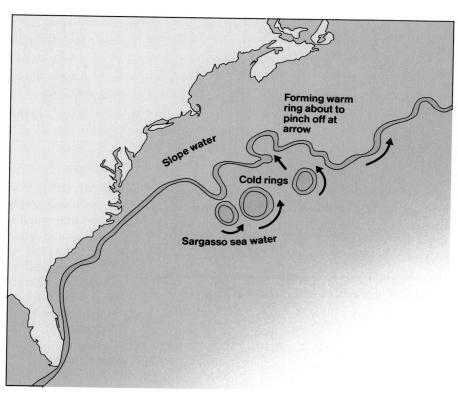

B.

FIGURE 8–17 Atlantic Ocean Subsurface Circulation, as Evidenced by Physical Properties of Water.

From Sverdrup, Johnson & Fleming, *The oceans: their physics, chemistry, and general biology,* © 1942, Renewal © 1970. Reprinted by permission of Prentice-Hall, Inc., Englewood Cliffs, New Jersey.

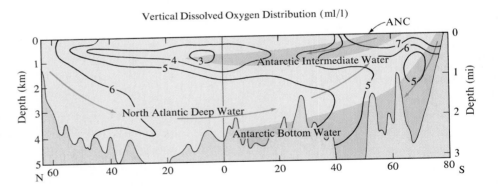

provide the explanation of these changes. We appear to be approaching such an understanding, but it may still be many years away.

PACIFIC OCEAN CIRCULATION

Surface Circulation

The surface circulation in the Pacific Ocean is generally similar to that which we described for the Atlantic Ocean, except that the Equatorial Countercurrent is much better developed in the Pacific Ocean than in

the Atlantic (fig. 8–18). We will first discuss the counterclockwise rotation in the South Pacific and the flow of an undercurrent present within the system.

Southeast trade winds drive the South Equatorial Current, which has a lateral dimension from about 10°S to 3°N latitude. Embedded within the South Equatorial Current and flowing in an easterly direction along the equator is a thin ribbonlike **Equatorial Undercurrent.** This undercurrent extends for over 6000 km (3726 mi) from the western Pacific to the Galápagos Islands. Only 0.2 km (0.12 mi) thick and approximately 300 km (186 mi) wide, the Equatorial Undercurrent at depths of 70 to 200 m (230 to 656 ft) has a volume transport of approximately 40 sv.

The South Equatorial Current becomes the western boundary current of the South Pacific gyre—the **East Australian Current**. This current, which is relatively weak for a western boundary current, joins the West Wind Drift that carries the water across the southern Pacific to be pushed up along the western coast of South America as the **Peru Current**, which completes the gyre by turning west near the equator to join the South Equatorial Current.

In the North Pacific, the North Equatorial Current flows from east to west between latitudes of 8° and 20°N. As the North Equatorial Current approaches the western boundary of the North Pacific, most of the water moves north to form the **Kuroshio Current**, the North Pacific counterpart of the Florida Current in the North Atlantic. Part of the North Equatorial Current water turns south and joins the Equatorial Countercurrent. After the Kuroshio Current, which has a maximum velocity of just under 9 km/h (5.6 mi/h), leaves the coast and begins to flow in an easterly direction, it becomes the **Kuroshio Extension** to a longitude of about 170°E. This segment of the North Pacific gyre corresponds to the Gulf Stream of the North Atlantic and has a volume transport of about 65 sv.

The Kuroshio Extension is met by the cold **Oyashio Current**, which carries water south from the Bering Sea and the Sea of Okhotsk. Beyond this juncture the flow becomes the **North Pacific Current** that moves toward the North American continent, where it divides. Part of the North Pacific Current turns southward as the **California Current**, which completes the gyre by joining the North Equatorial Current. The remainder flows north to become the **Alaskan gyre**.

Deep-Water Masses

Surface-temperature distribution with regard to latitude in the Pacific Ocean is similar to that observed in the Atlantic Ocean, but surface salinity in the Pacific is lower than in the Atlantic, with the South Pacific water being slightly more saline than that in the North Pacific (see fig. 8–9).

The Pacific Ocean bottom water is a mixture of the North Atlantic Deep Water and Antarctic Bottom Water. It is called the **Oceanic Common Water** and is introduced into the Pacific by the Antarctic Circumpolar Current. It moves very slowly toward the north with a volume transport of about 25 sv. The northward movement of this Antarctic mass can be identified by observing the temperature of the bottom water, which increases gradually from south to north. There is also a slight decrease in salinity and dissolved oxygen content from south to north (fig. 8–19).

In contrast to the condition of high volume interchange in the Atlantic Ocean, the rates and volume of water transport between the North and South Pacific are so low that it is difficult to accurately measure them. One of the main criteria for concluding that the deep waters in the Pacific Ocean move at very low rates and are old is the relatively low dissolved oxygen content of these waters. Compared to the Atlantic deep-water masses, where dissolved oxygen content falls within the range of 4.5 to 6.5 mL/L, the Pacific deep-water dissolved oxygen content ranges between 3.5 and 4.5 mL/L (fig. 8–19). The residence time of deep water in the Pacific Ocean is 510 years.

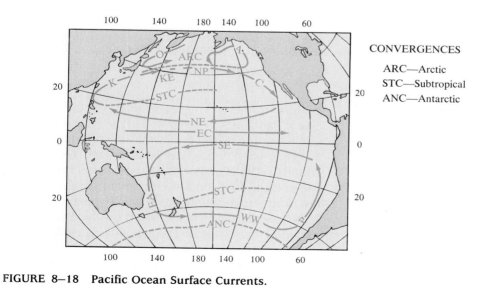

CONVERGENCES

ARC—Arctic
STC—Subtropical
ANC—Antarctic

CURRENTS

A—Alaskan
C—California
EA—East Australian
EC—Equatorial Counter
K—Kuroshio
KE—Kuroshio Extension
NE—North Equatorial
NP—North Pacific
O—Oyashio
P—Peru
SE—South Equatorial
WW—West Wind Drift

FIGURE 8–18 Pacific Ocean Surface Currents.
Base map courtesy of Open University Press.

FIGURE 8–19 Pacific Ocean Subsurface Circulation as Evidenced by Physical Properties of Water.

From Sverdrup, Johnson & Fleming, *The oceans: their physics, chemistry, and general biology*, © 1942, Renewal © 1970. Reprinted by permission of Prentice-Hall, Inc., Englewood Cliffs, New Jersey.

NPIW–North Pacific Intermediate Water   CW—Common Water

AAIW—Antarctic Intermediate Water

INDIAN OCEAN CIRCULATION

Surface Circulation

Surface circulation in the Indian Ocean, which extends only to about 20°N latitude, varies considerably from that which we have discussed for the Atlantic Ocean and Pacific Ocean. From November to March the equatorial circulation is similar to that of other oceans, with two westward-flowing equatorial currents separated by an **Equatorial Countercurrent**. In contrast to the wind systems in the Atlantic and Pacific oceans, which are shifted to the north of the geographical equator, we find that in the Indian Ocean the ITCZ is shifted to the south of the geographical equator. The Equatorial Countercurrent flows between 2° and 8°S latitude, bounded on the north by the North Equatorial Current, which extends as far as 10°N latitude, and on the south by the South Equatorial Current, which extends to 20°S latitude. During the Northern Hemisphere winter the typical northeast trade winds are developed and are referred to as the **northeast monsoon**. They are reinforced because rapid cooling of the air during the winter months over the Asian mainland has created a high-pressure cell that is forcing atmospheric masses off the continent out over the ocean, where the air pressure is lower (fig. 8–20).

The Asian mainland warms up faster than the oceanic water owing to the relatively lower heat capacity of continental crustal material compared to that of water. As a result, a low-pressure cell develops over the continent during summer and "sucks" in the air masses overlying the ocean. This force gives rise to the southwest monsoon, which may be thought of as a continuation of the southeast trade winds across the equator. During this season the North Equatorial Current disappears and is replaced by the Southwest Monsoon Current, which flows from west to east across the North Indian Ocean. In September or October, the northeast trade winds are re-established, and the North Equatorial Current reappears (fig. 8–20).

The surface circulation in the southern Indian Ocean is similar to the counterclockwise circulation observed in other southern oceans. During the time the northeast trade winds blow, the South Equatorial Current provides water for the Equatorial Countercurrent and the Agulhas Current, which flows south along the eastern coast of Africa and joins the West Wind Drift. Turning north out of the West Wind Drift is the West Australian Current, which completes the gyre by merging with the South Equatorial Current.

During the southwest monsoon, a northward flow from the equator along the coast of Africa, the Somali Current, develops with velocities approaching 4 km/h (2.5 mi/h).

Deep-Water Masses

Directly beneath the surface and near-surface water is the Antarctic Intermediate Water, identifiable by its low temperature and high dissolved oxygen content (fig. 8–21). The Red Sea Water is a high-salinity mass with low oxygen content that is fed by outflow of the Red Sea and descends to a depth between 1000 and 1500 m (3280 and 4920 ft) in the northern Indian Ocean.

The Oceanic Common Water of the Indian Ocean has an average temperature of 1.5°C (34.7°F) and an average salinity of 34.7‰. It enters the Indian Ocean at a rate of 20 sv, which results in a residence time for deep water in the Indian Ocean of 250 years.

FIGURE 8–20 Indian Ocean Surface Currents.
Base map courtesy of National Ocean Survey.

A. WINTER: November–March, Northeast monsoon wind season

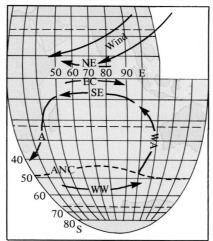

B. SUMMER: May–September, Southwest monsoon wind season

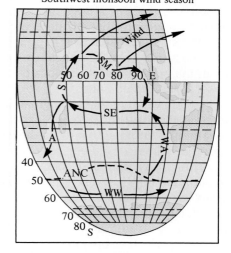

From November to March equatorial circulation in the Indian Ocean is similar to that of other oceans, except that the Equatorial Countercurrent is shifted south.

Low pressure over the mainland during summer draws the Southwest Monsoon winds over the North Indian Ocean. This wind produces the Southwest Monsoon Current which flows in an easterly direction and replaces the North Equatorial Current.

CURRENTS   A—Agulhas   EC—Equatorial Counter   NE—North Equatorial   S—Somali
SE—South Equatorial   SM—Southwest Monsoon   WA—West Australian   WW—West Wind Drift
ANC—Antarctic Convergence

FIGURE 8–21   Indian Ocean Subsurface Circulation as Evidenced by Physical Properties of Water.
From A. J. Clowes and G. E. R. Deacon, *The deep-water circulation of the Indian Ocean*, © 1935. Reprinted by permission of Nature, Macmillan Journals, and G. E. R. Deacon.

POWER FROM WINDS AND CURRENTS

The winds that transform solar energy into ocean currents have for centuries been used by society to drive its machines. Westerly winds off New England represent a near-shore wind supply that is reasonably sustained and contains a large amount of energy. There is some optimism that Offshore Windpower Systems (OWPS) such as that shown in figure 8–22 could provide electricity or hydrogen generated by electricity to meet the needs of large sections of the North Atlantic coast of the United States.

Many have considered the great amount of energy in the Florida–Gulf Stream Current System and dreamed of harnessing it to do the work of society. A group of scientists and engineers met in 1974 to consider this possibility and concluded that some 2000 megawatts (MW) of electricity could be recovered along the east coast of southern Florida. Devices proposed for extraction range from underwater "windmills" to a Water Low Velocity Energy Converter (WLVEC), which is operated by parachutes attached to a continuous belt. Again, calculations indicate such systems can be economically competitive. The Coriolis Program was proposed in 1973 by William J. Moulton of Tulane University. Hydroturbines with diameters of 170 m (558 ft) are envisioned that could generate 43 MW of electricity from the movement of ocean currents (fig. 8–23). An array of 242 units covering an area 30 km (18.6 mi) wide and 60 km (37 mi) long could generate 10,000 MW—the equivalent

of 130 million barrels of oil. Tests are now being conducted to establish an optimum size, identify suitable sites, and confirm engineering, economic, and environmental estimates. What remains is to see if the designs for ocean wind and current generation of electricity actually function as expected in the sometimes hostile marine environment.

FIGURE 8–22   Offshore Windpower System.
Courtesy of Woods Hole Oceanographic Institution.

FIGURE 8–23 Coriolis Program.

Buoyant Coriolis hydroturbines anchored in ocean current. It is estimated that an array of 242 such units placed in the Gulf Stream could provide 10% of the present electricity needs of Florida. Courtesy of U.S. Department of Energy.

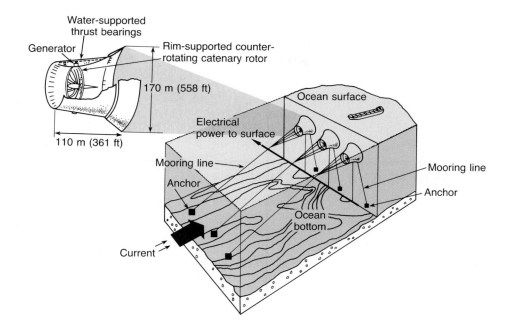

SUMMARY

Horizontal currents set in motion by wind systems are characteristic of the surface waters of the world ocean. According to the **Ekman spiral** concept, winds will set the surface waters in motion in a direction 45° to the right of the wind in the Northern Hemisphere, and the net water movement will be at right angles to the wind direction. As a result of the Ekman spiral phenomenon, water is pushed in toward the center of clockwise gyres in the Northern Hemisphere and counterclockwise gyres in the Southern Hemisphere, forming "hills." As water in the Northern Hemisphere runs downslope on the hills of water that result, the Coriolis effect causes it to turn right and into the clockwise flow pattern. As a result, a **geostrophic current** flowing parallel to the contours of the hill is maintained. The apex of the hill is always located to the west of the geographical center of the gyre. Using **dynamic topography**, oceanographers can map the surface currents in ocean basins by determining the average density of water columns at locations throughout the basin and mapping this distribution. The speed of the current will increase with increased slope of the hills, which will normally be steepest on the west side.

Vertical circulation produced by wind will be confined to the surface-water mass. An example is equatorial or coastal **upwelling**, which is created by the **Ekman transport** pulling wind-driven water masses away from the equator or coastal areas. Water will surface from a depth of no more than a few hundred meters to replace the surface water. An even shallower vertical movement of water is represented by the **Langmuir circulation** of long, shallow convection cells running parallel to the gentle, steady winds that create them. These cells may extend no deeper than 6 m (19.7 ft). The mixing of the water in the deepest ocean basins is caused by **thermohaline circulation**, which is initiated by the sinking of dense surface water in high latitudes. The **North Atlantic Deep Water** and the **Antarctic Bottom Water** carry most of the oxygen-bearing water to the deep basins of all the oceans to maintain their life-sustaining capacity.

The **Circumpolar Current** flows in a clockwise direction around the continent of Antarctica. It is the largest current in the world ocean, transporting over 130 **sverdrups** of water. The surface portion of this flow is often referred to as the **West Wind Drift** because it is driven by strong westerly winds. Two major deep-water masses form in Antarctic waters. The Antarctic Bottom Water, the densest water mass in the oceans, forms primarily in the **Weddell Sea** and sinks along the continental shelf into the Atlantic Ocean. Farther north, at the **Antarctic Convergence**, the low-salinity **Antarctic Intermediate Water** sinks to a depth of about 900 m (2952 ft). Sandwiched between these two masses is the North Atlantic Deep Water, rich in plant nutrients after hundreds of years in the deep ocean. The highest-velocity current in the oceans is the **Gulf Stream** as it flows along the southeast Atlantic coast. The best development of an equatorial countercurrent system is found in the Pacific, where large eastern movement of water is achieved by the **Pacific Countercurrent** and the **Equatorial Undercurrent**, a ribbonlike subsurface flow. The strong **Kuroshio Current** is the North Pacific counterpart of the Gulf Stream. The formation of deep-water masses is not pronounced in the Pacific Ocean, as there appears to be no deep sinking of surface water. The Indian Ocean circulation is dominated by the **monsoon** wind systems that blow out of the northeast in the winter and the southwest in the summer.

Continued investigation of the energy potential of ocean winds and currents may lead to development of a significant addition of **renewable** energy that can be exploited by society.

QUESTIONS AND EXERCISES

1. Compare the forces directly responsible for creating horizontal and deep vertical circulation in the oceans. What is the ultimate source of energy for driving both circulation systems?

2. On a base map of the world, plot the major surface circulation gyres of the oceans, the meteorological equators of each ocean, and the Subtropical, Arctic, and Antarctic Convergences. Superimpose the major wind belts of the world on the gyres. Label the currents using the symbols used in figures 8–15, 8–18, and 8–20.

3. Diagram and discuss how the Ekman transport produces the "hill" of water within major ocean gyres that causes geostrophic current flow. As a starting place on the diagram, use the prevailing wind belts, the trade winds, and the westerlies.

4. What causes the apex of these geostrophic "hills" to be offset to the west of the center of the ocean gyre systems?

5. Explain how oceanographers determine the dynamic topography of the ocean waters.

6. Describe the relationships among the wind, the surface current it creates, and the development of equatorial and coastal upwellings.

7. Discuss why thermohaline vertical circulation is driven by sinking of surface water that occurs only in high latitudes.

8. Name the two major deep-water masses and give the locations of their formation at the ocean's surface.

9. The largest current in the world ocean in terms of volume transport is the Antarctic Circumpolar Current. Explain why its surface portion is referred to as the West Wind Drift. What is its maximum volume transport as compared to the maximum volume transport measure for the Gulf Stream?

10. Why does the North Atlantic Deep Water enhance biological productivity in the Antarctic Zone as it surfaces between the Antarctic Intermediate Water and the Antarctic Bottom Water?

11. Observing the flow of Atlantic Ocean currents in figure 8–15, compare the Brazil Current to the Gulf Stream and offer some possible explanations for the fact that the Brazil Current has a much lower velocity and volume transport than the Gulf Stream.

12. Explain why the Gulf Stream eddies that develop northeast of the Gulf Stream rotate clockwise and have warm-water cores while those that develop to the southwest rotate counterclockwise and have a core of cold water.

13. The Antarctic Intermediate Water is identifiable throughout much of the South Atlantic on the basis of a temperature minimum, salinity minimum, and dissolved oxygen maximum. Why should it be colder, less salty, and contain more oxygen than the surface-water mass above it and the North Atlantic Deep Water below it?

14. What evidence is there that the bottom water in the Pacific Ocean has been away from the surface much longer than that of the Atlantic Ocean?

15. Discuss the changes in current flow in the North Indian Ocean and their relationship to the monsoon winds.

16. Where is the potential for using ocean windpower systems and ocean current power systems greatest, and why?

REFERENCES

Alayne, F., Street-Perrott, and Perrott, R. 1990. Abrupt climate fluctuations in the tropics: The influence of Atlantic Ocean circulation. *Nature* 343:6259, 607–12.

Gill, A. 1982. *Atmosphere-ocean dynamics*. Orlando, Florida: Academic Press.

Goldstein, R., Barnett, T., and Zebker, H. 1989. Remote sensing of ocean currents. *Science* 246:4935, 1282–86.

MacLeish, W., ed. 1974. Energy and the sea. *Oceanus* 17:52.

Montgomery, R. 1940. The present evidence of the importance of lateral mixing processes in the ocean. *American Meteorological Society Bulletin* 21:87–94.

Pedlosky, J. 1990. The dynamics of the oceanic subtropical gyres. *Science* 248:4935, 316–22.

Pickard, G. L. 1975. *Descriptive physical oceanography.* 2nd ed. New York: Pergamon Press.

Stommel, H. 1987. *A view of the sea.* Princeton, N.J.: Princeton University Press.

————.1958. The abyssal circulation. Letters to the editor. *Deep Sea Research.* 5. New York: Pergamon Press.

Stuiver, M.; Quay, P. D.; and Ostlund, H. G. 1983. Abyssal water carbon-14 distribution and the age of the world oceans. *Science* 219:4586, 849–51.

Sverdrup, H. U.; Johnson, M. W.; and Fleming, R. H. 1942. Renewal 1970. *The oceans.* Englewood Cliffs, N.J.: Prentice Hall.

SUGGESTED READING

Sea Frontiers

Frye, J. 1982. The ring story. 28:5, 258–67.
A discussion of the Gulf Stream, its rings, and how the rings are studied.

Miller, J. 1975. Barbados and the island-mass effect. 21:5, 268–72.
A discussion of the phenomenon by which waters around tropical islands are much more productive than the surface waters of the open ocean.

Smith, F. G. W. 1972. Measuring ocean movements. 18:3, 166–74.
Discussed are some practical problems related to current flow and some methods used to determine current direction, speed, and volume.

Smith, F. G. W., and Charlier, R. 1981. Turbines in the ocean. 27:5, 300–305.
A discussion of the potential of extracting energy from ocean currents is presented.

Sobey, E. 1982. What is sea level? 28:3, 136–42.
The factors that cause sea level to change are discussed.

Scientific American

Baker, D. J., Jr. 1970. Models of ocean circulation. 222:1, 114–21.
This article discusses observations of a model depicting a segment of the surface of the earth over which fluids move and helps explain geostrophic flow within ocean gyres. Some knowledge of basic physics is necessary for full comprehension of material presented.

Hollister, C. D., and Nowell, A. 1984. The dynamic abyss. 250:3, 42–53.
Submarine "storms" are associated with deep, cold currents flowing away from polar regions toward the equator.

Munk, W. 1955. The circulation of the oceans. 191:96–108.
A young physical oceanographer who has gone on to be highly honored by the scientific community presents his views of ocean circulation.

Stewart, R. W. 1969. The atmosphere and the ocean. 221:3, 76–105.
The exchange of energy between the atmosphere and ocean and the resulting phenomena, currents, and waves are covered in this readable, comprehensive article.

Stommel, H. 1955. The anatomy of the Atlantic. 190:30–35.
Another vintage article by a young oceanographer who went on to great achievements.

Webster, P. J. 1981. Monsoons. 245:5, 108–19.
The mechanism of the monsoons and their role in bringing water to half the earth's population is discussed.

Weibe, P. H. 1982. Rings of the Gulf Stream. 246:3, 60–79.
The biological implications of the large cold-water rings are considered.

9
WAVES

Waves are one of the most obvious phenomena of the ocean. Yet it was not until well into the nineteenth century that some understanding of what caused waves and how they behaved was developed. We will first look at the character of waves in general before proceeding to discuss oceanic waves. Wave phenomena, with the exception of electromagnetic waves, involve the transmission of energy and momentum by means of vibratory impulses through the various states of matter. Theoretically, the medium itself does not move in the direction the energy passes through. The particles that make up the medium simply oscillate in a back-and-forth or orbital pattern, transmitting energy from one particle to another.

Simple **progressive waves** (a phenomenon in which the wave form can be observed to move), shown in figure 9–1A, may be described as **longitudinal** and **transverse**. In longitudinal waves, such as sound waves, the particles that are in vibratory motion move back and forth in a direction parallel to the propagation of energy. Energy may be transmitted through all states of matter—gaseous, liquid, or solid—by longitudinal movement of particles.

Transverse wave phenomena involve the propagation of energy at right angles to the direction of particle vibration. An example of such a wave would be that created when one end of a rope is tied to a doorknob while the other end is moved up and down with the hand. A waveform is set up and progresses along the rope, and energy is transmitted from the motion of the hand to the doorknob. If one were to pick out a particular segment of the rope and watch it, the segment would be seen to move up and down at right angles to a line drawn from the doorknob to the hand that is putting energy into the rope. We generally consider that this type of wave can transmit energy only through solids, since it is only in a solid that particles are strongly attached to one another. Longitudinal and transverse waves are called **body waves**, as they transfer energy through a body of matter.

Waves on the ocean surface, as do all waves that transmit energy along an interface between any two fluids of different density, have particle movements that are neither longitudinal nor transverse. We may say more correctly that the movement of particles along such an interface involves components of both, since the particles move in circular orbits at the interface between the atmosphere and ocean. They are **orbital waves**, or **interface waves**.

Preceding page: Boats swept ashore by Hurricane Hugo, Charleston, South Carolina, September 1989. Photo from AP/Wide World Photos.

WAVE CHARACTERISTICS

When we observe the ocean surface, we see that there are waves of various sizes moving in various directions, resulting in a complex wave pattern that is constantly changing. We will, for the sake of introducing some characteristics that must be used to discuss waves, look at a much simpler form of an idealized wave representing the transmission of energy created by a single source and traveling along the ocean-atmosphere interface. Such a series of waves will have very uniform characteristics. We use such an idealized form in figure 9–1B.

As an idealized progressive wave passes a permanent marker, such as a pier piling, we will notice a succession of high parts of the waves, **crests**, separated by low parts, **troughs**. If we were to mark the water level on the piling when the troughs pass and then do the same for the crests, the vertical distance between the marks would be the **wave height**, H. The horizontal distance between corresponding points on successive waveforms, such as from crest to crest, is the **wavelength**, L. The ratio of H/L is **wave steepness**. The time that elapses during the passing of one wavelength is the **period**, T. Because the period is the time required for the passing of one wavelength, if either the wavelength or period of a wave is known, the other can be calculated, since $L(m) = 1.56 (m/s)T^2$ (fig. 9–2):

$$\text{Speed } (S) = \frac{L}{T}$$

For example:

$$\text{Speed } (S) = \frac{L}{T} = \frac{156 \text{ m}}{10 \text{ s}} = 15.6 \text{ m/s}$$

Another characteristic related to wavelength and speed is **frequency**, f. Frequency is the number of wavelengths that pass a fixed point per unit of time and is equal to $1/T$. If 6 wavelengths pass a point in 1 min, and we have the same wave system as in our previous example, then:

$$\text{Speed } (S) = Lf = 156 \text{ m} \times \frac{6}{\text{min}} = 936 \text{ m/min}$$

$$\frac{936 \text{ m}}{1 \text{ min}} \times \frac{1 \text{ min}}{60 \text{ s}} = 15.6 \text{ m/s}$$

Since the speed and wavelength of ocean waves are such that less than one wavelength passes a point per second, the preferred unit of time for scientific measurements, **period** (rather than frequency), is the

LONGITUDINAL WAVE
Particles (color) move back and forth in direction of energy transmission. These waves transit energy through all states of matter.

A.

TRANSVERSE WAVE
Particles (color) move back and forth at right angles to direction of energy transmission. These waves transmit energy only through solids.

ORBITAL WAVE
Particles (color) move in orbital path. These waves transmit energy along interface between two fluids of different density (Liquids and/or gases).

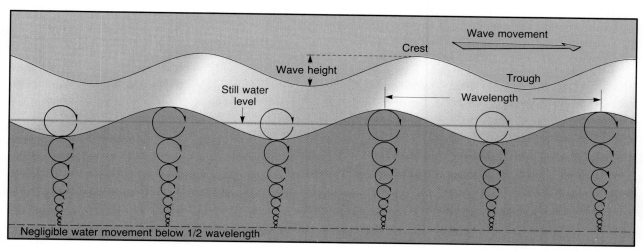

B.

FIGURE 9–1   Types of Progressive Waves.

A, types of progressive waves. B, simple sea wave and its parts. B from the Tasa Collection: *Shorelines.* Published by Merrill Publishing, Columbus, OH. Copyright © 1986 by Tasa Graphic Arts, Inc. All rights reserved.

FIGURE 9–2   Speed— Deep-Water Waves.

The theoretical relationships between wave speed, wavelength, and period for deep-water waves. Speed is equal to wavelength divided by period. $L(m) = 1.56T^2$

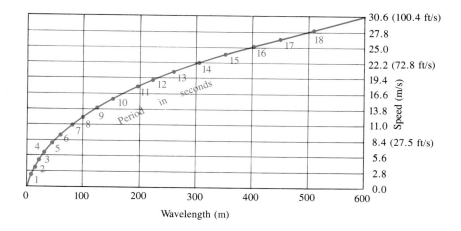

more practical measurement to use when calculating speed.

Returning to the particle motion of ocean waves, the circular orbits followed by the water particles at the surface have a diameter equal to the wave height. While a particle is in the crest of a passing wave, it is moving in the direction of energy propagation. While it is in the trough, it is moving in the opposite direction. The half of the orbit that is accomplished in the trough is at a lower speed than the crest half of the orbit. Therefore, there is a small net transport of water in the direction the waveform is moving. This condition results from the fact that particle speed decreases with increasing depth below the still-water line. Also, the diameters of particle orbits decrease with increased depth until particle motion associated with our idealized wave ceases at a depth of one-half wavelength, $L/2$.

Deep-Water Waves

The ocean waves with the characteristics just discussed belong to a category called **deep-water waves**.

Such waves travel across the ocean where the water depth *(d)* is greater than one-half the wavelength (fig. 9–3A). Included in deep-water waves are all wind-generated waves as they move across the open ocean. As can be seen by the speed calculations previously discussed, speed of deep-water waves is related to wavelength *(L)* [S (m/s) $= 1.25 \sqrt{L(m)}$] and period *(T)*. The easier of these characteristics to measure is period. Thus the following equation is most commonly used for computing wave speed (*g* is acceleration due to gravity).

$$\text{Speed (m/s)} = \frac{gT}{2\pi} = \frac{9.8 \text{ m/s}^2 \times T \text{ (s)}}{2 \times 3.14} = 1.56T$$

Shallow-Water Waves

Waves in which $d < L/20$ are classified as **shallow-water waves** (long waves). Included in this category are wind-generated waves that have moved into shallow nearshore areas; **tsunami** (seismic sea waves), generated by disturbances in the ocean floor; and **tide waves**, generated by gravitational attraction of the

A. DEEP-WATER WAVE   Wave profile and water-particle motions of a deep-water wave. Note the diminishing size of the orbits with increasing depth below the surface.

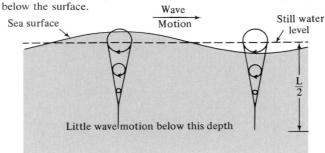

B. SHALLOW-WATER WAVE Motions of water particles in shallow-water waves. Water motion extends to ocean floor.

C. RELATIONSHIP OF WAVELENGTH TO WATER DEPTH

FIGURE 9–3   Deep-Water and Shallow-Water Waves.

sun and moon (fig. 9–3B). In these waves, the wave-length is very great relative to water depth, and speed is determined by water depth, d:

$$\text{Speed (m/s)} = \sqrt{gd} = 3.1 \sqrt{d(m)}$$

Particle motion in shallow-water waves is in a very flat elliptical orbit approaching horizontal oscillation, and the vertical component of particle motion decreases with increasing depth. The presence of shallow-water waves can be detected to the ocean bottom and therefore can affect the bottom.

Transitional Waves

Transitional waves have wavelengths greater than twice, but less than 20 times the water depth. The speed of transitional waves is determined partially by wavelength and partially by water depth (fig. 9–4). Deep-water waves generated by winds at the ocean surface usually have periods of 10–12 s. They maintain their periods after encountering shallow coastal water.

Since it is easier to measure the period of a wave than any of its other characteristics, Kinsman has used the system presented in figure 9–5 as a means

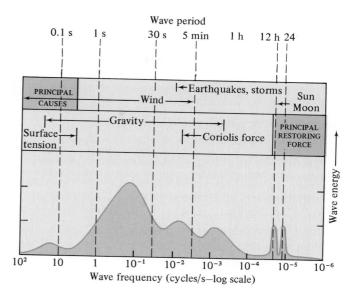

FIGURE 9–5   Distribution of Energy in Ocean Waves.
This figure shows that most of the energy possessed by ocean waves is in wind-generated waves with a period of about 10 seconds. The long period peak above the 10^{-3} represents tsunami, while the two sharp peaks to the right represent the tides with semidaily and daily periods. After Blair Kinsman, *Wind waves: Their generation and propagation on the ocean surface*, © 1965. Reprinted by permission of Prentice-Hall, Inc., Englewood Cliffs, New Jersey.

of showing the concentration of energy in ocean waves with periods ranging from less than 0.1 s to more than 1 day. The illustration also shows the principal causes of waves of different periods.

WIND-GENERATED WAVES

Sea

As the wind blows over the ocean surface, the pressure and stress deform the ocean surface into small rounded waves with wavelengths less than 1.74 cm (0.7 in.). These waves are called **capillary waves** (ripples) because the dominant **restoring force** that works to destroy them and smooth the ocean surface is surface tension (capillarity). Capillary waves characteristically have rounded crests and V-shaped troughs (fig. 9–6).

As capillary wave development increases, the sea surface takes on a rougher character, which allows the wind and ocean surface to interact more efficiently. As more energy is transferred to the ocean, **gravity waves** with lengths over 1.74 cm (0.7 in.) develop with a shape more like that of a sine curve. Because they reach sufficient height at this stage, the force of gravity becomes the dominant restoring force. The length of "young" waves in a sea is generally 15–35 times their height. As additional energy is

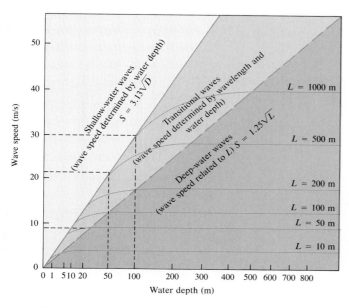

FIGURE 9–4   Wave Speed.
To determine the wave speed of shallow-water waves, move up from the water depth to the line marked shallow-water waves, then over to the wave-speed scale [*see* example for 50-m (164-ft) water depth]. To determine the wave speed of deep-water waves, continue the wavelength (L) line to the left until it intersects the wave-speed scale [*see* example for 50-m wavelength]. Combine these procedures for transitional waves (colored lines) [*see* example for 1000-m (3280-ft) wavelength and 100-m (328-ft) water depth]. Note that the transitional waves have a lower wave speed than they would have had as deep-water waves.

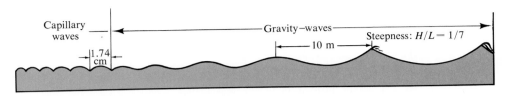

FIGURE 9–6   Capillary and Gravity Waves.
As energy is put into the ocean surface by wind, small rounded waves with V-shaped troughs develop (capillary waves). As the water gains energy, the waves increase in height and length. When they exceed 1.74 cm (0.7 in.) in length, they take on the shape of the sine curve and become gravity waves. Increased energy increases the steepness of the waves. The crests become pointed and the troughs rounded. As the steepness reaches 1/7, the waves become unstable, and whitecaps form as they "break." Not drawn to scale.

gained, the wave height increases more rapidly than wavelength, and the crests become pointed, while the troughs are rounded. When the steepness reaches 1/7, the velocity with which the waves travel is 1.2 times that of typical deep-water waves, or $S = 1.56T \times 1.2 = 1.87T$.

Energy imparted by the wind increases the magnitude of height, length, and speed. When the wave speed reaches that of the wind, neither of these characteristics can change because there is no net energy exchange, and the wave is at its maximum size.

The area in which wind-driven waves are generated is called **sea** and is characterized by a choppy, short wave structure with waves moving in many directions and having many different periods and lengths. The variety of wave periods and wavelengths is caused by frequently changing wind speed and direction. The factors that are important in increasing the amount of energy that waves obtain are (1) wind speed, (2) the duration—the time during which the wind blows in one direction, and (3) the **fetch**—the distance over which the wind blows in one direction.

Figure 9–7 shows the relationship among wind direction, fetch, and the sea.

Wave height is a characteristic that is directly related to the amount of energy possessed by the wave. Wave heights in a sea area are usually less than 2 m (6.6 ft), although it is not uncommon to observe waves with heights of 10 m (32.8 ft) and periods of 12 s. The largest wind-generated wave authentically measured was a 34-m (112-ft) wave with a period of 14.8 s seen in the North Pacific in February 1935 by the crew of the U.S. Navy tanker U.S.S. *Ramapo*. As sea waves gain energy, their steepness increases. When it reaches a critical value of 1/7, open ocean **breakers** called **whitecaps** form. The appearance of a sea surface as it changes from calm to the condition that results from hurricane-force winds is described in table 9–1.

See "Seasat Measures the Ocean Surface" on page 220 for a discussion of how wave height conditions in the world ocean can be related to the strength of the prevailing wind belts by means of data obtained from remote-sensing systems aboard satellites.

FIGURE 9–7   The Sea and Swell.
As wind blows across the "sea," wave size increases with increased wind speed, fetch, and duration. As waves advance beyond the "sea," they continue to advance across the ocean surface as "swell," free waves that are not driven by the wind but sustained by the energy they obtained in the "sea."

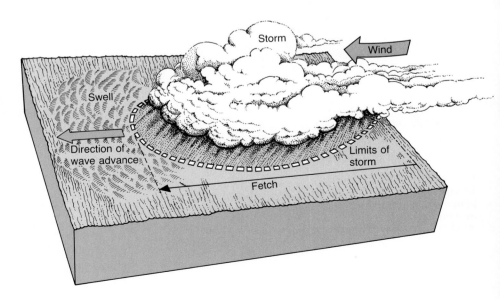

TABLE 9–1 Beaufort Wind Scale and the State of the Sea.

Beaufort Number	Descriptive Term	Speed m/s	Speed mi/h	Appearance of the Sea
0	Calm	—	—	Like a mirror
1	Light air	0.3–1.5	1–3	Ripples with the appearance of scales; no foam crests
2	Light breeze	1.6–3.3	4–7	Small wavelets; crests of glassy appearance, no breaking
3	Gentle breeze	3.4–5.4	8–12	Large wavelets; crests begin to break; scattered whitecaps
4	Moderate breeze	5.5–7.9	13–18	Small waves, becoming longer; numerous whitecaps
5	Fresh breeze	8.0–10.7	19–24	Moderate waves, taking longer form; many whitecaps; some spray
6	Strong breeze	10.8–13.8	25–31	Large waves begin to form; whitecaps everywhere; more spray
7	Near gale	13.9–17.1	32–38	Sea heaps up and white foam from breaking waves begins to be blown in streaks
8	Gale	17.2–20.7	39–46	Moderately high waves of greater length; edges of crests begin to break into spindrift; foam is blown in well-marked streaks
9	Strong gale	20.8–24.4	47–54	High waves; dense streaks of foam and sea begins to roll; spray may affect visibility
10	Storm	24.5–28.4	55–63	Very high waves with overhanging crests; foam is blown in dense white streaks, causing the sea to appear white; the rolling of the sea becomes heavy; visibility reduced
11	Violent storm	28.5–32.6	64–72	Exceptionally high waves (small and medium-sized ships might be for a time lost to view behind the waves); the sea is covered with white patches of foam; everywhere the edges of the wave crests are blown into froth; visibility further reduced
12	Hurricane	32.7–36.9	73–82	The air is filled with foam and spray; sea completely white with driving spray; visibility greatly reduced

After Bowditch, 1958.

For a given wind speed there is a maximum fetch and duration of wind beyond which the waves will not grow. When both maximum fetch and duration are reached for a given wind velocity, the sea is said to be **fully developed**. The reason it can grow no further is the waves are losing as much energy through the breaking of whitecaps as they are receiving from the wind.

Table 9–2 shows the fetch and duration of wind required to produce a fully developed sea for given wind speeds. Table 9–3 describes the fully developed sea at given wind speeds in terms of average height, length, and period of waves. Also, the average height of the highest 10% of the waves is given.

TABLE 9–2 Fetch and Duration Required to Produce Fully Developed Sea for a Given Wind Speed.

Wind Speed km/h (mi/h)	Fetch km (mi)	Duration h
20 (12)	24 (15)	2.75
30 (19)	77 (48)	7
40 (25)	176 (109)	11.5
50 (31)	380 (236)	18.5
60 (37)	660 (409)	27.5
70 (43)	1093 (678)	37.5
80 (50)	1682 (1043)	50
90 (56)	2446 (1517)	65.25

Swell

As waves generated in a sea area move toward the margins of the generating area, where wind speeds are lower, they eventually are moving with a speed greater than that of the wind. When this occurs, the steepness of the waves decreases, and they become long-crested waves called **swell**. The swell moves with little loss of energy over large stretches of the ocean's surface, transmitting energy away from the input area of the sea to where most of it will eventually be released along the margins of continents. As a part of this transfer of energy away from the sea area,

SEASAT MEASURES THE OCEAN SURFACE

It has long been the complaint of meteorologists who try to forecast the world's weather that there are never enough data available from the 71% of the earth's surface covered by ocean. They, of course, are joined in their desire to know what is going on "out there" by the seafarers who make their living on the ocean. Many a vessel, from the smallest fishing boat to massive oil tankers, has disappeared with all hands aboard beneath unexpected massive waves that may reach heights of 30 m (98 ft).

Most observations of conditions at sea are provided by ships moving through major shipping lanes—mostly in the Northern Hemisphere. This state of affairs changed for a 100-day period beginning June 28, 1978. On this date, the U.S. satellite *SEASAT* began measuring atmospheric liquid water, water vapor, sea-surface temperature, sea-surface topography, ocean wave height, and wind speed—all on a global basis. Unfortunately, the satellite power supply short-circuited and *SEASAT* stopped gathering data on October 10, 1978. However, the data obtained during the life of *SEASAT* have proven the instrumentation effective in providing useful data on all the atmospheric and oceanographic conditions it was designed to investigate.

The coverage of the world ocean achieved by *SEASAT* during August 9–11, 1978, is shown in figure 9A*A*. For comparison, the distribution of ship observations for the same three-day period is shown in Figure 9A*B*. It is clear that the satellite coverage is much more uniform. Averaging the data collected in strips of the earth's surface 2.5° of latitude by 7.5° of longitude throughout the entire 100-day period resulted in the maps of water vapor, wind speed, and wave height shown on figures 9B, 9C, and 9D.

Average Water Vapor High concentrations of atmospheric water vapor associated with the southwest monsoon of the northern Indian Ocean exceed 5 g/cm^2 of ocean surface (fig. 9B). High values are also associated with the Intertropical Convergence Zone, the doldrums, along the meteorological equator a few degrees north of the geographical equator. It is here that the northeast trade winds of the Northern Hemisphere and the southeast trade winds of the Southern Hemisphere converge and rise, producing a low-pressure belt of warm, moist air. Water-vapor concentrations decrease toward higher latitudes. The anomalous apparent high concentrations off Antarctica are thought to be caused by interference effects of ice in the Weddell Sea and Ross Sea.

The lower water-vapor levels in the Southern Hemisphere, compared to those in the same latitudes in the Northern Hemisphere, reflect the colder winter air with its lower capacity for holding water vapor.

Had water-vapor values been available over a longer period of time, it would probably have been possible to greatly increase our understanding of the role of water's latent heat of evaporation in transferring heat from the warm low-latitude ocean to the high-latitude atmosphere.

FIGURE 9A Three-Day Global Data Coverage Comparison.
A, from *SEASAT*, compares to *B*, conventional ship weather reports, for August 9–11, 1978.

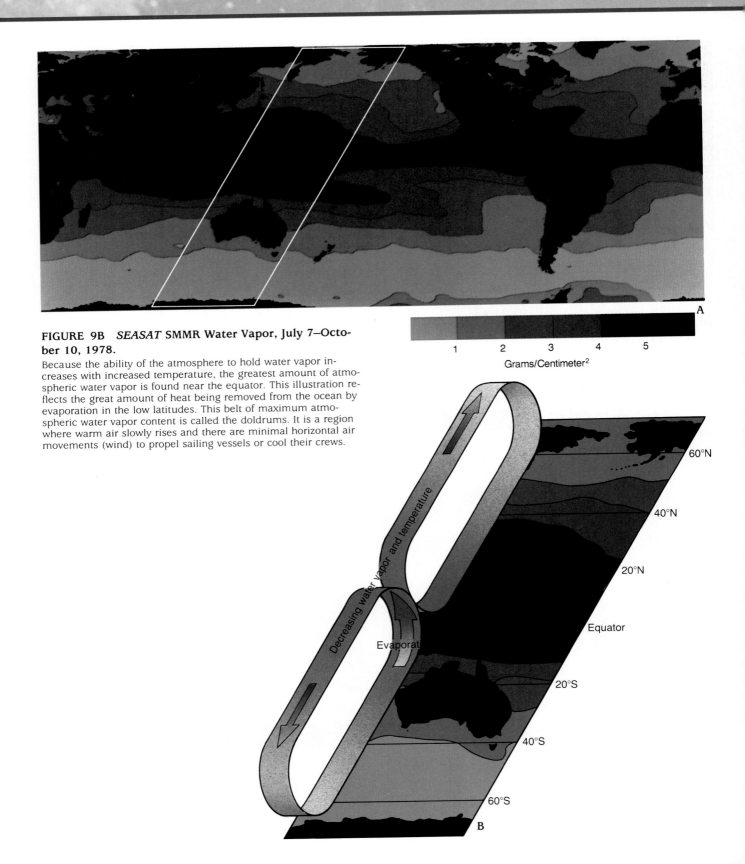

FIGURE 9B *SEASAT* SMMR Water Vapor, July 7–October 10, 1978.

Because the ability of the atmosphere to hold water vapor increases with increased temperature, the greatest amount of atmospheric water vapor is found near the equator. This illustration reflects the great amount of heat being removed from the ocean by evaporation in the low latitudes. This belt of maximum atmospheric water vapor content is called the doldrums. It is a region where warm air slowly rises and there are minimal horizontal air movements (wind) to propel sailing vessels or cool their crews.

Grams/Centimeter²

5.0   6.8   8.6   10.4
Meters/Second

A

FIGURE 9C *SEASAT* **Altimeter Wind Speed, July 7–October 10, 1978.**

Four of the earth's prevailing wind systems are the trade winds that blow from 30° latitudes toward the equator and the westerlies that blow from 30° toward 60° latitudes. Because there are fewer continents in the Southern Hemisphere, the westerly wind belt there contains the strongest year-round winds. The lowest wind speeds occur at the doldrums, where the air is rising, and the horse latitudes (30°N and S latitudes), where air descends. Wind speeds gradually increase away from these belts of minimum wind speed.

B

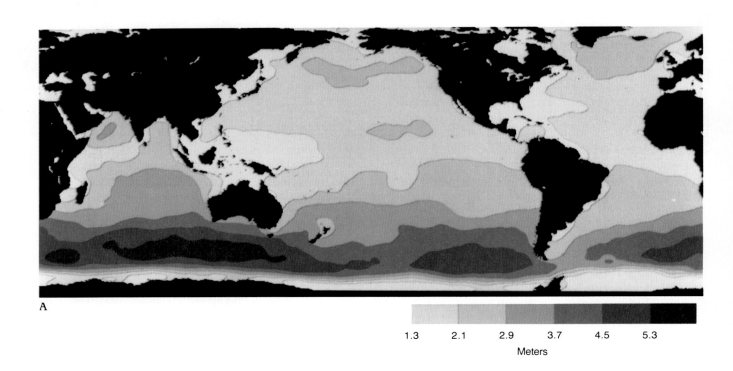

A

1.3   2.1   2.9   3.7   4.5   5.3

Meters

B

| Meters | .67 | 1.7 | 2.5 | 3.3 | 4.1 | 4.9 | 5.7 |
| (Feet) | (2.2) | (5.6) | (8.2) | (10.8) | (13.5) | (16.1) | (18.7) |

FIGURE 9D   *SEASAT* **Altimeter Wave Height, July 7–October 10, 1978.**

Because the westerly wind belt in the Southern Hemisphere reaches the highest average wind speed, this wind system produces the largest ocean waves. The largest waves occur in the southern Indian Ocean. The wave heights represented here are the averages for the largest one-third of the waves occurring at any location.

Average Wind Speed   The prevailing wind systems are evident in figure 9C. The northeast and southeast trade winds, with wind speeds within the 6.8 to 8.6 m/s (15 to 19 mi/hr) range in the Atlantic and Pacific oceans, are separated by the doldrums along the Intertropical Convergence Zone where horizontal movement (wind speed) drops below 5 m/s (11.2 mi/h). The highest wind speeds at trade wind latitudes occur in the Indian Ocean, where the southeast trade winds of the Southern Hemisphere and the southwest monsoon of the Northern Hemisphere summer season exceed 8.6 m/s (19 mi/h).

The horse latitudes (30° N and S) separating the trade winds from the westerlies also show up as low-speed wind belts. Here dense, dry air is descending in the lower atmosphere to produce high-pressure belts. The westerlies show up as high-speed winds in both hemispheres and are best developed between 50° and 60° latitudes. Within this southern latitudinal belt, the highest wind speeds are recorded. Winds averaging in excess of 10.4 m/s (23 mi/h) occur in the Southern Hemisphere winter storms. The high wind speeds observed within this belt were surprisingly steady, with over 68% of the observations falling between 8 and 12 m/s (17.9 and 26.8 mi/h).

The high wind speeds along Antarctica may represent the polar easterlies moving away from the high-pressure cell of the polar region.

Average Wave Height   Figure 9D gives the first accurate picture of global wave conditions. It actually presents an average of the one-third largest waves observed by the altimeter, the *significant wave height* (SWH). These values are relatively small (about 2 m) in the Northern Hemisphere, where summer wind speed is low. The largest waves are created by the high winter wind speed of the Southern Hemisphere westerlies. Average heights in excess of 4.5 m (14.8 ft) are found in the southern reaches of all oceans, with values in excess of 5.3 m (17.4 ft) observed in the southern Indian Ocean.

The narrowest range of wave conditions occurred at the Intertropical Convergence Zone, where weak wind conditions in the doldrums resulted in a wave field of which over 70% of the waves were less than 2 m (6.7 ft) high, and none exceeded a height of 4 m (13 ft). In contrast, the high speed of the Southern Hemisphere westerlies produced waves of which only 2% were less than 2 m (6.7 ft) high, and over 40% were more than 5 m (16.4 ft) high. In one southern Indian Ocean strip, the average significant wave height was 6.5 m (21.3 ft).

The data indicate that, although the probability of doing so is less than 1%, one might encounter waves as high as 8 m (26.2 ft) at a latitude of 52.5°N and waves of over 11 m (36.1 ft) height at 52.5°S.

Whether or not the data recorded during this three-and-a-half-month period are typical cannot be determined. The quality of the data obtained does, however, demonstrate the unique ability of satellites to measure atmospheric and sea-state conditions on a global basis. Having such data continuously available greatly increases the ability to forecast world weather and sea states.

With such a system of prediction, it is possible that many marine disasters could be prevented. The justification for taking the necessary steps to regain the ability to make such satellite wind and wave observations seems obvious.

Reference: Chelton, Dudley B.; Hussey, Kevin J.; and Parke, Michael E. 1981. Global satellite measurements of water vapour, wind speed and wave height. *Nature,* 294:5841, 529–532.

TABLE 9–3 Description of
Fully Developed Sea for a
Given Wind Speed.

Wind Speed km/h (mi/h)	Average Height m (ft)	Average Length m (ft)	Average Period s	Highest 10% Waves m (ft)
20 (12)	.33 (1)	10.6 (34.8)	3.2	.75 (2.5)
30 (19)	.88 (2.9)	22.2 (72.8)	4.6	2.1 (6.9)
40 (25)	1.8 (5.9)	39.7 (130.2)	6.2	3.9 (12.8)
50 (31)	3.2 (10.5)	61.8 (202.7)	7.7	6.8 (22.3)
60 (37)	5.1 (16.7)	89.2 (292.6)	9.1	10.5 (34.4)
70 (43)	7.4 (24.3)	121.4 (398.2)	10.8	15.3 (50.2)
80 (50)	10.3 (33.8)	158.6 (520.2)	12.4	21.4 (70.2)
90 (56)	13.9 (45.6)	201.6 (661.2)	13.9	28.4 (93.2)

the waves with the longer length, traveling at higher speed, leave the sea area first. They are followed by the shorter-length, lower-speed **wave trains** (groups of waves). This progression illustrates the principle of **wave dispersion** (the sorting of waves by wavelength). In the generating area, waves of many wavelengths are present. In deep water, wave speed is a function of wavelength, so the longer waves "outrun" the shorter ones. To illustrate the distance that can be traveled by swell without having its energy depleted, swell that originated from storms in the Antarctic has been recorded breaking along the Alaskan coast after traveling a distance of more than 10,000 km (6210 mi).

As a set of waves leaves a sea and becomes a swell wave train, the group moves across the ocean surface at only half the velocity of an individual wave in the group. Progressively, the leading wave disappears. There will, however, always be the same number of waves in the group. As the front wave disappears, a new wave replaces it at the back of the group.

Interference Patterns Since swells may be moving away from a number of storm areas in a given ocean, it is inevitable that the swell forms will run together and interfere with one another. This gives rise to one of the special features of wave motion— interference patterns. An interference pattern produced by the superposition of two or more wave systems will be the algebraic sum of the disturbance each wave would have produced individually. The result may be a larger or smaller trough or crest, depending on whether the individual disturbances are of the same or opposite sign, in or out of phase (fig. 9–8).

For instance, if swells are moving away from two storm areas and come together as shown in figure 9–8, the interference pattern may be constructive, de-

structive, or more likely mixed. **Constructive interference** results if, in an ideal case, wave trains with the same wavelength came together in phase, crest to crest and trough to trough. Adding the displacements that would result from the waves individually, we will see that the interference pattern produced will be a wave with the same length as the two wave systems that are converging but with a wave height that is equal to the sum of the wave heights of the individual wave systems.

Destructive interference results if the crests produced by the waves from one generating area coincide with the troughs produced by the waves from the second generating area. If the waves had identical characteristics, the algebraic sum of the crest plus the trough would be zero, and the waves would cancel each other out.

It is more likely that the two systems possess waves of different heights and lengths and would come together with both destructive and constructive interference. Thus, a more complex **mixed interference** pattern would develop. It is such interference that explains the occurrence of a sequence of high waves followed by a sequence of lower waves and other irregular wave distribution patterns observed as the swell approaches the margins of the continents.

Free and Forced Waves In the swell we see what may be referred to as a **free wave**, which is moving with the momentum and energy imparted to it in the sea area but is not experiencing a maintaining force that keeps it in motion. In the wave-generating area, free and forced waves are present. A **forced wave** is one that is maintained by force that has a periodicity coinciding with the period of the wave. This force would be the wind; owing to the variability of the wind, many wave systems in the sea area alternate between being forced waves and free waves.

Surf

Most waves that are generated in the sea area by the force of storm-velocity winds move across the ocean as swell. They release their energy at the margins of the continents in the **surf zone**, where the swell forms breakers. As the deep-water waves making up the swell move toward the margin of the continent over gradually shoaling water, they eventually encounter water depths which are less than one-half wavelength.

These shoaling depths interfere with the particle movement at the base of the wave, and the wave slows down. As one wave is slowed down, the following waveform, which is still moving at an unaffected speed, tends to "catch up" with the wave that is being slowed, thus reducing the wavelength. Wave height increases, and the crests become narrow and pointed while the troughs become wide curves, a form that was previously described for high-energy waves in the sea. The increase in wave height accompanied by a decrease in wavelength increases the steepness (H/L) of the waves. As the wave steepness reaches 1/7, the waves break as **surf**.

If the surf is composed of swell that has traveled from distant storms, breakers will develop relatively near shore in shallow water, the shoaling of which is primarily responsible for their breaking. By the time waves break, they have become shallow-water waves. The horizontal water motion associated with such waves translates water toward and away from the shore. The surf will be characterized by parallel lines of relatively uniform breakers. However, if the surf is composed of waves that have been generated by local wind, the waves may not have been sorted out into swell. The surf may be more nearly characterized by unstable, deep-water, high-energy waves with steepness already near 1/7. They will break shortly after feeling bottom some distance from shore and the surf will be rough and choppy with an irregular nature.

Ideally, when the water depth is about 1.3 times the breaker height, the crest of the wave breaks, producing surf. When the water depth becomes less than 1/20 the wavelength, the waves in the surf zone begin to behave as shallow-water waves. Particle motion is greatly interfered with by the bottom, and a very significant transport of water toward the shoreline occurs (fig. 9–9).

The breaking experienced in the surf results from the fact that the particle motion near the bottom of the wave has been interfered with greatly by contact with the ocean bottom, and this tends to slow down the waveform. However, the individual particles that are orbiting near the ocean surface have not been slowed down as much as the entire waveform because they have experienced no contact with the bottom. This causes the top of the waveform to lean toward the shore as the bottom of the waveform is held back. The slowing down of the entire waveform compared to the increasing speed of motion in the circular orbits near the ocean surface because of increasing wave height results in the water particles at the interface between the ocean and the atmosphere moving faster toward shore than the waveform itself.

This motion can be seen particularly well in **plunging breakers** (fig. 9–10), which have a curling crest that moves over an air pocket resulting from the fact that the curling particles have outrun the wave and there is nothing beneath them to support their motion. Plunging breakers form on moderately steep beach slopes. The more commonly observed breaker is the **spilling breaker** that results from a relatively gentle slope of the ocean bottom, which more gradually extracts the energy from the wave, producing a turbulent mass of air and water that runs down the front slope of the wave instead of a spectacular cresting curl. Because of the gradual extraction of energy of spilling breakers, they have a longer life span and give surfers a longer, if less exciting, ride than do the plunging breakers.

Recalling the particle motion of ocean waves in figure 9–3, we can see that in front of the crest the water particles are moving up into the crest. It is this force along with the buoyancy of the surfboard that helps maintain a surfer in front of a cresting wave. When this upward motion of water particles is interrupted by the wave passing over water that is too shallow to allow this movement to continue, the ride is over. A skillful surfer, by positioning his board properly on the wave front, can regulate the degree to which the gravitational forces that propel him exceed the buoyancy forces, and high speeds can be obtained while moving along the face of the breaking wave.

Wave Refraction

We previously discussed that waves begin to bunch up and wavelengths become shorter as swell begins to "feel bottom" upon approaching the shore. It is seldom that the swell will approach the shore at right angles. Some segment of the wave can be expected to feel bottom first and therefore be slowed down before the rest of the wave. This produces **refraction**, or bending, of the waves as they approach the shore. As shown in figure 9–11, an irregular shoreline might result from an irregular bottom topography that could

A.

FIGURE 9–8 Wave Interference.

A, as wave trains come together from different sea areas (1 and 2), three possible interference patterns may result (3). Very rarely, if the waves have the same length and come together in phase (crest to crest and trough to trough), totally constructive interference may occur. The algebraic sum of amplitudes produces a wave of the same length but of greater height. If two sets of waves have identical characteristics and come together 180° out of phase, the result will be no wave at all (destructive). More commonly, waves of different lengths and heights encounter one another and produce a mixed interference pattern. *B,* as swell from seas A and B cross, wave interference occurs. The waves from Sea A are larger than those from Sea B, and a varied pattern of interference occurs. If the waves come together crest to crest or trough to trough, *constructive interference* occurs and produces a higher crest or deeper trough. If the waves come together crest to trough, *destructive interference* occurs. Whichever represents the smaller displacement from the still-water line is subtracted from the other, and the resulting crest or trough is reduced in size. The overall pattern of interference includes many degrees of constructive and destructive interference and is called *mixed interference. C,* the individual crests and troughs of waves from Sea A and Sea B along line A–A' of part *B* are shown separately and in a pattern of mixed interference. Points along the line where constructive and destructive interference occurred are indicated by the letters C and D and the lines that run vertically through all three wave patterns.

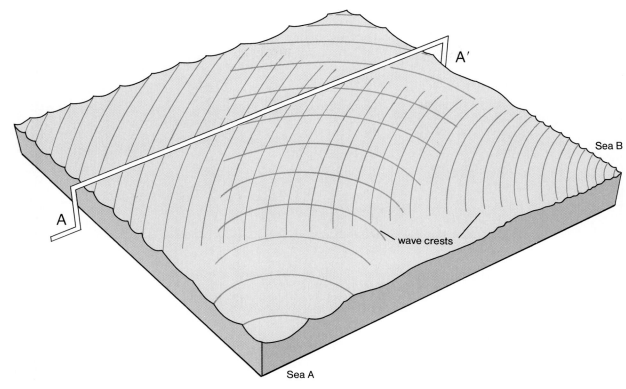

wave crests

Sea B

Sea A

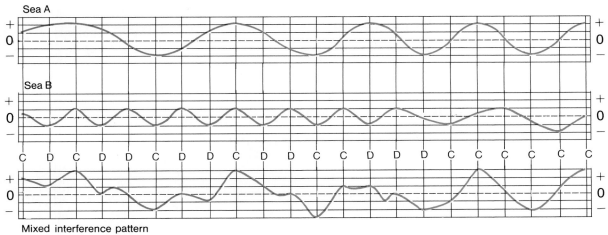

Sea A

Sea B

C D C D D C D D C D D C C D D D C C C C C C

Mixed interference pattern

C.

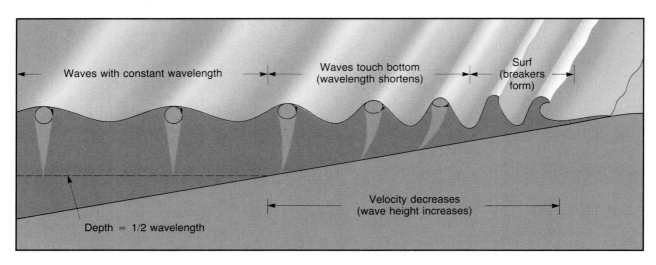

FIGURE 9–9 Surf Zone.

As waves approach the shore and encounter water depths less than one-half a wavelength ($L/2$), friction removes energy and the waves slow down, wavelength decreases, and wave height increases. When the water depth is $1.3H$, the wave reaches a steepness of $\frac{1}{7}$ and breaks on the shore. The Tasa Collection: *Shorelines.* Published by Merrill Publishing Co., Columbus, OH. Copyright © 1986 by Tasa Graphic Arts, Inc. All rights reserved.

FIGURE 9–10 Breakers.

A, spilling breakers. *B,* plunging breakers. Photos courtesy of Charles H. V. Ebert.

A.

B.

FIGURE 9–11   Refraction.
As the waves "feel bottom" first in the shallow areas off the headlands, they are slowed. The segments of the waves that move through the deeper water leading into the bay are not slowed until they are well into the bay. As a result, the waves are refracted (bent) so the release of wave energy is concentrated on the headlands. Erosion is active on the headlands, while deposition occurs in the bay, where the energy level is low. Orthogonal lines, spaced so that equal amounts of energy are between any two adjacent lines, help to show the distribution of energy along the shore. Adapted from Tarbuck, E. J., and Lutgens, F. K., *Earth science*, 5th ed. (Columbus: Merrill, 1988).

Orthogonal lines

slow down portions of a wave that was approaching the shore.

The effect of wave refraction is to unevenly distribute wave energy along the shoreline (fig. 9–11). Orthogonal lines are constructed perpendicular to the wave fronts and spaced in such a way that the amount of energy between lines is equal at all times. They are of great assistance in illustrating how energy is distributed along the shoreline by breaking waves. Orthogonals indicate the direction the wave is traveling and can be seen to converge on headlands jutting out into the ocean and diverge in bays. A concentration of energy is released against the headlands, while energy released in the bays is spread more thinly. This condition produces erosion on the headlands, while deposition may occur in the bays. The increased energy of the waves breaking on headlands is reflected in an increased wave height.

Wave Diffraction

Wave **diffraction** can be considered as the bending of waves around objects. It is this kind of movement that allows waves to move past barriers into harbors as energy moves laterally along the crest of the wave, as shown in figure 9–12. This bending is on a much smaller scale and is less easily explained than the bending discussed in refraction, which is a simple response to changes in velocity. Diffraction results because any point on a wave front can be a source from which energy can propagate in all directions.

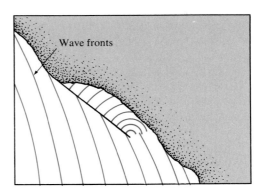

Wave fronts

FIGURE 9–12   Diffraction.
Diffraction is bending caused by waves passing an obstacle; it is not related to refraction. By diffraction, wave energy may spread to the most protected areas within a harbor.

Wave Reflection

Swell can be reflected back into the ocean with little loss of energy from a vertical barrier such as a sea wall. For this ideal **reflection** to occur without energy loss, the wave would have to strike the barrier at a right angle. Such a condition would be rare in nature. Less ideal reflections will nonetheless produce **standing waves**, which are the product of two waves of the same length moving in opposite directions.

The standing wave is a special interference pattern where no net momentum is carried because the waves are moving in opposite directions. The particles continue to move vertically and horizontally, but there is no more of the circular motion that we saw in the progressive wave. Standing waves are characterized by lines along which there is no vertical movement. There may be one or more such **nodes** or **nodal** lines. **Antinodes**, crests that alternately become troughs, represent the points of the greatest vertical movement within a standing wave (fig. 9–13). There is no particle motion when an antinode is at its greatest vertical displacement, and the maximum particle movement occurs when the water surface is level. At this time, the maximum movement of the water is in a horizontal direction directly beneath the nodal lines. The movement of water particles beneath the antinodes is entirely vertical. We will consider standing waves further when we discuss the tides in the following chapter, as there are certain conditions under which the development of standing waves has a significant effect on the tidal character of coastal regions.

Reflection of wind-generated waves from coastal barriers will be at an angle equal to the angle at which the wave approached the barrier, as shown in figure 9–14. This type of reflection may produce a small-scale interference pattern similar to those we previously discussed.

An outstanding example of reflection phenomena is the Wedge that develops west of the 400-m (1312-ft) jetty at Newport Harbor, California. As the incoming waves strike the jetty and are reflected, a constructive interference pattern develops. When the crests of the incoming waves merge with the crests of the reflected waves, plunging wedge-shaped breakers that may reach heights in excess of 8 m (26 ft) form. These waves present a fierce challenge to the most experienced body surfers. Attracting some who were not up to the challenge, the Wedge has killed and crippled many who have come to try it.

Storm Surge

Large cyclonic storms that develop over the ocean will, owing to their low pressure, produce a hill of water beneath them that moves with the storm across the open ocean. Additionally, as the storm approaches shallow water near shore, the portion of the hill over which the wind is blowing shoreward will produce a **storm surge**, a mass of wind-driven water that produces an increase in sea level that may be extremely destructive to low-lying coastal areas. Storm surges can be particularly destructive when accompanied by high tide. The coincidence of a storm surge with high tide in areas that have particularly high tides frequently produces major catastrophes, with great loss of life and property damage (fig. 9–15).

One of the most outstanding examples of storm surge is that which relates to the windows of a lighthouse atop a 90-m (295-ft) cliff at Dunnet Head, Scotland, being broken by stones tossed by waves breaking on the cliff. The lighthouse marks the western entrance to Pentland Firth, which connects the Atlantic Ocean with the North Sea between the Orkney Islands to the north and the Scottish mainland to the south.

TSUNAMI

The large waves referred to by the Japanese word **tsunami** originate as a result of disturbances within the earth's crust. We commonly hear them called *tidal waves,* which implies that they are related to the tides. They have no relationship to the tides, however. Tsunami are usually caused by fault movement, a displacement in the earth's crust along a fracture, that causes a sudden change in water level at the ocean surface. Secondary events, such as underwater avalanches, produced by the faulting may also produce tsunami (fig. 9–16). The ocean that is most plagued by the tsunami is the Pacific. It is ringed by a series of trenches that are unstable margins of lithospheric plates along which large-magnitude earthquakes occur.

FIGURE 9–13 Reflection— Standing Waves.

An example of water motion at quarter-period intervals. Water is motionless when antinodes reach maximum displacement. Water movement is maximum when the water is level. Movement is totally vertical beneath the antinodes, and maximum horizontal movement occurs beneath the node. The circular motion of particles in progressive waves does not exist in standing waves.

A.

B.

FIGURE 9–14   Reflection—The Wedge.
A, the Wedge, a wedge-shaped crest that may reach heights in excess of 8 m (26 ft), develops as a result of interference between incoming waves and reflected waves near the jetty protecting the entrance to Newport Harbor, California. B, a view of a wedge crest taken from the landward end of the jetty. The three dots in the water in front of the wave are the heads of body surfers waiting to catch the wave.

A.

B.

FIGURE 9–15   Storm Surge.
A, as a cyclonic storm in the Northern Hemisphere moves ashore, the low-pressure cell around which the storm winds blow and the onshore winds associated with the storm will produce a high-water phenomenon called a *storm surge*. B, low pressure and high onshore winds of Hurricane Camille produced a storm surge along the Mississippi coast in 1969. Photo courtesy of NOAA.

A.

B.

C.

FIGURE 9–16 Tsunami.

A, movement along a fault in the earth's crust sends energy to the ocean surface. This energy then is distributed laterally along the atmosphere-ocean interface in the form of a tsunami. The energy is transmitted across the open ocean by undetectable waves over 200 km (124 mi) long and about 0.5 m (1.6 ft) high. They release their energy after reaching the shore and developing heights that may be in excess of 30 m (98 ft). Vertical uplift of ocean floor pushes up the ocean water column above the fault. The energy released into the water column by the earthquake travels away from the affected area as a tsunami. *B* shows Sandy Beach on the Island of Oahu moments before a tsunami generated by an earthquake in the Aleutian Trench area struck the beach. Note that the beach was exposed as the shoreline moved out to sea. The arrow indicates the location of *C,* which shows the beach a few minutes later, after the tsunami arrived. A man escaping the wave is circled at left; several atuomobiles swept off the highway are circled at right. Photos by Y. Ishii. Reprinted by permission of *Honolulu Advertiser.*

One of the most destructive tsunami ever generated came from the greatest release of energy from the earth's interior observed during historical times. On August 27, 1883, the island of Krakatoa in the Sunda Strait, between Sumatra and Java, exploded and essentially disappeared. The sound of the explosion was heard 4800 km (2981 mi) away at Rodriguez Island in the western Indian Ocean, and the dust that ascended into the atmosphere circled the earth and produced unusual and beautiful sunsets for nearly one year. A tsunami that rose to heights of more than 30 m (98 ft) devastated the coastal region of the Sunda Strait, taking more than 36,000 lives. The energy carried by this wave was still detectable after crossing the Indian Ocean, passing the southern tip of Africa, and moving north in the Atlantic Ocean into the English Channel.

Since the wavelength of a typical tsunami will be in excess of 200 km (124 mi), it is obviously a shallow-water wave, the speed of which is determined by water depth. Moving at speeds well in excess of 700 km/h (435 mi/h) and with wave heights of approximately 0.5 m (1.6 ft) in the open ocean, tsunami are not readily observable until they reach shore. In shallow water, they are slowed down, and the water begins to pile up to form crests that may exceed heights of 30 m (98 ft).

Unlike the hurricane that represents a great hazard to ships at sea and may send them seeking the protection of a coastal harbor, a tsunami sends them from their coastal moorings into the open ocean. Until 1948, there was no possible warning of the coming of a tsunami that would be adequate to allow shippers to take the appropriate precautions to avoid the destructive waves. Since that time and as a result of a destructive wave that struck Hawaii in 1946, a tsunami warning system has been established throughout the Pacific Ocean. With every seismic disturbance that occurs beneath the ocean surface with the potential of producing a tsunami, observations are made at the closest tide-measuring stations to see if there is any indication that such a wave has been created. Should one be detected, warnings are sent to all the coastal regions that might conceivably encounter a destructive wave along with the time at which the wave is to be expected. This warning allows for the removal of ships from harbors before the waves arrive if the disturbance has occurred at a great enough distance.

Prior to such a warning system, the first notice that most observers had of an impending tsunami would be the rapid seaward recession of the shoreline. The recession would then be followed in a few minutes by the destructive wave (fig. 9–16B, C). Such a recession was observed in the port of Hilo, Hawaii, on April 1, 1946, as a result of an earthquake in the Aleutian Trench off the island of Unimak over 3000 km (1863 mi) away. The recession created by the formation of a trough preceding the first tsunami wave crest was followed by a wave that carried the waters to heights nearly 8 m (26 ft) above the normal high tide level. The tsunami also struck Scotch Cap, Alaska, on Unimak Island. The lighthouse on the island, the base of which stood 14 m (46 ft) above sea level, was totally destroyed by a wave that must have reached a height of 36 m (118 ft).

It was the wave produced by this disturbance in the Aleutian Trench that was recorded throughout the coastal regions of the Pacific Ocean at tide-recording stations and led to the development of what is now the International Tsunami Warning System (ITWS).

INTERNAL WAVES

We have to this point discussed waves that occur at an obvious density discontinuity existing between the atmosphere and the ocean. From our previous discussions we should be aware that there are also such discontinuities within the ocean water column itself. For instance, the pycnocline represents a density interface that is typically associated with the thermocline. Just as energy can be transmitted along the interface between the atmosphere and the ocean, it can also be transmitted as an interface wave along density interfaces beneath the ocean surface in what are called **internal waves**. Internal waves may have heights above 100 m (328 ft) (fig. 9–17). The greater the difference in density between the two fluids, the faster the waves will move.

There is still much to be learned about internal waves, but their existence is well documented, and it is thought that many causes for them may be identified. Internal waves are known to have periods related to tidal forces, indicating that these forces may be a significant cause. Underwater avalanches in the form of turbidity currents, wind stress, and energy put into the water by moving vessels may also be causes of internal waves. It is thought that the parallel slicks seen on the surface waters may overlie the troughs of internal waves. The slicks are caused by a film of surface debris accumulating and dampening surface waves.

Internal waves reach greater heights from a smaller energy input than do the waves resulting from very large energy input observed at the ocean's surface. This is because they move along interfaces, such as the pycnocline, across which the density dif-

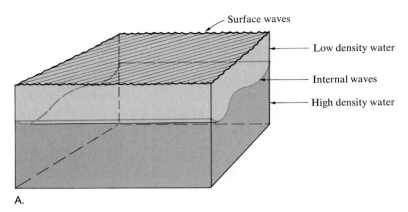

A.

FIGURE 9–17   Internal Wave.
A, a simple internal wave moving along the density interface below the ocean surface. *B, Seasat* Synthetic Aperture Radar image of internal waves in the Gulf of California. They are thought to be associated with tidal currents in the shallow water around the island of Angel de La Guarda located at the bottom. The image is about 100 km (62 mi) on each side.

B.

ference is considerably less than that which exists between the ocean surface and the atmosphere. They are thought to move as shallow-water waves at speeds considerably less than those of surface waves, with periods of 5–8 min and wavelengths of 0.6–0.9 km (0.37–0.56 mi).

POWER FROM WAVES

Extracting power from waves as they approach the shore is a possibility, although there are significant problems to be overcome. Owing to the refraction of waves, a large half-cone could be constructed to focus the energy of waves at the apex of the structure. Such a device might extract up to 10 MW of power per kilometer of shore, but could produce significant power only when large storm waves broke against it. Such a system would operate only as a power supplement, and a series of such structures along the shore would be required. Structures of this type could have a serious effect on longshore transport of sediment, which could lead to serious coastal erosion problems in areas deprived of sediment.

Along shores endowed with proper topography of the ocean floor, internal waves with their great heights may be effectively focused by refraction and induced to "break" against an energy-conversion device.

Two engineers at Lockheed Corporation have developed an interesting ocean wave energy device called **Dam-Atoll**, which is shown in figure 9–18. Waves enter the top of the unit, just at the ocean surface. Guide vanes at the opening cause the entering

water to spiral into a whirlpool within an 18-m (60-ft) central core. The swirling column of water turns a turbine, the unit's only moving part. It can provide a continuous electrical power output of 1 to 2 MW according to the inventors, Leslie S. Wirt and Duane L. Morrow. In particularly good wave areas, such as the Pacific Northwest, they believe it would be possible to anchor 500 to 1000 units that would provide power in quantities comparable to those provided by Hoover Dam. These units, 76 m (250 ft) in diameter and made of concrete, have many potential uses other than the generation of electricity. They could be used in cleaning up oil spills, protecting beaches from wave erosion, creating calm harbors in the open sea, and desalinating seawater through reverse osmosis.

Should economic conditions in the future enhance the appeal of converting wave energy to electrical energy, the coastal regions with the greatest potential for exploitation are shown in figure 9–19.

SUMMARY

Wave phenomena transmit energy through various states of matter by setting up patterns of oscillatory motion in the particles that make up the matter. **Progressive waves** may be described as **longitudinal, transverse,** or **orbital,** depending on the pattern of particle oscillation. Particles related to ocean waves move primarily in orbital paths.

Characteristics used to describe waves are **wavelength** (L), **wave height** (H), **period** (T), and **wave speed** (S). If water depth is greater than 1/2 wave-

FIGURE 9–18   Dam-Atoll.

Lockheed Corporation's Dam-Atoll designed to generate electricity from wave action. *A,* water entering at the surface spirals down an 18-m (60-ft) central cylinder to turn a turbine located at the bottom of the cylinder. Each unit is designed to produce from 1 to 2 MW. *B,* view from above. Courtesy of Lockheed Corporation.

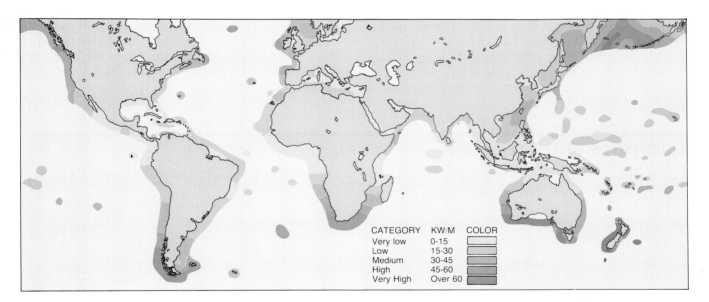

CATEGORY	KW/M	COLOR
Very low	0-15	
Low	15-30	
Medium	30-45	
High	45-60	
Very High	Over 60	

FIGURE 9–19   Global Coastal Wave Energy Resources.

This map, constructed from data available in the U. S. Navy Summary of Synoptic Meterological Observations (SSMO), shows the effect of west to east movement of storm systems in the temperate latitudes. The west coasts of continents are struck by larger waves than the east coasts. Also the larger waves appear to be associated with the westerly wind belts between 30° and 60° latitudes in both hemispheres. The significant wave height associated with the available wave power categories shown on the map are very low (1 m), low (2–2.5 m), medium (3 m), high (3.5 m), and very high (3.9 m). Adapted from *Sea frontiers* (July-August 1987) "Sea Secrets," International Oceanographic Foundation (Vol. 33, No. 4) pp. 260–261; produced by The National Climatic Data Center with support from the U.S. Department of Energy.

length, a progressive wave travels as a **deep-water wave** with a speed that is directly proportional to wavelength. If water depth is less than 1/20 wavelength, it will move as a **shallow-water wave**, the speed of which increases with increased water depth.

As wind-generated waves form in the **sea** area, **capillary waves** with rounded crests and wavelengths less than 1.74 cm (0.7 in.) form first. As the energy of the waves increases, **gravity waves** with increased wave speed, wavelength, and wave height form. Energy is transmitted from the sea area across the ocean by low, rounded waves called **swell**. The swell releases its energy in the **surf** that forms as the waves break in the shoaling water near shore. If the waves break on a relatively flat surface, the result is usually a **spilling breaker**, while breakers forming on steep slopes have spectacular curling crests and are called **plunging breakers**.

When swell approaches the shore, the segments of the waves that encounter shallow water first will be slowed, and those parts of the wave that have not yet been interfered with by shallow water will move ahead, causing the wave to refract, or bend. **Refraction** causes a concentration of wave energy on the headlands, while low-energy breakers are characteristically found in bays. **Reflection** of waves off sea walls or other barriers can cause an interference pattern called a **standing wave**. In standing waves,

crests do not move laterally as in progressive waves but form alternately with troughs at locations called **antinodes**. Separating the antinodes are locations where there is no vertical movement of the water—**nodes**.

During storms, the combination of low air pressure and onshore winds may produce **storm surge** that raises the water level at the shore many meters above normal sea level. Such surges are particularly destructive if they coincide with high tide.

Tsunami, or seismic sea waves, are generated by seismic disturbances beneath the ocean floor. Such waves have lengths in excess of 200 km (124 mi) and travel across the open ocean with undetectable heights of about 0.5 m (1.6 ft) at speeds in excess of 700 km/h (435 mi/h). On approaching shore, they may increase in height to over 30 m (98 ft). Single-wave tsunami have been known to cause millions of dollars worth of damage and take tens of thousands of lives. **Internal waves** are not well understood but are thought to form at density interfaces beneath the ocean surface, especially in connection with the pycnocline. They may be expected to have heights of up to 100 m (328 ft), with periods from 5 to 8 min.

Ocean waves can potentially be focused to produce hydroelectric power, but there are significant environmental considerations that will have to be incorporated into any such plan.

QUESTIONS AND EXERCISES

1. Discuss longitudinal, transverse, and orbital wave phenomena, including the states of matter in which they transmit energy.

2. Calculate the speed (S) for deep-water waves with the following characteristics:
   A. $L = 351$ m, $T = 15$ s
   B. $L = 351$ m, $f = 4$ waves/min
   Express speed (S) in meters per second (m/s).
   C. $T = 11$ s

3. Calculate the speed with which a shallow-water wave will travel across an ocean basin 4 km (2.5 mi) deep.

4. Describe the change in the shape of waves that occurs as they progress from capillary waves to increasingly larger gravity waves until they reach a steepness ratio of 1/7.
   A change in which variable will make gravity the dominant restoring force, H, L, S, T, or f?

5. What is the minimum fetch and duration of wind required to produce a fully developed sea with a wind speed of 45 km/h (27.9 mi/h)? What would be the av-

erage height, length, and period of waves in this fully developed sea?

6. Waves from separate sea areas move away as swell and produce an interference pattern when they come together. If the waves from Sea A and Sea B have wave heights of 1.5 m (5 ft) and 3.5 m (11.5 ft), respectively, what would be the height of waves resulting from constructive interference and destructive interference? Illustrate your answer with a diagram (refer to fig. 9–8).

7. Describe changes, if any, in wave speed (S), length (L), height (H), and period (T) that occur as waves move across shoaling water to break on the shore.

8. Using orthogonal lines, illustrate how wave energy can be distributed along the shore. Identify areas of high and low energy release.

9. On the basis of the fundamental characteristics of standing waves shown in figure 9–13, construct a similar diagram of a standing wave in which two nodes and three antinodes exist.

10. List three factors that may affect the height of storm surge. Make a diagram of a hurricane coming ashore from the south along an east-west shore and indicate the segment of shore along which you think the storm surge will reach maximum height. Explain why.

11. Why is it more likely that a tsunami will be generated by faults beneath the ocean along which vertical rather than horizontal movement has occurred?

12. How long will it take for a tsunami to travel across 2000 km (1242 mi) of ocean if the average depth of ocean is 4500 m (14,760 ft)?

13. What ocean depth would be required for a tsunami with a wavelength of 220 km (136 mi) to travel as a deep-water wave? Is it possible that such a wave could become a deep-water wave any place in the world ocean?

14. Why is the development of internal waves likely within the thermocline?

15. Discuss some environmental problems that might result from the development of facilities for the conversion of wave energy to electrical energy.

REFERENCES

Bowditch, N. 1958. *American Practical Navigator.* Rev. ed. H. O. Pub. 9. Washington, D.C.: U.S. Naval Oceanographic Office.

Gill A. E. 1982. *Atmosphere-ocean dynamics. International Geophysics Series,* Vol. 30. Orlando, Fla.: Academic Press, Inc.

Kinsman, B. 1965. *Wind waves: Their generation and propagation on the ocean surface.* Englewood Cliffs, N.J.: Prentice-Hall.

Melville, W., and Rapp, R. 1985. Momentum flux in breaking waves. *Nature* 317:6037, 514–16.

Pickard, G. L. 1975. *Descriptive physical oceanography: An introduction.* 2nd ed. New York: Pergamon Press.

Stewart, R. H. 1985. *Methods of satellite oceanography.* Berkeley, Calif.: University of California Press.

Sverdrup, H. U.; Johnson, M. W.; and Fleming, R. H. 1942. Renewal 1970. *The oceans: Their physics, chemistry, and general biology.* Englewood Cliffs, N. J.: Prentice-Hall.

van Arx, W. S. 1962. *An introduction to physical oceanography.* Reading, Mass.: Addison-Wesley.

SUGGESTED READING

Sea Frontiers

Barnes-Svarney, P. 1988. Tsunami: Following the deadly wave. 34:5, 256–63.
  The origin of tsunami, a history of major occurrences, and the methods used to detect them are covered.

Changery, M. J. 1987. Coastal wave energy. 33:4, 259–62.
  The wave energy resources of the world's shores are considered.

Ferrell, N. 1987. The tombstone twins: Lights at the top of the world. 33:5, 344–51.
  A short history of the two lighthouses on Unimak Island, Alaska.

Land, T. 1975. Freak killer waves. 21:3, 139–41.
  The British design a buoy that will gather data in areas where 30-m (98-ft) waves, which may be responsible for the loss of many ships, occur.

Mooney, M. J. 1975. Tragedy at Scotch Cap. 21:2, 84–90.
  A recounting of the events resulting from an earthquake off the Aleutians April 1, 1946. The resulting tsunami destroyed the lighthouse at Scotch Cap, Alaska.

Pararas-Carayannis, G. 1977. The International Tsunami Warning System. 23:1, 20–27.
  A discussion of the history and operations of the International Tsunami Warning System.

Robinson, J., P., Jr. 1976. Newfoundland's disaster of '29. 22:1, 44–51.
  A description of the destruction caused by a tsunami that struck Newfoundland on November 18, 1929.

———. 1976. Superwaves of southeast Africa. 22:2, 106–16.
  A discussion of the formation and destruction caused by large waves that strike ships off the southeast coast of South Africa.

Smail, J. 1982. Internal waves: The wake of sea monsters. 28:1, 16–22.
  An informative discussion of the causes of internal waves and their effect on surface ships and submarines.

Smail, J. R. 1986. The topsy-turvy world of capillary waves. 32:5, 331–37.
Capillary waves are clearly described, and their role in transmitting wind energy to the motion of waves and currents is discussed.

Smith, F. G. W. 1970. The simple wave. 16:4, 234–45.
This is a very readable explanation of the nature of ocean waves. It deals primarily with the characteristics of deep-water waves.

————. 1971. The real sea. 17:5, 298–311.
A comprehensive and readable discussion of wind-generated waves.

————. 1985. Bermuda mystery waves. 31:3, 160–63.
A discussion of the possible source of large waves that struck Bermuda on November 12, 1984.

Truby, J. D. 1971. Krakatoa—the killer wave. 17:3, 130–39.
The events leading up to the 1883 eruption of Krakatoa and the tsunami that followed are described.

Scientific American

Bascom, W. 1959. Ocean waves. 201:2, 89–97.
An informative discussion of the nature of wind-generated waves, tsunami, and tides.

Koehl, M. A. R. 1982. The interaction of moving water and sessile organisms. 274:6, 124–35.
A discussion of the adaptations of benthic shore-dwelling animals to the stresses of strong currents and breaking waves.

10
TIDES

Humans have undoubtedly observed the rise and fall of the tides since they first inhabited the coastal regions of the continents. But until Herodotus (450 B.C.) observed tides on the Mediterranean, there was no known record of them. The first Greek observers concluded that the tides were related to the motion of the moon since both followed a similar cyclic pattern, but it was not until **Sir Isaac Newton** (1642–1727) developed his universal law of gravitation that the tides could be explained adequately.

The tides are the ultimate manifestation of shallow-water wave phenomena, possessing lengths measured in thousands of kilometers and heights ranging from zero to more than 15 m (49 ft). The ocean tides generated by the gravitational attraction of the sun and moon on the mass of the ocean affect every particle of water from the surface to the deepest part of the ocean basin. Thus, the tide undoubtedly has a much more far-reaching effect on ocean phenomena than we can observe at the ocean surface.

TIDE-GENERATING FORCES

Law of Gravitation

Sir Isaac Newton published his *Philosophiae naturalis principia mathematica* in 1686 and stated in his preface, ". . . I derive from the celestial phenomena the forces of gravity with which bodies tend to the sun and several planets. Then from these forces, by other propositions which are also mathematical, I deduce the motions of the planets, the comets, the moon, and the sea." What followed is our first understanding of why tides behave as they do.

Newton's **law of gravitation** states that every particle of mass in the universe attracts every other particle of mass with a force that is proportional to the product of their masses and inversely proportional to the square of the distance between the masses. The greater the mass of the objects and the closer they are together, the greater will be the gravitational attraction. Mathematically this can be expressed as:

$$\text{Gravitational force} = G\, \frac{m_1 m_2}{r^2}$$

Here, G is the universal gravitational constant, m_1 and m_2 are the masses, and r is the distance between the masses. For spherical bodies, all of the masses can be considered to exist at the center of the sphere. Thus, r will always be the distance between the centers of bodies being considered.

Actually, tide-generating forces vary inversely as the cube of the distance from the center of the earth to the center of the tide-generating object, instead of varying inversely to the square of the distance as does the gravitational attraction. Therefore, the tide-generating force, although it is derived from the force of gravitational attraction, is not linearly proportional to it. Distance becomes a more highly weighted variable in the tide-generating forces than it is in gravitational attraction force:

$$\text{Tide-generating force} \propto \frac{m_1 m_2}{r^3}$$

Although the gravitational attraction between the earth and sun is over 177 times that between the earth and moon, the moon dominates the tides, as will be seen from the following comparison of the tide-generating force of the sun with that of the moon. Since the sun is 27 million times more massive than the moon, it should, solely on the basis of comparative masses, have a tide-generating force 27 million times greater than that of the moon. However, we must also consider the distance between the earth and the moon as compared to the distance between the earth and the sun. Since the sun is 390 times farther from the earth than the moon, its tide-generating force is reduced by 390^3, or about 59 million times compared to that of the moon. These conditions result in the sun's tide-generating force being $^{27}\!/_{59}$, or about 46%, that of the moon (fig. 10–1 and table 10–1).

The full explanation is complex, but we will attempt a brief explanation of how the tide-generating force decreases in proportion to the cube of the increase in distance between the particle of the earth that is affected and the center of the tide-generating bodies (moon and sun). We will confine our consideration to the moon, but a similar explanation applies to the sun.

The tidal pattern we see on the earth is primarily the result of the rotation of the **earth-moon system** around its center of mass. Figure 10–2A shows the path followed by the center of the earth as it rotates

Preceding page: View of Mont-Saint-Michel in the Gulf of Saint-Malo, France, where the tidal range exceeds 10 m (33 ft). At extreme low tide, the shoreline moves 15 km (9 mi) from the high-tide shoreline of the mainland. Mont-Saint-Michel is built on a granitic outcrop about 1.7 km (1 mi) from the high-tide shoreline of the mainland and is surrounded by ocean water during high tide. Photo courtesy of the French Government Tourist Office.

FIGURE 10–1 Relative Sizes of the Earth, Moon and Sun
The earth has a diameter of 12,682 km (7876 mi). The diameter of the moon is 3478 km (2160 mi), which is 0.27 that of the earth. The diameter of the sun is 1,392,000 km (864,432 mi), which is 109 times the diameter of the earth. The relative sizes are shown to scale. See table 10–1 for a comparison of the masses of the sun and moon and their distances from the earth. These factors are important in determining their tide-generating effect on the earth.

MOON	EARTH	SUN
3478 km (2160 mi) (0.27 × earth)	12,682 km (7876 mi)	1,392,000 km (864,432 mi) (109 × earth)

around the center of gravity for the earth-moon system, a point about 4700 km (2918 mi) from the center of the earth. It shows also that the **zenith** (point on the earth's surface closest to the moon) and **nadir** (point on the earth's surface farthest from the moon) follow identical circular paths. Actually, all points on the earth must follow circular paths identical to that of the center of the earth. If we divide the earth into millions of tiny particles of equal mass, each must have an identical **centripetal force** act on it to keep it in this circular path.

As an example of centripetal force, if you tie a string to a rock and swing the tethered rock around your head, the string pulls the rock toward your hand—a point in the center of the circular orbit the rock is following. The string is providing a centripetal (center-seeking) force to the rock. If the string breaks, the rock can no longer maintain its circular orbit and will fly off in a straight line tangent to the circle (fig. 10–2B).

Figure 10–3A shows that each particle of equal mass is located at a different distance and/or direction from the center of the moon. This causes the force of gravitational attraction between each particle and the moon, which must provide the required centripetal force to each particle, to be different. As a result of the differences between the required centripetal force and the gravitational force of attraction

TABLE 10–1 Relative Masses of Sun and Moon and Distances from Earth—Relative Tide-Generating Effect.

Tide-Generating Body	Distance from Earth (avg)	Mass (Metric tons)	Relative Tide-Generating Effect
Moon	384,835 (km) (238,483 mi)	7.3×10^{19}	Based on relative masses, the sun is 27 million times more massive than the moon and has 27 million times the tide-generating effect. However, since the sun is 390 times farther than the moon from the earth, its tide-generating effect is reduced by 390^3, or 59 million times.
Sun	149,758,000 km (93,016,845 mi)	2×10^{27}	

Moon
Earth   Distance of moon and sun from earth shown approximately to scale.   Sun

DETERMINATION OF TIDE-GENERATING FORCE OF SUN RELATIVE TO MOON

$$\text{Tide-generating force} \propto \frac{\text{Mass}}{(\text{Distance})^3} \propto \frac{\text{Sun—27 million times more mass}}{(\text{Sun—390 times farther away})^3}$$

$$(390)^3 = 59,000,000 \quad \text{Thus,} \quad \frac{27 \text{ million}}{59 \text{ million}} = 0.46 \text{ or } 46\%$$

The sun has 46% the tide-generating force of the moon.

FIGURE 10–2 Earth-Moon-System Rotation.

A, the dashed line through the center of the earth is the path of the earth's center as it moves around the common center of gravity of the earth-moon system. The circular paths followed by the nadir (point a) and zenith (point b) have the same radius as that followed by the earth's center. *B,* if you tie a string to a rock and swing it in a circle around your head, it stays in the circular orbit because the string exerts a centripetal (center-seeking) force on the rock. This force pulls the rock toward the center of the circle. If the string breaks with the rock at the position shown, the rock will fly off along a straight path tangent to the circle.

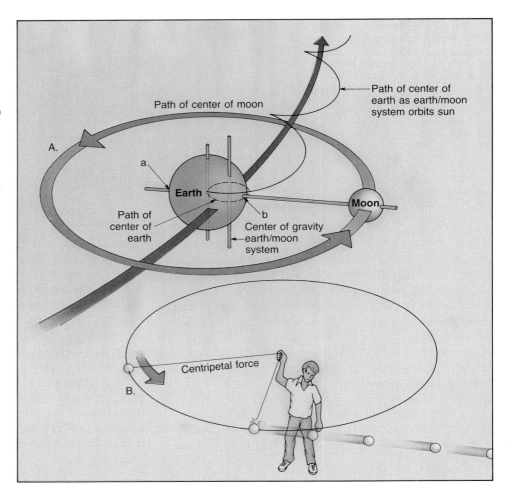

between each particle and the moon, small lateral (tangent to earth's surface) or tide-generating forces are produced to push the ocean's water into bulges at the zenith and the nadir.

A plane passed through the center of the earth and perpendicular to a line through the centers of the earth and moon intersects the earth's surface along a circle where tide-generating forces are zero. There is also no tide-generating force at the nadir and zenith. At all other points on the earth's surface there is a lateral component to the gravitational force between the particle it represents and the moon. It is these tiny lateral forces, the tide-generating forces, that push the water into the bulges at the zenith and nadir. The magnitude of the tide-generating forces increases from zero along the circle previously discussed to a maximum at an angle of 45° on either side of the circle and decreases again to zero at the zenith and nadir. These two bulges provide the framework within which the equilibrium tide operates (fig. 10–3B).

EQUILIBRIUM THEORY OF TIDES

We will now consider the kinds of tide observations that would be predictable on an earth with two tidal bulges, one toward the moon and one away from the moon as the earth rotates on its axis. In the following discussion, we assume an ideal ocean of uniform depth covers the earth and that there is no friction between the water in the ocean and the ocean floor. We will call this theoretical tide the **equilibrium tide.** Although this oversimplification will not give us an accurate means of predicting the types of tides that will occur throughout the earth's surface, it will give us the gross characteristics of tides in the world ocean. Only after we have established an understanding of the equilibrium theory will we attempt to consider the **dynamical theory** of tides that deals with changes in ocean depth, the existence of continents, and friction between the ocean water and ocean floor.

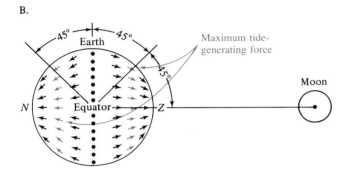

FIGURE 10–3 Tide-Generating Force.

A, the length and direction of the arrows in this figure represent the magnitude and direction of the forces related to each of the eight identical masses located at each of the points shown on the figure. *C arrows*—The centripetal force required to keep each particle in the identical circular orbit it follows as a result of the earth-moon rotation system. *G arrows*—The gravitational attraction between particles and the moon. This force provides the required centripetal force, but it is identical to it only for the particle at the center of the earth. It is in the required direction only for particles on a line connecting the centers of the earth and moon. For all particles on the half of the earth facing the moon, the force is larger than required, and for the particles on the half of the earth facing away from the moon, it is smaller. *Blue arrows*—For all particles except at the center of the earth, there is a resulting residual force because the gravitational force varies from the required centripetal force. This force is small—averaging about 10^{-7} of the magnitude of the earth's gravity. Therefore, where forces act perpendicular to the earth's surface, as does gravity, they do not have any tide-generating effect. However, where they have a significant horizontal component—tangent to the earth's surface—they aid in producing tidal bulges on the earth. Since there are no other large horizontal forces on the earth with which they must compete, these small, ever-present residual forces can push water across the earth's surface. *B,* along the intersection of the earth's surface and a plane running through the earth's center perpendicular to a line connecting the centers of the earth and moon, the residual forces are directed toward the center of the earth and have no horizontal component—tide-generating forces are zero (dots). There is also no horizontal component at the zenith (*Z*) and nadir (*N*) because the residual forces point away from the center of the earth. Elsewhere on the earth's surface, there are horizontal tide-generating forces. The magnitude of the tide-generating force varies. It reaches maximum values along two circles that lie 45° on either side of the previously described circle of zero tide-generating force.

The Rotating Earth

This ideal ocean is modified only by the tide-producing forces that cause bulges on opposite sides of the earth, as shown in figure 10–4. Let us assume that a stationary moon is directly above the equator so that the maximum bulge will occur on the equator on opposite sides of the earth. Since the earth requires 24 h for one complete rotation, an observer on the equator would experience two high tides per day. The time that would elapse between high tides, the **tidal period,** would be 12 h. An observer at any latitude north or south of the equator would experience a similar period, but the high tides would not be so high because the observer would be at the edge of the bulge rather than at the apex.

But high tides do not occur every 12 h on the earth's surface. This is due to the fact that the earth-moon system is rotating about its center of mass while the earth is rotating on its axis.

The **lunar day,** the time that elapses between successive passages of the moon across the meridian (longitude line) of an observer, must be somewhat longer than the solar day of 24 h. It is actually 24 h 50 min. If one observes the time at which the moon rises on successive nights, it can be seen to rise 50 min later each night. Figure 10–5 shows that this results from the fact that as the earth is making its rotation on its axis in 24 h, the moon has moved 12.2° to the east. The earth must rotate another 50 min to have the moon again on the meridian of the observer. Therefore, if we wish to divide the lunar day of 24 h 50 min into 24 lunar hours, each **lunar hour** will be

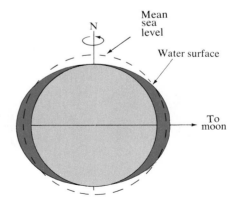

FIGURE 10–4 Equilibrium Tide.

Assuming an ocean of uniform depth covering the earth, and the moon located above the equator, the tide-generating forces will produce two bulges on the ocean surface. One will extend in the direction of the moon and the other away from the moon. As the earth rotates, all points on the earth except the poles will experience two high tides each day as the earth rotates beneath the two bulges.

FIGURE 10–5 The Lunar Day.
A lunar day is the time that elapses between successive appearances of the moon on the meridian directly over a stationary observer. As the earth rotates on its axis, the earth-moon-system rotation moves the moon in the same direction (to the east). During one complete rotation of the earth on its axis (the 24-h solar day), the moon moves east 12.2°, and the earth must rotate another 50 min to put the observer in line with the moon.

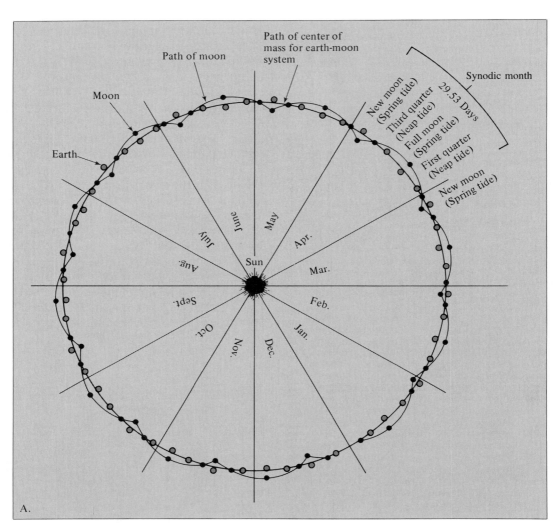

FIGURE 10–6 Sun-Moon-Earth Motion and Tides.
As the earth-moon system moves around the sun, the centers of each body follow a wavy path because they are also rotating about the center of mass of the earth-moon system (see fig. 10–2A). We can think of this system as a sledgehammer, with the earth being the heavy business end and the moon being a knob at the end of the handle. The "handle" that connects the earth and moon is gravity. If the hammer were thrown through space, the center of the knob (moon) at the end of the handle would follow an orbit around the center of mass with a much greater radius than that of the orbit of the center of the head of the hammer (earth). Therefore, the wobble of the moon is more pronounced and is shown in exaggerated form, A. During a synodic month, the moon moves from a position between the sun and the earth (new moon) to a position that puts the earth between the moon and the sun (full moon) and back to its original position.

about 1 h 2 min of solar time in length. With the knowledge that the moon completes a 360° revolution in 29.53 days, this 12.2° of eastward revolution for the moon can be computed by

$$\frac{360°}{29.53 \text{ days}} = 12.2° \text{ per day}$$

We have so far considered the effects of the earth's rotation and the revolution of the earth-moon system about its center of mass on the prediction of tides. We have ignored the effect of the sun. In the following discussion we will consider the combined effect of the moon and sun on the earth's tides.

Combined Effects of Sun and Moon

Figure 10–6 shows the path of the earth and moon as the earth-moon system revolves around the sun. It can be seen that approximately every 29½ days the moon is in the same phase. When the moon is between the earth and the sun it is said to be in **conjunction**, producing a **new moon**. The moon is in **opposition** when it is on the opposite side of the earth from the sun, causing a **full moon**. A **quarter moon** results when the moon is in **quadrature**, at right angles to the sun relative to the earth.

When the sun and moon are in opposition or conjunction, the tide-generating forces of the sun and

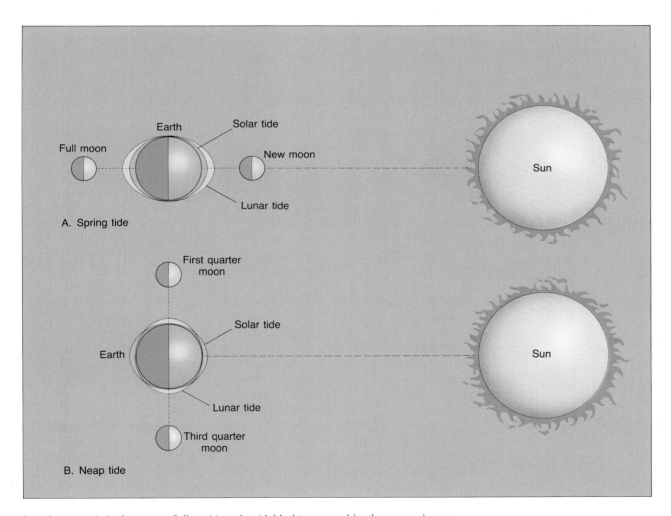

B, when the moon is in the new or full position, the tidal bulges created by the sun and moon are aligned, producing constructive interference and a larger bulge. When the moon is in the positions halfway between the new and full phases, the first and third quarters, the tidal bulge produced by the moon is at right angles to the bulge created by the sun. The bulges tend to "cancel" each other (destructive interference), and the resulting bulge is smaller. New moon and full moon phases produce spring tides with maximum tidal ranges, while the first and third quarter phases of the moon produce neap tides with minimal tidal ranges. *B,* from The Tasa Collection: *Shorelines.* Published by Merrill Publishing Co., Columbus, OH, Copyright © 1986, by Tasa Graphic Arts, Inc. All rights reserved.

GRUNION AND THE TIDES

Along the beaches of southern California and Baja California from March through September a very unusual fish-spawning behavior can be observed. Shortly after the maximum spring tide has occurred, small silvery fish come ashore to bury their fertilized eggs in the sand. They are the grunion *(Leuresthes tenuis),* slender little fish 12 to 15 cm (4.7 to 6 in.) in length. The name grunion comes from the Spanish *gruñón,* which means grunter. The early Spanish settlers gave the fish this name because of the faint noise they make during spawning.

The type of tide that occurs along southern California and Baja California beaches is a mixed tide. On most tidal days (24 h 50 min) there are two high and two low tides. There is usually a significant difference in the heights of the two high tides that occur each day. During the summer months, this higher high tide occurs at night. As the higher high tides become higher each night while the maximum spring-tide range is approached, sand is eroded from the beach (fig. 10A). After the maximum height of the spring tide has occurred, the higher high tide that occurs each night will be a little lower than the one of the previous night. During this sequence of decreasing heights of the tides as neap-tide conditions approach, sand is deposited on the beach. The grunion depend greatly on this pattern of beach sand deposition and erosion in their spawning process.

Grunion spawn only after each night's higher high tide has peaked on the three or four nights following the night of the occurrence of the highest spring high tide. This behavior assures that the eggs will be covered deeply by sand deposited by the receding high tides. The fertilized eggs buried in the sand are ready to hatch nine days after spawning. By this time, another spring tide is approaching, and the higher high tide that occurs each night will be higher than that of the previous night. This condition causes the beach sand to erode, exposing the eggs to agitation by waves that break ever higher on the beach. The eggs hatch about three minutes after being freed in the water. Tests done in laboratories have shown that the eggs will not hatch until agitated in a manner that simulates the agitation of the eroding waves.

The spawning begins as the grunion come ashore immediately following an appropriate high tide, and it may last from 1 to 3 hours. Spawning activity usually peaks about an hour after its start and may last from 30 minutes to an hour. Thousands of fish may be on the beach at this time. During a run, females move high on the beach. If no males are near, a female may return to the water without depositing her eggs. In the presence of males, she will drill her tail into the semifluid sand until only her head is visible. The female continues to twist, depositing her eggs 5 to 7 cm (2 to 3 in.) below the surface. The male curls around the female's body and deposits his milt against it (fig. 10B). The milt runs down the body of the female to fertilize the eggs. The spawning completed, both fish return to the water with the next wave.

As soon as the eggs are deposited, another group of eggs begins to form within the female. They will be deposited during the next spring-tide run. Larger females are capable of producing up to 3000 eggs for each series of

Spring tide   Neap tide   Spring tide

| 1 | 2 | 3 | 4 | 5 | 6 | 7 | 8 | 9 | 10 | 11 | 12 | 13 | 14 | 15 | 16 | 17 | 18 |

Days

Flood tides erode sand and free grunion eggs during higher high tide as maximum spring tidal range is approached.

Grunion deposit eggs in beach sand during early stages of the ebb of higher high tides on the three or four days following maximum spring tidal range.

Maximum spring tidal range

FIGURE 10A   Grunion and the Tidal Cycle
Eggs deposited in the beach during ebb tides on days 3, 4, and 5 will be ready to hatch when eroded by the flood tides of days 13, 14, 15, and 16. The lines separating days on the tide curve chart represent midnight.

FIGURE 10B   Grunion Spawning.
A male grunion wraps around the female that has burrowed into the sand to deposit her eggs beneath the surface. The male releases sperm-laden milt, which runs down the female's body to fertilize the eggs. Photo by Bill Beebe.

spawning runs, which are separated by the 2-week period between spring-tide occurrences. Early in the spawning season, only older fish spawn, but by May even the 1-year-old females are in spawning condition.

Young grunion grow rapidly and are about 12 cm (5 in.) long when they are a year old and ready for their first spawning. They usually live 2 or 3 years, but 4-year-olds have been recovered. The age of a grunion can be determined by the scales. After growing rapidly during the first year, they grow very slowly. There is no growth at all during the 6-month spawning season, which causes marks to form on each scale.

How the grunion are able to time their spawning behavior so precisely is not known. Some investigators believe the grunion are able to sense very small changes in the hydrostatic pressure caused by the changing level of the water associated with the tidal ebb and flow. Certainly some very dependable detection mechanism keeps the grunion accurately informed of the tidal conditions, because their survival depends on a spawning behavior precisely tuned to tidal motions.

moon are additive, and we experience maximum **tidal ranges**, the vertical difference between high and low tide. During quadrature, the tide-generating force of the sun is working at right angles to the tide-generating force of the moon, and we have minimum tidal range. The maximum tidal range condition that exists during the new and full moon phases is referred to as the **spring tide**, and during the quadrature phases we have **neap tide** (fig. 10–6). The time that elapses between successive spring tides (full moon and new moon) or neap tides (first and third quarters) is a **fortnight**, about two weeks.

In figure 10–7A we see the spring-tide condition associated with the new moon (the same conditions would exist with a full moon). As is the case in all the diagrams in the figure, the tide generated by the sun is indicated by a dash-and-dot line and the tide generated by the moon is indicated by a dashed line. The net tide of both tide-generating forces is depicted by the colored line. Since the moon and sun are on the same side of the earth and are more or less aligned along a line connecting the center of the earth with the center of the sun, maximum tide heights occur at midnight and noon. We have extremely high tides and extremely low tides under this arrangement of the earth, moon, and sun—a spring-tide condition.

Part B of the figure illustrates the tide that results from earth-moon-sun positions during quadrature. During third quarter, the tide-generating effect of the moon is maximum at 6:00 A.M. and 6:00 P.M., when it

FIGURE 10–7 Sun-Moon Tide Curves.

A, **spring tide.** During a new moon, the sun and moon tidal bulges are centered at the same longitude, so their effects are added to produce maximum high tides and minimum low tides. This produces maximum tidal range. *B,* **neap tide.** During a third-quarter moon, the sun and moon tidal bulges are at right angles to each other. Thus, the sun tide reduces the effect of the moon tide. For example, at 6 A.M. the observer is passing through the moon bulge and the sun trough. This produces a high tide in which the effect of the moon bulge is reduced by an amount equal to the sun trough. At 12 noon the moon trough is reduced by the sun bulge, so a reduced low tide occurs. This produces minimum tidal range. *C,* in a situation where the sun and moon bulges are separated by an angle of 45° the sun bulge is alternately added to and subtracted from the moon bulge. This produces a tidal range halfway between that of the spring and neap tides.

is at zenith and nadir, respectively, in relation to the observer. The sun is at nadir at 12:00 midnight and zenith at 12:00 noon. These are the times of maximum tide-generating effect of the sun. In this instance, we can see that the maximum lunar tides correspond with the minimum solar tides, producing a tide pattern with relatively small tidal range—a neap tide. The same conditions would exist when the moon is in its first quarter.

Figure 10–7C shows a situation that would develop halfway between neap-tide and spring-tide alignments of the earth, moon, and sun. For example, the moon is in a position halfway between the full moon and the quadrature that occurs at the end of the third quarter. When the moon is at zenith at 3:00 A.M., the observer still has to rotate 135° before the sun will be at zenith at 12:00 noon. The sun and moon are at nadir at 12:00 midnight and 3:00 P.M., respectively. The pattern is a little more complicated than that which occurs during spring and neap tides. Nevertheless, it can still be seen that the net tide curve, which shows a tidal range that is less than that of spring tides and greater than that of neap tides, is still determined by the combined effect of the solar and lunar tide-generating forces. "Grunion and the Tides," a special feature on page 248, discusses the close association between the spawning of these small fish and the spring neap-tide sequence.

Effects of Declination

Up to this point we have considered that the moon and the sun remained at all times above the equator, or at least we did not describe specifically any deviation from such a position. We now must consider the fact that this is not the case. As the earth revolves around the sun, its axis of rotation is tilted 23.5° from vertical relative to the **ecliptic**, the plane of the earth's orbital path about the sun. It is owing to this tilt that we experience the seasons—spring, summer, fall, and winter. To further complicate our consideration, the plane of the moon's orbit is at an angle of 5° to the ecliptic. This angular distance of the sun or moon above or below the equatorial plane of the earth is called **declination**.

The tilt of the earth's axis relative to its plane of revolution around the sun is shown in figure 10–8A. It can be seen that the tilted axis assumes a constant direction in space throughout the yearly cycle that includes the equinoxes and solstices. At the **vernal equinox**, which occurs about March 21, the sun is directly above the equator and is moving from the Southern Hemisphere to the Northern Hemisphere. On about June 21 the **summer solstice** occurs. At this

time the sun reaches its most northerly point in the sky, directly above the Tropic of Cancer, which is at 23.5°N latitude. Following this occurrence, the sun moves farther south in the sky each day, and on about September 23 it is directly above the equator again and produces the **autumnal equinox**. During the next three months the sun appears to be more southerly in the sky until the **winter solstice** on about December 22, when it is directly over the Tropic of Capricorn at 23.5°S latitude. Thus the sun may be found at declinations between 23.5° north and 23.5° south of the equator on a yearly cycle.

Since the plane of the moon's orbit intercepts the plane of the ecliptic at an angle of 5°, and since the plane of the moon's orbit **precesses**, or rotates, while maintaining its 5° angle with the ecliptic with a precessional cycle of 18.6 yr, we have a relatively complex consideration regarding the declination of the moon relative to the plane of the earth's equator. In part 1 of figure 10–8B, the declination of the moon's orbit relative to the earth's equator is 28.5°. The declination will change from 28.5° south to 28.5° north and back to 28.5° south of the equator in a period of one month. Part 2 of the illustration shows the relationship of the ecliptic, the plane of the moon's orbit, and the plane of the earth's equator after one-fourth of the precession, or 4.65 yr later. The maximum declination of the moon's orbit relative to the earth's equator still approaches 28.5°. However, in part 3, when one-half precession is completed, which is 9.3 yr after the condition observed in part 1, it can be observed that the maximum declination of the moon relative to the earth's equator is 18.5°.

On the basis of these considerations we must alter our previous concept of the predicted equilibrium tide. We must now expect that tidal bulges will rarely be aligned with the equator and will occur for the most part north and south of the equator. Since the moon is the dominant force that creates tides in the earth's oceans, we would suspect that the bulges would follow the moon as it moves on its monthly journey across the equator, being found at a maximum of 28.5° north and south of the equator (fig. 10–9).

Effects of Distance

Additional considerations that will affect the tide-generating force of the sun and moon on the earth are their changing distances from the earth. The earth ranges from **perihelion** with the sun, when the distance between the bodies is 148.5 million km (92.2 million mi) during the winter months in the Northern Hemisphere, to **aphelion**, when the distance between

FIGURE 10–8 Orbital Planes of Earth and Moon.

A, as the earth orbits the sun during one year, the axis of rotation is tilted 23.5° from perpendicular relative to the ecliptic. The sun shines down directly over the Tropic of Cancer (23.5°N) on the day of the summer solstice, June 21. Three months later the sun is directly above the equator (0°) during the autumnal equinox, September 23. The sun is directly over the Tropic of Capricorn (23.5°S) during the winter solstice, December 22, and returns to a position above the equator on the vernal equinox, March 21. Three months later, the yearly orbit is completed, and the sun is again directly over the Tropic of Cancer. *B,* the plane of the moon's orbit (gray plane) is tilted at an angle of 5° relative to the plane of the ecliptic (yellow plane) and rotates with a clockwise precession that has a period of 18.6 years. In *B1,* the declination of the moon's orbit is the sum of the angle of intersection of the plane of the earth's equatorial plane (blue plane) with the plane of the ecliptic (23.5°) plus the angle of intersection between the plane of the moon's orbit and the ecliptic (5°). This produces the maximum declination of the moon relative to the earth's equator of 28.5°. *B2* shows the positions of the planes 4.65 years later when the moon has achieved one-fourth of its precessional rotation. *B3* shows the relative positions of the planes after 9.3 years or one-half of the precession. The maximum declination of the moon relative to the earth's equator is now 18.5°, or 23.5° less 5°. From C. Hauge, Tides, currents, and waves, *California Geology,* July 1972. Reprinted by permission of the author.

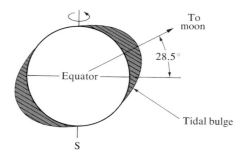

FIGURE 10–9   Maximum Declination of Tidal Bulges from Equator.

The center of the tidal bulges may lie at any latitude from the equator to a maximum of 28.5° on either side of the equator.

the bodies is 152.2 million km (94.5 million mi) during the summer months. The moon moves from **perigee**, its closest approach to the earth, of 375,200 km (233,000 mi) to **apogee** of 405,800 km (252,000 mi) and back to perigee during a period of 27½ days—the **anomalistic month**. (See fig. 10–10.)

Because of these movements, spring tides have greater ranges during the Northern Hemisphere winter than in the summer. Also, as the result of changing distance between the earth and moon we will find that tidal ranges will become greater at perigee each anomalistic month. Keeping in mind that tide-generating forces vary inversely with the cube of the distance from the center of the earth to the center of the tide-generating body, it can be appreciated that these changes resulting from the elliptical nature of the earth's orbit around the sun and the moon's orbit around the earth can be readily observed in the tides.

Equilibrium Tide Prediction

To consider the tidal patterns we would predict for an idealized water-covered earth, let us return our atten-

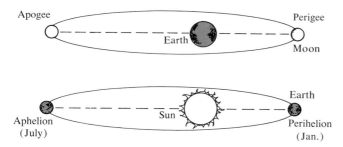

FIGURE 10–10   The Effects of Elliptical Orbits.

The moon moves from perigee of 375,200 km (233,000 mi) to apogee of 405,800 km (252,000 mi) (top). Greater tidal ranges are experienced during perigean tides. Perihelion brings the earth within 148,500,000 km (92,200,000 mi) of the sun. Aphelion distance is 152,200,000 km (94,500,000 mi) (bottom). Greater tidal ranges are experienced during perihelion tides.

tion to the effect of declination. Let us assume that the declination of the moon, which will determine the alignment of the tidal bulges, is 28° north of the equator. If we position an observer at this latitude on a permanent point, the observation of the tides that occurs here will be different from observations at the equator.

If the observations begin when the moon is directly over the observer's head, high tide will be recorded. Six lunar hours later a low tide will be recorded, followed by another high tide that will be much lower than the initial high tide (fig. 10–11A–D). At the end of a 24-lunar-hour period, the observer will have passed through a complete lunar-day cycle of two low tides and two high tides. Thus, the predicted period of the equilibrium tide is **semidiurnal**, half a lunar day. A representative curve of the type of tide observed can be seen in figure 10–11E. Tide curves showing the heights of the same tides during one lunar day at the equator and at 28°S latitude are also provided. The difference in heights of successive high tides or successive low tides that occurs at a given location as a result of the declination of the moon and the sun relative to the earth's equator is called the **diurnal inequality**. When the moon is at its maximum northern declination and again about two weeks later when it is at its maximum southern declination, the diurnal inequality reaches a maximum. Since the moon is generally above the tropics at these times, we call these tides **tropical tides**. Separating the tropical tides are tides that have a minimum diurnal inequality when the moon is over the equator, **equatorial tides**.

To summarize the tides that we might predict as a result of the tide-generating forces of the moon and sun on an earth covered with a uniform depth of water for any point not on the equator, we would predict that the equilibrium tide would generally be observed at any location to have two high tides and two low tides per lunar day. Except for the rare occasions when the sun and moon are simultaneously above the equator, we would expect that neither the two high tides nor the two low tides would be of the same height because of the changing declination of the moon and the sun. We would expect yearly and monthly cycles of tidal range related to the changing distances of the earth from the sun and moon. Lastly, we would expect that each fortnight, half a lunar month, we would experience spring tides that would be separated by neap tides (fig.10–11F).

Considering the great number of variables that are involved in predicting tides, it is interesting to consider when the conditions might be right to produce the maximum possible tide-generating force. The

FIGURE 10–11 Predicted Equilibrium Tides.

F, from Anikouchine and Sternberg. *The world ocean: An introduction to oceanography,* © 1973. Reprinted by permission of Prentice-Hall, Englewood Cliffs, New Jersey.

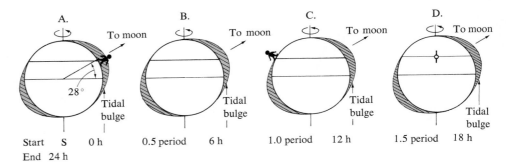

A. At start of lunar day the observer is at zenith in the center of the tidal bulge. He experiences a high tide. *B.* Six lunar hours later (0.5 period) the observer experiences low tide while located on the back side of diagram. *C.* After twelve lunar hours (1.0 period or 0.5 lunar day) the observer again experiences high tide. It is much lower than the first high tide, however, because the observer is now passing through the edge of the tidal bulge. *D.* Observer experiences low tide again eighteen lunar hours (1.5 period) after start. At the end of one lunar day (24 hours) the observer returns to *A* and experiences a high high tide.

Tide curves for 28°N, 0°, and 28°S latitudes when the declination of the moon is 28° N (all curves for same longitude). Note that tide curves for 28°N and 28°S have identical highs and lows, but are out of phase by 12 hours. This results from the fact that the bulges in the two hemispheres occur on opposite sides of the earth.

Along with the unequal heights for the two high and low tidal extremes occurring each lunar day, we expect to observe spring and neap tides. Thus the tide curve for the month of June 1969 demonstrates the general character of the predicted equilibrium tide.

maximum tidal range will be produced when the sun is at perihelion and in conjunction or opposition with the moon at perigee and when both the sun and moon have a zero declination. This condition occurs only once each 1600 years, and the next occurrence is predicted for A.D. 3300.

An indication of the significance of even a near coincidence of the earth's perihelion with the perigee of the moon during a spring tide was made woefully clear during the January 1983 storms in the North Pacific. The introduction of slow-moving low-pressure cells that developed over the Aleutians caused strong northwest winds to blow across the ocean from Kamchatka to the U.S. coast. Averaging about 50 km/h (31 mi/h), the winds produced a near fully developed 3-m (10-ft) swell along the coast from Oregon to Baja California (see tables 9–2 and 9–3). The storm surge from this condition would have been trouble enough under average conditions, but the situation was made worse by the 2.25-m (7.4-ft) high spring tides and a slight increase in sea level along the coast that resulted from the influx of warm water driven north by the 1982–83 El Niño (see "El Niño–Southern Oscillation Events," p. 160).

The unusually high tides occurred because the earth was still near perihelion (January 2) when the moon reached perigee on January 28, 1983. Some of the largest waves came ashore January 26–28, when over $100 million in damage was done (fig. 10–12). Twenty-five homes were destroyed, over 3500 homes were seriously damaged, and many commercial and municipal piers collapsed. At least a dozen lives were lost. With each such occurrence, we learn more of the cost of development on and near the shore. If the predicted increased greenhouse effect does occur, destructive storms of this type will occur more frequently than in the past because of the increased sea level resulting from the water expansion caused by increased sea-surface temperature. Should melting of glacial ice occur also, the threat of such destructive storms would greatly increase.

A.

B.

FIGURE 10–12 High Tides of January 1983.
A, January 1983 storm winds blow uninterrupted across the North Pacific Ocean from Kamchatka to the U.S. coast. B, homes threatened by storm waves and unusually high tides on January 27, 1983, at Stinson Beach north of San Francisco. Wide World Photos.

DYNAMICAL THEORY OF TIDES

In our previous discussion of the equilibrium tide, we considered that the tidal bulges were directed at the moon and away from the moon on opposite sides of the earth. Maintaining this relationship with the moon as the earth rotates beneath the moon, the bulges (or wave crests), which would be separated by a distance of half the earth's circumference (about 20,000 km, or 12,420 mi), would be moving across the earth at a speed of more than 1600 km/h (994 mi/h). We previously stated that the tides were an extreme example of shallow-water waves, whose velocity is proportional to the square root of the water depth. In order for the tidal wave to travel at 1600 km/h, the depth of the idealized ocean would have to be 22 km (13.7 mi). Since the average depth of the ocean is less than 3.9 km (2.4 mi), the tidal bulges move at speeds slightly under 700 km/h (435 mi/h). The actual position of the bulges relative to the tide-generating forces is determined by a balance that is reached between these forces and the shallow-water wave movement of the bulges provided by the depth of the ocean.

We must also consider the effect of the continents, which interrupt the free movement of the tidal bulges across the ideal unobstructed ocean surface we considered to exist in our discussion of the equilibrium tide. The ocean basins between continents have set up within them free oscillatory waves whose character modifies the forced astronomical tidal waves that develop within the basin. It is impossible for us to explain here the various tidal phenomena that occur throughout the world. For instance, high tide rarely occurs at the time the moon is at zenith, and the amount of time that elapses between the passing of the moon and the occurrence of high tide varies from place to place as a result of the many factors that determine the characteristic of the tide at any given location.

In order to develop a maximum understanding of tides, a combination of mathematical analysis and observation is required. When we consider the **harmonic analysis**, a mathematical approach to the study of tides which embraces all the tide-generating variables that possess a **periodicity** (cyclic pattern), there are almost 400 tide-generating variables, called **partial tides**, identified. The actual tide observed at any given location is the combined effect of all the partial tides at that point. To simplify things somewhat, a relatively accurate model of the actual tide can be computed considering only the seven major partial tides. They are listed with their characteristics in figure 10–13. Combining the periods of each of the partial tides with the amplitudes and phases that can be obtained from observation, relatively accurate predictions of the tide at any location can be made. The observations must have been made throughout a period of at least 18.6 yr to make the predictions as accurate as possible, as this is the period of the precession of the plane of the moon's orbit through the ecliptic (fig. 10–8).

Types of Tides

It was previously mentioned that everywhere on the earth we should expect two high tides and two low tides of either equal or unequal heights during a lunar day. Owing to modifications resulting from varying depths, sizes, and shapes of ocean basins, the tide predicted by the equilibrium theory is replaced in many parts of the world. In reality one finds either a diurnal (daily), semidiurnal (twice daily), or mixed tide (fig. 10–14).

The **diurnal tide** is characterized by a single high and low water each lunar day. These tides are common in the Gulf of Mexico and along the coast of Southeast Asia. Such tides have a tidal period of 24 h 50 min.

The **semidiurnal tide** has two high and two low waters each lunar day, and the heights of successive high waters and successive low waters are approximately the same. Since tides are always getting

A. The seven most important partial tides.

	Symbol	Period in solar hours	Amplitude $M_2 = 100$	Description
Semidiurnal tides	M_2	12.42	100.00	Main lunar (semi-diurnal) constituent
	S_2	12.00	46.6	Main solar (semi-diurnal) constituent
	N	12.66	19.1	Lunar constituent due to monthly variation in moon's distance
	K_2	11.97	12.7	Soli-lunar constituent due to changes in declination of sun and moon throughout their orbital cycle
Diurnal tides	K_1	23.93	58.4	Soli-lunar constituent
	O	25.82	41.5	Main lunar (diurnal) constituent
	P	24.07	19.3	Main solar (diurnal) constituent

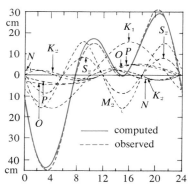

B. Partial tides, computed tide, and observed tide at Pula, Yugoslavia (January 6, 1909). Note close fit of computed and observed tides.

FIGURE 10–13   Partial Tides.

The semidiurnal and diurnal partial tides described in *A* are shown in tide curve form in *B*, along with the computed and observed tides that resulted from their combined effects at Pula, Yugoslavia. *A* and *B* from A. Defant, *Ebb and flow,* © 1958. Reprinted by permission of The University of Michigan Press, Ann Arbor.

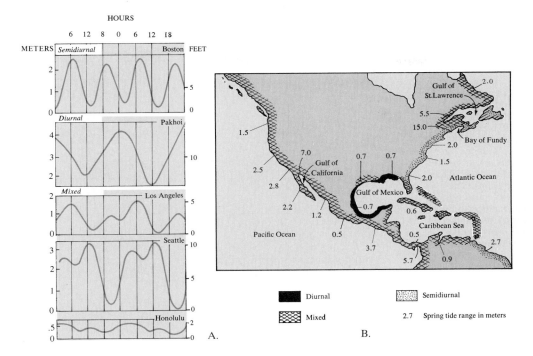

FIGURE 10–14 Types of Tides.

A, types of tides. In a **semidiurnal** (twice daily) type of tide, there are two highs and lows during each tidal day and the heights of each successive high and low are about the same. In the **diurnal** (daily) type of tide, there is only one high and one low each tidal day. In the **mixed** type of tide, both diurnal and semidiurnal effects are detectable and the tide is characterized by a large difference in the high-water heights, the low-water heights, or both, during one tidal day. Even though a tide at a place can be identified as one of these types, it still may pass through stages of one or both of the other types. For example, the tide along the North Carolina coast changes from semidiurnal to mixed, but is predominantly semidiurnal. *B,* map showing the types of tides that have been observed along portions of the coasts of North and South America. The numbers give the spring tide range in meters and are therefore near the maximum tidal range that can be expected. Storm waves, lower barometric pressure, ocean currents, and the alignment of additional celestial bodies could increase the range. After C. Hauge, Tides, currents, and waves, *California Geology,* July 1972. Reprinted by permission of the author.

higher or lower at any location, owing to the spring neap-tide sequence, successive high tides and successive low tides can never be exactly the same at any location. Semidiurnal tides are common along the Atlantic coast of the United States. The tidal period is 12 h 25 min.

The **mixed tide** may have characteristics of both diurnal and semidiurnal tides. The diurnal inequality discussed earlier is a characteristic of this tide, as successive high tides and/or low tides will have significantly different heights. Mixed tides usually have a tidal period of 12 h 25 min, which is a semidiurnal characteristic, but will also possess diurnal periods for a few days of each tidal month. This is the tide that is most common throughout the world and the type that is found along the Pacific coast of the United States. Note that the tide curve for Los Angeles, shown in figure 10–15, is mixed because it is semidiurnal in period during all days except September 15 and 16, when the period is diurnal.

Tides in Lakes

Before discussing the open ocean, we will consider the effects of the tide-generating forces on a body of water in a small rectangular closed basin. In general, lake basins are too small to have any appreciable tide setup by the tide-producing forces, but such forces may be significant where the long axis of the basin runs parallel to lines of latitude. In any such basin, very small standing waves will be generated with a period equal to that of the tide-generating force. This wave is referred to as a **forced standing wave** (fig. 10–16).

Of much greater importance is the **free standing wave** initiated by strong winds at the surface or by some seismic disturbance that may be set up in the basin (fig. 10–16). The period of a free standing wave is determined by the length and depth of the basin, and this period is termed the **characteristic period** for the basin. If the characteristic period of the free wave

FIGURE 10–15 Tidal Curves for Three Locations.

The declinations of the moon and the sun, phase of the moon, and position of the moon in its orbit as noted across the top of the chart all contribute to the tidal variations that are depicted. Additional contributing factors include the position of the earth in its orbit around the sun and the configuration of the sea bottom and basin boundaries. Some names of water levels are listed along the right margin. Those that are used as the chart datum for a place are marked with an asterisk. MHWS, mean high water springs—the average height of the high water of the spring tides; MHW, mean high water—the average height of all the high tides at a place; MLW, mean low water—the average height of all the low tides at a place; MLWS, mean low water springs—the average height of all low waters of the spring tides; MHHW, mean higher high water—the average height of the higher high tides at a place where the tide is the mixed type and displays an inequality during a tidal day; MLLW, mean lower low water—the average height of the lower low tides at a place where the tide is of the mixed type. From C. Hauge. Tides, currents, and waves, *California Geology,* July 1972. Reprinted by permission of the author.

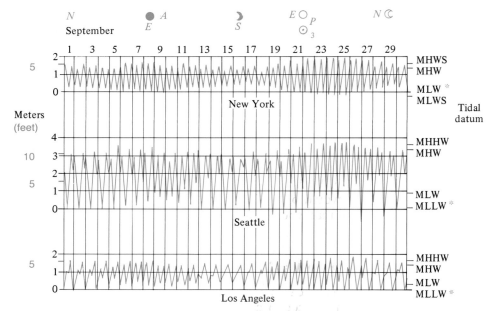

●, new moon; ☽, first quarter; ○, full moon; ☾, last quarter; *E*, moon on the equator; *N, S*, moon farthest north or south of the equator; *A, P*, moon in apogee or perigee; ₁⊙₃, sun at autumnal equinox; * , chart datum.

FIGURE 10–16 Tides in Lakes.

A. A forced standing wave generated by tide-generating forces. It will have a period of 12 h 25 min.

B. A free standing wave (seiche) generated by atmospheric disturbance. Its period is determined by the size and shape of the basin of the lake. The period (*T*) is related to length and depth of the lake as is shown by the equation:

$$T = \frac{2L}{\sqrt{gh}}$$

L = length of lake (m)
g = acceleration due to gravity (9.8 m/s²)
h = depth of lake (m)

If the lake has dimensions that produce a period for the seiche that is approximately equal to the forced tide generated wave, they will be resonant and produce greater displacements at the antinodes.

is very nearly that of the period of a forced wave resulting from the tide-generating force, the oscillations may reinforce each other, and **resonance tides** with greater range may develop as they do in Lake Ontario.

Free oscillations of the type described above are common on large lakes and were given the name **seiche** by the inhabitants of the region near Lake Geneva, Switzerland. For seiches that have a single nodal line, the formula for the period in figure 10–16 gives a close approximation to the actual period that develops. For rectangular basins with two and three nodal lines the periods would be approximately one-half and one-third those that would be computed for a mononodal seiche.

Tides in Narrow Open Basins

As discussed above, even under conditions of resonance between free and forced oscillations in the small closed bodies of water, the maximum tides are not great. Seldom do they reach ranges in excess of a few centimeters. By contrast, if we consider similarly sized basins that are open at one end to the ocean, we find that the tides can become large in comparison. Why the difference?

To help answer this question, let us consider a rectangular bay aligned east-west with one end open to the ocean (fig. 10–17). Even though one end of the basin is open, we will still have to consider the effect of free standing waves developing much as we might conceive of the standing waves that develop in an organ pipe that has one end open. Without attempting to explain the mathematics, it is a fact that free standing waves can be reflected from the open end of the basin.

At the open end of the bay, the water level must always be the same as the height of the ocean at the location. Because of this, tidal range will be of greater magnitude than it is in closed basins, which have relatively small tidal ranges. If we think of the free standing wave that forms as a result of reflection of wave energy from the closed end of the basin as superimposed on the forced wave produced by tidal forces, we can see that combining the energy of both waves could produce a standing wave with increased amplitude within the embayment (fig. 10–17).

Tides in Broad Open Basins

In large, broad basins, a manifestation of the earth's rotation causes a rotary movement of the wave crest around the margins of the embayment. Since there is a significant horizontal transport of water associated

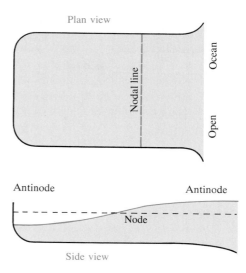

FIGURE 10–17   Tides in Seas, Bays, and Gulfs.
In seas, bays, and gulfs the forced standing waves will have a greater height than those created in lakes. This is because the height of the tide at the open end of the basin must be the same as the open ocean. Therefore, the development of a resonant condition between the free and forced standing waves in such basins will produce much greater displacements at the antinodes.

with standing waves, the Coriolis effect becomes important in embayments that are wide and will thus allow lateral movement of the water.

Let us assume an embayment in the Northern Hemisphere that has developed a mononodal standing wave is experiencing the development of a crest at its closed end, as illustrated in figure 10–18A. As the water begins to move toward the open end of the embayment (10–18B), the Coriolis effect diverts the water to the right, forcing it to pile up along the left-hand side of the embayment as we look at it from the open end. One-half period after the crest at the closed end of the basin begins to fall, a crest will form at the open end (10–18C). As the water moves toward the closed end of the basin, again the Coriolis effect will pile it up along the right-hand margin of the embayment (10–18D). If a basin is wide, this produces an **amphidromic point**, a point of no tidal range, around which the crest of the wave rotates in a counterclockwise direction instead of in a mononodal line as might have developed in a narrow basin.

Radiating from the amphidromic point and connecting points at which high water will occur simultaneously are **cotidal lines**. Figure 10–18E shows the cotidal lines at one-hour intervals for a tide in a rectangular basin with a period of 12 lunar hours. The cotidal lines are designated by Roman numerals that show that the crest of the tide wave moves in a counterclockwise direction. The crest makes one complete

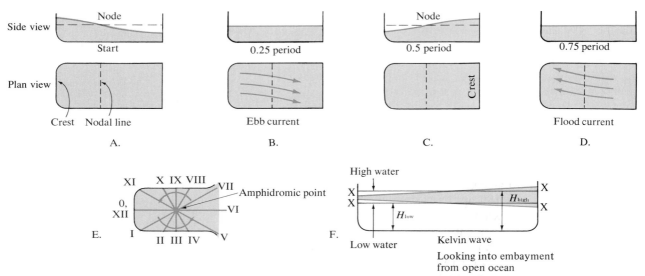

A–D. As the crest forms in the head of the bay there is no horizontal water movement. In *B* the surface is in an equilibrium (flat) condition and water moves horizontally toward the open ocean. The Coriolis force causes the water to veer to the right in the Northern Hemisphere. After the crest forms at the mouth of the bay (*C*), the same effect causes water to veer toward the opposite side of the bay when water moves toward the head of the bay in *D*.

E. In bays with sufficient width, the nodal line of the standing wave disappears and is replaced by an amphidromic point, a point of no tide. The standing wave has taken on a progressive character, and the crest moves in a counterclockwise path around the bay. The lines labeled with Roman numerals are cotidal lines that connect all points experiencing high tide at the indicated lunar hour. The tide crest rotates around the amphidromic point, and tidal range increases with increasing distance along the cotidal lines from the amphidromic point.

F. Kelvin wave—Along with the effects shown in *A–E*, the Coriolis force acting on a progressive wave moving along any channel or embayment produces a tilted water surface instead of a flat high and low water at tidal extremes. This results in a greater tidal range on one side of a basin than the other. On the right side, the tidal range is $H_{high} - H_{low} + 2x$. On the left side the range is $H_{high} - H_{low} - 2x$.

FIGURE 10–18   Tides in Wider Embayments.

rotation around the amphidromic point during the tidal period. The greater width of the basin combined with the Coriolis effect prevents the perfect back-and-forth reflection needed to produce pure standing waves. In wide basins a progressive wave moves around the basin in a counterclockwise direction. Figure 10–19 shows the cotidal rotations that have replaced the equilibrium-theory bulges in the world ocean.

Figure 10–18F shows that as progressive waves move through broad basins there will be a much greater tidal range on the right-hand side of the basin, looking into the basin, than on the left. Assuming a theoretical high- and low-tide level for the system and a tilt in the surface of the wave caused by the Coriolis effect as it moves into and out of the basin, it can be seen that along the right-hand side of

the basin the tidal range is equal to the theoretical tidal range plus $2x$. The x value results from the Coriolis force tilting the wave, a **Kelvin wave**, as it moves through the basin. On the left-hand side of the basin the tidal range, as a result of this tilting, will be equal to a theoretical difference between high and low tide of minus $2x$.

Coastal Tide Currents

The current that accompanies the slowly turning tide crest in a Northern Hemisphere basin will turn in a counterclockwise direction, producing a **rotary current** in the open portion of the basin. Near shore, except where the shore is straight and steeply sloping, the rotary current is changed to an alternating or reversing current that moves in and out rather than along the coast as would a rotary current. These reversing currents are of the greatest concern to navigators, since they reach velocities of 44 km/h (27.6 mi/h) in restricted channels between islands of coastal British Columbia. Velocities of the rotating currents in the open ocean are usually well below 1 km/h (0.62 mi/h).

The reversing tidal current that develops throughout a lunar day for a mixed tide is depicted in figure 10–20. Beginning at high tide, the current velocity is zero because the water has just reached its highest stage and is momentarily to begin its outward flow. Following this **high slack water**, the lowering of the tide begins, and the **ebb current** velocity increases and reaches a maximum about 3 lunar hours after high slack water. The velocity decreases and eventually reaches zero again at the first **low slack water**. Following the change in current velocity associated with the tidal phase throughout the day, it can be seen that the maximum current velocity is reached midway through the ebb current that occurs between the higher high water and lower low water.

The illustration shows that the lower low water (LLW) is below the **tide datum** of the chart, the zero mark. How can the tide be less than zero? The datum that is commonly used for mixed tides is the mean lower low water, the average height of the lower low tides at the locality. Since the lower low tide that we are recording is below the average lower low tide for this locality, this tide will have a negative value. In areas where mixed tides are not observed, the tide datum is very commonly the mean low tide recorded at that place—the average low tide. Thus most tide extremes recorded, even low tides, will have positive values (*see* fig. 10–15). Only during spring tides are negative low tides observed.

TIDES OBSERVED THROUGHOUT THE WORLD

Bay of Fundy

With a length of 258 km (160 mi), the Bay of Fundy has an extremely wide opening into the Atlantic Ocean. The bay splits into two narrow basins at its northern end, Chignecto Bay and Minas Basin (fig. 10–21). The period of free oscillation in the Bay of Fundy is very nearly that of the tidal period. This con-

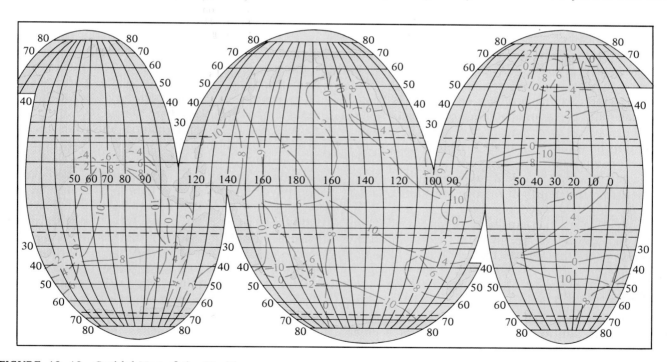

FIGURE 10–19   Cotidal Map of the World.
Contour numbers indicate times of M_2 high tide in lunar hours after the moon has crossed the Greenwich Meridian. Tidal ranges generally increase with increasing distance along cotidal lines away from the amphidromic points. Where cotidal lines terminate at both ends in amphidromic points, maximum tidal range will be near the midpoints of the lines. Base map courtesy of National Ocean Survey. After von Arx, 1962; original by H. Poincaré, 1910, Leçons de Mécanique Céleste, Gauther-Crofts, Vol. 3.

FIGURE 10–20   Reversing Current.

Note that tidal current velocity is zero at high and low tidal extremes. Maximum tidal current velocity occurs midway between tidal extremes.

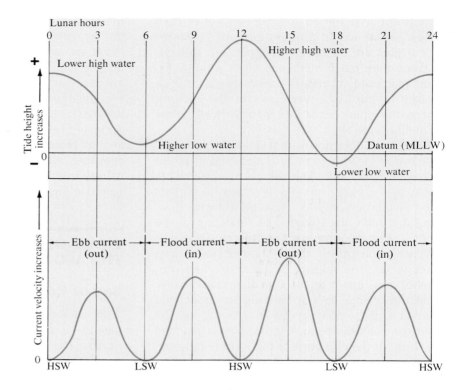

HSW—High slack water (velocity zero)

LSW—Low slack water (velocity zero)

dition brings about resonance, which along with the narrowing of the bay toward the north end and the shoaling in that direction, produces maximum tidal ranges in the extreme northern end of Minas Basin. The maximum perigean tidal range of about 17 m (56 ft) occurs at the north end of Minas Basin, and the minimum tidal range of about 2 m (6.6 ft) is found at the opening into the bay. There are no nodes within the bay, and the tidal range progressively increases from the mouth of the bay northward. Along with the standing wave phenomena, a progressive component exists that causes the tidal range on the south shore to be greater than that experienced on the north shore (see fig. 10–18F).

North Sea

The North Sea is a very broad basin that is relatively shallow and open to the Atlantic Ocean for a great distance along its northern margin. Bounded on the west by the British coast, and to the south and east by the coast of northern Europe, the North Sea experiences maximum tidal ranges approaching 5 m (16 ft). As the semidiurnal tide in the form of a progressive wave enters the North Sea from the Atlantic Ocean, the maximum tidal range within the sea is

produced along the Scottish and English coasts. In contrast to the 3 to 4 m (9.8 to 13.1 ft) tidal range seen along the western coast of the North Sea, the tides recorded at the southern tip of Norway are frequently of a range less than 25 cm (10 in.).

Owing to the earth's rotation, the mass of water that moves with the progressive wave is deflected to the right as it enters the North Sea. The wave is reflected off the coast of Belgium and the Netherlands and travels north with the right-hand deflection, this time pushing the mass of moving water against the coasts of Denmark and Norway as the wave moves toward the Atlantic Ocean. The reason tidal ranges are so much greater along the coast of Scotland and England is twofold. First, the wave possesses more energy as it moves in from the Atlantic Ocean. As the wave moves southward toward the European coast it gradually loses energy to the gently shoaling bottom, and the reflected wave contains considerably less energy than that possessed by the southbound wave. Secondly, the Kelvin wave that develops in any situation where a progressive wave travels through a broad channel has contributed to the effect. The effect of the loss of energy as the wave moves into and out of the North Sea and the tilting of the Kelvin wave can be seen in the corange lines of figure 10–22A.

A.

B.

FIGURE 10–21 Bay of Fundy.

The largest tidal range known in the world occurs at the north end of the Minas Basin in the Bay of Fundy. Because of its dimensions, this bay has a natural free-standing-wave period about equal to that of the forced tide wave. This combined with the fact that the bay narrows and becomes shallower toward its head causes a maximum tidal range at the northern end of 17 m (56 ft). High tide (A) and low tide (B) in Minas Basin. Photos courtesy of Nova Scotia Department of Tourism.

FIGURE 10–22 North Sea Tide.

In A the cotidal lines show a much higher tidal range along the coast of the British Isles than along the coasts of Denmark and Norway. Also, the amphidromic points are offset to the east, because the tide wave enters the North Sea from the North Atlantic, and the incoming Kelvin wave is much more energetic than the outgoing Kelvin wave that moves north along the coasts of Denmark and Norway. Much of the incoming wave energy is lost as it moves into the shallow water at the south end of the North Sea. B shows the theoretical pattern that would develop without the energy loss described in a bay the size and shape of the North Sea. A and B from A. Defant, *Ebb and flow,* © 1958. Reprinted by permission of the University of Michigan Press, Ann Arbor.

A. Cotidal and corange lines of the North Sea M_2 tide, ———: time of high water after moon's transit through Greenwich meridian; – – –: mean corange lines.

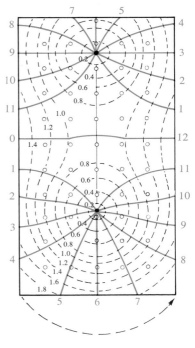

B. Theoretical cotidal lines and corange lines of the semi-diurnal tide in a bay whose length is twice its breadth. The arrow shows the direction of the earth's rotation.

FIGURE 10–23 River Bores.
As the tidal crest moves upriver it develops a steep forward slope through resistance to its advance by the river flowing to the ocean. Such crests (bores) may reach heights of 5 m (16.4 ft) and move at speeds up to 22 km/h (13.6 mi/h).

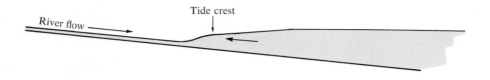

Theoretically, for a basin with the dimensions of the North Sea, the tidal pattern should be approximately that shown in figure 10–22B. Two rotary tides should develop around amphidromic points, one in the north end and one in the south end of the basin. The tidal wave should travel from the north along the western margin, around the southern end of the basin, which is closed, then back along the eastern side to the north. The actual pattern that develops in the North Sea is shown in figure 10–22A and can be seen to be similar to the predicted tidal picture, with the exception that the amphidromic points are offset considerably to the east.

Tides in Rivers

The Amazon River probably possesses the longest estuarine stretch that is affected by oceanic tides. Tides can be measured as far as 800 km (546 mi) from the mouth of the Amazon, although the effects are quite small at this distance. Tidal waves that move up river mouths lose their energy owing to the decreasing depth of water and the flow of the river water against the tide during the flood interval. As the wave moves up the river it develops a steep front (fig. 10–23). This front produces a rapidly rising tide. An extreme development of this type produces a **tidal bore** in which a very steep wave front surges up the river. In the Amazon it is called *pororoca* and appears as a waterfall up to 5 m (16.4 ft) in height moving upstream at speeds up to 22 km/h (13.6 mi/h). Other rivers that experience bores are the Chientang in China, where they may reach 8 m (26 ft); the Petitcodiac in New Brunswick, Canada; the Seine in France; and the Trent in England.

TIDES AS A SOURCE OF POWER

The history of human efforts to harness the energy of the tides dates back at least to the Middle Ages. Interest in pursuing this renewable energy source waned during the time of cheap fossil fuel availability. But today, as we realize that the end of cheap fuel

could arrive at any time, there is increased interest in assessing the prospects of generating electricity by using tidal energy.

Some Basic Considerations

The most obvious benefit of using the tides to generate electrical power would be in reduced operating costs as compared to the conventional thermal power plants that require radioactive isotopes or fossil fuels. Even though the initial cost of the tidal power-generating plant may be higher, there would be no ongoing fuel bill.

A negative consideration involves the periodicity of the tides. Power would be generated throughout only a portion of a 24-h day unless special design features were included in the construction of the facility. Because the tides operate on a lunar period and our energy demand operates on a solar period, the energy available through generating power from the tides would only accidentally coincide with need.

Generators must run at constant speed, and we must use the flow of the tidal current in two directions. It is therefore obvious that we would benefit by using turbine blades whose angle could be varied and reversed. Adjusting the pitch of the turbine blades would set them flatter when the head of water was greatest and increase their angle as the head of water decreased, so that the generators could be maintained at a constant speed. The maximum head would usually not be great, so the flow channels constructed through dams would have to be of small diameter. This would requires a series of small channels and small turbines, which would be less efficient than one large machine.

La Rance Power Plant

A successful tidal power plant was constructed in the estuary of **La Rance River** off the English Channel in France. The estuary shown in figure 10–24 has a surface area of approximately 23 km² (8.9 mi²), and the tidal range at La Rance reaches a maximum of 13.4 m (44 ft). Usable tidal energy is proportional to the area of the basin and to the square of the amplitude of the tide. A barrier was built across the estuary a

one tide cycle = 12 h 25 min

Side toward sea

rising tide

falling tide

turbine-generator unit

Side toward basin

FIGURE 10–24 La Rance Tidal Power Plant at St. Malo, France.
Block diagram shows the barrier between the open ocean to the right and the La Rance estuary to the left. The relative water levels during rising and falling oceanic tides are also shown. Photo courtesy of Phototeque/Electricite de France.

little over 3 km (1.9 mi) upstream, where it is 760 m (2493 ft) wide to protect it from storm waves. The deepest water ranges from just over 12 m (39.4 ft) at low tide to more than 25 m (82 ft) at high tide. To allow water to flow through the barrier when the generating units are shut down, sluices (artificial channels) with adjustable vanes 10 m (33 ft) high and 15 m (49 ft) wide have been built into the barrier. Twenty-four generating units are located in conduits below the power plant. The bottoms of the conduits containing the turbine-generator units are 10 m (33 ft) below the surface of the water at the lowest tide. The conduits containing the units are 53 m (174 ft)

long, the cross-sectional area at each end is about 93 m² (995 ft²), and each unit can generate 10,000 kW at 3.5 kV.

The plant generates electricity only when sufficient head exists between the pool and the ocean—about one-half of the tidal period. Annual power production of about 540 million kWh without pumping can be increased to 670 million kWh by using the turbine-generators as pumps at the proper times to increase the head of water.

Although some engineers think a tide-generating plant proposed for **Passamaquoddy Bay** near the United States–Canadian border at the south end of

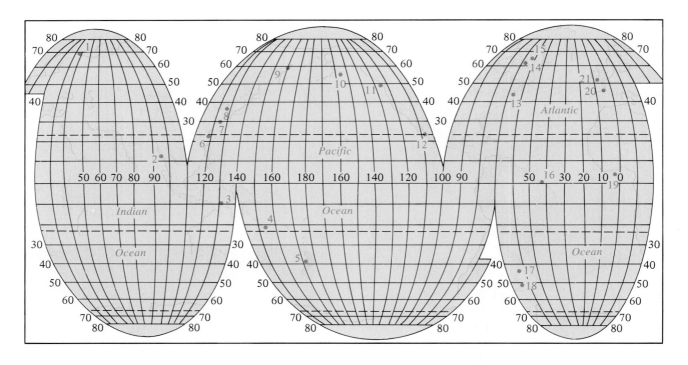

1. Mezan/Kislaya	8. Asan Bay	15. Frobisher Bay
2. Rangoon	9. Sea of Okhotsk	16. Sao Luis
3. Darwin	10. Cook Inlet	17. Golfo San Jorge
4. Broad Sound	11. Strait of Georgia	18. Straits of Magellan
5. Aukland	12. Gulf of California	19. Abidjan
6. Amoy	13. Bay of Fundy/Passamaquoddy Bay	20. Rance River/Chausey I.
7. Shanghai	14. Ungava Bay	21. Severn River

FIGURE 10–25   Sites with Major Potential for Tidal Power Generation.
Base map courtesy of National Ocean Survey.

the Bay of Fundy could be made to generate electricity constantly, others who have studied the project are less optimistic. Potentially, the usable tidal energy seems great compared with La Rance because the volume of flow is about 117 times greater than that at La Rance. Whether or not tide-generating units are ever constructed on a large scale, this potential source of energy will receive increased attention as the cost of generating electricity by conventional means increases. Figure 10–25 shows the locations of some sites around the world that have potential for tidal generation of electrical power.

SUMMARY

The tides of the earth are derived from the gravitational attractions of the sun and moon. The moon has about twice the tide-generating effect of the sun. Small horizontal forces tend to push water into two bulges, one at the earth's **zenith** and one at the **nadir**

relative to the **tide-generating body**—the sun or moon. Since the moon bulges are dominant, the tides we observe on earth have periods dominated by lunar motions and modified by the changing positions of the sun bulges.

If the earth were a uniform sphere covered with an ocean of uniform depth, and if we could ignore some of the considerations of physical motion, the tides on earth would be those predicted by the **equilibrium theory** of tides. Such a tide would have a **tidal period** of 12 h 25 min, or half a lunar day. A tide with maximum **tidal range** would occur each **new moon** and **full moon**, and tides with minimum range would occur with the **first and third quarter** phases of the moon. These would be the **spring** and **neap** tides, respectively. Since the moon may be as much as 28.5° north or south of the equator on a monthly cycle and the sun is directly over the equator only two times per year, the tidal bulges would usually be located so as to create two high tides of unequal height per lunar day. The same may be said for the low tides. Tidal ranges are greater when the earth is at **perihelion** in its orbit around the sun and when the moon is at **perigee** in its orbit around the earth because the tide-generating bodies are then closest to the earth.

Since the earth has an irregular surface with continents dividing the world ocean into irregularly shaped basins, the tides we actually observe on earth are explained by the **dynamical theory** of tides. The basic types of tides observed on the earth are a **diurnal tide** with a period of one lunar day, a **semidiurnal tide** with a period of half a lunar day (like that predicted for the equilibrium tide), and a **mixed tide** with characteristics of both. Mixed tides are usually dominated by semidiurnal periods and display a significant **diurnal inequality**. The diurnal inequalities are greatest when the moon is over the tropics and least when it is over the equator.

All basins have a characteristic free standing wave, or **seiche**, depending on their length and depth. It usually is not very high for small basins, but if it is in phase with the forced standing wave created by the tide-generating bodies, the height can be significant.

If the basin is wide enough and in the Northern Hemisphere, the standing wave may be converted to a progressive wave rotating in a counterclockwise direction around an **amphidromic point**, a point of zero tidal range. Tidal currents follow this rotary pattern in open-ocean basins but are converted to reversing currents at the margins of continents. The maximum velocity of reversing currents occurs during ebb and flood currents when the water is halfway between high and low standing waters. The effects of shoaling and narrowing on the tide can be observed in the **Bay of Fundy**. The development of multiple amphidromic points can be observed in the **North Sea**. **Tidal bores**, tide waves that force their way up rivers, are common in such rivers as the Amazon, Chientang, Seine, and Trent.

Since tides can be used to generate power without the requirement of fossil or nuclear fuel, the possibility of constructing such generating plants has always been attractive to engineers. One such plant is located in the estuary of **La Rance River** in France and is operating satisfactorily.

QUESTIONS AND EXERCISES

1. Explain why the sun's influence on the earth's tides is only 46% that of the moon, even though the sun exerts a gravitational force on the earth 177 times greater than that of the moon.

2. Describe how the centripetal force *required* to keep particles of the earth rotating in their identical orbits within the earth-moon system varies from the centripetal force *provided* by the gravitational attraction between the particles and the moon. Note: Consider strength and direction of the force.

3. Construct a diagram of vector arrows to show how the horizontal tide-generating force results from the difference in the *required* centripetal force, C, for a particle on the earth's surface at 45° from the zenith, and the centripetal force *provided* by the gravitational attraction between the moon and that particle, G.

4. Discuss why the length of the lunar day is 24 h 50 min of solar time.

5. Explain why the maximum tidal range (spring tide) occurs during new and full moon phases and the minimum tidal range (neap tide) with quadratures.

6. Discuss the length of cycle and degree of declination of the moon and sun relative to the earth's equator. Include a discussion of the effects of precession of the plane of the moon's orbit through the ecliptic.

7. Describe the effects of the declination of the moon and sun on the world ocean tides.

8. Diagram the moon's orbit around the earth and the earth's orbit around the sun. Label the positions on the orbits at which the moon and sun are closest to and farthest from the earth, stating the terms used to identify them. Discuss the effects that the moon and earth being in these positions have on the earth's tides.

9. Define tropical and equatorial tides. Include the concept of diurnal inequality (*see* the days of September 1, 2, 22, and 23 in fig. 10–15).

10. Describe the period and diurnal inequality of the following: diurnal tide, semidiurnal tide, and mixed tide.

11. What forces produce forced and free standing waves in lakes and narrow ocean embayments?

12. In narrow embayments, standing waves may develop as a result of ocean tides entering from the ocean. In wider embayments, a rotating progressive wave may develop instead. Discuss how the increased width of the embayment allows this rotary wave to develop.

13. Describe the velocity of tidal currents in relationship to high and low tidal extremes.

14. Using a diagram, explain how development of a Kelvin wave could cause the tidal range on the south side of the Bay of Fundy to be greater than that on the north side.

REFERENCES

Clancy, E. P. 1969. *The tides: Pulse of the earth*. Garden City, N.Y.: Doubleday.

Defant, A. 1958. *Ebb and flow: The tides of earth, air and water*. Ann Arbor: University of Michigan Press.

Gill, A. E. 1982. *Atmosphere and ocean dynamics*. International Geophysics Series, Vol. 30. Orlando, Florida: Academic Press.

Pond, S., and Pickard, G. L. 1978. *Introductory dynamic oceanography*. Oxford: Pergamon Press.

Sverdrup, H. U.; Johnson, M. W.; and Fleming, R. H. 1942. Renewal 1970. *The oceans: Their physics, chemistry, and biology*. Englewood Cliffs, N.J.: Prentice-Hall.

von Arx, W. S. 1962. *An introduction to physical oceanography*. Reading, Mass.: Addison-Wesley.

SUGGESTED READING

Sea Frontiers

Sobey, J. C. 1982. What is sea level? 28:3, 136–42.
 The role of tides and other factors in changing the level of the ocean surface is discussed.

Zerbe, W. B. 1973. Alexander and the bore. 19:4, 203–8.
 An account of Alexander the Great's encounter with a tidal bore on the Indus River.

Scientific American

Goldreich, P. 1972. Tides and the earth-moon system. 226:4, 42–57.
 The tide-generating force of the sun and moon on the earth and the effect of transfer of angular momentum from the earth to the moon as a result of tidal friction are discussed. Also considered are theories of lunar origin.

Greenberg, D. A. 1987. Modeling tidal power. 257: 5, 128–31.
 The effects of using the large tidal ranges experienced in the Bay of Fundy to generate electricity are modeled on a computer.

Lynch, D. K. 1982. Tidal bores. 246:4, 146–57.
 The tidal bore phenomenon is explained.

11
THE SHORE

Throughout our history, humans have been attracted to the coasts because of the moderate climates, food, and recreational opportunities they provide as well as the commercial benefits associated with nearness to the sea. The push toward the coasts has continued until 80% of the U.S. population lives within easy access of these regions. The coastal regions are undergoing increasing levels of stress that threaten to destroy this national resource to the degree that future generations may not have the opportunity to enjoy it. Before discussing the coastal region and its processes, I want to discuss a situation that has contributed to increasing the magnitude of this problem in hopes of stirring your interest in bringing about an improvement in the situation.

One of the saddest comments on government's ability to meet its responsibilities for protecting the health and safety of the citizenry derives from the federal government's development of the National Flood Insurance Program (NFIP) in 1968. Developed in response to Hurricane Betsy, which killed 75 people and did $1.4 million in property damage, the original goal of NFIP was to encourage people to build beyond the flood-threatened areas in return for federally subsidized flood insurance in case some very unusual circumstance resulted in flood damage to their property. The goal was to reduce deaths and property damage from flooding. By the time the lobbyists for the banking, construction, and real estate industries had their fears alleviated by legislators, the program turned into a vast subsidy in support of overdevelopment of the U.S. coastal region.

Following passage of the bill, coastal development boomed. Compared to the ten years prior to passage of the flood insurance bill, when hurricanes killed 186 people and did $2.2 billion in damage, there were 411 deaths and $4.7 billion in property damage during the ten years following its passage. And the problem continues to get worse. Studies conducted by the General Accounting Office in 1983 into the actuarial soundness of the program showed that many premiums should be increased by 800%. As a result, only new properties were subject to higher rates.

In addition to the flood insurance paid out to help owners replace structures built in locations that are clearly in danger of damage by storms, the federal government subsidizes the development of these areas by paying most of the cost of infrastructure development. After the development gets into trouble, the federal government then assumes from 50% to 70% of the cost of erosion-control programs. It is clearly a case of the U.S. taxpayer subsidizing imprudent development decisions that benefit a small number of individuals.

Congress has passed the **Coastal Barrier Rezoning Act**, which excludes construction in undeveloped coastal areas from receiving federal assistance, and has encouraged states to develop setback regulations that realistically incorporate the threat of storm damage into all coastal building programs. Yet, if the National Academy of Sciences warning of possible accelerated sea level rise in coming years is realized, the problem with existing structures could become immense in as few as 10 years.

What can be done to change the direction of such an unsound public policy? What general guidelines should we follow in developing a national program for conservative use of our coastal areas? Of course, politics is the major hurdle, but understanding the environment in question is an important first step. In the following pages we will discuss some of the features of the coast and shore, the processes that modify them, and some examples of the effects of human interference with these processes.

CLASSIFICATION OF COASTS

According to the classification of coasts developed by Francis P. Shepard, all coasts can be divided into two basic categories. **Primary coasts** are young features that have been formed by nonmarine processes; **secondary coasts** have aged enough that physical or biological marine processes have obliterated the nonmarine character of the coast. Table 11–1 shows the nonmarine processes that produce primary coasts. **Land erosion coasts** such as **drowned rivers** (Chesapeake Bay) or **drowned glacial erosion coasts** (Puget Sound) have been formed as a result of an ocean that began rising about 18,000 years ago as the last Pleistocene ice advance began to wane. The rising water invading these nonmarine landforms reached its present level about 3000 years ago, and sea level has been relatively stable since then. **Subaerial deposition coasts, glacial deposition coasts, volcanic coasts, coasts shaped by earth movements,** and **ice coasts** all reflect coastlines that have recently been formed by nonmarine geological processes (fig. 11–1).

With time, exposure of primary coasts to the action of ocean waves will destroy all evidence of the nonmarine process that produced them, converting them to secondary coasts. For instance, an irregular primary coast resulting from earth movements may be eroded sufficiently by wave action to produce a rela-

Preceding page: Wave action modifying the shoreline of Hawaii. Photo by Stephen Trimble.

270

TABLE 11–1   Classification of Coasts.

I. Coasts Shaped by Nonmarine Processes (Primary Coasts)
  A. Land erosion coasts
    1. Drowned rivers
    2. Drowned glacial-erosion coasts
      a. Fjord (narrow)
      b. Trough (wide)
  B. Subaerial deposition coasts
    1. River deposition coasts
      a. Deltas
      b. Alluvial plains
    2. Glacial-deposition coasts
      a. Moraines
      b. Drumlins
    3. Wind deposition coasts
      a. Dunes
      b. Sand flats
    4. Landslide coasts
  C. Volcanic coasts
    1. Lava flow coasts
    2. Tephra coasts
    3. Coasts formed by volcanic collapse or explosion
  D. Coasts shaped by earth movements
    1. Faults
    2. Folds
    3. Sedimentary extrusions
      a. Mud lumps
      b. Salt domes
  E. Ice coasts
II. Coasts Shaped by Marine Processes or Marine Organisms (Secondary Coasts)
  A. Wave erosion coasts
    1. Straightened coasts
    2. Irregular coasts
  B. Marine deposition coasts (prograded by waves, currents)
    1. Barrier coasts
      a. Sand beaches (single ridge)
      b. Sand islands (multiple ridges, dunes)
      c. Sand spits (connected to mainland)
      d. Bay barriers
    2. Cuspate forelands (large projecting points)
    3. Beach plains
    4. Mud flats, salt marshes (no breaking waves)
  C. Coasts formed by biological activity
    1. Coral reef, algae (in the tropics)
    2. Oyster reefs
    3. Mangrove coasts
    4. Marsh grass
    5. Serpulid reefs

After Francis P. Shepard, *Submarine geology,* 3rd ed. Copyright © 1973 by Harper & Row. Reprinted by permission.

tively straight cliffed coast if bedrock is of uniform strength. If the resistance of the bedrock varies along the coast, it will become irregular. Much of the Pacific Coast of the United States is classified as a **wave erosion coast.** From Massachusetts south, most of the Atlantic Coast is classified as a **marine deposition coast.** Coasts formed by marine biological activity are quite restricted along the U.S. shoreline; some of them are discussed elsewhere in the text.

GENERAL DESCRIPTION OF THE COASTAL REGION

Moving from the oceans onto the continent, one encounters the **shore,** which is the zone that lies between low tide and the highest elevations on the continent that are affected by storm waves. The **coast** extends from the landward limit of the shore inland as far as features that seem to be related to marine processes can be found. The width of the coast may vary from less than 1 km (0.6 mi) to many tens of kilometers. As the waves beat against the shore they cause erosion, producing sediment that will be transported along the shore and deposited in the low-energy areas.

Wave Erosion Coasts

The landward limit of a shore that is dominated by erosion will normally be marked by a cliff. The **coastline,** which marks the boundary between the shore and the coast, will be a line along the cliff, connecting points at which the highest effective wave action takes place.

The shore is divided into the **foreshore,** that portion exposed at low tide and submerged at high tide, and the **backshore,** which extends from the normal high-tide shoreline to the coastline. The **shoreline** migrates back and forth with the tide and is the water's edge. The **nearshore** zone is that region between the low-tide shoreline and breakers. Beyond the low-tide breakers is the **offshore** zone (fig. 11–2).

The **beach** consists of the wave-worked sediment that moves along the shore. It may extend from the coastline to the low-tide line of breakers (fig. 11–2).

Wave Erosion   Owing to refraction (bending of waves discussed in chapter 9), wave energy is concentrated on **headlands** that jut out from the continent, while the amount of energy reaching the shore in **bays** is reduced. As the waves concentrate their energy on the headlands, erosion occurs and the

A. Drowned river coast—Chesapeake Bay

B. Deposition coast—Mississippi delta

C. Lava flow coast—Hawaii

D. Fault movement coast—Alaska

FIGURE 11–1   Examples of Primary and Secondary Coasts.

Primary coasts: A, land erosion coast—drowned river. Chesapeake Bay formed when the rising ocean flooded the valley of the Susquehanna River and its tributaries (courtesy Earth Satellite Corporation). *B,* subaerial deposition coast—delta. Satellite view of the Mississippi River Delta clearly shows the characteristic "bird's foot" shape (NASA). *C,* volcanic coast—lava flow. Lava from the Kilauea east rift flowing into the sea changes the contour of the island of Hawaii (U.S. Geological Survey). *D,* coast shaped by earth movements—faults. Fault movements associated with the Alas-

E. Fault movement coast—San Francisco Bay

F. Wave erosion coast—Santa Cruz

G. Barrier coast—Martha's Vineyard

H. Mangrove coast—Galápagos Islands

kan earthquake of 1964 uplifted the wave-cut bench to become a wave-cut terrace. (U.S. Geological Survey). *E,* coast shaped by earth movements—faults. Satellite view of San Francisco Bay shows a coastline controlled by vertical movements of fault blocks within the San Andreas Fault system (NASA). *Secondary coasts: F,* wave erosion coast—irregular. View of coast along the island of Santa Cruz, California (photo by James R. McCullagh). *G,* marine deposition coast—barrier coast. Spit and bay barrier along the coast of Martha's Vineyard, Massachusetts (USDA-ASCS). *H,* coast formed by biological activity—mangrove coast. Mangrove trees on Isabela Island in the Galápagos extend the coast seaward (photo by James R. McCullagh).

FIGURE 11–2 Landforms and Terminology of Coastal Regions.

The coastline marks the most landward evidence of direct erosion by ocean waves. It separates the shore from the coast. The shore extends from the coastline to the low-tide shoreline (water's edge). It is divided into the backshore, above the high-tide shoreline, which is covered with water only during storms, and the foreshore (intertidal or littoral zone). Never exposed but affected by waves that touch bottom is the nearshore that extends seaward to the low-tide breaker line. The greatest amount of sediment transport as beach deposit occurs within the shore and nearshore zones. Beyond the nearshore lies the offshore region, where depths are such that waves rarely affect the bottom.

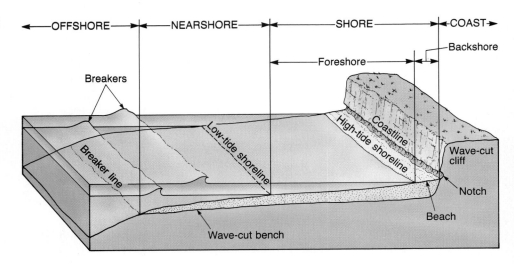

shoreline retreats. The greatest concentration of wave energy, on a day-to-day basis, is in the foreshore region. However, during the rare periods when storm waves batter the shore, more erosion may occur across the entire shore in one day than may be achieved by average wave conditions in a year.

The cliff shown in figure 11–2 is referred to as a **wave-cut cliff**, which has been produced by wave action cutting away its base. The cliff develops as the upper portions collapse after being undermined by wave action. The undermining may be evident in the form of a notch at the base of the cliff that may be characterized by **sea caves**. Such caves are most commonly cut into hard sedimentary rock. Further wave action eroding the softer portions of the rock outcrops may develop the caves into openings running through the headlands, **sea arches**. Continued erosion and crumbling of arches will produce **stacks**. Such remnants rise from the relatively smooth **wave-cut bench** cut into the bedrock by wave erosion (fig. 11–3). The rate of erosion by waves is determined by a number of variables:

1. One of the most important variables is the **degree of exposure** of the coastal region to the open ocean. Coasts that are fully exposed receive higher-energy wave action and are likely to have rugged cliffs in areas of high topographic relief.
2. The **tidal range** is also an important variable affecting the amount of wave erosion. Given the same amount of wave energy, a region with a small tidal range will erode much more

rapidly than one with a large tidal range that allows the wave energy to be spread over a much broader shore. Although high-velocity tidal currents may develop in areas where a large tidal range exists, these currents are of limited importance in eroding the coastline.

3. The **composition of coastal bedrock** is very significant. Crystalline igneous rocks, such as granite, as well as metamorphic and hard sedimentary rocks, are relatively resistant and generally produce rugged shoreline topography. Weak sedimentary rocks, such as sandstone and shale, are more easily eroded, and a gentler topography associated with more extensive beach deposits is produced by their erosion.

Regardless of the rate of erosion, all coastal regions follow the same developmental path. As long as there is no change in the elevation of the landmass relative to the ocean surface, the cliffs will continue to retreat until the beaches widen sufficiently to prevent waves from reaching them, and the eroded material will be carried from the high-energy areas and deposited in the low-energy areas.

Marine Depositional Coasts

The coastal erosion we have just discussed, as well as erosion being carried on by running water inland, produces large amounts of sediment that must be distributed along the continental margin. As waves strike the shore at an angle, they set up a longshore movement of water—the **longshore current**. The velocity of the longshore current increases with increasing beach slope, angle of breakers with the beach, wave height, and decreasing wave period.

This current moves parallel to the shore between the shoreline and the breaker line, carrying with it the materials that make up the beach. At the landward margin of the surf zone the **swash**, a thin sheet of water, moves sediment onto the exposed beach at an angle, but the force of gravity causes the backwash to carry the sediment straight down the beach face. As a result, the pebbles and sand grains that are transported by the swash move in a zigzag pattern along the shore in the same direction as the longshore current within the surf zone (fig. 11–4). The net annual longshore current is south along the Atlantic and Pacific shores.

Longshore drift is a term that has been applied to the movement of sediment by the processes just described. The amount of longshore drift in any coastal region is determined by an equilibrium between erosional and depositional forces. Any interference with the movement of sediment along the shore will destroy this equilibrium and result in a new erosional and depositional pattern determined by the nature of the interference.

Rip Currents As longshore-current water moves onto the shore it must eventually run back into the ocean. This backwash of water finds its way into the open ocean as a thin sheet flow across the ocean bottom or in local **rip currents** that occur perpendicular to or at an angle to the coast where topographic lows or other conditions allow their formation. Rip currents may be less than 25 m (82 ft) wide and can attain velocities of 7 to 8 km/h (4 to 5 mi/h) but do not travel far from shore before they break up. If a light moderate swell is breaking, numerous rip currents moderate in size and velocity may develop. A heavy swell will usually produce fewer, more concentrated rips (fig. 11–4).

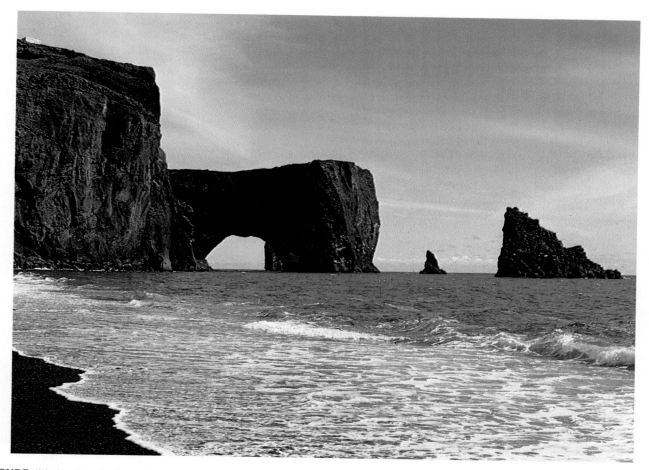

FIGURE 11–3   Sea Arch and Sea Stack along the Coast of Iceland.
Photo by Bruce F. Molnia, Terraphotographics/BPS.

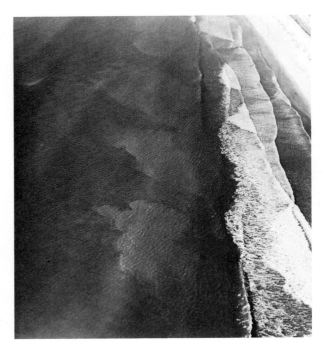

FIGURE 11–4   Longshore Currents and Rip Currents.
As waves approach shore from a southerly direction, they produce a longshore current that flows north. Water in the current follows a zigzag path as waves push it up the beach slope from the direction of approach. The water runs back down the slope under the influence of gravity. Water of the longshore current finds its way offshore by passing through topographic lows as strong seaward flows. These rip currents are visible in the photograph as jets of water that appear light in color because of the turbidity resulting from sediment they have resuspended from the ocean floor. Photo courtesy of Scripps Institute of Oceanography, University of California/San Diego.

Beach Composition   The material found in a beach deposit will depend on the source of the sediment that is locally available or transported by the longshore drift. In areas where the sediment is provided by coastal mountains, the beaches will be composed of minerals contained in the rocks of those mountains and may be of relatively coarse texture. If the sediment is provided primarily by rivers that drain lowland areas, sediment that reaches the coastal regions will normally be finer in texture; in many cases mud flats develop along the shore because only clay- and silt-sized particles are emptied into the ocean. In low-relief low-latitude areas such as southern Florida, where there are no mountains or other sources of rock-forming minerals nearby, most of the material of the beaches is derived from the remains of the organisms that live in the coastal waters. Beaches in these areas will be composed predominantly of shell fragments and the remains of microscopic animals, particularly foraminifers. Many volcanic island beaches in the open ocean will be composed of dark-colored fragments of the basaltic lava that makes up the islands or of coarse fragments of coral debris from the

reefs that develop around the margins of islands in low latitudes.

Beach Slope   The slope of beaches is closely related to the size of the particles of which they are composed. Waves washing onto the beach carry sediment, thereby increasing the slope of the beach. If the backwash returns as much sediment as the waves carried in, the beach has reached equilibrium and will not steepen. A beach composed of fine-grained sand that is relatively angular will have a gently sloping, firm surface. Because these small grains interlock closely, little of the swash sinks down among the grains. Most of it runs back down the slope to the ocean, possessing enough energy to maintain equilibrium on a gentle slope. The backshores of such beaches are usually nearly horizontal.

Beaches composed of coarse sands or pebbles will usually contain more-rounded particles that are more loosely packed. The swash can quickly percolate into such deposits when it moves up the beach slope. Deposition of particles by the swash will continue until the beach slope becomes steep enough that the backwash running to the ocean has sufficient energy to maintain equilibrium. Such beaches are usually much less firm than beaches composed of finer material, and the backshores will slope significantly toward the coastline (table 11–2).

Special Geomorphic Features Numerous depositional features that are partially or wholly separated from the shore are deposited by the longshore drift and other processes that are not well understood. A **spit** is a linear ridge of sediment attached at one end to land. The other end of the deposit points in the direction of longshore drift and ends in the open water. Spits are simply extensions of beaches into the

deeper water near the mouth of a bay. The open-water end of the spit will normally curve into the bay as a result of current action.

If tidal currents or the currents initiated by river runoff are too weak to keep the mouth of the bay open, the spit may eventually extend across the bay and tie to the mainland to completely separate the bay from the open ocean. The spit then has become a **bay barrier**. A **tombolo** is a sand ridge that connects an island with another island or the mainland. Tombolos usually are aligned at a high angle to the main shore. This is because they are usually sand deposits laid down in the wave-energy shadow of the island; thus, their alignment is at right angles to the average direction of wave approach (fig. 11–5).

Barrier islands are long offshore deposits of sand lying parallel to the coast. The origin of a barrier island is complex, and it may be that there are several explanations for its existence. It appears, however, that many such structures developed during the rise in sea level that began with the last melting of the glaciers some 18,000 years ago.

Barrier islands are nearly continuous along the Atlantic Coast of the United States. They extend around Florida and along the Gulf of Mexico coast, where they may be found well past the Mexican border. Barrier islands may attain lengths in excess of 100 km (62 mi) and have widths of several kilometers. Examples of such features are Fire Island off the New York coast and Padre Island off the coast of Texas.

A typical barrier island has the following physiographic features from the ocean to the lagoon behind

TABLE 11–2 The Relationship of Particle Size to Beach Slope.

Wentworth Particle Size	(mm)	Mean Slope of Beach
Cobble	256	24°
Pebble	64	17°
Granule	4	11°
Very coarse sand	2	9°
Coarse sand	1	7°
Medium sand	0.5	5°
Fine sand	0.25	3°
Very fine sand	0.125	1°
	0.063	

SOURCE: After Table 9, "Average beach face slopes compared to sediment diameters" from *Submarine geology*, 3rd ed. by Francis P. Shepard, p. 127. Copyright © 1948, 1963, 1973 by Francis P. Shepard. Reprinted by permission of Harper & Row, Publishers, Inc.

FIGURE 11–5 Coastal Depositional Features.

A *tombolo* is a deposit that connects an island to the mainland or another island. A *spit* is a deposit that extends from land into open water. A spit that is extended across a bay and closes the bay off from open water is a *bay barrier*. A *barrier island* is a long deposit separated from the mainland by a lagoon.

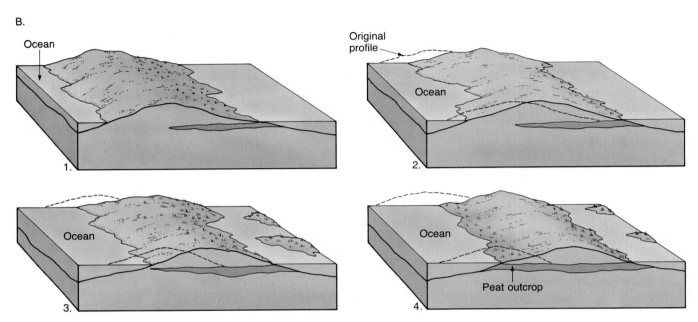

FIGURE 11–6 Barrier Islands.

A, cross section through a barrier island. The major physiographic zones of a barrier island are the
ocean beach, dunes, barrier flat, and *high* and *low salt marsh.* MSL is mean sea level. The peat bed
represents ancient marsh environments that have been covered by the island as it migrates toward
the mainland with rising sea level. *B,* (1) barrier island before overwash, with salt marsh development on mainland side; (2) overwash erodes seaward margin of barrier island, cuts inlets, and
carries sand into the lagoon, covering much of the existing marsh; (3) new marsh forms in the
lagoon that has retreated toward the mainland; (4) with continued rise in sea level and migration
of the barrier island toward the mainland, peat formed from previous marsh deposits may be exposed by erosion of the foreshore of the ocean beach.

it: (1) ocean beach, (2) dunes, (3) barrier flat, and (4)
salt marsh (fig. 11–6A). The **ocean beach** is typical of
the beach environment discussed earlier. During the
summer, as gentle waves carry sand to the beach, it
widens and becomes steeper. Higher-energy winter
waves carry sand offshore and produce a narrow,
gently sloping beach.

Winds blow sand inland during dry periods to produce **dunes,** which are stabilized by dune grasses
that can withstand salt spray and burial by sand.
Dunes are the estuary's primary protection against
excessive flooding during storm-driven high tides.
Numerous passes exist through the dunes, particu-

larly along the southeast Atlantic Coast, where dunes
are less well developed than to the north.

Behind the dunes the **barrier flat** forms as the result of deposition of sand driven through the passes
during storms. These flats are quickly colonized by
grasses. If for some reason the frequency of overwash
by storms decreases, the grasses will successively be
replaced by thickets, woodlands, and forests.

Salt marshes typically lie inland of the barrier flat.
They are divided into the **low marsh,** extending from
about mean sea level to the high neap-tide line, and
the **high marsh,** extending to the highest spring-tide
line. The low marsh is by far the most productive part

of the salt marsh. New marsh is formed as overwash carries sediment into the lagoon, filling portions so they are intermittently exposed by the tides. Marshes may be poorly developed on parts of the island that are far from flood-tide inlets. Their development is greatly restricted behind barrier islands where artificial dune enhancement and inlet filling prevent overwashing and flooding.

Because of the gradual rise in sea level relative to the eastern North American coast, barrier islands are migrating landward, a fact clearly visible to those who build structures on these islands. Evidence for such migration can be seen in peat deposits cored beneath the barrier islands. Since the only barrier island environment in which peat forms is the salt marsh, the island must have moved inland over previous marsh development.

A possible cycle of salt marsh formation and destruction by landward migration of barrier islands is presented in figure 11–6B.

Deltas   Some rivers carry more sediment than can be distributed by the longshore current. Such rivers develop a **delta deposit** as sediment settles out at the river mouth. One of the largest such features is that produced by the Mississippi River. Deltas are fertile, flat areas that are subject to periodic flooding.

Delta formation begins when a river has filled its estuary with sediment. Once the delta forms, it grows through the distribution of sediment by **distributaries,** branching channels that radiate out over the delta. Distributaries lengthen as they deposit sediment and produce fingerlike extensions to the delta. When the fingers get too long, they become choked with sediment. At this point, a flood may easily cause a shift in the distributary course and provide sediment to the low-lying areas between the fingers.

In areas where depositional processes are dominant over coastal erosion and transportation processes, the "bird foot" **Mississippi-type delta** results. Where erosion and transportation processes exert a significant influence, the shoreline of the delta will be smoothed to a gentle curve like that of the **Nile Delta** (fig. 11–7A). The Nile Delta is presently eroding owing to the entrapment of sediment behind the Aswan High Dam, which was completed in 1964. Erosion is

FIGURE 11–7   Deltas.
A, photographed from *Gemini IV.* The relatively smooth, curved surface of the Nile Delta can be seen. The Mediterranean Sea is to the left, and the northern Red Sea is in the upper right. Photo courtesy of NASA. *B,* digitate structure of the Mississippi River Delta results from low-energy environment. Location of present main channel and older main channels (Atchafalaya River and Bayou Lafourche) show how river shifts position with time.

A.

B.

only one of the negative effects of the construction of this dam. This erosion, however, may have made possible the discovery of the ruins of ancient Alexandria beneath the Mediterranean waters at the edge of the delta. The problem of reduced availability of sediment may become very significant along the eastern end of the delta, where subsidence has occurred at a rate of 0.5 cm (0.2 in.)/yr for 7500 years.

CHANGING LEVELS OF THE SHORELINE

Along some coasts there are flat platforms, called **marine terraces,** backed by cliffs. **Stranded beach** deposits and other evidence of marine processes may exist many meters above the present shoreline. These features characterize **shorelines of emergence** that have reached their present positions relative to the existing shoreline by an uplift of the continent, by the lowering of sea level, or by a combination of the two.

In other areas one may find beneath the water overlying the continental shelf **drowned beaches** and **submerged dune topography.** These features along with the drowned river mouths along the present shoreline indicate **shorelines of submergence.** This submergence must have been caused by a subsidence of the continent, a rise in sea level, or a combination of the two.

Attempts to determine the causes of the change in relative level of the ocean and the continent have not met with great success. Whether the shoreline has become submerged because of a rising sea level or a subsiding continent in a particular region cannot be determined by examining the coastal features in that area, since both processes produce the same end results (fig. 11–8).

Tectonic and Isostatic Movements

Changes in sea level relative to the continent may occur because of movement of the land—tectonic movement. Such movement includes large-scale uplift or subsidence of large portions of continents or ocean basins or more localized deformation of the continental crust involving folding, faulting, and tilting. The earth's crust also responds isostatically to the accumulation or removal of heavy loads of ice, sediment, or lava.

There is evidence that during the last 2.5 to 3 million years at least four major accumulations of glacial ice developed in high latitudes. Although Antarctica is still covered by a very large glacial accumulation, much of the ice cover that once existed in northern Asia, Europe, and North America has disappeared. The most recent period of melting began about 18,000 years ago. Accumulations of ice that were up to 3 km (2 mi) in thickness have disappeared from northern Canada and Scandinavia. While these areas were beneath the thick ice sheet, they were pushed down and are still in the process of recovery after the melting of the ice. Some geologists believe that by the time the **isostatic rebound** is finished, the floor of Hudson Bay, which is now about 150 m (492 ft) deep, will be above sea level. There is evidence in the Gulf of Bothnia, between Sweden and Finland, of 275 m (902 ft) of isostatic rebound during the last 18,000 years. Generally, tectonic and isostatic changes in the level of the shoreline are confined to a segment of the shoreline of a given continent.

Eustatic Movements

Changes in the level of the shoreline that can be measured on a worldwide basis because they are caused by the increase or decrease of water volume in the ocean or the capacity of the ocean basin are termed **eustatic.** This term refers to the highly idealized situation in which all of the continents remain static while the sea rises or falls. Small changes in sea level could be created by the formation or the destruction of large inland lakes. Probably more important is the locking into or the release from continental glaciers of the earth's water during glacial and interglacial stages.

Changes in sea-floor spreading rates can change sea level. Fast spreading produces larger rises like the East Pacific Rise that will displace more water than slow-spreading ridges like the Mid-Atlantic Ridge. Thus, fast spreading produces a rise in sea level.

FIGURE 11–8 Evidence of Changing Levels of the Shoreline.
Marine terraces resulting from ancient sea cliffs and wave-cut benches being exposed above present sea level mark ancient shorelines, as do drowned beaches that lie below sea level.

During the Pleistocene Epoch, when the previously mentioned glacial advances were occurring, the amount of water in the ocean basin fluctuated considerably. Since the climate was colder during ice advances, we might account for some of the lowering of the shoreline by the contraction of the ocean volume as its temperature decreased. It has been calculated that for every 1°C (1.8°F) decrease in the mean temperature of the ocean water, sea level would drop 2 m (6.6 ft). Temperature indications derived from study of fossils from Pleistocene ocean sediments suggest the ocean surface temperature may have been as much as 5°C (9°F) lower than at present. Therefore, contraction of the ocean water may have lowered sea level by about 10 m (33 ft).

Although it is difficult to say with certainty what the range of shoreline fluctuation was during the Pleistocene, there is cause to believe that the shoreline was at least 120 m (394 ft) below the present shoreline. It is also estimated that if all the remaining glacial ice on earth were to melt, sea level would rise by another 60 m (197 ft). This would give a minimum possible range of sea level during the Pleistocene of 180 m (590 ft), most of which must be explained through the capture and release of the earth's water by glaciers. Such changes in sea level are termed **glacioeustatic oscillations.**

During the last 18,000 years the ocean volume has been increasing owing to the expansion of the water resulting from warmer temperature and the melting of polar sea ice and glacial ice. Since the combination of tectonic and eustatic changes in sea level may be very complex and difficult to identify, it is hard to classify coastal regions as purely emergent or submergent. Most coastal areas show evidence of having experienced both submergence and emergence in the recent past. It is believed, however, that sea level has not risen significantly as a result of melting glacial ice during the last 3000 years.

There is at present much discussion resulting from the combined gleanings from historical research. There is clear evidence that atmospheric carbon dioxide has increased by 15% since 1958, when monitoring began. There has also been an apparent global warming of about 0.5°C (0.9°F) and eustatic rise in sea level of 10 cm (4 in.) since 1880. Are all of these observations tied together and the result of an increased greenhouse effect? Regardless of the answer to this question, which is at present not known, we have come to realize that we cannot dominate nature and must attempt to live within it. To do otherwise is futile. Some of the consequences of a rising sea level for U.S. coastal communities are discussed in the feature "Can We Learn to Live with our Changing Shores?" on page 282.

THE UNITED STATES COASTS

The most dramatic examples of sea level change during the last 3000 years, have been due to tectonic processes. To understand the underlying tectonic cause of emergence or submergence, we must return to the concept of global plate tectonics.

Atlantic-type (Passive) Margins

Considering only the coasts of North America, we find the Atlantic Coast is subsiding and the Pacific Coast is emerging. Beginning with the breakup of Pangaea and the presently forming Atlantic Ocean, this emergence-submergence pattern can be readily understood. It was previously discussed that the lithosphere thickened, the ocean deepened, and heat flow decreased as the lithospheric plates cooled with age or with increasing distance from the spreading centers. As a result of this process, the ocean floor of this passive margin, the Atlantic-type margin, is thought to have subsided about 3 km (2 mi) over the last 150 million years.

Erosion of the continent produced sediment that has accumulated to a maximum thickness of 15 km (9.3 mi) along the Atlantic Coast. This thick deposit of sediment has been made possible by the plastic asthenosphere allowing the lithosphere to flex downward as the sediment burden is increased. It is this thick sedimentary wedge underlying the continental shelf, slope, and rise that has been exploited for its large petroleum reserves in the Gulf of Mexico.

These two processes account for the subsidence of the Atlantic and Gulf coasts. Although the rate of thermal subsidence decreases with time and the rate of subsidence due to sediment loading depends on sediment supply, it is not reversed unless the tectonic or isostatic conditions change.

Pacific-type (Active) Margins

In contrast to the passive and subsiding Atlantic-type margin, the Pacific-type margin, or active margin, is the scene of intense tectonic activity. Along such margins, the eruption of volcanoes and shock of earthquakes are common. The thick, broad sediment wedge characteristic of the Atlantic Coast is not well developed. Instead of being the product of the creation of a new ocean basin, the Pacific-type margin is the scene of lithospheric plate destruction.

The characteristic alignment of mountain ranges parallel to the continental margin can be seen. In all cases of continental margins associated with plate convergence, the compressive forces produce uplift.

Recent studies have shown that there has been a 10-cm (4-in.) rise in sea level during the last century (fig. 11A). Investigators believe most of this rise comes from thermal expansion of ocean water as a result of increasing surface temperatures. This temperature increase may well be the result of increasing concentrations of CO_2 in the atmosphere. A continued rise of sea level by as much as another 30 cm (12 in.) *could* occur in the next 70 years. Should temperatures rise significantly, the Antarctic ice sheet could begin to melt and increase the rate of rise in sea level.

If, as some investigators predict, sea level does continue to rise, it could create major problems for low-lying coastal areas. Fresh groundwater supplies could become salty, and the threat of damage from storm surge could increase dramatically. Many coastal communities along the Atlantic and Gulf coast shores would be seriously threatened because, in addition to the possible danger from a rise in sea level, these coastal areas have been subsiding at an average rate of 3 cm/1000 yr (1.2 in./1000 yr) for the last 150 million years.

The city of New Orleans, Louisiana, provides us with an example of the problems that may face many low-lying coastal cities. Although the city is now protected from direct ocean waves by 48 km (30 mi) of salt marsh and has a levee and sea wall to keep out water from the Mississippi River and Lake Pontchartrain, respectively, a rising sea could be devastating to the city (fig. 11B). The subsiding marshes are getting little sediment to build them up because they are separated from the flow of the Mississippi River by levees. Should the marshes become submerged, the ocean would be at New Orleans' door. This is a serious problem because parts of the city are already 2 m (6.6 ft) below sea level.

To keep the city dry, 22 pumps 4.3 m (14 ft) in diameter and 12 pumps 3.7 m (12 ft) in diameter must pump a minimum of 113,599 m^3 (4,010,045 ft^3) of water per day. In time, with continued subsidence of the land and rise of sea level, maintaining the pumping rate necessary to keep the city from flooding could become increasingly difficult. Many other coastal cities, from Brownsville, Texas, to Long Island, New York, would also suffer damage from encroachment of the ocean (fig. 11C).

Although the Pacific Coast is rising tectonically, open exposure to the storm waves of the Pacific Ocean results in heavy erosion and property damage there, too.

A major part of the problem along the Atlantic and Gulf coasts has resulted from the development of barrier islands, which undergo natural shoreward migration as the sea rises relative to the continent along the Atlantic and Gulf coasts. Sea walls have been built to stabilize the islands, but they are only a temporary solution to protecting developed areas and actually accelerate the rate of beach erosion. Unless sand is continually replenished on the seaward side of the sea wall, the beaches will erode and the sea walls will tumble into the ocean.

FIGURE 11A   Sea-Level Change from 1880 to 1980.

Tide-gauge data averaged over 5-year periods show that global mean sea level has increased about 10 cm (4 in.) over the last 100 years. After V. Gornits, S. Lebedeff, and J. Hausen, *Science* 215, pp. 1611–14, 1982.

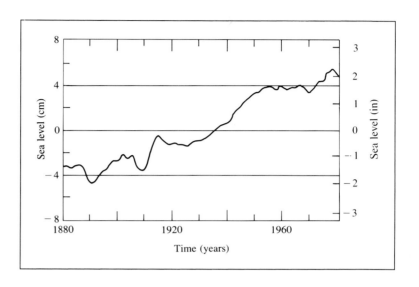

FIGURE 11B   New Orleans Area.

New Orleans, parts of which lie 2 m (6.6 ft) below sea level, is protected from direct inundation by the waters of the Gulf of Mexico by salt marsh to the south and east. The waters of the Mississippi River and Lake Pontchartrain are kept out of the city by a levee and seawall, respectively.

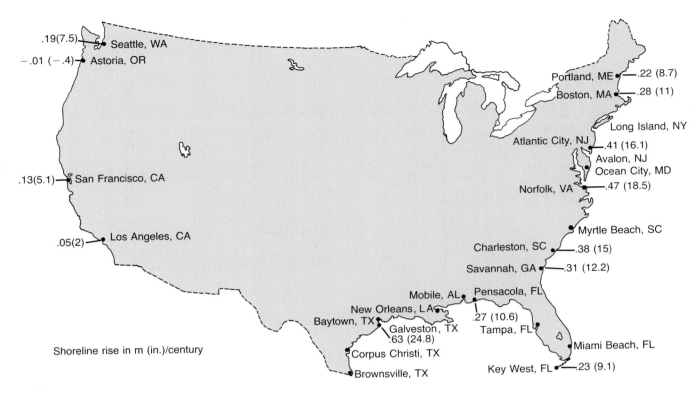

FIGURE 11C Coastal Cities Subject to Damage by an Encroaching Sea.

The general problem of maintenance of coastal construction is becoming a very sensitive issue because a large amount of public money is being spent to protect private property. In a sense, all areas lived in and built upon by humans are on loan from nature, but in the case of barrier islands that loan is very short-term. As the cost of protecting developed areas increases, a point may be reached where the public decides it is time for the private interests to assume the expense of protecting their investments or take their losses.

Whether or not sea level continues to rise, there is no doubt that most of the Atlantic and Gulf coasts will continue to subside. At the very minimum, it seems prudent for the development policy of our coastal regions to include a prohibition against further development of barrier islands. A very cautious approach to the development of mainland coasts should also be followed. Such a policy may be very distasteful to many of us who love to be near the ocean. However, if the reported rise in sea level continues, there may be no alternative.

Along the California coast, the process has been further complicated with the development of the San Andreas Fault, a transform fault. This fault has made possible the local subsidence of the depositional basins in the Los Angeles area. But on the broader scale, emergence along Pacific-type margins is the characteristic condition.

UNITED STATES COASTAL CONDITIONS

Whether the dominant process along a coast is accretion or erosion depends on the combined effect of the variables we have been discussing—degree of exposure, tidal range, composition of coastal bedrock, tectonic or isostatic subsidence, or emergence and eustatic sea level change.

In its 1971 publication, "A Report on the National Shoreline Study," the U.S. Army Corps of Engineers found that of the nation's 135,870 km (84,240 mi) of coastline, 33,011 km (20,500 mi), or over 24%, were "seriously eroding." Subsequent studies supported by the U.S. Geological Survey along the shores of the contiguous coastal states and Alaska produced the rates of shoreline change presented in tables 11–3 and 11–4 and figure 11–9. These tables reflect conditions as they existed between 1979 and 1983. Table 11–3 shows the mean rate of change for each coastal state.

The Atlantic Coast

Most of the Atlantic coastline is exposed to storm waves from the open ocean, but barrier island development from Massachusetts south provides protection for the mainland from large storm waves. Tidal ranges generally increase from less than 1 m (3.3 ft) along the Florida coast to more than 2 m (6.5 ft) in Maine. Bedrock for most of Florida is a resistant limestone. From Florida north through New Jersey, the bedrock is composed primarily of poorly consolidated sedimentary rocks formed in the recent geologic past. These rocks do not provide great resistance to erosion and are sources of sand that is deposited as barrier islands and other depositional features common along the coast. From New York north there is considerable evidence of continental glaciers having affected the coastal region directly. Many coastal features, including Long Island and Cape Cod, are the products of glacial deposition. They are moraines that were deposited when the glaciers melted.

North of Cape Hatteras the coast is subject to very high energy conditions during fall and winter when

TABLE 11–3 Rates of accretioin (+) and Erosion (−) along Oceanic Coasts and in Bays of the United States (Contiguous and Alaska).

Region	Change (cm/yr)
Atlantic Coast	−0.8
Maine	−0.4
New Hampshire	−0.5
Massachusetts	−0.9
Rhode Island	−0.5
New York	+0.1
New Jersey	−1.0
Delaware	+0.1
Maryland	−1.5
Virginia	−4.2
North Carolina	−0.6
South Carolina	−2.0
Georgia	+0.7
Florida	−0.1
Gulf of Mexico	−1.8
Florida	−0.4
Alabama	−1.1
Mississippi	−0.6
Louisiana	−4.2
Texas	−1.2
Pacific Coast	−0.0
California	−0.1
Oregon	−0.1
Washington	+0.5
Alaska	−2.4
Delaware Bay	−1.6
Chesapeake Bay	−0.7

TABLE 11–4 Rates of Accretion (+) and Erosion (−) for Coastal Landform Types.

Region	Change (m/yr)
Mud Flats	−2.0
Florida	−0.1
Louisiana–Texas	−2.1
Gulf of Mexico	−1.9
Sand Beaches	−0.8
Maine–Massachusetts	−0.7
Massachusetts–New Jersey	−1.3
Atlantic Coast	−1.0
Gulf Coast	−0.4
Pacific Coast	−0.3
Barrier Islands	−0.8
Louisiana–Texas	−0.8
Florida–Louisiana	−0.5
Gulf of Mexico	−0.6
Maine–New York	+0.3
New York–North Carolina	−1.5
North Carolina–Florida	−0.4
Atlantic Coast	−0.8
Rock Shorelines	+0.8
Atlantic Coast	+1.0
Pacific Coast	−0.5

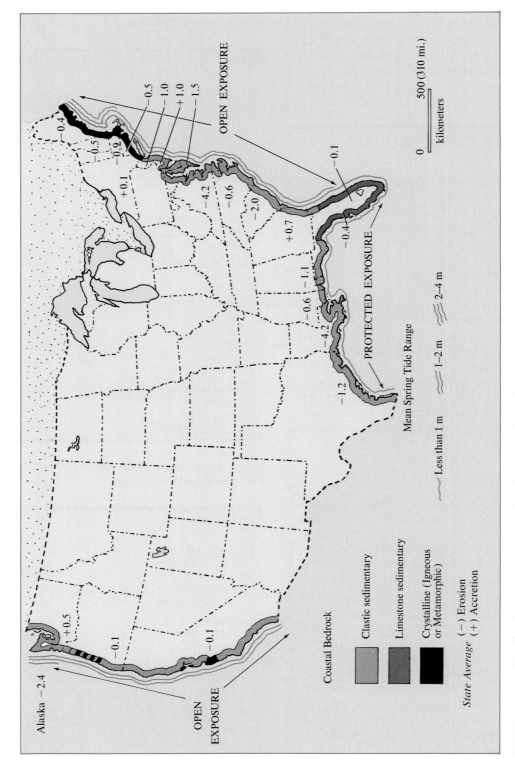

FIGURE 11–9 Distribution of Conditions Related to Coastal Erosion and Rates of
Erosion vs. Accretion.

the "nor'easters" blow out of the North Atlantic. The great energy of these storms is manifested in up to 6-m (20-ft) waves with a 1-m (3.3-ft) rise in sea level that follow the low pressure as it moves northward. Such high-energy conditions will significantly change coastlines that are predominantly depositional and may be expected to cause considerable erosion.

Most of the Atlantic coast appears to be experiencing a sea level rise of about 0.3 m (1 ft) per century. This condition may be reversed in northern Maine because of isostatic rebound since removal of the continental ice sheet.

The Atlantic Coast has an average annual rate of erosion of 0.8 m (2.6 ft). Virginia leads the way with a 4.2-m (13.7-ft) loss per year, which is largely confined to barrier islands. Erosion rates for Chesapeake Bay are about average for the Atlantic coast, but Delaware Bay shows average erosion rates about twice (1.6 m/yr) the Atlantic coast average (fig. 11–10). Seventy-nine percent of the observations made along the Atlantic coast showed some degree of erosion. In spite of the fact there are serious erosion problems in these states as well, Delaware, Georgia, and New York display accreting coasts.

FIGURE 11–10 Locations of Major Bays along the Atlantic Coast.

The Gulf Coast

The Louisiana-Texas coastline is dominated by the Mississippi River Delta, which is being deposited in a microtidal and generally low-energy environment. The tidal range is normally less than 1 m (3.3 ft) and, with the exception of the hurricane season, wave energy is generally low. Tectonic subsidence is general throughout the Gulf Coast, and the average rate of sea level rise is similar to that of the southeast Atlantic coast, about 0.3 m (1 ft) per century.

The average rate of erosion is 1.8 m (6 ft) per year in the Gulf Coast. The Mississippi River Delta is experiencing the highest rate of erosion, which results in the state of Louisiana losing an average of 4.2 m (13.7 ft) per year. Aided by dredging of barge channels through marshlands, Louisiana has lost more than one million acres of delta since 1900 and is losing it now at a rate in excess of 130 km^2 (50 mi^2) per year. Although all Gulf states show a net loss of land, and the Gulf Coast has a higher rate of erosion than the Atlantic Coast, only 63% is receding because of erosion. This reflects the heavy erosion loss in the delta area.

The Pacific Coast

Along the Pacific Coast, there is open exposure to large storm waves, with a tidal range mostly between 1 and 2 m. The bedrock varies from the predominant relatively weak marine deposits, which are geologically young, to local outcrops of granite, metamorphic, and volcanic rocks that are more resistant to erosion. Tectonically, the coast is rising (fig. 11–11), but, with the exception of a segment along the coast of Oregon and the Alaskan coast, sea level still shows at least small rates of rise. High-energy waves may strike the coast in winter, with 1-m waves being normal. Frequently, the wave height will increase to 2 m, and a few times per year 6-m (20-ft) waves hammer the shore.

During the winter, many beaches have the sand eroded from the foreshore area by high-energy storm waves. The exposed beaches, which are composed primarily of pebbles and boulders during the winter months, regain their sand as low-energy summer waves beat more gently against the shore. Damming of many of the rivers along the Pacific Coast for flood control and hydroelectric power generation has reduced the amount of sediment for the longshore transport. As a result, some areas experience severe threats of erosion.

FIGURE 11–11   Marine (Wave-cut) Terraces on San Clemente Island South of Los Angeles, California.
Once at sea level, the highest terraces are now about 400 m (1320 ft) above it. Photo by John S. Shelton.

With an average erosion rate of only 0.005 m (0.016 ft)/yr, and only 30% showing erosion loss, the Pacific Coast appears to be under much less of an erosion threat than the Atlantic and Gulf coasts. Yet, all along the coast there are areas that are experiencing high rates of erosion. Only Washington shows accretion, while Alaska is losing an average of 2.4 m (7.9 ft)/yr. The long, protected Washington shoreline within Puget Sound helps skew the Pacific Coast figures (fig. 11–12). Although the average rage of erosion for California is only 0.1 m (0.33 ft)/yr, over 80% of the coast is experiencing erosion with local rates of up to 0.6 m (2 ft)/yr.

Erosion Rates and Landform Types

The variability we see in erosion rates along coasts is closely correlated to coastal landform types (table 11–4). Deltas and mud flats composed of fine-grained sediment erode most rapidly, with mean erosion rates of 2 m (6.6 ft)/yr. This is not surprising, and it accounts for the fact the Gulf Coast is eroding at a greater rate than other coasts. Coarser deposits of sandy beaches and barrier islands erode more slowly (0.8 m/yr). The rock shore of Maine, which is experiencing isostatic rebound, shows great resistance to erosion, with an accretion rate of 1 m/yr. On the Pacific Coast, the rocks are not so durable and erode at an average rate of 0.5 m/yr.

Effects of Artificial Barriers

Jetties that have been constructed to protect harbor entrances from wave action are the most common barriers to the longshore transport of sediment. Many examples of the destruction resulting from the con-

FIGURE 11–12   Features of Special Interest along the Pacific Coast.

struction of such features along the shore can be cited. Along the southern California coast, excessive beach erosion related to the development of local harbor facilities is especially common. The longshore drift in California is predominantly from north to south, and the construction of a jetty results in the entrapment of sediment on the north side, causing increased erosion on the south side. A jetty constructed as a **breakwater** to protect harbor waters at Santa Barbara had a very significant effect upon the equilibrium established in this coastal region. The barrier created by the breakwater on the west side of the harbor caused an accumulation of sand that was moving in an easterly direction along this portion of the California coast, which trends generally east-west. The beach to the west of the harbor continued to grow until finally the sand moved around the breakwater and began to fill in the harbor (fig. 11–13).

While deposition was occurring at an abnormal rate to the west, erosion was proceeding at an alarming rate east of the harbor. The energy available east of the harbor was no greater than it had been previously, but the sand that had formerly moved down the coast to replace sand removed was no longer available owing to its entrapment behind the breakwater. To compensate for this deficiency of sand

downcurrent from the harbor, dredging had to be initiated within the harbor. The dredging keeps the harbor from filling in, and the sand is pumped down the coast so it can re-enter the longshore drift and replenish the eroded beach. The dredging operation has stabilized the situation, but at a considerable expense. It seems obvious that any time human beings interfere with natural processes in the coastal region, they will have to provide the energy needed to replace that which they have misdirected through modification of the shore environment.

Along many coasts, small jettylike structures called **groins** have been constructed at intervals along the beach (fig. 11–14). They have helped beach development by causing sand to accumulate on the

FIGURE 11–13 Santa Barbara Harbor.

Construction of a breakwater to the west of Santa Barbara Harbor interfered with the eastward-moving longshore drift, creating a broad beach. Sand being deposited against the breakwater was no longer available to replace sand being removed by wave erosion to the east. As the beach extended around the breakwater into the harbor, dredging operations were initiated to keep the harbor open and to put the sand back into the longshore drift. This helped reduce the coastal erosion east of the harbor. U.S. Coast and Geodetic Survey Chart 5161.

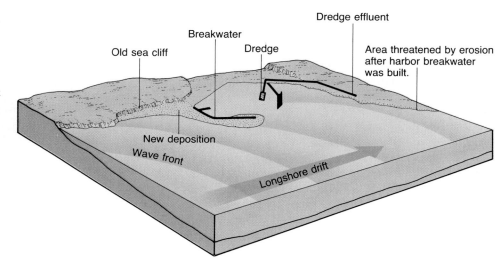

FIGURE 11–14 Jetties and Groins.
Jetties and groins trap sand that would otherwise be moved down the beach by wave action. From Tarbuck, E. J., and Lutgens, F. K., *Earth science,* 5th ed. (Columbus: Merrill, 1988). Reprinted by permission.

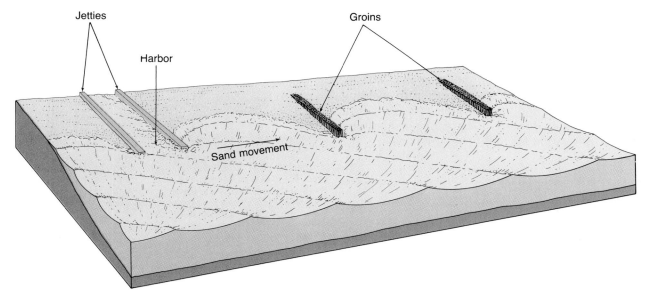

north sides of these structures. Being shorter than jetties built to protect harbors, the groins eventually allow the sand to migrate around their ends. An equilibrium may be reached that allows sufficient sand transport along the coast before excessive erosion occurs downcurrent from the last groin. Although some serious erosional problems in these regions have developed, construction of a series of groins usually results in much less beach erosion than the construction of one large jetty.

One of the most destructive of structures built to "stabilize" eroding shores is the **seawall**. Instead of extending out into the longshore current as do groins and jetties, they are built parallel to the shore to protect developments landward of them from the action of ocean waves. Once waves begin breaking against the seawalls, turbulence generated by the abrupt release of wave energy quickly erodes the sediment on their seaward sides, causing them to collapse into the surf (fig. 11–15). Where they have been used to protect property on barrier islands, the seaward slope of the island beach is steepened and the rate of erosion increased.

Although most state laws say beaches belong to the public, government actions allow the owners of shore property to destroy beaches in an effort to provide short-term protection to their property. To add to the public insult, taxpayers subsidize property owners in many shore developments by helping to rebuild their storm-damaged properties at great public expense through the programs discussed earlier in the chapter.

An interesting possible alternative to the solid structures used to protect local shore areas from ocean waves is the **tethered-float breakwater** developed at Scripps Institution of Oceanography (fig. 11–16). Tested successfully in San Diego Harbor, these inverted pendulums protect against coastal wave erosion by extracting energy from passing waves without stopping the transport of sediment. If they prove durable enough, they would certainly be a welcome substitute for traditional breakwaters.

SUMMARY

Since passing the **National Flood Insurance Program (NFIP)** in 1968, the federal government has encouraged development of stretches of our shores subject to damage by storm waves. The direction of this and complementary programs must be reversed if the situation is be improved. One helpful piece of legislation in the direction was the **Coastal Barrier Rezoning Act**, which excludes construction in undeveloped coastal areas from receiving federal assistance.

FIGURE 11–15 Seawalls.
When a seawall is built on a beach to protect beachfront property, the first large storm will remove the beach from the seaward side of the wall and steepen its seaward slope. Eventually, the wall is undermined and falls into the sea, and the property is lost as the oversteepened beach slope advances landward in an effort to reestablish its natural slope angle.

Coasts can be classified either **primary coasts** or **secondary coasts**. Primary coasts include all coasts whose character has been determined by nonmarine processes and include **land erosion coasts, subaerial deposition coasts, glacial deposition coasts, volcanic coasts, coasts shaped by earth movements**, and **ice coasts**. Secondary coasts include all coasts whose

FIGURE 11-16 Tethered-float Breakwaters.

A, deployment of tethered-float breakwater at San Diego Harbor. Photograph courtesy of Scripps Institute of Oceanography, University of California, San Diego. *B,* artist's sketch of a tethered breakwater system incorporating 1.5-m (4.8-ft) spheres of steel placed 1.5 m apart. The spheres are anchored to the ocean floor and held just beneath the ocean surface. This system has the advantage over traditional breakwater structures in that it removes energy from waves but does not interfere with the transport of sediment along the shore.

A.

B.

nonmarine character has been destroyed by marine processes and include **wave erosion coasts, marine deposition coasts,** and **coasts formed by marine biological activity.**

The region of contact between the oceans and the continents is marked by the **shore,** lying between the lowest low tides and the highest elevation on the continents affected by storm waves. The **coast** extends inland from the shore as far as land features related to marine processes can be found. The shore is divided into the **foreshore,** extending from low tide to high tide, and the **backshore,** extending beyond the high-tide line to the **coastline** that separates the shore from the coast. Seaward of the low-tide line are the **nearshore zone,** extending to the breaker line, and the **offshore zone** beyond.

Wave erosion of the shore produces a **wave-cut cliff** that constantly retreats, leaving behind such features as **wave-cut benches, sea caves, sea arches,** and **stacks.** The rate of wave erosion increases with increased **exposure** of the shore to the open ocean, decreasing **tidal range,** and decreasing **strength of bedrock.**

As waves break at an angle to the shore, a **longshore current** is set up that produces a **longshore drift** of sediment along the shore. Deposition of sediment transported by the longshore current may produce such features as **beaches; spits,** deposits attached at one end to land; **tombolos,** sand ridges connecting an island to another island or the mainland; and **barrier islands,** long offshore deposits lying parallel to the shore. From ocean side to lagoon side, barrier islands commonly exhibit the following divisions: **ocean beach, dunes, barrier flat,** and **salt marsh. Deltas** form at the mouths of rivers that carry more sediment to the ocean than can be distributed by the longshore current.

A drop in sea level may be indicated by ancient wave-cut cliffs and **stranded beaches** well above the present shoreline, and a rise in sea level may be indicated by submerged **relict beaches,** wave-cut cliffs, and **drowned river valleys.** Such changes in sea level may result from **tectonic** processes causing local movement of the land mass or from **eustatic** processes changing the amount of water in the oceans or the capacity of ocean basins, thus causing worldwide changes in sea level. Melting of continental ice caps during the last 18,000 years has caused a **glacioeustatic rise** in sea level of about 120 m (394 ft).

Along the Atlantic Coast, tidal range varies from less than 1 m (3.3 ft) along the Florida coast to over 2 m (6.5 ft) along the coast of Maine. Sea level is rising about 0.3 m (1 ft) per century along most of the coast except for Maine, where isostatic rebound keeps pace with the rising water level. The average annual rate of erosion is 0.8 m (2.6 ft), although accretion is dominant along the coasts of Delaware, Georgia, and New York. Along the Gulf Coast, wave energy levels are generally low and the tidal range is less than 1 m. Sea level is rising at a rate of 0.3 per century, and the average rate of erosion is 1.8 m (6 ft) per year. The Mississippi River Delta is eroding at a rate of 4.2 m (13.7 ft) per year, resulting in an annual loss of wetlands set at 130 km^2 (50 mi^2). Along the Pacific Coast high energy winter waves and a 1 to 2 m (3.3 to 6.6 ft) tidal range are found. The average rate of erosion is only 0.005 m (0.016 ft) per year; because of the protected conditions of Puget Sound, Washington shows a net accretion. There are, however, many areas where the problem of erosion is significant. Based on coastal landform types, mud flats erode more rapidly than sand beaches, barrier islands, or rocky shores.

Structures constructed along the shore, such as **jetties** to protect harbors and **groins** used to widen beaches, will trap sediment on the upcurrent side, but erosion may then become a problem downcurrent. A possibly less damaging replacement for fixed breakwater structures is the **tethered-float breakwater.**

QUESTIONS AND EXERCISES

1. Why has the National Flood Insurance Program had a negative impact on the nation's shores?
2. Define the characteristics of the two major categories of coast classification and list the subcategories of each.
3. To help you reinforce your knowledge of the shore, construct and label your own diagram similar to that in figure 11–2.
4. Discuss the formation of such erosional features as sea cliffs, sea caves, sea arches, and stacks.
5. List and discuss three factors in the rate of wave erosion.
6. What variables affect the velocity of the longshore current?
7. What is the longshore drift, and how is it related to the longshore current?
8. Discuss the composition of beaches and the relationship of beach slope to particle size.

9. List and define the depositional features spit, tombolo, bay barrier, and barrier island. Include discussion of how some barrier islands develop peat deposits running through them from ocean shore to marsh.

10. Discuss why some rivers have deltas and others do not. Also include the factors that determine whether a "bird foot" or a smoothly curved Nile-type delta will form.

11. Compare the causes and effects of tectonic vs. eustatic changes in sea level.

12. List the two basic processes by which coasts advance seaward, and list their counterparts that lead to coastal retreat.

13. Describe the tectonic and depositional processes causing subsidence along Atlantic-type margins.

14. How does the Pacific-type margin differ from the Atlantic-type margin?

15. Discuss the Atlantic Coast, Gulf Coast, and Pacific Coast by describing the conditions and features of emergence-submergence and erosion-deposition that are characteristic of each.

16. Describe the effect on erosion and deposition caused by putting a structure such as a breakwater or jetty across the longshore current and drift.

REFERENCES

Bird, E. C. F. 1985. *Coastline changes: A global review.* Chichester, U. K.: John Wiley & Sons.

Burk, K. 1979. The edges of the ocean: An introduction. *Oceanus* 22–3:2–9.

Coates, R., ed. 1973. Coastal geomorphology. *Publications in Geomorphology.* Binghamton, N.Y.: State University of New York.

Ingmanson, D. E., and Wallace, W. J. 1973. *Oceanology: An introduction.* Belmont, Calif.: Wadsworth.

Kuhn, G. G., and Shepard, F. P. 1984. *Sea cliffs, beaches, and coastal valleys of San Diego County: Some amazing histories and some horrifying implications.* Berkeley: University of California Press.

Leatherman, S. P. 1983. Barrier dynamics and landward migration with Holocene sea-level rise. *Nature* 301:5899, 415–17.

May, S. K., Kimball, H., Grandy, N., and Dolan, R. 1982. The Coastal Erosion Information System. *Shore Beach* 50, 19–26.

Peltier, W. R., and Tushingham, A. M. 1989. Global sea level rise and the greenhouse effect: Might they be connected? *Science* 244:4906, 806–10.

Shepard, F. P. 1973. *Submarine geology,* 3rd ed. New York: Harper & Row.

_____. 1977. *Geological oceanography.* New York: Crane, Russak & Company.

SUGGESTED READING

Sea Frontiers

Carr, A. P. 1974. The ever-changing sea level. 20:2, 77–83.
A discussion of the causes of sea-level change is very well presented.

Emiliani, C. 1976. The great flood. 22:5, 256–70.
An interesting discussion of the possible relationship of the rise in sea level resulting from the melting of glaciers 11,000–8,000 years ago and biblical and other ancient accounts of a great flood.

Feazel, C. 1987. The rise and fall of Neptune's kingdom. 33:2, 4–11.
A discussion of factors that change the level of the sea, including atmospheric conditions, currents, climate, ocean topography, and sea-floor spreading.

Fulton, K. 1981. Coastal retreat. 27:2, 82–88.
The problems of coastal erosion along the southern California coast are considered.

Grasso, A. 1974. Capitola Beach. 20:3, 146–51.
The destruction of the beach of Capitola, California, shortly after the construction of a harbor by the U.S. Army Corps of Engineers at Santa Cruz to the north.

Mahoney, H. R. 1979. Imperiled sea frontier—barrier beaches of the east coast. 25:6, 329–37.
The natural alteration of barrier beaches is considered.

Schumberth, C. J. 1971. Long Island's ocean beaches. 17:6, 350–62.
This is a very informative article on the nature of barrier islands. The specific problems observed on the Long Island barriers serve as examples.

Wanless, H. R. 1989. The inundation of our coastlines: Past, present and future with a focus on South Florida. 35:5, 264–7.
This is an informative overview of the evidence that is useful in evaluating the past and future movements of the shorelines.

Westgate, J. W. 1983. Beachfront roulette. 29:2, 104–109.
The problems related to the development of barrier islands are discussed.

Scientific American

Bascom, W. 1960. Beaches. 203:2, 80–97.
A comprehensive consideration of the relationship of beach processes, both large- and small-scale, to release of energy by waves.

Broecker, W. S., and Denton, G. H. 1990. What drives glacial cycles? 262:1, 48–107.
The most up-to-date information is included in this consideration of the causes of glacial cycles.

Dolan, R., and Lins, H. 1987. Beaches and barrier islands. 257:1, 68–77.
A discussion dealing with the ultimate futility of trying to develop and protect structures on beaches and barrier islands.

Fairbridge, R. W. 1960. The changing level of the sea. 202:5, 70–79.
A discussion of what is known of the causes of changing level of the sea, which seems to be related mostly to the formation and melting of glaciers and changes in the ocean floor.

12

THE COASTAL WATERS AND MARGINAL SEAS

In this chapter we will consider the increasingly stressed estuary and wetland environments, which are among the most biologically productive earth environments, and some classic examples of marginal seas. We will concentrate on the physical processes and conditions present in these features because they foster the high levels of biological productivity in coastal waters.

COASTAL WATERS

The primary difference between the coastal waters and the open ocean is that of depth. Generally, changes in the nature of water with respect to distance and time are much greater in the shallower coastal waters. Because of the shallowness of the coastal waters, river runoff and tidal currents have a very significant effect on the nature of coastal water.

Salinity

River runoff has the direct effect of reducing salinity of the surface layer in areas where mixing is not significant and throughout the water column where mixing does occur. In areas where the precipitation on the landmass is predominantly rain, the runoff of the rivers will be at the maximum during the season of maximum precipitation. However, if the runoff is fed to a great extent by the melting of snow and ice, the season of maximum runoff will always be the summer. In general, salinity will be lower in coastal regions than in the open ocean owing to the runoff of fresh water from the continents (fig. 12–1A, C).

Counteracting the effect of runoff in some coastal regions will be the presence of prevailing offshore winds that will usually have lost most of their moisture over the continent. These winds evaporate considerable quantities of water as they move across the surface of the coastal waters. The increase in the evaporation rate in these areas tends to increase the surface salinity, as shown in figure 12–1B.

Temperature

In coastal regions of the ocean where the water is relatively shallow, very great ranges in temperature may

Preceding page: Satellite image of a portion of the East Coast showing the land erosion coastline of Chesapeake Bay, an estuary created when the lower portion of the Susquehanna River valley and its tributary valleys were submerged by the rise in sea level that followed the end of the last Ice Age. Courtesy of Earth Satellite Corporation.

occur on a yearly basis. Sea ice forms in many of the high-latitude coastal areas in which temperatures are determined by the freezing point of the water, which will generally be above $-2°C$ (28.4°F). Maximum surface temperature in low-latitude coastal water may approach 45°C (113°F) in areas where the coastal water is somewhat restricted in its circulation with the open ocean and protected from strong mixing. The seasonal change in temperature can be most easily detected in the coastal regions of the midlatitudes, where surface temperatures are at a minimum in the winter and reach maximum values in the late summer.

Figure 12–1 shows how strong thermoclines may develop in areas where mixing does not occur. Very high-temperature surface water may form a relatively thin layer. Mixing reduces the surface temperature by distributing the heat through a greater vertical column of water, thus pushing the thermocline deeper and making it less pronounced. One major factor in mixing coastal water is the effect of tidal currents, which can have a considerable influence on the vertical mixing of shallow water near the coast. Prevailing winds can also have a significant effect on surface temperatures if they blow from the continent. These air masses we previously mentioned in regard to salinity will usually be of relatively high temperature during the summer and cause an increase in the surface temperature and rate of evaporation of ocean water. They will be of much lower temperature than the ocean surface during the winter and will cause a heat loss and cooling of the surface water near shore.

Coastal Geostrophic Currents

The geostrophic effect that was discussed in association with the current gyres in chapter 8 also develops in the coastal waters. The causes are basically the same. Wind blowing parallel to the coast in a direction that would cause water to pile up along the shore under the influence of the Coriolis effect produces a situation in which that water must eventually, under the influence of gravity, run back down the slope toward the open ocean. As the water runs down the slope away from the shore, the Coriolis effect causes it to veer to the north on the western coast and to the south on the eastern coast of continents in the Northern Hemisphere.

Another condition that produces geostrophic flow along the margins of a continent is the runoff of large quantities of fresh water that gradually mix with the oceanic water. This produces a surface slope of water away from the shore. The seaward slope is associated with salinity and density gradients, as both increase

FIGURE 12–1   Temperature and Salinity Changes in the Coastal Ocean.
A, freshwater runoff is not mixed into water column and forms surface layer of low salinity. A well-developed halocline occurs. *B,* warm, dry offshore winds cause a high rate of evaporation that may offset the effects of runoff, producing a halocline with a gradient that is the reverse of that seen in *A. C,* runoff is mixed with deeper water, producing an isohaline water column of generally lower salinity than the open ocean. *D,* in high latitudes where sea ice is forming or thawing throughout the year, the temperature of coastal water will remain uniformly near the freezing point. *E,* water in shallow, low-latitude coastal regions protected from free circulation with the open ocean may develop a high-temperature isothermal condition. *F,* in the mid-latitudes, coastal water will undergo a significant warming during summer. A strong seasonal thermocline may develop. *G,* winter may produce a layer of low-temperature water at the surface. The thermocline shown may develop; however, cooling may cause the surface water to sink because of increased density. This would result in a well-mixed isothermal water column. Mixing due to strong winds may drive the thermoclines shown in *F* and *G* deeper and may even cause mixing of the entire water column, producing an isothermal condition.

seaward (fig. 12–2). These variable currents, which depend upon the wind and the amount of runoff for their strength, are bounded on the ocean side by the more steady boundary currents of the open-ocean gyres.

These local geostrophic currents frequently flow in the opposite direction of the boundary current, as is the case with the **Davidson Current** that develops along the coast of Washington and Oregon during the winter. A very great amount of precipitation occurs in the Pacific Northwest during the winter months, and the winds that generally blow out of the southwest are the strongest during these same months. Their

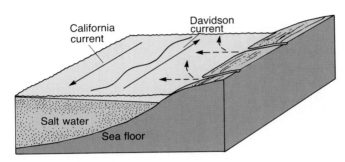

FIGURE 12–2   Davidson Current.
During the winter rainy season, fresh runoff water produces a seaward slope away from low-salinity surface water near the shore. A surface flow of low-salinity water from the shore toward the open ocean occurs. As the flow of surface water is acted on by the Coriolis effect, it veers right. This creates a north-flowing current (Davidson Current) along the coast of Washington and Oregon between the shore and the southbound California Current.

combined effect produces a relatively strong northward-flowing geostrophic current between the southward-flowing California Current, which is part of the open-ocean circulation, and the continent of North America.

ESTUARIES

Estuaries (fig. 12–3) are semienclosed coastal bodies of water in which the ocean water is significantly diluted by fresh water from land runoff. The mouths of large rivers throughout the world form the most economically significant estuaries, since many serve as ports and centers of ocean commerce. Many estuaries support important commercial fisheries as well, and the environmental changes that are occurring as a result of the commercial importance of the estuaries will be a major concern as the future development of these areas is planned. Many bays, inlets, gulfs, and sounds may be considered estuaries on the basis of their structure.

Origin of Estuaries

Essentially all estuaries in existence today owe their origin to the fact that in the last 18,000 years sea level has been raised approximately 120 m (394 ft) owing to the melting of much of the major continental glaciers that covered portions of North America, Europe, and Asia during the Pleistocene Epoch, more com-

FIGURE 12–3   Classification of Estuaries on the Basis of Origin.
Four classes of estuaries are coastal plain (e.g., Chesapeake Bay, Maryland and Virginia), fjord (e.g., Strait of Juan de Fuca, Washington), bar-built (e.g., Laguna Madre, Texas), and tectonic (e.g., San Francisco Bay, California).

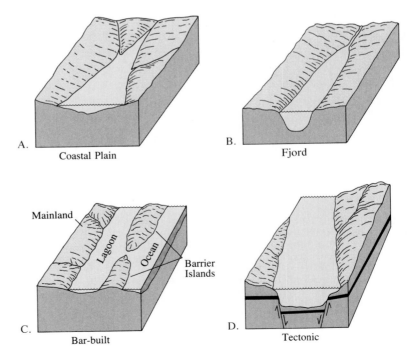

monly referred to as the Ice Age. Four major classes of estuaries can be identified on the basis of their origin (fig. 12–3):

1. **Coastal plain estuaries** were formed as the rising sea level caused the oceans to invade the existing river valleys. These estuaries are sometimes referred to as *drowned river valleys.* Chesapeake Bay is an example.

2. **Fjords** are glaciated valleys that are U-shaped with steep walls. They usually have a glacial deposit forming a sill near the ocean entrance. Fjords are common along Norwegian, Canadian and New Zealand coasts.

3. **Bar-built estuaries** are shallow estuaries separated from the open ocean by bars composed of sand deposited parallel to the coast by wave action. Lagoons separating barrier islands from the mainland are bar-built estuaries.

4. **Tectonic estuaries** are produced by faulting or folding, which causes a restricted downdropped area into which rivers flow. San Francisco Bay is in part a tectonic estuary.

Water Mixing in Estuaries

Generally, the freshwater runoff that flows into an estuary moves as an upper layer of low-density water across the estuary toward the open ocean. An inflow from the ocean takes place below the upper layer, and mixing takes place at the contact between these water masses. Stommel has classified estuaries on the basis of the degree of mixing as determined by the distribution of water properties into the following categories, which are shown in figure 12–4:

1. Vertically mixed—shallow, low-volume estuaries where the net flow always proceeds from the river at the head of the estuary toward the

mouth. Salinity at any point in the estuary will be uniform from the surface to the bottom owing to the even mixing of the river water with ocean water by eddy diffusion at all depths. Salinity increases from the head to the mouth of the estuary.

2. Slightly stratified—a relatively shallow estuary in which the salinity increases from the head to the mouth at any depth. Two basic water layers can be identified: the less saline upper water provided by the river and the deeper marine water separated by a zone of mixing. It is in this type of estuary that we begin to see the typical estuarine circulation pattern develop, as there is a net surface flow of low-salinity water into the ocean and a net subsurface flow of marine water toward the head of the estuary.

3. Highly stratified—typical of deep estuaries in which the salinity in the upper layer increases from the head of the estuary to the mouth, where it reaches a value close to that of open-ocean water. The deep-water layer, however, has a rather uniform marine salinity at any depth throughout the length of the estuary. The net flow of the two layers is similar to that described for the slightly stratified estuary, except that the mixing that occurs at the interface of the upper water and the lower water is such that the net movement is from the deep-water mass into the upper water. The less saline surface water does not seem to dilute the deep-water mass and simply moves from the head toward the mouth of the estuary, having its salinity increased as water from the deep mass joins it. Relatively strong haloclines develop in such estuaries at the contact between the upper and lower water masses. It is not unusual for these haloclines to reach magni-

FIGURE 12–4 Classification of Estuaries on the Basis of Degree of Mixing.

Note that due to the Coriolis effect, surface freshwater flow toward the mouth of the estuary extends seaward much farther along the right-hand shore as one faces seaward. The opposite side of the estuary experiences a greater marine influence. In most estuaries, the marine flow of water into them is a subsurface flow.

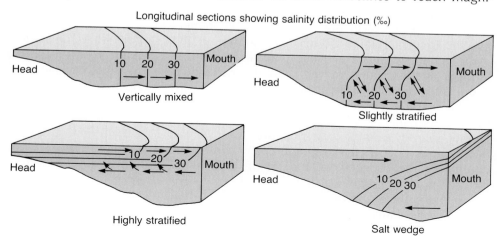

Longitudinal sections showing salinity distribution (‰)

tudes approaching 20‰ near the head of the estuary at times of maximum river flow.

4. **Salt wedge**—resulting where a saline wedge of water intrudes from the ocean below that of the river water; typical of the mouths of deep, large-volume-transport rivers. There is no horizontal salinity gradient at the surface in these deep estuaries, the water being essentially fresh throughout the length of, and even beyond, the estuary. There is, however, a horizontal salinity gradient at depth and a very pronounced vertical salinity gradient manifested as a strong halocline at any station throughout the length of the estuary. This halocline will be shallower and more highly developed near the mouth of the estuary.

The mixing patterns described often cannot be applied to an estuary as a whole. Mixing within an estuary may change with longitudinal distance, season, or tidal conditions.

Estuaries and Human Activities

The overriding problem regarding the health of estuaries is that they are not only of great value in their contribution to natural ecosystems that have evolved within them over millennia, but they also enhance the efficiency of economic activities that are not directly dependent upon the health of the estuary. These economic activities include shipping, logging, manufacturing, waste disposal, and others that can potentially damage the estuarine environment.

Even in areas where population pressure is still modest, estuaries can be severely damaged. The Columbia River estuary, which forms most of the border between Washington and Oregon, is flushed by tides that drive a salt wedge 42 km (26 mi) upstream and raise its water level by over 3.5 m (12 ft) (fig. 12–5). When the tide falls, up to 28,000 m^3/s (1,000,000 ft^3/s) of fresh water push out hundreds of kilometers into the Pacific Ocean to aid returning salmon to find their "home" waters. In the late nineteenth century, farmers and dairymen moved onto the rich floodplains and eventually diked them to prevent the inevitable natural flooding. Now these lands are removed from the natural ecosystem and, ironically, neither are they suitable for **agriculture**. The river has been a great boon to the **logging** industry, which has dominated the economy of the region throughout most of its history. The river has survived some terrible insults inflicted by the logging industry, but the **hydroelectric dams** have dealt the system a serious

blow. Many of these dams did not include salmon ladders, thereby closing off many of the previous spawning grounds. As is the case with most interactions between nature and humans, these structures were almost essential to the development of the region, but a high price was paid. At present, in an effort to diversify the local economy, the development of **shipping facilities** is occurring, with the accompanying dredging and increased potential for pollution. If such problems have developed at such sparsely populated areas as the Columbia River estuary, what must be the conditions at more highly stressed estuaries?

Chesapeake Bay is a classic example of a coastal plain estuary, produced by the drowning of the Susquehanna River and its tributaries. Most of the fresh water entering the bay enters along the western margin via rivers draining the slopes of the Appalachian Mountains. Also contributing to this pattern of salinity distribution is the Coriolis effect, acting on marine water entering the bay from the south, and the southward flow of river water (fig. 12–6). The area has experienced intense pressure from human activities.

With maximum river flow in the spring, a strong halocline (pycnocline) develops, preventing mixing of the fresh surface water and saltier deep water. Beneath the pycnocline, which can be as shallow as 5 m (16 ft), waters may become **anoxic** (without oxygen) from May through August as dead organic matter decays in the deep water. Major kills of commercially important blue crab and oysters as well as other bottom-dwelling organisms occur during these events. For unknown reasons, the degree of stratification and kills of bottom-dwelling animals have increased over the last 40 years. It is, however, suggested that increased nutrients have been added to the bay as a result of human activities. The increased microscopic plant production resulting from this could cause the deposition of increased amounts of organic particulates on the bottom.

That much is yet to be learned about estuarine circulation is illustrated by the following observations made over an extended period in Chesapeake Bay. Just the opposite of the expected pattern, intervals of upstream surface flow accompanied by downstream deep flow have been recorded. Also, periods when all flow was upstream as well as periods of total downstream flow have been observed. Even more complex patterns known to develop involve a surface and bottom flow in one direction separated by a middepth flow in the opposite direction or landward flows along the shores and seaward flows in the central portions of the estuaries.

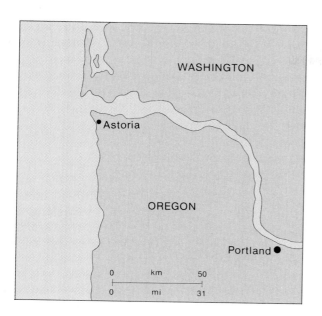

FIGURE 12–5   Columbia River Estuary.

This large-volume salt-wedge estuary has suffered more from interference with its flow by dams and diked floodplains than from chemical pollution. Like the Mississippi River, the large volume of its flow will aid it in fighting off the effects of chemical pollution.

FIGURE 12–6   Chesapeake Bay.

A, location map and average surface salinity distribution. *B*, comparison of dissolved oxygen levels for the summers of 1950 and 1980. Deeper Midbay waters are anoxic in 1980. After Officer et al., 1984.

A.

B.   Distance from the mouth of bay (Km)

Estuaries and the Continental Shelf Plankton Community

Studies of four estuarine environments and their influence on the continental shelf plankton community conducted between Charleston, South Carolina, and Jacksonville, Florida, revealed similar conditions (fig. 12–7A). As estuarine phytoplankton (microscopic photosynthetic cells) are carried seaward by the estuarine flow, the rate of photosynthetic production increased sixfold because of improved water clarity. Beyond a distance of 5 km (3 mi), production gradually dropped across the shallow shelf as the nutrient-poor Gulf Stream waters were approached (fig. 12–7B). Seasonal changes in photosynthetic production were similar in the estuaries, nearshore shallow shelf waters (less than 20 m depth), and deeper offshore shelf waters. Photosynthetic productivity in all regions peaked in late summer, as did the abundance of zooplankton (microscopic animal populations), including fish eggs and larvae (fig. 12–7C). Because these populations have different reproduction rates and would not be expected to peak together, it appears that the similar patterns of changes in photosynthetic production and the abundance of populations of animal plankton may be explained by the fact all were produced in the estuaries and flushed into the shelf waters by estuarine outflow. If this is true, these estuaries have a major influence on the plankton populations and the overall ecology of the continental shelf.

Understanding the dynamics of estuarine circulation is essential if desired advances are to be made in our ability to meaningfully describe (1) parameters of water quality such as dissolved oxygen and coliform bacteria, (2) suspended particulate matter and bottom sediment transport, and (3) biological activity, particularly in relation to microscopic plants and animals, fish eggs, and larvae.

WETLANDS

Bordering estuaries and other shore areas protected from the open ocean are **wetlands**, biologically productive strips of land delicately in tune with natural shore processes. They are of two types: **salt marshes** and **mangrove swamps**. Both are intermittently submerged by ocean water and are characterized by oxygen-poor mud and peat deposits. Marshes, characteristically inhabited by a variety of grasses, are known to occur from the equator to latitudes as high as 65°. Mangrove trees are restricted to latitudes below 30° (fig. 12–8). Once mangroves colonize an area, they can normally outgrow and replace marsh grasses.

With our strong desire to live near the oceans, marsh and mangrove wetlands have suffered from systematic filling and development for housing, industry, and agriculture.

Research is showing that wetlands have a very high economic value when left alone. Salt marshes are believed to serve as nursery grounds for over half the species of commercially important fishes in the southeastern United States. Other fishes such as flounder and bluefish use them for feeding and overwintering, and fisheries of oysters, scallops, clams, and fishes such as eels and smelt are located directly in the marshes.

More than half of the nation's wetlands have been lost. Of the original 215 million acres of wetlands that once existed in the conterminous United States, only about 90 million acres remain. To help prevent the loss of remaining wetlands, the Environmental Protection Agency established an Office of Wetlands Protection (OWP) in 1986. At that time, wetlands were being lost to development at a rate of 300,000 acres per year. The OWP will actively pursue enforcement of regulations against wetlands pollution and identification of the most valuable wetlands that can be protected or restored.

A very important characteristic of wetlands is their ability to remove inorganic nitrogen compounds and metals from groundwater polluted by land sources. Most of the removal is probably achieved through adsorption on clay-sized particles. Some of the nitrogen compounds trapped in the sediment are decomposed by denitrifying bacteria, releasing the nitrogen to the atmosphere as nitrogen gas. Much of the remaining compounds is used for plant production in this environment, which is one of the most productive in the world. With the death of the plants, the organic nitrogen compounds are either incorporated into the sediment and converted to peat or are broken up and become food for bacteria, fungi, or detritus-feeding shell- and fin fish.

LAGOONS

Landward of the barrier island salt marshes lie protected, shallow bodies of water called **lagoons**. Because of very restricted circulation between lagoons and the ocean, three distinct zones can usually be identified within a lagoon. There will usually be a

A. Study Area

B. Photosynthetic Production
(carbon produced—mg/m³/hr)

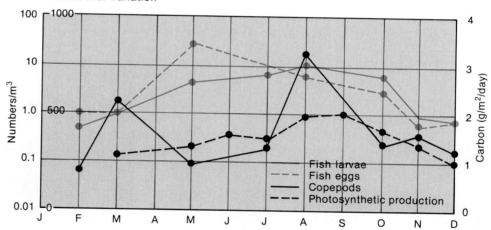

C. Seasonal Variation

Fish larvae
Fish eggs
Copepods
Photosynthetic production

FIGURE 12–7   Estuaries and Shelf Waters.

A, along the southeast U.S. coast, transects from estuaries across the continental shelf were conducted to study photosynthetic production and plankton abundance at the following locations: Charleston, S.C.; Savannah, Ga.; Sapelo Island, Ga., and Jacksonville, Fla. Along this coast, relatively small-volume rivers flow onto a gently sloping continental shelf at least 100 km (62 mi) wide. *B,* this curve simulates average photosynthetic production pattern along the transects. Zero on the distance scale represents the nearshore shelf waters adjacent to the mouth of the estuary. Patterns of zooplankton abundance are similar. *C,* seasonal changes in photosynthetic production, copepods, an important population of small plankton that eat photosynthetic cells, fish eggs, and fish larvae. A similar pattern for all these quite different life forms indicates they may be produced in the estuary and flushed together into the shelf waters by estuarine outflow. After Turner, R. E., et al., 1979.

A.

B.

C.

FIGURE 12–8   World Distribution of Salt Marshes and Mangrove Swamps.

A, note dominance of mangrove swamps in low latitudes and salt marshes in high latitudes. Base map courtesy of National Ocean Survey. *B*, salt marsh. Photograph by Michael DiSpezio. *C*, mangrove swamp. Photograph by T. W. Vaughan, U.S. Geological Survey.

freshwater zone near the mouths of rivers that flow into the lagoon, a **transitional zone** of brackish water (water with a salinity between that of fresh water and ocean water) and a **saltwater zone** close to the entrance, where the maximum tidal effects within the lagoon can be observed. Moving away from the saltwater zone near the mouth of the lagoon, the tidal effects diminish and are usually undetectable in the freshwater region that is well protected from tidal effects. Figure 12–9 shows the general characteristics of a typical lagoon.

The salinity of lagoons is also determined by factors other than position relative to the mouth of a lagoon. In latitudes with seasonal variations in temperature and precipitation, ocean water will flow through the entrance during a warm, dry summer to compensate for the volume of water that is lost through evaporation. This flow results in an increased salinity within the lagoon. Lagoons may actually become hypersaline in arid regions where the inflow of seawater is not sufficient to keep pace with evaporation that is taking place at the surface of the lagoon. During the rainy season the lagoon will become much less saline as the amount of freshwater runoff increases.

A **hypersaline lagoon** along the coast of Texas between Corpus Christi and the mouth of the Rio Grande is **Laguna Madre**, which is a long, narrow body of water protected from the open ocean by Padre Island, a 160-km (100-mi)-long offshore island. Much of the lagoon, which probably formed about 6000 years ago as sea level was approaching its present height, is less than 1 m (3.3 ft) in depth. The tidal range of the Gulf of Mexico in this area is about 0.5 m (1.6 ft), and the inlets, one of which is shown in figure 12–10, at each end of the barrier island are quite small. There is, therefore, very little tidal interchange between the lagoon and the open ocean.

The shallowness of the water in the lagoon makes possible a very great seasonal range of temperature and salinity in this semiarid region. The water temperatures are high in the summer and may fall below 5°C (41°F) in the winter. Salinities range from 2‰ to over 100‰ because of infrequent local storms that provide large volumes of fresh water in a short period of time and high evaporation that keeps the salinity generally well above 50‰. Because even the salt-tolerant marsh grasses cannot withstand such high salinities, the marsh is replaced by an open sand beach on Padre Island. At the inlets, the inflow from the ocean is as a surface wedge over the more dense water of the lagoon. In turn, the outflow from the lagoon occurs as a subsurface flow, just the opposite of the circulation described for estuaries.

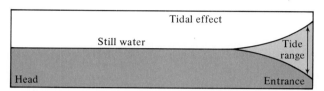

FIGURE 12–9 Lagoons.
Typical configuration, salinity conditions, and tide conditions of a lagoon.

FIGURE 12–10 Laguna Madre.
Typical summer surface salinity distribution (‰).

FIGURE 12–11 General Bathymetry and Subseas of the Mediterranean Sea.

From *Encyclopedia of oceanography,* edited by Rhodes Fairbridge, © 1966. Reprinted by permission of Dowden, Hutchinson, & Ross, Inc., Stroudsburg, Pa.

MARGINAL SEAS

At the margins of the ocean bodies are relatively large bodies of water, **marginal seas**, that have varying degrees of restriction in the circulation that exists between them and the open oceans. We will now discuss the circulation that exists between some of the major marginal seas and the open-ocean bodies they border and the water characteristics that develop within these seas as a result of these circulation patterns.

Marginal Seas of the Atlantic Ocean

Mediterranean Sea The Mediterranean Sea is an east-west-trending body of water that has a very irregular coastline dividing it into subseas with separate circulation patterns. Bounded by Europe and Asia Minor on the north and east and Africa to the south, it is surrounded by land except for a narrow connection with the Atlantic Ocean through the **Strait of Gibraltar** and the very narrow **Bosphorus** to the Black Sea. A man-made canal, the Suez Canal, connects the Mediterranean Sea to the Red Sea. The Mediterranean is divided into two major basins separated by a sill with a 400-m (1312-ft) depth, extending from Sicily to the coast of Tunisia. To the north of Sicily, strong currents run between Sicily and the Italian mainland through the Strait of Messina. The much broader connection between Sicily and Tunisia is the Strait of Sicily (fig. 12–11).

Atlantic water enters the Mediterranean through the Strait of Gibraltar as a surface flow that is carried in to replace water evaporated at a high rate in the very arid eastern end of the Mediterranean Sea. The

water level in the eastern Mediterranean is generally 15 cm (6 in) lower than at the Strait of Gibraltar. This surface flow follows the northern coast of Africa throughout the length of the Mediterranean and spreads north across the sea (fig. 12–12).

The Atlantic water that finds its way to the northern coast of the eastern basin is cooled by evaporation in the Adriatic Sea and sinks to the bottom with a tem-

Surface flow ⟶
Intermediate flow (200–600 m; 650–2,000 ft) ----➤

A. SURFACE AND INTERMEDIATE CIRCULATION

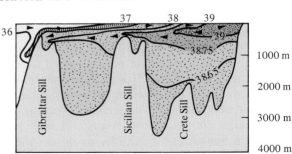

B. VERTICAL DISTRIBUTION OF SALINITY (°/₀₀)
Most of the mass of Mediterranean Water has salinity ranging between 38 °/₀₀ and 39 °/₀₀. Maximum salinities in excess of 39 °/₀₀ are found in the surface waters at the east end of the sea. After sinking and moving as Intermediate Water toward the Atlantic Ocean, salinity is reduced to about 37.3 °/₀₀ as it flows over the sill at Gibraltar.

FIGURE 12–12 Mediterranean Circulation.

perature of 13.2°C (55.8°F) and a salinity of 38.6‰. The water apparently is not trapped in these deep basins because of the formation of new bottom water each winter. This conclusion is reached on the basis of the relatively high dissolved oxygen content that may reach 4.7 ML/L in the western basin and 5.0 ML/L in the eastern basin.

The remaining Atlantic water continues eastward to Cyprus, where during the winter it sinks to form the **Mediterranean Intermediate Water**, with a temperature of 15°C (59°F) and a salinity of 39.1‰. This water flows west along the North African coast at a depth ranging from 200 to 600 m (656 to 1968 ft) and passes into the North Atlantic as a subsurface flow through the Strait of Gibraltar. By the time it passes through the strait, its temperature has dropped to 13°C (55.4°F) and the salinity is 37.3‰. It is still much denser than water at this depth in the Atlantic Ocean, so it moves down the continental slope until it reaches a depth of approximately 1000 m (3280 ft), where it encounters Atlantic water of the same density. It is at this level that it spreads out in all directions into the Atlantic water and becomes the **Mediterranean water mass**, detectable over a broad area because of its high salinity.

The circulation between the Mediterranean Sea and the Atlantic Ocean is typical of closed restricted basins in areas where evaporation exceeds precipitation. In such situations the restricted basins will always lose water at a very high rate from the surface through evaporation, and this water will have to be replaced by surface inflow from the open ocean. The evaporation of the water that flows in from the open ocean increases its salinity to very high values, causing eventual sinking and the return to the open ocean as a subsurface flow. Since the Mediterranean Sea is the classic area for this type of circulation it is given the name **Mediterranean circulation**. You will recognize it as being opposite the pattern we observed to be characteristic of estuaries, where the surface flow goes into the open ocean and subsurface flow into the estuary. In areas where **estuarine circulation** develops between the marginal body of water and the ocean, the availability of fresh water from runoff and precipitation is greater than the rate of water loss to evaporation.

The "American Mediterranean"
The "American Mediterranean" includes the four basins of the Caribbean Sea and the Gulf of Mexico. The **Caribbean Sea** is separated from the Atlantic Ocean by an island arc called the **Antillean Chain**. Composed of the islands of Cuba, Hispaniola, Puerto Rico, and Jamaica, the **Greater Antilles** constitute most of the northern boundary of the Caribbean Sea. The **Lesser Antilles** extend from the Virgin Islands in an arc to the continental shelf of South America near the island of Trinidad. The deepest connection between the Caribbean and the Atlantic Ocean is through **Anegada Passage** east of the Virgin Islands. This passage has a maximum depth near 2300 m (7544 ft). Other channels through which communication occurs have depths ranging between 1500 and 2000 m (4920 and 6560 ft). The Caribbean Sea is divided into four major basins from east to west—Venezuela, Colombia, Cayman, and Yucatán—all of which reach depths in excess of 4000 m (13,120 ft) (fig. 12–13).

The **Gulf of Mexico**, or the Mexican Basin of the American Mediterranean, is much less complex than the Caribbean Sea in that it is a simple structure with broad continental shelf surrounding much of its margin. It slopes off into a relatively broad basin that reaches a maximum depth in excess of 3600 m (11,800 ft). The Gulf of Mexico is connected to the Caribbean Sea by the **Yucatán Strait**, which reaches a maximum depth of 1900 m (6232 ft). The only connection with the Atlantic Ocean is through the **Straits of Florida**, which reach depths approaching 1000 m (3280 ft).

Surface Currents. Entering through the Lesser Antilles, where sill depths less than 1000 m generally prevail, the **Guiana Current**, which represents the portion of the South Equatorial Current that moves northwest along the Guiana Coast of South America, brings water with a temperature between 26° and 28°C (78.8° and 82.4°F) and a salinity between 35.0‰ and 36.5‰ into the Caribbean Sea. This relatively thin mass of water passes into the Caribbean Sea through the shallow channels north and south of **St. Lucia Island** and mixes in a 1:3 ratio with North Atlantic Water. Becoming the Caribbean Current, the axis of which ranges between 200 and 300 km (124 and 186 mi) north of the Venezuelan coast, the surface flow continues generally west over the deepest portion of the Caribbean Sea. It turns north and passes through the Yucatán Strait to the Gulf of Mexico. Surface velocities as high as 4.5 km/h (2.8 mi/h) have been observed in the main axis of the Caribbean Current, but the maximum surface velocities usually average less than 2 km/h (1.2 mi/h). The current can be detected as deep as 1500 m (4920 ft) in the eastern basins. Volume transport into the Caribbean Sea by the Caribbean Current is about 30 sv.

The easterly component of the trade winds blowing along the coast of Venezuela and Colombia sets up a surface flow away from the coast that produces a shallow upwelling. Most of the water rising to the sur-

FIGURE 12–13 "American Mediterranean" Surface Currents and Bathymetry.

Vertical salinity profiles for numbered stations are shown in figure 12–15.

Deeper than 4000 meters

face comes from depths of less than 250 m (820 ft). The upwelling water contains high concentrations of nutrients and low temperatures that result in relatively high biological productivity in the coastal region.

The Caribbean Current passes through the Yucatán Strait, causing a "dome" of water in the Gulf of Mexico that stands 10 cm (4 in.) higher than the Atlantic water southeast of Florida. This hydraulic head forces an intense flow known as the **Florida Current** to pass through the Straits of Florida. It joins with water carried north by the **Antilles Current** and flows north along the coast of Florida (fig. 12–13). The **Loop Current** is a significant feature of the Gulf of Mexico's surface circulation. Figure 12–14 shows the relationship of the current flow direction to the topography of the 20°C (68°F) **isotherm** surface in the southeastern Gulf of Mexico. Much of the surface water enters the Gulf of Mexico through the Yucatán Strait, loops around the temperature "bowl" in a clockwise flow and heads toward the Straits of Florida.

Water Masses in the Caribbean Sea. There are four readily identifiable water masses in the Caribbean Sea, two of which are relatively warm surface masses found above 200 m (656 ft) depth. The two deeper masses are characterized by lower temperature. The

salinity of the **Caribbean Surface Water** mass is determined by the rate of evaporation versus the amount of precipitation and runoff. It generally ranges above 36‰ during the winter, partially because of the upwelling of high-salinity water. The surface salinity decreases to the north to values generally less than 35.5‰.

Extending from the southeast to the northwest of the Caribbean Sea near the Yucatán Strait is a thin, sheetlike, high-salinity layer—the **Subtropical Underwater**. Located at depths as shallow as 50 m (164 ft) in the southeast, it dips to a depth of 200 m near the Yucatán Strait. The maximum salinity within this sheet seems to follow the axis of flow for the Caribbean Current and exceeds 37‰ in the Yucatán Strait. Salinity decreases away from the flow axis.

Also following the main flow axis is a low-salinity water mass, the **Intermediate Water**, located directly beneath the Subtropical Underwater. A salinity minimum that falls below 34.7‰ in the southeast and becomes less detectable at the Yucatán Strait identifies this mass (fig. 12–15).

Entering the Caribbean primarily through the **Anegada Passage** between the Virgin Islands and the Leeward Islands of the Lesser Antilles and through the **Windward Passage** between Cuba and Hispaniola is **North Atlantic Deep Water**. Characterized by a salin-

FIGURE 12–14 Loop Current.

A, the position and strength of the Loop Current can be deduced from the depth (m) to the 20°C (68°F) temperature surface. Flowing clockwise, its velocity increases with increased slope of the temperature surface. What may be a clockwise-rotating eddy is indicated northwest of the main loop current. Courtesy of NOAA. B, block diagram showing flow of Loop Current in a clockwise pattern around the 10-cm-high surface "dome" of water that directly overlies the "bowl" representing the depth of the 20°C temperature surface. The vertical scale is greatly exaggerated. C, NOAA 9 satellite infrared image of sea-surface temperature in the Gulf of Mexico. The 20°C temperature surface that forms the "bowl" in B intersects the ocean surface where the two lightest shades of blue meet. The Loop Current loops clockwise around the warm water that "fills" the "bowl," then moves off through the Straits of Florida. Note that the warm water that forms the "dome" reaches temperatures in excess of 26°C (79°F) (red). Courtesy of NOAA.

FIGURE 12–15 "American Mediterranean" Water Masses. The cores of the Subtropical Underwater and the Intermediate Water can be identified on vertical salinity profiles as maxima and minima, respectively, at the stations 1, 2, 3, and 4. The locations of the stations are indicated on figure 12–13. The Caribbean Surface Water salinity is generally above 36‰ while the Bottom Water salinity is uniformly between 34.9 and 35.0‰. The saltier surface water does not sink because of its high temperature.

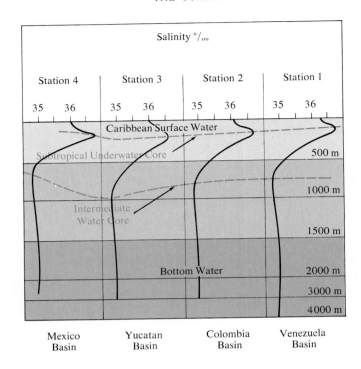

ity slightly less than 35‰ and temperature just above 2°C (35.6°F), this water spreads as Bottom Water and can be identified by an oxygen maximum that reaches values in excess of 5 ML/L.

Water Masses in the Gulf of Mexico. After passing through the Yucatán Strait, the surface-water characteristics can be identified down to a depth of 90 m (295 ft) during the winter and 125 m (410 ft) during the summer months, this being the depth through which seasonal temperature changes extend. Generally, the surface temperature just off the Yucatán coast ranges from 24° to 27°C (75° to 81°F), while that along the northern Gulf coast ranges between 18° and 21°C (64° and 70°F).

The salinity of the surface water generally ranges between 36.0‰ and 36.3‰. Runoff from the Mississippi River significantly decreases salinity to depths of 50 m (164 ft) and as far as 150 km (93 mi) from the coast. The influence of the runoff can be identified by salinity values as low as 25.0‰ near the surface.

The Subtropical Underwater is still identifiable as salinity maximum north of the Yucatán Strait at a depth of 100 to 200 m (328 to 656 ft). The boundary between the upper water, containing the surface water and Subtropical Underwater, and the deep water is marked by the 16°C (61°F) isotherm, which is at a depth of approximately 200 m.

The Intermediate Water, which enters the Gulf through the Yucatán Channel below 850 m (2788 ft), can be identified as a salinity minimum throughout the Gulf of Mexico. The core of the Intermediate Water can be identified throughout the Gulf at depths as shallow as 550 m (1804 ft), but its identity is lost passing through the Straits of Florida. The salinity and temperature increase very slightly to the bottom of the basin.

Marginal Seas of the Pacific Ocean

Gulf of California This northwest-southeast-trending sea extends from the Tropic of Cancer at its open south end to the mouth of the **Colorado River**, which is the main river system draining into it (fig. 12–16). Some half-dozen smaller rivers also empty into the Gulf from the east, carrying water from the Sierra Madre Occidental across a broad coastal plain that forms the western margin of the Gulf. From this coastal plain, which is on the **North American Plate**, a continental shelf extends into the Gulf to distances from 5 to 50 km (3 to 30 mi). The shelf terminates at an average depth somewhat in excess of 100 m (328 ft). This depositional feature does not exist on the western side of the Gulf, which is characterized by steep rocky slopes and is on the **Pacific Plate**. The Colorado River has developed a significant delta at the northern end, with depths rarely exceeding 200 m

FIGURE 12–16 Gulf of California Bathymetry and Surface Circulation.

A, basins increase in depth from the 980 m (3214 ft) depth of the Delfin Basin in the north to more than 3700 m (12,136 ft) in the Pescadero Basin. Winds that reverse on a seasonal basis produce upwelling on the east side of the Gulf in the winter and on the west side during the summer. From *Encyclopedia of oceanography,* edited by Rhodes Fairbridge, © 1966. Reprinted by permission of Dowden, Hutchinson, & Ross, Inc., Stroudsburg, Pa. *B,* pagodalike structure that vents 315°C (599°F) water atop sediment mound that covers a spreading center of the Guaymas Basin. *Alvin's* "arm" and vestimentiferan tube worms can be seen in the foreground. Photo by Robert Brown, Scripps Institution of Oceanography, University of California, San Diego.

A.

B.

(656 ft). Two exceptions are 1500-m (4920-ft) and 550-m (1804-ft) basins on the west and east side, respectively, of Angel de la Guarda Island. The depth gradually increases to the south through a series of basins, the deepest of which is 3700 m (12,136 ft). The basins are separated by sills that generally extend from 100 to 400 m (328 to 1312 ft) above the floors of the basin.

The sills may represent east-west-trending offsets of segments of spreading centers that are responsible for separating Baja California from the Mexican mainland. The character of the Gulf has changed somewhat since the completion of Hoover Dam in 1935. Until that construction was completed, the Colorado River provided an annual average flow of almost 18 billion m³ of water, which carried 161 million tn of sediment. Since 1935 the average annual flow has been reduced to less than 8 billion m³, which carries about 15 million tn of sediment. The drainage systems on the east side of the basin have relatively small flow volumes, as many of them are intermittent owing to the arid nature of their drainage basins.

Surface winds control the surface circulation, which has a seasonal pattern. A low-pressure system located over the northern end of the peninsula develops southerly summer winds that drive the surface water from the Pacific into the Gulf. This flow produces upwelling along the steep rocky coast of the peninsula. During the winter months, the low-pressure system is found on the mainland east of the Gulf, and the winter winds are northerly. The wind change produces upwelling along the mainland side of the Gulf, so there is a rich plankton bloom of diatoms and dinoflagellates throughout the year. This high plant productivity supports a significant zooplankton population that ultimately supports a large fish population.

The tidal range increases from about 1 m (3.3 ft) in the south to more than 10 m (33 ft) during spring tides at the mouth of the Colorado River. The tidal currents that develop in the north, along with convective mixing, produce an isothermal water column during the winter. Temperatures may drop as low as 16°C (61°F), as opposed to summer surface temperatures that may reach 30°C (86°F). A high rate of evaporation produces a marked stratification, with surface water being warmer and more saline. The water below the thermocline in the central southern portion of the Gulf possesses an oxygen minimum as low as 0.01 ML/L between the depths of 400 m (1312 ft) and 800 m (2624 ft).

A joint Mexican-American expedition identified a warm-spring biological community in the **Guaymas Basin** during the summer of 1980. In 1982 the biological community was observed and sampled during a dive of the submersible *Alvin*. Although they are covered by sediment, the vents are on a segment of a spreading center (fig. 12–16 *B*).

Bering Sea   This sea on the northern margin of the Pacific Ocean extends to latitude 66°N and has the shape of a triangle with a curved base represented by the **Aleutian Islands** (fig. 12–17). A broad continental shelf with depths less than 200 m (656 ft) extends off the Siberian and Alaskan coasts, but most of the rest of the basin is found to have depths in excess of 1000 m (3280 ft). The deep basin is located in the western half of the sea, and fully 90% of the total area of the sea is either less than 200 m in depth or more than

1000 m in depth. Except where it is cut by the **Bering Canyon** at the end of the Alaska Peninsula, the continental shelf slopes off very abruptly at 4° to 5° into the deep basin.

Circulation.   The major surface flow of water into the Bering Sea occurs between the **Komandorski Islands** of Russia and **Attu Island**, the most westerly of the Aleutian Islands belonging to Alaska. Here the **Alaskan Stream**, flowing in a westerly direction south of the Aleutian chain, converges with northward-moving water of the western Pacific and flows through the passages into the Bering Sea. A counterclockwise gyre is set up north of the Komandorski Islands, while a clockwise rotation develops to the north of the eastern end of the American islands. The main flow passes on to the east until it reaches the broad Alaskan shelf, then follows this shelf edge to the north. A portion of this northward flow passes between St. Lawrence Island and the east Siberian coast before crossing the Bering Strait into the Arctic Ocean.

Tidal currents dominate the shallow Alaskan shelf region, but a persistent northward-flowing current resulting from the runoff of fresh water from the Alaskan coast flows at speeds up to 10 km/h (6 mi/h) along the coast and through the Bering Strait. The Bering Sea possesses a large fisheries resource. For a discussion of how the character of this resource is related to physical processes of the continental shelf,

FIGURE 12–17 Bering Sea Bathymetry and Surface Circulation.

From *Encyclopedia of oceanography,* edited by Rhodes Fairbridge, © 1966. Reprinted by permission of Dowden, Hutchinson, & Ross, Inc., Stroudsburg, Pa.

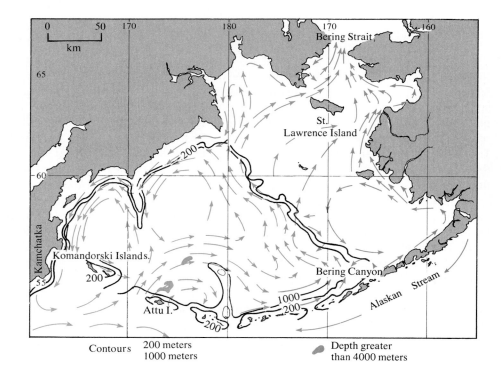

see "PROBES: Marine Science Steps Forward," on page 314.

Marginal Seas of the Indian Ocean

Red Sea The Red Sea extends over 1900 km (1180 mi) north of the narrow **strait of Bab-el-Mandeb** (Gate of Tears) to the northern tip of the **Gulf of Suez**, the western branch of the Red Sea that separates the Sinai Peninsula from the African mainland. Forming the eastern boundary of the Sinai Peninsula is the **Gulf of Aqaba**, which is an eastern branch at the northern end of the Red Sea (fig. 12–18). Extending from 12° to 30°N latitude, the Red Sea lies in a highly arid region and is characterized by surface waters of unusually high temperature and salinity. The sea is bordered by broad reef-covered shelves no more than 50 m (164 ft) deep that drop off sharply to a depth of about 500 m (1640 ft) to another gently sloping surface, which eventually leads into a deep central trough where depths are generally in excess of 1500 m (4920 ft) and extend to depths in excess of 2300 m (7545 ft) in the central region.

Geological evidence indicates the Red Sea has formed primarily during the past 20 million years, although the rifting may have begun 180 million years ago as a result of the Arabian Peninsula separating from the African mainland through the processes of sea-floor spreading.

Circulation. Sill depth at the strait of Bab-el-Mandeb is only 125 m (410 ft) compared to a maximum depth over part of the Red Sea that is in excess of 1000 m (3280 ft). Across this shallow sill, the basic circulation seems to be dominated by the high rate of evaporation, which exceeds 200 cm/yr (79 in./yr). The surface loss in the Red Sea must be replaced by Indian Ocean water from the Gulf of Aden that enters as a surface flow. As this surface flow moves north, its density is increased as evaporation increases salinity. The dense water sinks and returns as a subsurface flow to the sill and out into the Gulf of Aden. This outflowing warm saline water sinks rapidly until it finds its equilibrium depth and then spreads out into the Indian Ocean. This pattern of circulation (fig. 12–19) is similar to that of the Mediterranean Sea.

Because of the arid condition of the region, the surface water in the Red Sea reaches a salinity of 42.5‰ and a temperature of 30°C (86°F) during the summer months. Below a depth of 200 m (656 ft), a very uniform mass of water extends to the bottom throughout most of the Red Sea. This deep-water

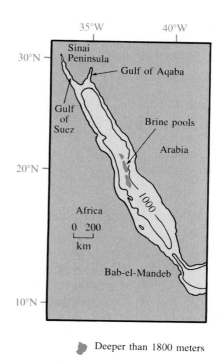

Deeper than 1800 meters

FIGURE 12–18 Bathymetry of the Red Sea.

mass has a temperature of 21.7°C (71°F) and a salinity of 40.6‰.

In 1966, investigators aboard the Woods Hole Oceanographic Institution vessel *Chain* studied a series of deep basins in the central Red Sea region that had been previously noted to contain extremely high-salinity and high-temperature water masses. Between 21°15′ and 21°30′N latitude, two major basins were found: the **Discovery Deep** to the south and the **Atlantis II Deep** to the north. These and other similar basins possess brine with temperatures in excess of 36°C (96.8°F) and salinity ranging as high as 257‰. These brines, because of their high salinity, have densities great enough to keep them in their basins and prevent them from mixing with the overlying Red

FIGURE 12–19 Circulation between the Red Sea and the Indian Ocean.

PROBES: MARINE SCIENCE STEPS FORWARD

The ultimate goal of marine scientists is to fully understand the biological and physical resources of the oceans. With this understanding will come the ability to develop levels of resource exploitation that will not excessively degrade the marine environment. In the case of biological resources, this understanding must take into account the relationships of organisms to their physical environment and to other organisms that share that environment. In a word, we must understand the *ecology* of the resource.

One study of the physical processes that lead to an abundance of animal life on the continental shelf of the Bering Sea goes by the acronym PROBES. The name of the project is Processes and Resources of the Bering Sea Shelf. Directed out of the University of Alaska by Professor John J. Goering, the study group includes many scientists from institutions throughout the nation. PROBES began in 1976 with the hypothesis that the physical processes present on the shelf lead to a pelagic food web dominating the outer shelf and a benthic food web on the middle shelf.

The Bering Sea is the third largest sea in the world, exceeded only by the Mediterranean and South China Seas. About 5% of the world's fishery is taken from the waters overlying the continental shelf which represents 45% of the sea. Although most of the shelf lies in U.S. water, American fishermen have concentrated on the king and tanner crabs and salmon. The large stocks of Alaska pollack and yellow-fin sole are exploited primarily by Japan, Korea, and Russia.

The continental shelf of the Bering Sea is divided into three domains by fronts where water masses moving in on- and offshore directions converge (Fig. 12A). The *coastal domain* extends seaward to a depth of about 50 m and ranges in width from 80 to 150 km. The *middle domain* extends seaward to near the 100 m depth, which occurs at distances from 350 to 400 km from shore. The *outer domain* extends to the shelf break, at a depth of about 170 m, and occurs about 500 km from shore.

The coastal domain is vertically mixed by tidal currents and wind waves. Freshwater runoff and salt water are well mixed although the surface water is slightly less salty than deeper water (Fig. 12B). The middle domain has enough water depth to separate the effects of tidal mixing (most effective on bottom) and wind wave mixing (most effective on top). This produces two distinct layers of water. A high-salinity, low-temperature (predominantly marine) layer underlies a warmer, less-saline surface layer. The greater water depth present in the outer domain produces a three-layered water structure. The top and bottom layers are similar to those of the middle domain although they are slightly more saline. Separating these layers is a layer of mixing where alternating thin laminae of offshore and shelf water can be identified.

Upwelling and enhanced biological productivity are associated with all of the fronts that separate shelf domains. This enhancement is most pronounced at the middle front, between the middle and outer domains (Fig. 12C). The patterns of phytoplankton and zooplankton distribution are organized around the inner, middle, and outer fronts.

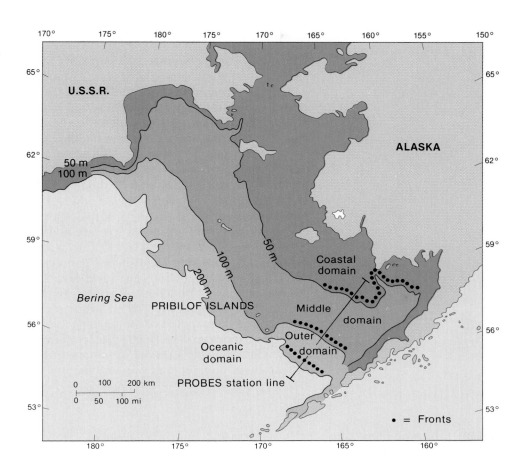

FIGURE 12A Physical Domains of the Bering Sea Shelf.

Primary productivity within the middle domain is about 400 grams of carbon per square meter per year, or about twice that of the outer domain (200 $gC/m^3/yr$). Most of the outer-domain phytoplankton is grazed by relatively large copepods and krill. These large grazers are prevented from entering the middle domain by their sensitivity to changes in physical water properties that occur at the middle front (Fig. 12D). The large outer-shelf copepods reproduce only once a year off the shelf in deep water, where food is scarce during winter. Juveniles migrate into the surface waters before the spring bloom begins (usually in May) to take advantage of the available food. They graze heavily on the phytoplankton and keep the bloom from attaining a large biomass.

The large crustaceans of the outer shelf are accompanied by smaller species of copepods that inflict much smaller grazing loss on the phytoplankton than do the larger forms. These smaller forms are found across the entire shelf and are the dominant grazers of the middle domain. The smaller copepods produce several broods per year, but reproduction does not occur until food becomes available, after the spring bloom has developed. Thus a large spring bloom is able to develop in the middle domain, and these small copepods never harvest a large amount of the phytoplankton production in the middle domain.

Because of the continual heavy grazing by the larger outer-domain herbivores, the phytoplankton biomass is kept at a low level. Little of this productivity is made available to bottom dwellers. On the other hand, the large phytoplankton biomass that goes unharvested by the smaller herbivores in the middle domain is made available to the benthos. Three times as much phytoplankton remains ungrazed in the middle domain as in the outer domain. As a result of the different phytoplankton productivity and zooplankton grazing patterns of the outer and middle domains, the outer domain has a predominantly pelagic food web and the middle domain has a primarily benthic food web. The Alaskan pollack, sablefish, and other pelagic fish are found in greater concentrations near the shelf break of the outer domain, while the benthic tanner crab, king crab, and yellow-fin sole are middle-domain fisheries.

The results of PROBES will help to provide an ecological basis for the management of Bering Sea fisheries and fisheries located on high-latitude continental shelves throughout the world ocean. A much greater understanding of the role of oceanic fronts in the ecology of shelf communities has been a major achievement of this study. PROBES represents a major step in the direction of increased understanding of the biological resources of the world ocean. As might be expected, the investigation confirmed the hypothesis on which the study was initiated, but more important is the fact it clearly identified *why* the pelagic outer-domain food web and benthic middle-domain food web exist.

FIGURE 12B Water Structure in Southeast Bering Sea.

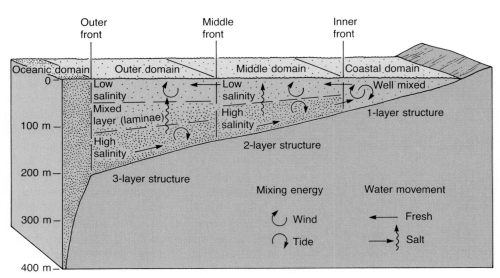

FIGURE 12C Nutrient Supply from Upwelling.

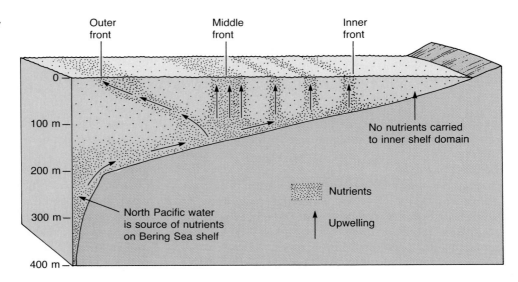

Outer front

Middle front

Inner front

0

100 m

200 m

300 m

400 m

No nutrients carried to inner shelf domain

North Pacific water is source of nutrients on Bering Sea shelf

Nutrients

Upwelling

FIGURE 12D Areas of Large Phytoplankton Biomass.

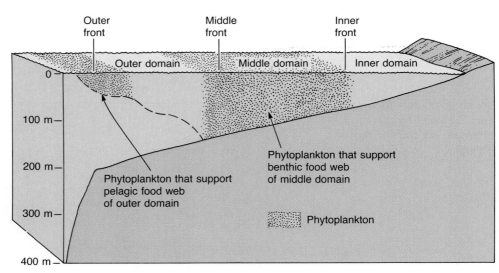

Outer front

Middle front

Inner front

Outer domain

Middle domain

Inner domain

0

100 m

200 m

300 m

400 m

Phytoplankton that support pelagic food web of outer domain

Phytoplankton that support benthic food web of middle domain

Phytoplankton

Sea water. The top of the brine accumulations is slightly below 2000 m (6650 ft).

Water may enter the porous oceanic crust beneath the Red Sea, be heated, and then dissolve the salts and metals that are found in the brines. Sediments found in the central deeps occupied by the brine possess concentrations of salts and metals that give them a great potential for economic development. Quite probably the enrichment in the crust beneath the sediment is also great and will enhance the economic potential of this area. These were the first **hydrothermal springs** to be discovered. This discovery provided the impetus to initiate the search that led to the discovery of the hydrothermal vents in the eastern Pacific.

Arabian Sea

The Arabian Sea is the northward extension of the Indian Ocean between Africa and India (fig. 12–20).

Circulation. Surface currents in the Arabian Sea are dominated by the monsoon winds that blow from the northeast from November until March, when the **southwest monsoon** begins to develop. The air carried onto the continent during this southwest monsoon contains large quantities of water and produces heavy precipitation in the coastal regions during this season.

During the **northeast monsoon** the surface current moves south along the west coast of India and turns west at about 10°N latitude, where some of the surface water flows into the Gulf of Aden and the rest turns south along the Somali coast and converges with the North Equatorial Current. When the southwest monsoon begins, the **North Equatorial Current** disappears, and a portion of the **South Equatorial Current** flows north along the Somali coast as the **Somali Current**. This strong seasonal current that flows with velocities in excess of 11 km/h (7 mi/h) continues along the coast of Arabia and India in a clockwise pattern until it reaches 10°N latitude. Here it joins the **Southwest Monsoon Current**, which has replaced the North Equatorial Current. Because of the alignment of winds relative to the African and Arabian coasts, significant upwelling occurs during the southwest monsoon.

Water Masses. Surface salinities north of 5°N latitude in the Arabian Sea are generally above 36‰ during the northeast monsoon. The Somali coastal region surface salinity may fall below 35.5‰ because of dilution by the South Equatorial Current and upwelling during the southwest monsoon. During the rainy season, surface salinities of less than 35‰ can be found as a result of dilution due to precipitation and runoff. Surface temperatures in the central ocean reach a maximum of 28°C (82°F) in June and a minimum temperature of 24°C (75°F) in February.

Below 200 m (656 ft), the salinity decreases to values near 35‰ until an abrupt salinity maximum of 35.4‰ to 35.5‰ develops at depths above 1000 m (3280 ft). This salinity maximum represents the flows

FIGURE 12–20 Bathymetry and Surface Circulation of the Arabian Sea and the Bay of Bengal.

Direction of Surface Flow

Summer—SW Monsoon – – – →
Winter—NE Monsoon ⸻→

of water into the Arabian Sea from the Persian Gulf and the Red Sea. The temperature of the **Red Sea Water** ranges from 9° to 10°C (48° to 50°F). The Red Sea Water can also be identified on the basis of 0.45 ML/L concentration of dissolved oxygen at about 790 m (2591 ft) off the Somali coast. Hydrogen sulfide has actually been observed on the continental slope at depths where this oxygen minimum occurs in the northern Arabian Sea.

Bay of Bengal The Bay of Bengal is bounded by India and the island of Sri Lanka on the west and by the Malay Peninsula and the Andaman-Nicobar Island Ridge to the east (fig. 12–20). A significant continental influence is exerted on this body of water by the runoff from the Ganges and Brahmaputra rivers at the extreme north end of the bay.

The currents in the Bay of Bengal are also dominated by the monsoon wind system. During the southwest monsoon, a clockwise rotation is established within the bay, and this circulation is accompanied by upwelling along the east Indian coast. With the development of the northeast monsoon in November, the circulation reverses and forms a counterclockwise gyre. At Chittagong in southeast Bangladesh, a seasonal change in the sea level of 1.2 m (4 ft) occurs as a result of monsoon wind changes.

The surface salinity in the Bay of Bengal seldom exceeds 34‰. During the southwest monsoon, particularly during late summer when the rainfall is the greatest, the runoff from the Ganges, Brahmaputra, and other rivers along the coast of Burma and India dilute the water and reduce surface salinity. Values as low as 18‰ may be observed at the extreme north end of the bay. The major influence of the dilution is observed along the Indian coast; it is less significant away from shore.

Of all the nations on the earth, Bangladesh is under the greatest threat if a significant sea level rise occurs. Over 80% of this nation is built on delta, and it is regularly devastated by storm surges that occur during the southwest monsoon. Any rise in sea level can only make things worse.

SUMMARY

Because of the shallow depth of the coastal ocean, river runoff, tidal currents, and seasonal changes in solar radiation, physical characteristics of the water such as temperature and salinity vary over a greater range than in the open ocean. **Coastal geostrophic** currents are produced as a result of runoff of fresh water and winds blowing in the proper direction along the coast. **Rip currents** and **longshore currents** are produced by waves breaking on the shore.

Estuaries are semienclosed bodies of water where freshwater runoff from the land and ocean water mix. They are classified on the basis of origin as **coastal plain, fjord, bar-built,** and **tectonic.** Mixing patterns range from that of a **vertically mixed** shallow estuary through **slightly stratified** and **highly stratified** to the **salt wedge** characteristic of large-volume river mouths. The typical pattern of **estuarine circulation** consists of a surface flow of low-salinity water toward the mouth and a subsurface flow of marine water toward the head.

Wetlands are of two basic types: **salt marsh** and tropical **mangrove swamp.** These biologically productive regions are alternately covered and exposed by the tide. Marshes are also important in their ability to adsorb land-derived pollutants before they reach the ocean. Despite their ecological importance, wetlands are fast disappearing from our coasts as they are encroached upon by human activities.

Some **lagoons** have very restricted circulation with the ocean and may exhibit a great range of salinity and temperature conditions as a result of seasonal change.

Marginal seas are relatively large bodies of marine water overlying the continental margins. They are shallower than and have varying degrees of restricted circulation with the open ocean. The **Mediterranean Sea** has a classic circulation characteristic of restricted bodies of water in areas where evaporation greatly exceeds precipitation. It is the reverse of estuarine circulation. The "**American Mediterranean**" includes the **Caribbean Sea** and **Gulf of Mexico.** The major surface currents are the **Caribbean Current** flowing in from the southeast and the **Florida Current** flowing out between Cuba and Florida.

The **Gulf of California** experiences an extreme tidal range at the north end and has high productivity resulting from seasonal upwelling on opposite sides of the Gulf on a seasonal basis. An extensive shelf on the northeast side and a deep basin to the southwest are characteristic of the **Bering Sea** lying north of the Aleutian Islands. Many important commercial fisheries exist in this sea.

The formation of a new sea is seen in the **Red Sea** between Africa and Arabia. Mineral-rich brine basins are found in its median rift. The **Arabian Sea** to the west of India and the **Bay of Bengal** to the east have circulation patterns determined by the **monsoon winds.**

QUESTIONS AND EXERCISES

1. For coastal oceans where deep mixing does not occur, discuss the effect offshore winds and freshwater runoff have on salinity distribution. How will the winter and summer seasons affect the temperature distribution in the water column?

2. How does coastal runoff of low-salinity water produce a longshore geostrophic current?

3. Describe the difference between vertically mixed and salt wedge estuaries in terms of salinity distribution, depth, and volume of river flow. Which displays the more classic estuarine circulation pattern?

4. Name the two types of wetland environments and the latitude ranges where each will likely develop. How do wetlands contribute to the biology of the oceans and the cleansing of polluted river water?

5. What factors lead to a wide seasonal range of salinity in Laguna Madre?

6. Describe the circulation between the Atlantic Ocean and the Mediterranean Sea, and explain how and why it differs from estuarine circulation.

7. How does the flow of the Loop Current relate to the slope of the 20°C (68°F) temperature surface?

8. On what basis can the Subtropical Underwater and the Subtropical Intermediate Water be identified in the "American Mediterranean"?

9. Describe how coastal upwelling in the Gulf of California is related to seasonal winds.

10. Discuss the unusual depth distribution of the Bering Sea and its relationship to biological productivity.

11. Compare the circulation between the Red Sea and the Indian Ocean to that between the Mediterranean Sea and the Atlantic Ocean. Discuss why they are or are not similar.

12. Explain why the Red Sea water that flows into the Indian Ocean at a depth of 125 m (410 ft) sinks to 1000 m (3280 ft) before spreading throughout the Arabian Sea.

13. Describe the relationship between the surface circulation and monsoon winds in the Bay of Bengal.

REFERENCES

Cherfas, C. 1990. The fringe of the ocean—under siege from land. *Science* 248:4952, 163–65.

Fairbridge, R. W., ed. 1966. *Encyclopedia of oceanography.* New York: Van Nostrand Reinhold.

Molinari, R. L.; Baig, S.; Behringer, D. W.; Maul, G. A.; and Legeckis, R. 1977. Winter intrusions of the Loop Current. *Science* 198: 505–6.

Officer, C. B. 1976. Physical oceanography of estuaries. *Oceanus* 19:5, 3–9.

Officer, C. B., Briggs, R. B., Taft, J. L., Tyler, M. A., and Boynton, W. R. 1984. Chesapeake Bay anoxia: Origin, development, and significance. *Science* 223:4631, 22–27.

Pickard, G. L. 1964. *Descriptive physical oceanography: An introduction.* New York: Macmillan.

Stommel, H. M., ed. 1950. Proceedings of the colloquium on "The flushing of estuaries." Woods Hole: Woods Hole Oceanographic Institution.

Valiela, I., and Vince, S. 1976. Green borders of the sea. *Oceanus* 19:5, 10–17.

SUGGESTED READING

Sea Frontiers

Baird, T. M. 1983. Life in the high marsh. 29:6, 335–41.
The ecology of the salt marsh is discussed.

Baker, R. D. 1972. Dangerous shore currents. 18:3, 138–43.
A discussion of the hazards to swimmers of various types of currents found near the shore.

de Castro, G. 1974. The Baltic—to be or not to be? 20:5, 269–73.
A review of the pollution threat to life in the Baltic Sea and what has been done to solve the problem.

Edwards, L. 1982. Oyster reefs: Valuable to more than oysters. 28:1, 23–25.
The value of oyster reefs to various estuarine life forms is detailed.

Heidorn, K. C. 1975. Land and sea breezes. 21:6, 340–43.
A discussion of the causes of land and sea breezes.

Iverson, E. S., and Jory, D. E. 1990. Arms race on the grass flats. 35:5, 304–11.

The arsenal of weapons used by crustaceans, turtles and octopuses to subdue their molluscan prey on the marsh grass flats is revealed.

Olson, D. B. 1990. Monsoons and the Arabian Sea. 36:1, 34–41.

The monsoons of the Arabian Sea are discussed in terms of their effect on coastal inhabitants and the physical conditions that cause them.

Osing, O. 1974. The meeting of the seas. 20:1, 21–24.

A description of the conditions at the northern tip of Denmark, where the North Sea and the Baltic Sea meet.

Sefton, N. 1981. Middle world of the mangrove. 27:5, 267–73.

The life forms associated with Cayman Island mangrove swamps are described.

Smith, F. 1982. When the Mediterranean went dry. 28:2, 66–73.

The role of global plate tectonics in causing the Mediterranean Sea to dry up some 5 million years ago.

Scientific American

Degens, E. T., and Ross, D. A. 1970. The Red Sea hot brines. 222:4, 32–53.

Describes the hot brine pools in the Red Sea. The source of brines and the economic potential resulting from metal enrichment associated with the brine deposits are considered.

Hsu, K. H. 1972. When the Mediterranean dried up. 227:6, 26–45.

Evidence is presented that shows the Mediterranean Sea was a dry basin 6 million years ago.

13
THE MARINE ENVIRONMENT

In chapter 2 we discussed some evidence for believing that life originated in the oceans. Primary among these considerations was the fact that the ocean constitutes the major part of the water environment. Considering the organisms' need for water, we can see that life would be totally impossible on the continents were it not for the hydrologic cycle that provides water that runs across and percolates through their surfaces, and is temporarily stored in lakes and rivers before it returns to the oceans. Although marine organisms rarely want for water, the success of individual organisms living in the oceans depends on their ability to adjust to many variables that we will consider in this chapter.

GENERAL CONDITIONS

The fact that representatives of all the animal *phyla* live in the marine environment attests to the ocean's being a far more fit biological realm than land. In fact, all phyla are thought to have originated in the ocean, and five are exclusively marine. Although it may seem inconsistent at this point, the fact that only simple one-celled plants inhabit the open oceans while the more complex and highly developed plants are restricted to its margins and the continents is further evidence pointing to the unique fitness of the oceans as a biological home.

The ocean environment is far more stable than the terrestrial environment; as a result, the organisms that live in the oceans have generally not developed highly specialized regulatory systems to combat sudden changes that might occur within their environment. They are, therefore, affected to various degrees by quite small changes in salinity, temperature, turbidity, and other environmental variables.

Water constitutes over 80% of the mass of **protoplasm**, the substance of living matter. Over 65% of the weight of a human being and 95% of that of the jellyfish is accounted for by the presence of water (table 13–1). Water carries dissolved within it the gases and minerals needed by organisms and is itself one of the raw materials required by plants in the production of food through photosynthesis. Land plants and animals have developed very complex systems and devices to distribute water throughout their bodies and prevent it from being lost. The threat of **desiccation** (drying out) through atmospheric exposure does not exist for the inhabitants of the open ocean, as they are abundantly supplied with water.

Preceding page: A sculpin, *Cottid orthonopias,* at San Clemente Island. Photo by James R. McCullagh.

TABLE 13–1 Water Content of Organisms.

Organism	Water Content (%)
Human	65
Herring	67
Lobster	79
Jellyfish	95

Support

Another necessity of plants and animals is support. Land plants have become complex multicellular organisms with vast root systems that fasten them securely to the earth. A number of support systems are used in the animal kingdom, but each requires some combination of appendages that must totally support the land animal's weight.

In the ocean this is not the case. The water of the oceans, as it so lavishly bathes organisms with the gases and nutrients they need, serves as their basic support as well. Organisms that live in the open ocean depend primarily upon buoyancy and frictional resistance to sinking to maintain them in their desired positions. This is not to say that they do not have problems maintaining their positions within the ocean; some of the special adaptations that are developed toward increasing their efficiency in this respect will be discussed in this and succeeding chapters.

Effects of Salinity

Marine animals are significantly affected by relatively small changes in their environmental conditions. One highly important condition is change of salinity. Some oysters that live in estuarine environments at the mouths of rivers are capable of withstanding a considerable range of salinity. During floods the salinity will be extremely low. On a daily basis, as the tides force ocean water into the river mouth and draw it out again, the salinity changes considerably. The oysters, and most other organisms that inhabit coastal regions, have a tolerance for a wide range of salinity conditions. They are **euryhaline**. By contrast, other marine organisms, particularly those that inhabit the open ocean, can withstand only very small changes in salinity; these organisms are **stenohaline**.

Certain plants and animals cause important changes in the amount of dissolved material in ocean water by extracting minerals to construct hard parts of their bodies that serve as protective covering. The compounds primarily used for this purpose are silica (SiO_2) and calcium carbonate ($CaCO_3$). The silica is used by the most important plant population in the ocean, the **diatoms**, and the microscopic animals called **radiolarians**. Calcium carbonate is used by the **foraminifers**, most members of the phylum Mollusca, corals, and some algae that secrete a calcium carbonate skeletal structure.

Cell membranes are semipermeable and will allow the passage of molecules of some substances while screening out others. By **diffusion**, cells may take in the nutrient substances they need from the surrounding fluid medium that contains them in high concentration. The nutrient compounds, to which the cell wall is permeable, pass from the surrounding medium into the cell, where they are in lesser concentrations (fig. 13–1).

The waste materials that a cell must dispose of after it has utilized the energy held in these nutrients are passed out of the cell by the same method. As the concentration of waste materials becomes greater within the cell than within the fluid medium surrounding the cell, these materials will pass out of this area of high concentration within the cell into the surrounding fluid. The waste products will then be carried away by the circulating fluid that services the cells in higher animals or by the surrounding water medium that bathes the simple one-celled organisms of the oceans.

Osmotic Pressure When aqueous solutions of unequal salinity are separated by a water-permeable membrane, there will be diffusion of water molecules through the membrane into the more concentrated solution. This process is called **osmosis** (fig. 13–2). Such solutions can be compared in terms of their osmotic pressures. The **osmotic pressure** of a solution is that which must be applied to keep pure water that is separated from it by a water-permeable membrane from passing through the membrane and diluting the solution. The higher the salinity of the solution, the higher its osmotic pressure. We can compare the osmotic pressure of solutions by comparing their concentrations of dissolved solids—their salinity. This comparison readily is done by determining their freezing points, since the addition of dissolved solids to an aqueous solution lowers its freezing point (fig. 13–3). This effect is due to the interference of the dissolved particles with the formation of ice crystals. The lower the freezing point, the higher the osmotic pressure of a solution.

If the salinity of body fluid and the external medium are equal, they are said to be **isotonic** and have equal osmotic pressure. No net transfer of water will occur through the membrane. If the external fluid

A.   DIFFUSION

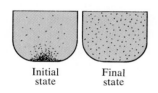

Initial   Final
state   state

If a substance soluble in water is placed in a pile on the bottom of a beaker of water, it will eventually become evenly distributed through random molecular motion. Diffusion means molecules of a substance move from areas of high concentration of the substance to areas of low concentration.

B.

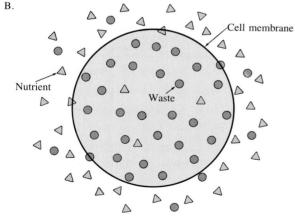

Nutrients are in high concentration outside the cell and diffuse into the cell through the cell membrane. Waste is in higher concentration inside the cell and diffuses out of the cell through the cell membrane.

FIGURE 13–1 Diffusion.

OSMOSIS

Hypotonic Hypertonic   Isotonic

Separating two water solutions of different salinities by a membrane that will allow the passage of water molecules but not the molecules of dissolved substances sets the stage for osmosis to occur. Water molecules will achieve a net motion by diffusion through the membrane into the more concentrated (hypertonic) solution from the less concentrated (hypotonic) solution. Osmosis will not produce a net movement of water through the membrane when the salinity of the two solutions is the same. They are then isotonic.

FIGURE 13–2 Osmosis—A Diffusion Process.

FIGURE 13–3 Osmotic Pressure and Freezing vs. Salinity.
Osmotic pressure increases and the freezing point decreases with increased salinity.

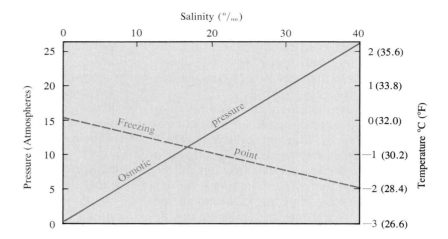

medium has a lower osmotic pressure and lower salinity than the body fluid within the cells of an organism, water will pass through the cell walls into the cells. This organism is **hypertonic** relative to the external medium. Should the salinity or osmotic pressure within the cells of an organism be less than that of the external medium, water from the cells will pass through the cell membranes into the external medium. This organism is described as **hypotonic** relative to the external medium.

In an oversimplification, we will consider osmosis as simply a diffusion process. If we consider the relative concentration of water molecules on either side of the semipermeable membrane, we will see that the net transfer of water molecules is from the side where the greatest concentration of water molecules exists through the membrane to the side that contains a lesser concentration of water molecules. The processes of water molecules moving through the semipermeable membrane (osmosis), nutrient molecules moving from the external medium where they

are more concentrated into the cell where they will be used to maintain the cell, and the movement of molecules of waste materials from the cell into the surrounding medium that carries them away are all going on at the same time.

We must keep in mind that during this process molecules of all of the substances in the system are passing through the membrane in both directions. Nevertheless, there will be a net transport of molecules of a given substance from the side on which they are most highly concentrated to the side where the concentration is less.

In the case of the marine invertebrates, the body fluids and the external medium in which they are immersed are in a nearly isotonic state. Since no significant difference exists, they do not have to develop special mechanisms to maintain their body fluids at a proper concentration. Thus they have an advantage over their freshwater relatives, whose body fluids are hypertonic in relation to the low-salinity external medium (table 13–2).

TABLE 13–2 Osmotic Pressure of Body Fluids in Various Organisms.
A comparison of the freezing points of the body fluids of some marine and freshwater organisms and the waters in which they live. The lower the freezing point of the fluid, the higher the salinity and osmotic pressure. Marine invertebrates and sharks are essentially isotonic, while most marine vertebrates are hypotonic. Essentially all freshwater forms are hypertonic. The freezing points of ocean water corresponding to invertebrate specimens vary because the specimens were taken from coastal waters, where water salinity varied.

	Marine Animals			Freshwater Animals	
	Freezing Points (°C)			Freezing Points (°C)	
Animals	Body Fluid	Water	Animals	Body Fluid	Water
		INVERTEBRATES			
Annelid worm	−1.70	−1.70	Mussel	−0.15	−0.02
Mussel	−2.75	−2.12	Water flea	−0.20	−0.02
Octopus	−2.15	−2.13	Crayfish	−0.80	−0.02
Lobster	−1.80	−1.80			
Crab	−1.87	−1.87			
		VERTEBRATES			
Cod	−0.74	−1.90	Eel	−0.62	−0.02
Shark	−1.89	−1.90	Carp	−0.50	−0.02
Turtle	−0.50	−1.90	Water snake	−0.50	−0.02
Seal	−0.50	−1.90	Manatee	−0.50	−0.02

Marine fish have body fluids slightly more than one-third as saline as ocean water. This level may be the result of their having evolved in low-salinity coastal waters. They are, therefore, hypotonic in relation to the surrounding medium. This difference produces a problem in that the saltwater fish, without some means of regulation, would be continually losing water from its body fluids to the surrounding ocean environment. This loss is counteracted by the fact that marine fish drink ocean water and secrete the salts through "chloride cells" located in the gills (fig. 13–4).

Freshwater fish are hypertonic in relation to the very dilute external medium in which they live. The osmotic pressure of the body fluids of such fish may be 20 to 30 times greater than that of the fresh water that surrounds them. Their problem is to avoid taking into their body cells large quantities of water that would eventually rupture the cell walls. In order to meet this problem, the freshwater fish does not drink water. Its cells have the capacity to absorb salt. The water that may be gained as a result of body fluids being greatly hypertonic is disposed of with large volumes of very dilute urine (fig. 13–4).

We should all be aware that we cannot drink salt water when at sea because the osmotic pressure of our body fluids is about one-fourth that of ocean water. To drink ocean water would cause a drastic loss

of water through the wall of the digestive tract by osmosis, causing eventual dehydration.

Availability of Nutrients

The distribution of life throughout the ocean's breadth and depth basically depends on the availability of plant nutrients. In areas where the conditions are right for the supply of large quantities of nutrient materials, marine populations will reach their greatest concentration. Where are these areas in which we find the greatest concentration of biomass? To answer this question, we must consider the sources of nutrients.

Previously, we discussed the role played by water in eroding the continents, carrying the eroded material to the oceans, and depositing it as sediment at the margins of the continents. During this process, the water will dissolve inorganic compounds in the form of **nitrates** and **phosphates**, that serve as the basic nutrient supply for plants. Because the sea bottom is relatively shallow near the continents, even nutrients that have settled out there can be returned to the surface waters by the actions of waves and tides. Through the process of photosynthesis, plants combine these with carbon dioxide and water to produce the carbohydrates that store the energy on which the rest of the biological community depends.

If the continents are the major sources of these inorganic nutrients, then we might expect that the greatest concentrations of marine life will be found at their margins. This is the case. As we go to the open sea from the margins of the continents, we will travel through waters containing progressively lower concentrations of marine life. This situation is basically due to the vastness of the world's oceans and the vertical descent of nutrients over the great distance between the open ocean and the coastal regions where the nutrients are supplied.

Availability of Solar Radiation

To make use of the available nutrients, plants must have another requirement met. Photosynthesis cannot proceed without the appropriate energy provided by solar radiation. The radiation from the sun penetrates the atmosphere quite readily so plants on the continents would seem to have an advantage over those in the ocean, as there is hardly ever any shortage of solar radiation to drive the photosynthetic reactions for these land-based plants.

By contrast, the ocean water does represent a significant barrier to the penetration of solar radiation, although in the clearest water solar energy may be

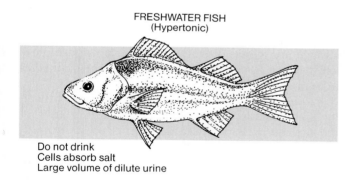

FRESHWATER FISH
(Hypertonic)

Do not drink
Cells absorb salt
Large volume of dilute urine

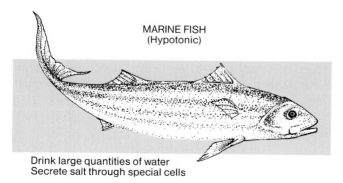

MARINE FISH
(Hypotonic)

Drink large quantities of water
Secrete salt through special cells

FIGURE 13–4   Freshwater and Marine Fish.

detected to depths in excess of about 1 km (0.6 mi). The amount of solar energy reaching these depths is extremely small and would not be sufficient to support the photosynthetic process. Photosynthesis in the ocean is commonly restricted to a very thin layer, approximately the uppermost 100 m (328 ft). Near the coast, this zone will be much thinner because the water contains more suspended material that will restrict the light penetration. In the clearest ocean water, the base of the photosynthetic zone may be at a depth as great as 200 m (656 ft).

Considering these two factors necessary for photosynthesis, the supply of nutrients and the presence of solar radiation, we find that in the open ocean away from the continental margins the column of water having solar energy available extends deeper, but that this column contains only a small concentration of nutrients. Conversely, in the coastal regions, where the water is more turbid, light penetrates to much shallower depths, but the nutrient supply is quite rich. Since the coastal zone is by far the most productive zone, we can see manifested in this comparison the overriding importance of nutrient availability to life in the oceans.

Margins of the Continents

It has been mentioned that the stable condition of the ocean environment made it ideal for the continuation of life processes, but actually the richest concentration of marine organisms is found in the very margins of the oceans, where the conditions are the least stable. When we consider the physical conditions that would appear at first to be deterrents to the establishment of life, we will see that many are present in the coastal regions.

The water depths are shallow, allowing seasonal variations in temperature and salinity that are much greater than would be found in the open ocean. A varying thickness of water column exists in the near-shore region as a result of the tidal movements that periodically cover and uncover a thin strip along the margins of the continents, where there is a constant beating of the surf. The surf condition represents a sudden release of energy that has been carried for great distances across the open ocean.

The development of such a great concentration of biomass in an environment containing so many factors that would seem to inhibit the development of life serves to highlight the importance of the evolutionary process in developing new species by natural selection. Over the eras of geologic time, new life forms have developed to fit into all biological niches.

Many of these forms have adapted to live under conditions that would seem deleterious to the conservation of life. This fact attests to the great range of physical conditions within which life can exist so long as those basic requirements for the production of the food supply are met.

We find that along continental margins there are areas where life is more abundant than others. What are the characteristics of these areas that result in such an uneven distribution of life along the continental margins? Again, we need consider only those basic requirements for the production of food for the answer. If we measure the various properties of the water along the coast of the continents and compare those measurements from place to place, we will find that the areas containing the greatest concentrations of biomass are those where the water temperatures are lower. Because of the low temperature, it is possible for the water to maintain greater amounts of the important gases, oxygen and carbon dioxide, in solution than would be possible in warmer waters. Of particular importance is the increased availability of carbon dioxide, and we see again an example of the greater availability of the basic requirements of plants affecting the distribution of life in the oceans.

Upwelling   In certain areas of the coastal margins we find an additional factor that enhances the conditions for life—upwelling. Upwelling occurs most commonly along the western margins of continents where surface currents are moving toward the equator. It consists of a flow of subsurface water to the surface and grows directly out of the relationship existing between the wind direction and the continental margin.

In the case of northward-flowing currents (driven by southeasterly winds) along the western margin of the Southern Hemisphere continents or southward-flowing currents (driven by northeasterly winds) along the western margin of Northern Hemisphere continents, the Ekman transport tends to move the surface water away from the coast. As the surface water moves away from the continents, water will surface from depths of 200 to 1000 m (656 to 3280 ft) to replace it. This water that surfaces from depths below those where photosynthesis occurs is rich in plant nutrients because there are no plants in these deeper waters to utilize them. This constant replenishing of nutrients at the surface enhances the conditions for life in these areas. Upwelling water is usually of low temperature, giving such water the additional benefit of having a high capacity for dissolved gases (fig. 13–5).

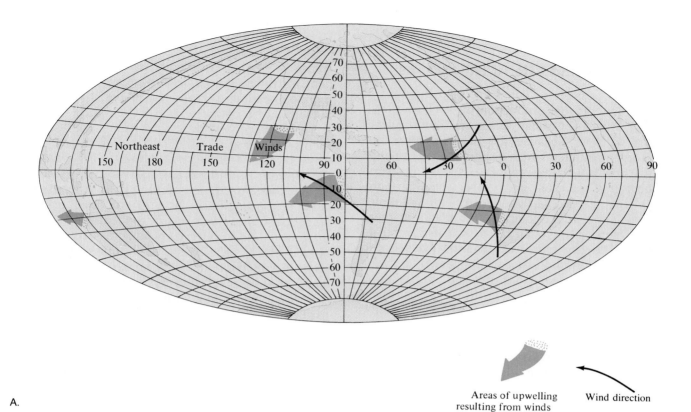

A.

Areas of upwelling
resulting from winds

Wind direction

B.

C.

FIGURE 13–5 Upwelling.

A, the coastal winds usually drive currrents south along the western margins of continents in the Northern Hemisphere, while they
are usually driven north in the Southern Hemisphere. The Ekman
transport moves water away from the western margins of the continents. Base map courtesy of National Ocean Survey. *B,* as the
surface water moves away from the continent, water from depths
of 200–1000 m (656–3280 ft) rises to replace it. Because this water comes from below the photosynthetic zone, it is cold and rich
in nutrients. *C,* in this false-color CZCS image from the *Nimbus-7*
satellite, wind-driven upwelling along the coast of northwest Africa results in high phytoplankton biomass and productivity. Highest productivity is indicated by colors of red and orange, decreasing through yellow, green, and blue. Imagine courtesy of NASA/
Goddard Space Flight Center.

Water Color and Life in the Oceans

It is usually possible to determine visually areas of high and low organic production in the ocean from the color of water. Coastal waters are almost always greenish in color because they contain more dissolved organic matter, which disperses solar radiation in such a way that the wavelengths most scattered are those that represent greenish or yellowish light. This condition is also partly the result of the presence of yellow-green microscopic marine plants in these coastal waters. In the open ocean, where particulate matter is relatively scarce and marine life exists in low concentration, the water will appear blue owing to the size of the water molecules and the scattering they effect on the solar radiation. This is similar to the process that produces the apparent blueness of the skies.

Green color in water usually indicates the presence of a lush biological population such as might correspond to the jungle environments on the continents. The deep indigo blue of the open oceans, particularly between the Tropics of Cancer and Capricorn, usually indicates an area that lacks abundant life and could be considered a biological desert.

Size

Earlier in this chapter we made reference to the fact that marine plants were quite simple in comparison to the specialized forms we find throughout the continents. How can we explain the fact that marine plants have not followed the path of aggregation (becoming multicellular)?

Assuming the availability of nutrients, the major requirements of marine plants are that they maintain themselves within the uppermost, sunlit, layers of ocean water to carry on photosynthesis. They must also take in nutrients from the surrounding waters and expel waste materials as efficiently as possible. These requirements lead us to a consideration of size and shape. The ease with which marine plants maintain their positions in the upper layers of the ocean is closely related to the ratio that exists between surface area and body mass. The greater the surface area per unit of body mass, the easier it will be for an organism to maintain its position because there is more surface per unit of body mass in frictional contact with the surrounding water. Greater surface area per unit of body mass amounts to a greater resistance to sinking per unit of body mass. An examination of figure 13–6 will help you understand that this ratio of surface area per unit of mass increases with a de-

crease in size. It can readily be seen that it would benefit the one-celled plants making up the bulk of marine plant life to be as small as possible. They are, in fact, microscopic.

The efficiency with which the plants take in the nutrients from the surrounding water and expel their waste through their cell membranes is also related to the amount of surface area per unit of mass. The greater the surface area per unit of mass, the easier it is for the organism to carry on these functions. Since both processes are dependent on diffusion, there would be a point at which increased size would reduce the surface area to body mass ratio to the point that the cell could not be serviced and would die. It is a direct result of this condition that we find cells in all plants and animals to be microscopic, regardless of the size of the organism. Each cell must take in nutrients and dispose of wastes by diffusion.

Examining diatoms carefully, we see that it is quite common for them, as well as other members of the microscopic marine community, to have long needle-like extensions from their siliceous **tests**, or protective coverings. These extensions serve to further increase the ratio of surface area to the mass of the organism. The increased ratio of surface area to body mass allows more frictional resistance against the tendency to sink.

Another contrivance used by the small plants and animals to increase their ability to stay in the upper layers of the ocean is to produce an internal drop of oil, which tends to decrease their overall density and increase their buoyancy.

In spite of these adaptations to increase the ability to float, the organisms still have a density in excess of the density of water and therefore tend to sink, if ever so slowly. This is not a serious handicap, since they are so small and are carried quite readily along with the water movements. In these surface layers, there is considerable turbulent action from the mixing that results from wind action on the surface. This turbulence tends to dominate the movement of these small organisms to the point that they are able to spend adequate time bathed in the solar radiation to carry on their role of producing the energy needed by the other members of the marine biological community.

Viscosity

We are familiar with the fact that liquids flow with varying degrees of ease, and we know that syrup flows less readily than water. This condition is indicative of the higher viscosity of syrup as compared to

FIGURE 13–6 The Importance of Size.

Cube A has ten times as much surface area per unit of volume as cube B. If both cubes are composed of the same substance, their masses are proportional to their volumes, since mass = volume × density. If cubes A and B were plankton, cube A would have ten times as much frictional resistance to sinking per unit of mass as cube B. It could stay afloat by exerting far less energy than cube B. Cube A, were it a planktonic alga, could also take in nutrients and dispose of waste through the cell wall ten times as efficiently as an alga with the dimensions of cube B.

CUBE A
Linear dimension—1 cm

→|1 cm|←
Area—6 cm²
Volume—1 cm³

Ratio of surface area to volume:

$$\frac{6 \text{ cm}^2}{1 \text{ cm}^3}$$

Cube A has 6 units of surface area per unit of volume.

CUBE B
Linear dimension—10 cm

Area—600 cm²
Volume—1000 cm³

Ratio of surface area to volume:

$$\frac{600 \text{ cm}^2}{1000 \text{ cm}^3} = \frac{0.6 \text{ cm}^2}{1 \text{ cm}^3}$$

Cube B has 0.6 units of surface area per unit of volume.

|———————10 cm———————|

water. **Viscosity,** defined as internal resistance to flow, is a characteristic of all fluids.

The viscosity of ocean water is affected by two variables—temperature and salinity. Viscosity of ocean water increases with an increase in salinity but, like density, is affected even more by temperature. A decrease in temperature increases viscosity. As a result of this temperature relationship, we find that floating plants and animals in colder waters have less need for extensions to aid them in floating than those that occupy warmer waters. It has been observed that members of the same species of floating crustaceans will be very ornate, with featherlike appendages, where they occupy warmer waters; these appendages are missing in the colder, more viscous environments. It appears then that high viscosity would benefit the floating members of the marine biocommunity because it would make it easier for them to maintain their positions in the surface layers.

With increasing size of organisms and a change in the mode of life, viscosity ceases to enhance the ability of an organism to pursue life and becomes an obstacle instead. This is particularly true of the large organisms that swim freely in the open ocean. They must pursue prey and displace water to move ahead. The more rapidly it swims, the greater the stress that is created on the organism. Not only must water be displaced ahead of the animal, but water must move in behind it to occupy the space that it has vacated. The latter is one of the more important considerations in streamlining.

The familiar fusiform shape of the free-swimming fish and the marine mammals manifests the type of

adaptations that must be achieved in streamlining an organism to move with a minimum of stress through the water. We see that a common shape achieved to meet this need is a laterally flattened body presenting a small cross section at the anterior end with a gradually tapering posterior. This form, which is characteristic of the bony fishes, reduces stress that results from movement through the water and makes it possible for this movement to be achieved with a minimum of effort (fig. 13–7).

A body with a low degree of streamlining. Its broad structure causes too much resistance area in front and too much wake region behind.

Wake

Teardrop shape (fusiform) front is well rounded and the body tapers gradually to the rear, producing a small wake region.

Wake

FIGURE 13–7 Streamlining.

Due to the viscosity of water, any body moving rapidly through it must be designed properly to produce as little stress as possible as it displaces the water through which it moves. After the water has moved past the body, it must fill in behind the body with as little eddy action as possible.

Temperature

Of all the conditions we can measure in the ocean, there are probably no physical characteristics we have discussed that are more important to life in the oceans than temperature. It has been stated that the marine environment is much more stable than that of the land. A comparison of temperature ranges that exist in the sea to those that are found on land will serve to exemplify this condition.

The minimum temperature observed in the sea is never much less than −2°C (28.4°F), and the maximum seldom exceeds 29°C (84.2°F). This value compares to the continental temperature range from a low of −88°C (−127°F) to a high of about 58°C (136°F). The temperature found on the continent varies considerably on a seasonal and daily basis. Daily variations at the ocean surface rarely exceed more than 0.2°C or 0.3°C (0.4° or 0.5°F), although they may be as great as 2°C or 3°C (4° or 5°F) in shallower coastal waters. Annual temperature variations are also small. They range from 2°C (4°F) at the equator to 8°C (14°F) at 35° to 45° latitude and decrease again in the higher latitudes. Annual variations of temperature in shallower coastal areas may be as high as 15°C (27°F).

The temperature variations we are discussing as characteristic of surface waters are reduced in magnitude with increasing depth of observation. In the deep oceans, daily or seasonal variation in temperature becomes a consideration of little or no importance. Throughout the deeper parts of the ocean the temperature remains uniformly low. Where the depth is in excess of 1.5 km (0.9 mi), temperatures will hover at or below 3°C (37°F), regardless of the latitude.

We previously discussed the fact that a decrease in temperature will increase density, viscosity, and the capacity of water to hold gases in solutions. All of these changes have significant effects on the organisms inhabiting the ocean. A direct consequence of the increased capacity of water to contain dissolved gases is seen in the vast plant communities that develop in the high latitudes during the summer seasons, when solar energy is available to carry on photosynthesis.

It has been observed that floating organisms are larger individually in colder waters than in warmer waters of the tropics, although the tropical populations appear to be characterized by a larger number of species. However, the total biomass of floating organisms in the colder high-latitude planktonic environments greatly exceeds that of the warmer tropics. This is possibly the result of greater nutrient availability in the cooler high-latitude waters. The fact that organisms in the tropics are smaller than those observed in the higher latitudes could quite possibly be related to the lower viscosity and density found in the lower-latitude waters. Because they are smaller, these tropical species can expose more surface area per unit of body mass, and they are also characterized by ornate plumage to further increase surface area per unit of body mass (fig. 13–8). These plumose adaptations are strikingly absent in the larger cold-water species.

Increases in temperature increase the rate of biological activity, which more than doubles with an increase in the temperature of 10°C (18°F). Tropical organisms apparently grow faster, have a shorter life expectancy, and reproduce earlier and more frequently than their counterparts in the colder waters.

There are some species of animals that can successfully live only in cooler waters and others that can do so only in warmer waters. Many of these can withstand only a very small change in temperature and are called **stenothermal**. Other varieties appar-

FIGURE 13–8 Water Temperature and Appendages.
A, copepod *(Oithona)* that displays the ornate plumage characteristic of warm-water varieties. *B,* copepod *(Calanus)* that displays the less ornate appendages found on temperate and cold-water forms.

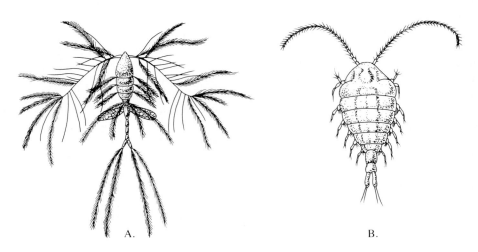

A.   B.

ently are little affected by temperature change and can withstand changes over a large range. These are classified as **eurythermal**. Stenothermal organisms are found predominantly in the open ocean and at the greater depths where it is very unlikely that large ranges of temperature will occur. The eurythermal organisms are more characteristic of the shallow coastal waters, where the largest ranges of temperature are found, and the surface waters of the open ocean. See "Gulf Stream Rings and Marine Life" on page 336 for an example of the effect of changing temperature on marine organisms.

DIVISIONS OF THE MARINE ENVIRONMENT

We can readily divide the marine environment into two basic units: the ocean water itself is the **pelagic environment**, and the ocean bottom constitutes the **benthic environment** (fig. 13–9).

Pelagic Environment

The pelagic environment is further divided into two provinces. The **neritic province**, which extends from the shore seaward, includes all water overlying an ocean bottom less than 200 m (656 ft) in depth. Beyond this 200 m depth the **oceanic province** is found.

The oceanic province, which includes water with a very great range in depth from the surface to the bottom of the deepest ocean trenches, is further subdivided on the basis of physical conditions. The oceanic province is subdivided into the **epipelagic zone** from the surface to a depth of 200 m, the **mesopelagic zone** from 200 to 1000 m (3280 ft), the **bathypelagic zone** from 1000 to 4000 m (13,120 ft), and the **abyssopelagic zone**, which includes all of the deepest parts of the ocean below 4000 m depth.

An important factor in determining the distribution of life in the oceanic province and the boundaries of biozones within it is the availability of light. The **euphotic zone** extends from the surface to a depth where there is enough light to support photosynthesis, rarely to more than a depth of 100 m (328 ft). Below this zone, there are small but measurable quantities of light within the **disphotic zone** to a depth of about 1000 m. Below this depth there is no light in the **aphotic zone** (fig. 13–9A).

The upper epipelagic zone is the only oceanic biozone in which there is sufficient light to support photosynthesis. The boundary between it and the meso-

pelagic zone (200 m) is also the approximate depth at which the level of dissolved oxygen begins to decrease significantly. This decrease in oxygen results because no plants are found beneath about 150 m (492 ft) and the dead organic tissue (detritus) descending from the biologically productive upper waters is undergoing decomposition by bacterial oxidation (fig. 13–9B). Nutrient content of the water also increases abruptly below 200 m (fig. 13–9B). This depth also serves as the approximate bottom of the mixed layer, seasonal thermocline, and surface-water mass.

Within the mesopelagic zone, a **dissolved oxygen minimum** occurs at a depth of about 700 to 1000 m (2296 to 3280 ft). The intermediate water masses that move horizontally in this depth range often possess the highest levels of plant nutrient content in the ocean. The base of the permanent thermocline and the boundary between the disphotic and aphotic zones characteristically occur at the 1000-m boundary between the mesopelagic and bathypelagic zones.

Within the mesopelagic zone, we see evidence of animals sensing the presence of light in the phenomenon known as the **deep scattering layer (DSL)**. During sonar equipment tests by the U.S. Navy early in World War II, a sound-reflecting surface that changed depth on a daily basis was observed. The variation indicated a reflecting mass at a depth of 100 to 200 m (328 to 656 ft) during the night that sank to depths as great as 900 m (2952 ft) during the day. After considerable investigation it was determined that this echo, the deep scattering layer, was produced by masses of migrating marine life that moved closer to the surface at night and then to a greater depth during the day. The DSL appeared to be a response to the changing intensity of solar radiation (fig. 13–10).

The use of plankton nets and submersibles has indicated that the DSL contains layers of siphonophores, copepods, euphausids, cephalopods, and small fish. Tests that have been done using fish with swim bladders suggest that they may be the primary cause of the reflection since a very small concentration of such fish is sufficient to produce such a phenomenon. Fish, however, are predators and if their presence is responsible for the echoes, the organisms on which they prey are probably responsible for the cyclic nature of their movement. The smaller organisms responding to a changing intensity of light could trigger the vertical movements.

The mesopelagic zone is also inhabited by species of fish that have unusually large and sensitive eyes, capable of detecting light levels 100 times less than those which the human eye could detect. Another important inhabitant of this zone is the bioluminescent group, especially shrimp, squid, and fish. Approxi-

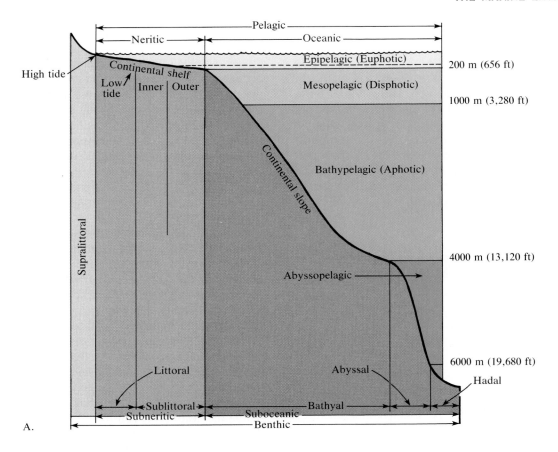

A.

FIGURE 13–9 Oceanic Biozones.

A, distribution of biozones of the pelagic and benthic environments. B, distribution of oxygen (O_2) and plant nutrient phosphate (PO_4) in the water column, mid- to low latitudes. Oxygen is abundant in surface water owing to mixing with the atmosphere and plant photosynthesis. Nutrient content is low in surface water owing to uptake by plants. Oxygen content decreases and nutrient content increases abruptly below the euphotic zone. An oxygen minimum and nutrient maximum are recorded at or near the base of the mesopelagic zone. Nutrient levels remain high to the bottom while oxygen content increases with depth as deep- and bottom-water masses carry oxygen into the deep ocean.

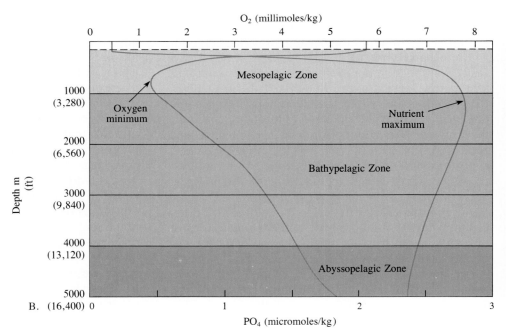

FIGURE 13–10 Deep Scattering Layer.

The deep scattering layer, which scatters and reflects sonar signals well above the bottom, may be caused by euphausids that grow to lengths greater than 2 cm (0.8 in.) and lantern fish (myctophid) that reach lengths of 7 cm (2.8 in.). They are predators that feed on smaller planktonic organisms migrating vertically in the water column on a daily cycle.

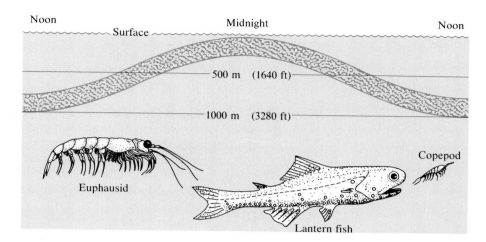

mately 80% of the inhabitants carry light-producing **photophores**. These are glandular cells containing luminous bacteria surrounded by dark pigments. Some contain lenses to amplify the radiation. This cold light is produced by a chemical process involving the compound luciferin. The molecules of luciferin are excited and emit photons of light in the presence of the enzyme luciferase and oxygen. Only a 1% loss of energy is required to produce the illumination. This system is similar to that of the firefly.

The aphotic bathypelagic zone and abyssopelagic zone represent more than 75% of the living space in the oceanic province. In this region of total darkness many totally blind fish exist. Very bizarre small predaceous species make up the total fish population. Many species of shrimp that normally feed on detritus become predators at these depths, where the food supply has been greatly reduced from what is available in the shallower waters. The animals that live in the aphotic bathypelagic and abyssopelagic zones feed mostly upon one another and have developed impressive warning devices and unusual apparatus to make them more efficient predators. They are characterized by small expandable bodies, extremely large mouths relative to body size, and very efficient sets of teeth (fig. 13–11).

Aside from the depth of light penetration, the primary physical criteria used in defining the boundaries of the bathypelagic zone are the tops and bottoms of deep and bottom-water masses. Oxygen content increases with depth, as these water masses carry oxygen from the cold surface waters where they formed to the deep ocean. The abyssopelagic zone is the realm of the bottom-water masses, which commonly are moving in the direction opposite to that of the overlying deep-water masses of the bathypelagic zone.

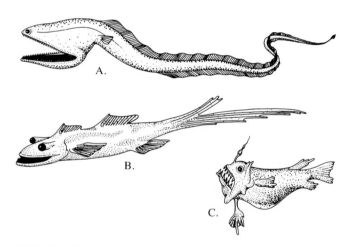

FIGURE 13–11   Some Examples of Deep-Sea Fish.

A, *Eurypharynx pelecanoides*. Length, 50–60 cm (20–24 in.). B, *Gigantura chuni*. Length, to 12 cm (5 in.). C, *Linophryne bicornis*. Length, 5–8 cm (2–3 in.).

Benthic Environment

The benthic environment can be subdivided into two larger units, which will then be subdivided on the basis of various criteria (fig. 13–9). These are the **subneritic province**, which extends from the spring high-tide shoreline to a depth of 200 m (656 ft) (approximately the continental shelf), and the **suboceanic province**, which includes all the benthic environment below the 200-m depth.

A transitional region above the high-tide line is the **supralittoral zone**. The supralittoral zone, commonly called the **spray zone**, is covered with water only during periods of extremely high tides and when tsunami or large storm waves break on the shore.

The intertidal zone is classified as the **littoral zone**, and from low tide to 200 m, the subneritic is called

GULF STREAM RINGS AND MARINE LIFE

The Gulf Stream separates two distinct water masses. To the north lies the Slope Water with an average temperature of less than 10°C (50°F), and to the south lies the Sargasso Sea water with temperatures ranging from 15°C (59°F) to as high as 25°C (77°F). The average width of the meandering stream is perhaps 100 km (62 mi), and periodically, the meanders cut themselves off and move as individual rings (eddies) on either side of the Gulf Stream (figs. 13A and 8–16).

When a meander breaks off and forms a ring (fig. 13B) on the north side of the Gulf Stream, it may be up to 200 km (124 mi) in diameter and have a lens of warm Sargasso Sea water trapped inside it. The warm water may go to a depth of 1500 m (4920 ft). Although of great variety, the concentration of life in this "warm ring" is meager and has not been studied in detail. Conversely, the "cold rings" that break off on the south side may be up to 300 km (186 mi)

FIGURE 13A Gulf Stream Rings.

Path of Gulf Stream separates cold, biologically productive, Slope Water on the north from the warm, less productive Sargasso Sea water to the south. As the Gulf Stream meanders, some meanders pinch off to form rings—warm rings of Sargasso Sea water enter the cold slope water to the north and cold rings of Slope Water enter the Sargasso Sea to the south. The arrows at the center of the eddies point in the direction of their movement. Base map courtesy of National Ocean Survey.

FIGURE 13B Temperature Structure of Warm Ring and Cold Ring.

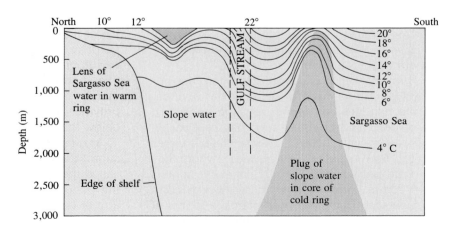

in diameter and contain within them a "plug" of cold Slope Water that extends to the ocean floor. The depth of the ocean in this region may be in excess of 4000 m (13,120 ft). This cold water contains a much smaller number of species but a much greater biomass than the Sargasso Sea water trapped in the warm rings.

Because of their greater concentration of marine life, the cold rings have been studied extensively by marine biologists. In the following paragraphs, we will describe the life history of the cold rings and the responses of the Slope Water biotic community to physical changes within the ring structure.

Although most cold rings have the energy to remain intact for from 2 to 4 years, the average life of a ring is thought to be about 1½ years. Most cold rings break away from the Gulf Stream between 60° and 75°W longitude and move southwest at an average speed of 4 km/day (2.5 mi/day) through the Sargasso Sea. They rejoin the Gulf Stream off the Florida coast north of the Bahamas.

While the rings are detached, the cold core water is modified. It warms up and becomes more saline, and the depth of the oxygen minimum drops from the 150 m (492 ft) typical of Slope Water to around 800 m (2624 ft), which is characteristic of the Sargasso Sea. A striking feature of this change is the collapse of the dome of cold water extending up into the ring. Figure 13C shows the difference between the temperature structure in a newly formed cold ring and one that has completed its trip across the Sargasso Sea.

As the cold-water environment in the ring decays, the Slope Water life forms in the core become stressed. Most severely affected are the planktonic forms that do not have the ability to swim away to a more favorable environment. The phytoplankton population in the core of newly formed rings is most abundant at a depth of about 30 m (98 ft), as is indicated by maximum chlorophyll concentrations of up to 3.0 micrograms/liter of water. Toward the flanks of the rings, the phytoplankton concentration decreases and is at a maximum at a depth of about 60 m (197 ft). This condition is typical of the Sargasso Sea. Observations in one ring showed a rapid decrease in chlorophyll content. At an age of 7 months, the ring contained only an eighth of the chlorophyll it had contained four months earlier. This rapid reduction in phytoplankton activity results because nutrient supply cannot be replenished in the trapped Slope Water.

The zooplankton population of a young ring core is quite similar in species abundance and depth range to that of Slope Water. It includes some eu-

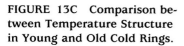

FIGURE 13C Comparison between Temperature Structure in Young and Old Cold Rings.

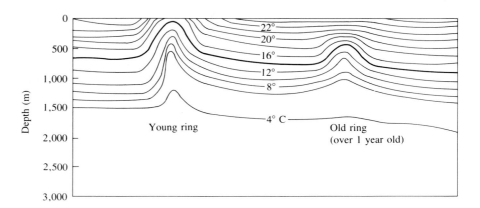

FIGURE 13D   Slope-Water Euphausid *Euphausia krohnii.*

Courtesy of Scripps Institution of Oceanography, University of California, San Diego.

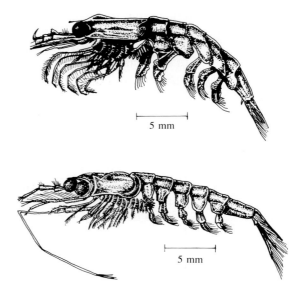

5 mm

FIGURE 13E   Slope-Water Shrimplike Crustacean *Nematoscelis megalops.*

Courtesy of Scripps Institution of Oceanography, Univerity of California, San Diego.

5 mm

phausids (small omniverous shrimplike crustaceans) and a herbivorous snail that is found only in the Slope Water. The euphausids (fig. 13D) and other herbivorous zooplankton populations are greatly reduced by the time the rings are six months old because of the rapid decline in phytoplankton.

One of the shrimplike crustaceans, *Nematoscelis megalops,* prefers temperatures ranging between 8° and 12°C (46° and 54°F) (fig. 13E). This population lives almost entirely above a depth of 300 m (984 ft) in the Slope Water when it is initially trapped in the ring. As the core water warms from the top and the isotherms (lines of equal temperature) sink, the depth range of *N. megalops* also descends. As the population sinks deeper into the ring, it becomes separated from the higher concentration of food near the surface. By the time the ring is 6 months old, most of the *N. megalops* population will be found between 300 and 800 m (2624 ft) depth. They begin to starve. Resulting physiological and biochemical changes include a declining rate of respiration, increased water content, and reduced lipid, carbon, and nitrogen content. In a 17-month-old-ring, no *N. megalops* remained. The same fate may await many of the cold-water species trapped in the ring.

Some of the Slope Water life forms may be left behind as the ring moves across the Sargasso Sea. Since the annulus of Gulf Stream water around the cold core extends only to a depth of 1000 to 1500 m (3280 to 4920 ft), deep-living forms may escape beneath this barrier. It is possible that some *N. megalops,* as well as a number of carnivorous species, may be removed from the core water in this way.

As the cold core decays, Sargasso Sea populations move into the ring structure. They gradually replace the Slope Water populations as conditions become favorable to them.

Rings similar to those formed along the Gulf Stream are formed in association with other major ocean current systems. Continued study of these systems will help increase our understanding of the physical processes of the oceans and the relationships of marine organisms to the processes.

References: Weibe, P. 1976. The biology of cold-core rings. *Oceanus* 19:3, 69–76. Wiebe, P. 1982. Rings of the Gulf Stream. *Scientific American* 246:3, 60–70.

the **sublittoral zone** or shallow subtidal zone. The **inner sublittoral** includes that part of the sublittoral to a depth of approximately 50 m (164 ft). This seaward limit will vary considerably because it is determined by that depth at which we find no plants growing attached to the ocean bottom. Its extent is determined primarily by the intersection of the euphotic zone and the ocean floor. In areas of high turbidity the inner sublittoral will have its seaward limit at shallower depths owing to the decreased penetration of solar radiation, while in areas where the water is unusually clear it may stand well below the 50-m mark. The **outer sublittoral** includes that portion of the sublittoral zone from the seaward limit of the inner sublittoral to a depth of 200 m or the shelf break, the seaward edge of the continental shelf.

With increased depth in the suboceanic system, we find the **bathyal zone** extending from the depth of 200 to 4000 m (656 to 13,120 ft) and corresponding generally to that geomorphic province found beyond the continental shelf—the continental slope. From a depth of 4000 to 6000 m (13,120 to 19,680 ft) stretches the **abyssal zone**, which includes in excess of 80% of the benthic environment. A restricted environment, including all depths below 6000 m, found only in the trenches along the margins of continents, is the **hadal** zone.

Once we have reached the outer sublittoral benthic region, we will see no more attached plants as we go from there into the deep-ocean basin. All of the photosynthesis seaward of the inner sublittoral zone is carried on by the microscopic algae floating in the neritic province above the outer sublittoral zone and in the epipelagic zone of the oceanic province. The *base,* or seaward limit, of the inner sublittoral can be said to coincide with the base of the **euphotic zone,** which is that zone of water near the ocean surface into which enough light penetrates to support photosynthesis. Currents will generally be stronger across the outer sublittoral bottom than along the upper continental slope of the bathyal zone, and sediments will be coarser on the continental shelf than on the continental slope.

Change in the sediment type of the continental slope to a depth of about 4000 m are subtle. Throughout much of the ocean, the calcium carbonate compensation depth occurs at about this depth, so sediments in the bathyal zone may contain considerable calcium carbonate, while those below will contain little or none. In general, the deposits of neritic sediment found around the continents begin to be replaced by oceanic sediment below 4000 m.

The ocean floor representing the abyssal zone is covered by soft oceanic sediment, primarily abyssal clay. The tracks and burrows of animals that live in this sediment are frequently recorded in bottom photographs. Figure 13–12 shows some typical examples of relatively large deep-sea fauna.

The hadal zone below 6000 m is primarily ocean trenches. Isolation in these deep, linear depressions allows the development of faunal assemblages found only in these trenches.

FIGURE 13–12 Benthic Organisms.
Deep-sea cucumber, brittle stars, and sea urchin. Photograph by Fred Grassle, Woods Hole Oceanographic Institution.

DISTRIBUTION OF LIFE IN THE OCEANS

It is difficult to describe the degree to which the sea is inhabited because of its immense space and the paucity of our knowledge of it. We do know, however, that some populations fluctuate greatly each season, and this fact increases the difficulty of describing with numbers the extent to which the marine environment is populated. Although it is perhaps meaningless to estimate the number of individual organisms in the ocean, we may derive some means of comparing the marine and terrestrial environments by comparing the number of marine species to the number of land species.

Well over 1.2 million species of animals are known, and only about 17% of these live in the ocean. Many biologists believe there may be from 3 to 10 million unnamed and undescribed animals living on earth. A large number of these may well inhabit the oceans. Yet, the following theory may explain why we can expect fewer species of marine animals to exist than terrestrial animals.

If the ocean is such a prime habitat for life, and if life originated in this environment, why do we now see such a small percentage of the world's animal species living there? This lesser number may well result because the marine environment is more stable than the terrestrial environment. The relatively uniform conditions of the open ocean do not produce pressures for adaptation. Also, once we get below the surface layers of the ocean, the temperatures are not only stable but also relatively low. Chemical reactions are retarded by this lower temperature; this, in turn, reduces the tendency for variation to occur.

If we are to consider briefly the great variety of species of organisms found on the continents, we can assume that this development was the product of an environment more diverse than is found in the ocean, one possessing many opportunities for natural selection to produce new species to inhabit new niches within this terrestrial environment. At least 75% of all land animals are insects that have evolved species capable of inhabiting very restricted environmental niches.

In describing the distribution of life within the ocean, we can get a general idea of the relative concentrations in the various marine environments if we consider that of the approximately 200,000 known animal species inhabiting the marine environment, only 4000 (about 2%) live in the pelagic environment. The remainder inhabit the ocean floor (fig. 13–13).

Before proceeding further in our discussion of the organisms of the oceans, we need to define some terms to describe these organisms on the basis of the portion of the ocean they inhabit and the means by which they move. We will examine these groups more closely in the following three chapters.

Plankton

The **plankton** include all those organisms that drift with ocean currents. This does not mean that all are without the ability to move. Many plankters have this capacity but are either able to move only weakly or are so restricted to vertical movement that they cannot determine their horizontal positions within the ocean. The plants that follow this life style in the upper layers of the ocean are called **phytoplankton**, and the animals are **zooplankton** (fig. 13–14). It has recently been discovered that free-living bacteria are much more abundant in the plankton community

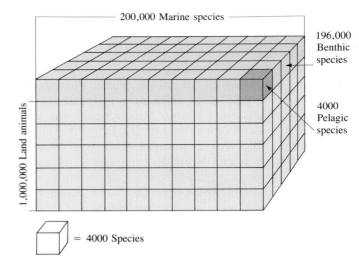

FIGURE 13–13   Distribution of Animal Species on Earth.
The cube represents the 1,200,000 species of animals living on the earth. The blue-shaded portion indicates the proportion (17%) that live in the oceans. Of the marine species, 98% live on or in the ocean floor.

than they were previously thought to be. Having an average dimension of only 0.5 μm (0.0002 in.), the **bacterioplankton** were missed in earlier studies because of their small size.

In considering the plankton in more detail, we will try to develop some basic understanding of this portion of marine life that many of us have never seen or imagined. The fact is, most of the biomass of the earth is found adrift in the oceans as plankton. The volume of the earth's space inhabited by animals that either drift or swim greatly exceeds that occupied by all animals that live on land or the ocean's bottom.

The plankton range in size from larger animals and plants such as jellyfish, salps, and *Sargassum* (**macroplankton**, 2 to 20 cm, or 0.8 to 8 in.) to bacteria that are too small to be filtered from the water by a silk net and must be removed by other types of microfilters (**picoplankton**, 0.2 to 2 μm, or 0.00008 to 0.0008 in.).

An additional scheme for classifying plankton is based on the portion of their life cycle spent within the plankton community. Those organisms such as planktonic diatoms and copepods, which spend their entire life as plankton, are **holoplankton**. Many of the organisms we normally consider as *nekton* or *benthos* because they spend their adult life in one or the other of these living modes actually spend a portion of their life cycle as plankton. Many of the nekton and very nearly all of the benthos during their larval stages make the plankton community their home. These organisms, which as adults sink to the bottom to be-

FIGURE 13–14 Phytoplankton and Zooplankton.

Maximum dimensions in parentheses. **Phytoplankton** *A* and *B*, Coccolithophoridae (15 μm, or 0.006 in.). *C–L*, diatoms (80 μm, or 0.032 in.): *C, Corethron; D, Asteromphalus; E, Rhizosolenia; F, Coscinodiscus; G, Biddulphia favus; H, Chaetoceras; I, Licmophora; J, Thalassiorsira; K, Biddulphia mobiliensis; L, Eucampia.* *M–S*, dinoflagellates (100 μm, or 0.04 in.): *M, Ceratium recticulatum; N, Gonyaulax scrippsae; O, Gymnodinium; P, Gonyaulax triacantha; Q, Dynophysis; R, Ceratium bucephalum; S, Peridinium.* **Zooplankton** *A,* fish egg (1 mm, or 0.04 in.); *B,* fish larva (5 cm, or 2 in.); *C,* radiolarian (.5 mm, 0.2 in.); *D,* foraminifer (1 mm, or 0.04 in.); *E,* jellyfish (30 m, or 98 ft); *F,* arrowworms (3 cm, or 1.2 in.); *G* and *H,* copepods (5 mm or 0.2 in.); *I,* salp (10 cm, or 4 in.); *J, Doliolum; K,* jellyfish (30 m, or 98 ft); *L,* worm larva (1 mm, or 0.04 in.); *M,* fish larva (5 cm, or 2 in.); *N,* tintinnid (50 μm, or 0.02 in.); *O,* foraminifer (1 mm, or 0.04 in.); *P,* dinoflagellate *(Noctiluca)* (100 μm, or 0.04 in.).

PHYTOPLANKTON

ZOOPLANKTON

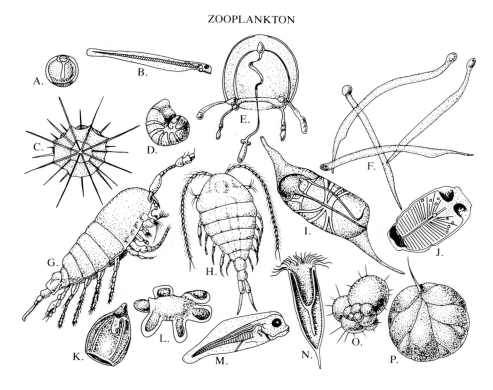

come members of the benthos or begin to swim freely as members of the nekton, are **meroplankton**.

Nekton

The **nekton** include all those animals capable of moving independently of the ocean currents. Such animals are not only capable of determining their own positions within relatively small areas of the ocean but are also capable, in many cases, of long migrations. Included in the nekton are most adult fish and squid, marine mammals, and marine reptiles (figs. 13–15 and 13–16).

The freely moving nekton, with some exceptions, are not able to move at will throughout the breadth of the ocean. By the presence of invisible barriers they are effectively limited in their lateral range. Gradual changes in temperature, salinity, viscosity, and availability of nutrients create impenetrable barriers. The death of large numbers of fish has frequently been caused by temporary lateral shifts of masses of ocean water. Vertical range is normally determined by pressure for all fish that contain air bladders and mammals that belong to the nekton.

Fish appear to be everywhere but are normally considered to be more abundant near continents and islands and in colder waters. Some fish, such as the salmon, ascend rivers to spawn. Many eels do just the reverse, growing to maturity in fresh water, then descending the streams to breed in the great depths of the ocean.

Benthos

The **benthos** live on or in the ocean bottom. There are those called **infauna** that live buried in the sand, shells, or mud. Those that live attached to rocks or move over the surface of the ocean bottom are the **epifauna**. Benthos such as some shrimp and demersal flounder that live in the bottom but move with relative ease through the water above the ocean floor are the **nektobenthos** (fig. 13–17 and 13–18).

FIGURE 13–15 Nekton.

Not drawn to scale; typical maximum dimension is in parentheses. *A*, bluefin tuna (2 m, or 6.6 ft); *B*, bottlenosed dolphin (4 m, or 13 ft); *C*, nurse shark (3 m, or 10 ft); *D*, barracuda (1 m, or 3.3 ft); *E*, striped bass (0.5 m, or 1.6 ft); *F*, sardine (15 cm, or 6 in.); *G*, deep-ocean fish (8 cm, or 3 in.); *H*, squid (1 m, or 3.3 ft); *I*, angler fish (5 cm, or 2 in.); *J*, lantern fish (8 cm, or 3 in.); *K*, gulper (15 cm, or 6 in.).

A.

FIGURE 13–16 Nekton.

A, this pelagic octopus is a strange and fearsome-looking creature as it jets through the open ocean. Thank goodness it is small enough to fit into the palm of your hand. This specimen was observed off the Kona coast of Hawaii. *B,* the identity of this fish, photographed at a depth of 130 ft, remains one of the sea's secrets. Photos © Christopher Newbert.

B.

FIGURE 13–17   Benthos—Some Intertidal and Shallow Subtidal Forms.
Not drawn to scale; typical maximum dimension is in parentheses. *A,* sand dollar (8 cm, or 3 in.);
B, clam (30 cm, or 12 in.); *C,* crab (30 cm, or 12 in.); *D,* abalone (30 cm, or 12 in.); *E,* sea urchins
(15 cm, or 6 in.); *F,* sea anemones (30 cm, or 12 in.); *G,* brittle star (20 cm, or 8 in.); *H,* sponge (30
cm, or 12 in.); *I,* acorn barnacles (2.5 cm, or 1 in.); *J,* snail (2 cm, or 0.8 in.); *K,* mussels (25 cm, or
10 in.); *L,* gooseneck barnacles (8 cm, or 3 in.); *M,* sea star (30 cm, or 12 in.); *N,* brain coral (50
cm, or 20 in.); *O,* sea cucumber (30 cm, or 12 in.); *P,* lamp shell (10 cm, or 4 in.); *Q,* sea lily (10
cm, or 4 in.); *R,* sea squirt (10 cm, or 4 in.).

FIGURE 13–18 Benthos.
Tiny and brilliantly colored feather-duster worms use their tentacled spirals to pluck microscopic plankton and detritus from the gently flowing currents. Like this specimen from the Coral Sea, they live in calcareous tubes tucked safely away in crevices of coral reefs or rocky shores. Photo © Christopher Newbert.

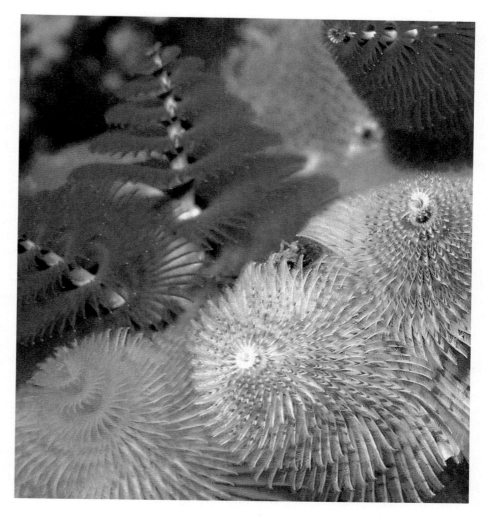

The littoral and inner sublittoral are the only zones in which we find the macroscopic algae attached to the bottom because they are the only benthic zones to which sufficient light can penetrate. There is a great diversity of physical and nutritive conditions in these zones. Animal species have developed in great numbers within this nearshore benthic community as a result of the variations existing within the habitat. As one moves across the bottom from the littoral into the deeper benthic environments, one commonly observes that an inverse relationship exists between the distance from shore and the number of benthic species that can be found. Some studies, however, indicate a local increase in species diversity at the shelf break.

Throughout most of the benthic environment, animals live in a region of perpetual darkness, where photosynthetic production cannot occur. They must feed on each other or on whatever outside nutrients fall from the productive zone near the surface. The deep-sea bottom is an environment characterized by coldness, stillness, and darkness. Under these conditions, it would be expected that life would move at a relatively slow pace. For those animals that move around on the bottom, streamlining, which is very important to the nekton, is of little importance.

The organisms that live in the deep sea normally have quite a wide range of distribution because the physical conditions for life do not vary greatly, even over great distances, on the deep-ocean floor. A few species appear to be extremely tolerant of pressure changes in that members of the same species may be found in the littoral province and at depths of several kilometers.

The 1977 discovery of the first hydrothermal vent biocommunity in the Galápagos Rift has shown us that high concentrations of deep-ocean benthos are possible. See "Life on the Deep-Ocean Floor," on page 430, for an interview with Dr. Robert Hessler, one of the foremost authorities on deep-ocean biology. He presents his views on the nature, distribution, and origin of life on the deep-ocean floor.

SUMMARY

The relatively stable marine environment is thought to have given rise to all phyla of organisms. Those organisms that have established themselves in the terrestrial realm have had to develop complex systems for support and for acquiring and retaining water.

Although the relative proportions of the constituents of salinity in the ocean and in the body fluids of the organisms are often very nearly the same, the salinity of one may differ greatly from that of the other. If the body fluids of an organism and ocean water are separated by a membrane that allows water molecules to pass, problems requiring adaptation may develop. Marine invertebrates and sharks are essentially **isotonic**, having body fluids with a salinity similar to that of ocean water. They rarely face adaptation problems. Most marine vertebrates are **hypotonic**, having body fluids with a salinity lower than that of ocean water, and tend to lose water through **osmosis**, the passing of water molecules by **diffusion** from a region in which they are in higher concentration through a semipermeable membrane into a region where they are in lower concentration. Freshwater organisms are essentially all **hypertonic**, having body fluids much higher in salinity than the water in which they live, so they must compensate for a tendency to take water into their cells through osmosis.

For life to flourish in any environment, there must be a sufficient food supply. The basic producers of food are plants, so the requirements of plants must be met if food is to be plentiful. The **availability of nutrients** and **solar radiation** make plant life possible. Since solar radiation is available only in surface water of the ocean, plant life is restricted to a thin layer of surface water usually no more than 100 m (328 ft) deep. Nutrients derived ultimately from the continents are much more abundant near continental features. Although much is yet to be learned about the distribution of life in the oceans, it appears that the **biomass concentration** of the oceans decreases away from the continents and with increased depth. The color of the oceans ranges from green in highly productive regions to blue in areas of low productivity.

The plants that must stay in surface water to receive sunlight and the small animals that feed on plants do not have effective means of locomotion. They depend, therefore, on their **small size** and other adaptations to give them a **high ratio of surface area per unit of body mass**, which results in a greater frictional resistance to sinking. Large animals that swim freely face an altogether different problem and generally have **streamlined bodies** to reduce frictional resistance to motion.

Compared to life in colder regions, organisms living in warm water tend to be individually smaller, comprise a greater number of species, and constitute a much smaller total biomass. Warm-water organisms also tend to live shorter lives and reproduce earlier and more frequently than their cold-water counterparts.

The marine environment is divided into two basic units—the **pelagic** (water) and the **benthic** (bottom) environments. These regions, which are further divided primarily on the basis of depth, are inhabited by organisms we can classify into three categories on the basis of life style. These categories are the **plankton**, or free-floating forms with little power of locomotion; the **nekton**, or free swimmers; and the **benthos**, or bottom dwellers.

QUESTIONS AND EXERCISES

1. Discuss the major differences between marine plants and land plants, and explain land plants' need for greater complexity.

2. Define the terms euryhaline, stenohaline, eurythermal, and stenothermal. Where in the marine environment will organisms displaying a well-developed degree of each characteristic be found?

3. Describe the relationships among osmotic pressure, salinity, and freezing point of a solution.

4. What is the problem requiring osmotic regulation that is faced by a hypotonic fish in the ocean? How have these animals adapted to meet this problem?

5. An important variable in determining the distribution of life in the oceans is the availability of nutrients. What are the relationships among the continents, nutrients, and the concentration of life in the oceans?

6. Another important determinant of plant productivity is the availability of solar radiation. Why is biological productivity relatively low in the tropical open ocean where the penetration of sunlight is greatest?

7. Discuss the characteristics of the coastal waters where unusually high concentrations of marine life are found.

8. What factors create the color difference between coastal waters and the less productive open-ocean water?

9. Compare the ability to resist sinking of an organism with an average linear dimension of 1 cm (0.4 in.) to that of an organism with an average linear dimension of 5 cm (2 in.). Discuss some adaptations other than size used by organisms to increase their resistance to sinking.

10. Changes in water temperature significantly affect the density, viscosity of water, and ability of water to hold gases in solution. Discuss how decreased water temperature changes these variables and may affect marine life.

11. Describe how higher water temperatures in the tropics may account for the greater number of species in these regions compared to low-temperature, high-latitude areas.

12. Construct a table listing the subdivisions of the benthic and pelagic environments and the physical factors used in assigning their boundaries.

13. Describe the vertical distribution of oxygen and nutrients in the oceanic province, and discuss the factors that are responsible for this distribution.

14. Discuss the probable cause and composition of the deep scattering layer.

15. List the relative number of species of animals found in the terrestrial, pelagic, and benthic environments, and discuss the factors that may account for this distribution.

16. Describe the lifestyles of plankton, nekton, and benthos. Why do plankton account for a relatively larger percentage of the biomass of the oceans than the benthos and nekton?

17. List the subdivisions of plankton and benthos and the criteria used for assigning individual species to each.

REFERENCES

Borgese, E. M., and Ginsburg, N. 1988. *Ocean Yearbook 7* Chicago: The University of Chicago Press.

Coker, R. E. 1962. *This great and wide sea: An introduction to oceanography and marine biology.* New York: Harper and Row.

Darwin, C. 1859. *On the origin of species by means of natural selection, or the preservation of favoured races in the struggle for life.* London: John Murray.

Hedpeth, J., and Hinton, S. 1961. *Common seashore life of southern California.* Healdsburg, Calif.: Naturegraph.

Isaacs, J. D. 1969. The nature of oceanic life. *Scientific American* 221:65–79.

May, R. M. 1988. How many species are there on Earth? *Science* 241:4872, 1441–48.

Sieburth, J. M. N. 1979. *Sea microbes.* New York: Oxford University Press.

Sumich, J. L. 1976. *An introduction to the biology of marine life.* Dubuque, Iowa: Wm. C. Brown.

Sverdrup, H.; Johnson, M.; and Fleming, R. 1942. Renewal 1970. *The oceans.* Englewood Cliffs, N.J.: Prentice-Hall.

Thorson, G. 1971. *Life in the sea.* New York: McGraw-Hill.

SUGGESTED READING

Sea Frontiers

Burton, R. 1977. Antarctica: Rich around the edges. 23:5, 287–95.
   The high level of biological productivity around the continent of Antarctica is the topic.

Gruber, M. 1970. Patterns of marine life. 16:4 194–205.
   Many varieties of life in the ocean are discussed in terms of how their form and size fit them for life in a particular environmental niche.

Hammer, R. M. 1974. Pelagic adaptations. 16:1, 2–12.
   A comprehensive discussion of the adaptations of pelagic organisms to reduce the energy required to maintain their positions in the open ocean.

Patterson, S. 1975. To be seen or not to be seen. 21:1, 14–20.
   A discussion of the possible role of color in the protection and behavior of tropical fishes.

Perrine D. 1987. The strange case of the freshwater marine fishes. 33:2, 114–19.
Explains how marine crevalle jacks are able to inhabit the fresh waters of Crystal River, Florida.

Schellenger, K. 1974. Marine life of the Galápagos. 20:6, 322–32.
A discussion of the unique life forms of the Galápagos Islands, 950 km (589 mi) from South America.

Thresher, R. 1975. A place to live. 21:5, 258–67.
An interesting discussion of how bottom-dwelling animals compete for space on the ocean floor.

Williams, L. B., and Williams, E. H., Jr. 1988. Coral reef bleaching: Current crisis, future warning. 34:2, 80–87.
Corals and related reef animals underwent "bleaching" along the Pacific Central American coast in 1983 and in the Caribbean Sea in 1987. Possible causes are discussed.

Scientific American

Denton, E. 1960. The buoyancy of marine animals. 203:1, 118–29.
The means by which some marine animals reduce the energy expenditure required to live in the ocean water far above the ocean floor are discussed.

Eastman, J. T., and DeVries, A. L. 1986. Antarctic fishes. 255:5, 106–14
Explains how one group of fish survived when the Antarctic turned cold.

Horn, M. H., and Gibson, R. N. 1988. Intertidal fishes. 258:1, 64–71.
Intertidal fishes have undergone remarkable adaptation to survive this physically harsh environment.

Isaacs, J. D., and Schwartzlose, R. A. 1975. Active animals of the deep-sea floor. 233:4, 84–91.
A surprisingly large population of large fishes on the deep-sea floor is suggested by automatic cameras dropped to the ocean bottom.

Isaacs, J. D. 1969. The nature of oceanic life. 221:3, 146–65.
A well-developed survey of the conditions for life in the ocean as they relate to the variety and distribution of marine life forms.

Palmer, J. D. 1975. Biological clock and the tidal zone. 232:2, 70–79.
This article investigates the mechanism of biological clocks set to the rhythm of the tides which are found in organisms from diatoms to crabs.

Partridge, B. L. 1982. The structure and function of fish schools. 246:6, 114–23.
Schooling benefits and the means by which fish maintain contact with the school are considered.

Vogel, S. 1978. Organisms that capture currents. 239:2, 128–39.
The manner in which sponges use ocean currents is an important part of this discussion.

14

BIOLOGICAL PRODUCTIVITY— ENERGY TRANSFER

The major primary producers of the oceans, marine algae, capture most of the energy used to support the marine biological community. When we think of marine plants, most of us undoubtedly first consider the large macroscopic plants we see growing near shore in many coastal areas. These large plants, however, play only a minor part in the production of energy for the ocean population as a whole. Instead, marine organisms depend primarily on the small planktonic varieties of marine plant life that inhabit the uppermost sunlit water of the world's oceans. They are not so obvious. All are microscopic, but they are scattered throughout the breadth of the ocean surface layer and represent the largest biomass community in the marine environment—phytoplankton.

In this chapter we will first consider briefly the classification of organisms. We will then discuss the composition of the marine algae, the ecological factors related to plant and bacterial productivity, and the passing of stored chemical energy on to other marine organisms within food webs.

TAXONOMIC CLASSIFICATION

All living things belong to one of the five **kingdoms** shown in figure 14–1 (*see* appendix VI). The simplest of all organisms are classified as **Monera**. These organisms are single-celled and have their nuclear material spread throughout the cell. Included in this kingdom are the cyanobacteria (blue-green algae) and heterotrophic bacteria.

Representing a higher stage of evolutionary development is the kingdom **Protista**. It includes those organisms that are single-celled but have their nuclear material contained within a nuclear sheath. Here we find the rest of the algae (simple plants) and the single-celled animals, Protozoa.

Kingdom **Mycota**, the fungi, appears to be poorly represented in the oceans. Fewer than 1% of the known 50,000 species are found there. Fungi are found throughout the marine environment, but they

are much more common in the intertidal zone in a mutualistic relationship with cyanobacteria or green algae. In this relationship, called **lichen**, the fungus provides a protective covering that retains water during periods of exposure, while the algae provides food for the fungus through photosynthesis. Other fungi function primarily as decomposers in the marine ecosystem: they remineralize organic matter.

Though there is one more kingdom of marine plants—**Metaphyta**, the multicelled plants—its members are mostly restricted to the shallow coastal margins of the ocean and are not an important portion of the marine productive community. They do play a very important role as producers within restricted communities, such as mangrove swamps and salt marshes.

The kingdom **Metazoa** consists of multicelled animals. Metazoans range in complexity from the simple sponges to the vertebrates (animals with backbones).

The kingdoms are divided into increasingly specific groupings (**taxa**). The system of taxonomic classification introduced by Carl von Linné (Linnaeus) in 1758 includes the following major categories:

> Kingdom
> Phylum
> Class
> Order
> Family
> Genus
> Species

Every organism's scientific name includes its genus and species. For example, the toothed whale called the common dolphin is the species *Delphinus delphis.*

MACROSCOPIC PLANTS

As we discuss the marine plants, we will first consider those groups with which we are most familiar. These are the attached forms of macroscopic algae and Metaphyta found in shallow waters along the margins of the ocean (fig. 14–2).

Phylum Phaeophyta (Brown Algae)

The largest members of the marine plant community that are attached to the bottom wherever suitable substrate is available in the littoral and inner sublittoral zones belong to the **Phaeophyta**, brown algae. The dominant pigment in these brown algae is **fucoxanthin** and the color may range from a very light brown to black. The brown algae occur primarily in temperate and cold-water areas. Their sizes range

Preceding page: This false-color image of chlorophyll pigment concentration in the northwest Atlantic Ocean shows the low level of biological productivity of the Gulf Stream in blue. Increasing levels of productivity are shown as green, yellow, red, and dark red. A striking feature of this image is the low level of productivity indicated by the blue color of the warm-core ring south of Cape Cod. Note that it is surrounded by more-productive green nearshore water. The white is cloud cover. Image from the Coastal Zone Color Scanner (CZCS) *Nimbus 7* satellite, courtesy NASA/Goddard Space Flight Center.

FIGURE 14–1 The Five Kingdoms of Organisms.
Monera—single cells without a nucleus; Protista—single cells with a nucleus; Mycota—the fungi; Metaphyta—complex, many-celled plants; Metazoa—complex, many-celled animals.

from the small black encrusting patch of the *Ralfsia,* found in the upper and middle intertidal zones where it is exposed to an extensive threat of desiccation and may become crisp and dry in the sun without dying, to the *Pelagophycus* (bull kelp), which may grow to the surface from small holdfasts (attachment structures) in water depths in excess of 30 m (98 ft). A genus of brown algae, *Sargassum,* may be attached to the bottom of nearshore waters in a belt between the subtropics or float freely at the ocean's surface. The floating variety found in the Sargasso Sea southeast of Bermuda is buoyed up by small grape-sized air bladders.

Phylum Chlorophyta (Green Algae)

The predominant forms of algae found in freshwater environments belong to the Chlorophyta, green algae, and are not well represented in the ocean. Most species are intertidal, or grow in shallow waters of bays. Because of the pigment **chlorophyll** they range in color from yellow green to a very dark green, but most are grass green in color. They grow only to mod-

erate size, seldom exceeding 30 cm (12 in.) in the largest dimension. Forms range from finely branched filaments to flat thin sheets. The various species of the genus *Ulva* (sea lettuce), a thin membranous sheet two cell layers thick, may be found widely scattered throughout cold-water areas (fig. 14–2A). The genus *Codium* (sponge weed) is a dichotomously branched form that can be in excess of 6 m (20 ft) in length and is more commonly found in warm waters (fig. 14–2D).

Three genera of green algae, *Acetabularia, Halimeda,* and *Penicillus,* secrete calcium carbonate structures and are important contributors to marine sediment in warm waters.

Phylum Rhodophyta (Red Algae)

Red algae belong to the **Rhodophyta** and are the most abundant and widespread of the marine macroscopic algae. Over 4000 species are found, many of them attached, from the very highest intertidal levels to the outer edge of the inner sublittoral zone. They are very rare in fresh water. The red algae range in

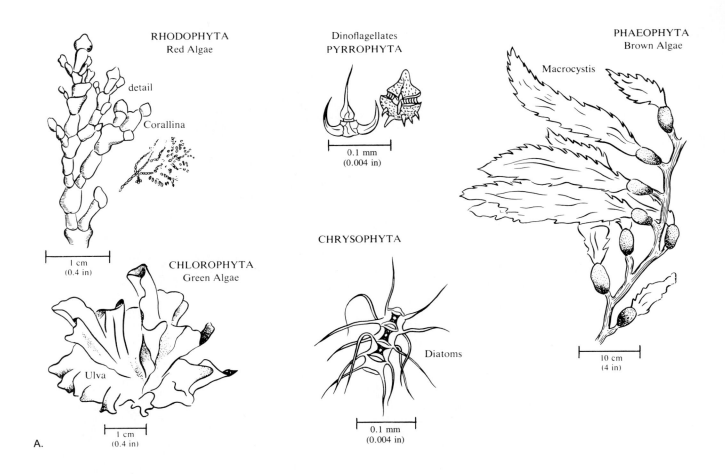

RHODOPHYTA
Red Algae

detail

Corallina

1 cm
(0.4 in)

Dinoflagellates
PYRROPHYTA

0.1 mm
(0.004 in)

CHRYSOPHYTA

Diatoms

0.1 mm
(0.004 in)

PHAEOPHYTA
Brown Algae

Macrocystis

10 cm
(4 in)

CHLOROPHYTA
Green Algae

Ulva

1 cm
(0.4 in)

A.

B.

C.

FIGURE 14–2   Algae.

A, detailed drawing of several algae. Line drawing by Phil David Weatherly. *B,* encrusting red algae, *Lithothamnion,* with a sea star, *Henricia.* Photo by Shane Anderson. *C.* brown algae, *Laminaria;* oarweed at low tide. Photo by D. J. Wrobel, Monterrey Bay Aquarium/BPS. *D,* green algae, *Codium fragile.* The prominent fingerlike plant is called *sponge weed. E,* photomicrograph of *Gonyaulax polyedra* magnified 1100 times. *Gonyaulax* is a large genus of phosphorescent marine dinoflagellates which, in great abundance, cause what is known along the shoreline as red tide. Photo from Scripps Institution of Oceanography, University of California, San Diego. *F,* coccoliths, disc-shaped $CaCO_3$ plates that cover coccolithophore cells. Diameter is 0.06 mm. Courtesy of Deep Sea Drilling Project, Scripps Institution of Oceanography, University of California, San Diego.

D.

E.

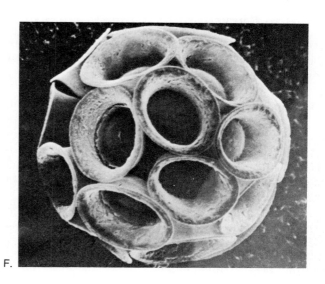

F.

size from just visible to the naked eye to lengths of up to 3 m (10 ft). While Rhodophyta are found in both warm- and cold-water areas, the warm-water varieties are relatively small. The characteristic pigment of the red algae is **phycoerythrin**. The color of the red algae will vary considerably depending on their depth in the intertidal or inner sublittoral zones. In the upper well-lighted areas it may be green to black or purplish in color, and it changes through a brown to a pinkish red in the deeper-water zones where light concentrations lessen. Although the bulk of marine plant productivity is believed to occur above water depths where the amount of light is reduced to 1% of that available at the surface (approximately 100 m, or 330 ft) a red alga has been observed growing at a depth of 268 m (880 ft) on a seamount near San Sal-

vador, Bahamas. Available light at this sighting was thought to be only 0.05% of the light available at the ocean's surface.

Phylum Spermatophyta (Seed-Bearing Plants)

The only Metaphyta observed in the marine environment belong to the highest group of plants, the seed-bearing **Spermatophyta**. Two seed-bearing plants commonly found in the marine environment are *Zostera* (eelgrass) and *Phyllospadix* (surf grass). *Zostera*, a grasslike plant with true roots, is found primarily in quiet waters of bays and estuaries from the low-tide zone down to a depth of some 6 m (20 ft). *Phyllospadix* prefers the high-energy environment of an exposed rocky coast and can be found from the intertidal zones down to a depth of 15 m (50 ft). Both of these plants are considered to be important sources of the detrital food for the marine animals that inhabit their environment. Found in salt marshes are grasses belonging mostly to the genus *Spartina,* while mangrove swamps contain primarily the mangrove genera *Rhizophora* and *Avicennia.*

MICROSCOPIC PLANTS

The following plants include all the members of the important phytoplankton that produce in excess of 99% of the food supply for marine animals. They are primarily free-drifting forms, although some will live on the bottom in the nearshore environment.

Phylum Chrysophyta (Golden Algae)

Containing the yellow pigment **carotene** (or carotin), these microscopic plants store food in the form of *leucosin* (a carbohydrate) and oils.

Diatoms   The diatoms are cells contained in a shell, or **frustule**, composed of opaline silica ($SiO_2 \cdot nH_2O$). These silica housings are important geologically because they accumulate on the ocean bottom and produce a siliceous sediment, **diatomite**. Some deposits of diatomite on land are mined and used primarily in the manufacture of filtering devices.

The frustule of the diatom is similar in structure to a microscopic pill box. The top and the bottom of the box are called *valves;* the larger of the valves is the **epitheca** and the smaller valve is the **hypotheca**. The protoplasm of the plant is contained within this housing and exchanges nutrients and waste products with the surrounding water through slits or pores in the valves.

The reproduction of diatoms is by simple cell division. To achieve this division the valves of the frustule must separate. Each valve then serves as one-half the housing for each newly formed cell. A new valve must form to complete the enclosure of each daughter cell, and each of these newly formed halves forms as a hypotheca.

As can be seen in figure 14–3, this process leads to the formation of smaller and smaller cells. Eventually, the newly formed daughter cell is going to approach a size so small that if it were decreased further, abnormalities would result. At this point, a change in the process occurs, namely, the formation of an **auxospore**. The auxospore forms between the two separating valves within a membrane that allows the auxospore to grow into a full-sized diatom so that the process of reproduction by the splitting of the frustule can commence again.

Coccolithophores   Coccolithophores are for the most part flagellum-bearing organisms (whiplike locomotive structures) covered with small calcareous plates ($CaCO_3$), **coccoliths**. The name of the group means *bearers of coccoliths.* The individual plates are about the size of a bacterium, and the entire organism is too small to be captured in plankton nets. Coccolithophores are included in the nanoplankton, which have dimensions of less than

FIGURE 14–3   Diatom Reproduction.

Diatom cell division. As the epitheca and hypotheca of a diatom separate, each becomes the epitheca of a new cell. The new frustule is completed by the generation of a new hypotheca by each new diatom. Following the arrow through the formation of three generations of diatoms, it can be seen that some cells will be crowded into ever smaller frustules. When the size of the frustule becomes critically small, an auxospore forms to allow growth of a larger cell.

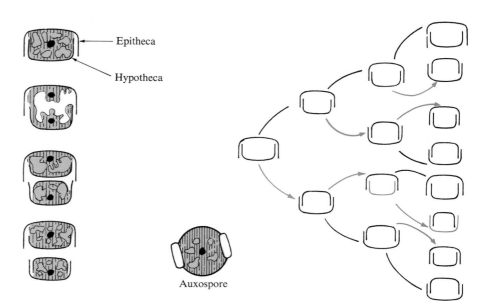

Epitheca

Hypotheca

Auxospore

0.06 mm (0.002 in.). The coccolithophores contribute significantly to calcareous deposits in all of the temperate and warmer oceans.

Phylum Pyrrophyta (Dinoflagellate Algae)

A second group of phytoplankton important in marine productivity is the **dinoflagellates**. Possessing **flagella** for locomotion, they have a slight capacity to move into more favorable areas for plant productivity. The dinoflagellates are not important geologically because many have no protective covering. Those that do have coverings made of cellulose, which is easily decomposed by bacterial action when the organism dies. Many of the 1100 species are luminescent.

Red tides result from conditions where up to 2 million dinoflagellates may be found in one liter (1.1 qt) of water. Mainly responsible for this red tide phenomenon are the genera *Ptychodiscus* and *Gonyaulax,* which produce water-soluble toxins. *Gonyaulax* toxin is not poisonous to shellfish but concentrates in their tissue and is poisonous to humans who eat the shellfish. *Ptychodiscus* toxin kills fish and shellfish. April through September are particularly dangerous months. In most areas there is a quarantine against the taking of shellfish that feed on these microscopic organisms and concentrate the poisons that they secrete to levels that are dangerous to humans. *Ptychodiscus breve* is an important contributor to Gulf of Mexico red tides. *Gonyaulax* is an important participant in the cooler waters of the New England and West Coast areas.

A potentially tragic epidemic of **paralytic shellfish poisoning (PSP)** from a *Gonyaulax tamarensis* red tide occurred in the coastal waters of Massachusetts in the fall of 1972. Fortunately, no deaths occurred, although 30 cases of poisoning were reported. Symptoms of PSP are similar to those of drunkenness, including incoherent speech, uncoordinated movement, dizziness, and nausea. Documented cases throughout the world include 300 deaths and 1750 nonfatal cases. There is no known antidote for the toxin, which attacks the human nervous system, but the critical period usually passes in 24 hours.

PRIMARY PRODUCTIVITY

Primary productivity is defined as the amount of carbon fixed by autotrophic organisms through the synthesis of organic matter from inorganic compounds such as CO_2 and H_2O using energy derived from solar radiation or chemical reactions. The major process through which primary productivity occurs is thought to be photosynthesis. Although new knowledge of the role of chemosynthesis in supporting hydrothermal vent communities on oceanic spreading centers has emerged, chemosynthesis is much less significant in overall marine primary production than is photosynthesis. Because of the far greater importance of photosynthesis in the overall productivity of the oceans and our greater knowledge of the factors that affect marine photosynthesis, the following considerations will relate primarily to this process.

Photosynthetic Productivity

The total amount of organic matter produced by photosynthesis per unit of time represents the **gross primary production** of the oceans. Plants will use some of this organic matter for their own maintenance through respiration. That which remains is the **net primary production**, which is manifested as growth and reproduction products. It is the net primary production that supports the heterotrophic marine populations—animals and bacteria.

In the 1920s, the **Gran Method** of measuring primary productivity was developed based on the fact that oxygen is produced by photosynthesis in proportion to the amount of organic carbon synthesized. The method involves having equal quantities of phytoplankton in each of a series of bottles, all of which contain the same amount of dissolved oxygen. The bottles are then arranged in pairs, one being transparent and the other totally opaque. These pairs are suspended on a hydrographic line through the euphotic zone, where they are left for a specific period of time. After the bottles are brought to the surface, the oxygen concentration is determined for each bottle.

Photosynthetic activities, which are confined to the **transparent bottles**, will add oxygen to the water, whereas respiration will reduce the oxygen content in both the *transparent* and *opaque bottles.* Increased oxygen concentration in the transparent bottles is proportional to the amount of photosynthesis that has occurred minus the oxygen consumed by plant respiration and thus represents the net primary production. Decreased oxygen content within the opaque bottles corresponds to the respiration rate. For any depth, an assessment of gross primary production can be made by adding the oxygen gain in the clear bottles to the oxygen loss in the opaque bottles.

The depth at which the oxygen production and the oxygen consumption are equal is called the **oxygen compensation depth**, which represents that light intensity below which plants do not survive. Since respiration goes on at all times, during the daylight hours plants must produce, through photosynthesis, biomass in excess of that which is consumed by respiration in any 24-h period if the total biomass of the community is to increase (fig. 14–4). An analogy can be made with the common paycheck: Gross primary production (Gross pay earned) = Oxygen change in clear bottle (Take-home pay) + Oxygen loss in dark bottle (Income tax withheld).

During the 1950s, a method involving the use of **radioactive carbon** (C^{14}) was developed. It has been refined and is currently the most often used method for determining marine primary productivity. The procedure is similar to that described above, because it involves the use of a series of paired clear and opaque bottles. Each bottle contains identical phytoplankton samples, equal amounts of CO_2 containing carbon-14, and equal amounts of CO_2 containing stable carbon.

The phytoplankton sample is filtered from each bottle after the system has been suspended in the ocean for a sufficient period of time. The amount of beta-radiation emitted by each sample is measured by a radiation counting device, and the rate of assimilation of carbon-14 is computed from these measurements. High levels of radioactive emission indicate high levels of productivity.

A third method used in studies of marine primary productivity is the measurement of **chlorophyll a** content of living phytoplankton samples taken from ocean surface waters. Although this method is not as precise as the measurement of net primary productivity by use of carbon-14, there is a direct relationship between the amount of chlorophyll *a* in a phytoplankton sample from a given volume of ocean water and gross primary productivity.

The special feature "Comparison of Spring Blooms in Shelf, Slope, Warm-Core Rings, and Gulf Stream Waters," on page 364, discusses a method being developed to help extrapolate local measurements of chlorophyll content into a global estimate of biological productivity.

Distribution of Photosynthetic Productivity

Primary photosynthetic productivity of the oceans varies from about 0.1 gram of carbon per square meter per day ($gC/m^2/d$) in the open ocean to over $10gC/m^2/d$ in highly productive coastal areas. This variability is the result of the uneven distribution of nutrients throughout the photosynthetic zone and seasonal changes in the availability of solar energy. The causes of this variability are discussed below.

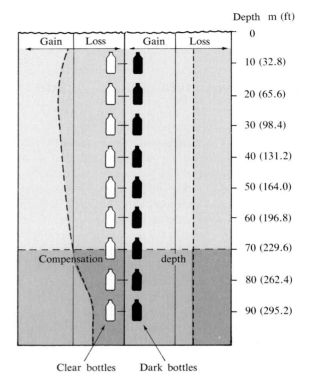

FIGURE 14–4   Oxygen Compensation Depth.
Biological assay of phytoplankton production and consumption of oxygen (O_2) shows to what depth light penetrates with sufficient intensity to sustain photosynthesis at a level sufficient to increase phytoplankton mass. This occurs only above the compensation depth, the depth at which oxygen production by photosynthesis is equal to consumption through respiration. The change in oxygen content in the clear bottles represents a net photosynthesis resulting from gross photosynthesis less the consumption of oxygen by respiration. The loss of oxygen in the dark bottles gives a value for the amount of respiration that occurred in the clear bottles. This amount added to the net photosynthesis that occurred in the clear bottles gives a value for the gross photosynthesis in the clear bottles.

Temperature Stratification and Nutrient Supply

Thermoclines are relatively permanent in low-latitude areas, while a seasonal thermocline may be present during the summer months in the mid-latitudes. They are generally nonexistent in high-latitude areas. In the following discussion of productivity, we will see that the presence or absence of the thermocline will have an important effect on nutrient availability.

In considering the availability of solar radiation and nutrients, we can rather conveniently divide the oceans into polar, temperate, and tropical regions. We will concern ourselves primarily with the nature of the open ocean in all three regions. The polar region will be discussed first as the productivity considerations are simplest there.

Polar Region   As an example of productivity on a seasonal basis in a polar sea, we will consider the **Barents Sea** off the northern coast of Europe, where there is peak diatom productivity during the month of May that tapers off through July (fig. 14–5A). This brings about a peak period of zooplankton development, consisting primarily of small crustaceans of the genus *Calanus* (fig. 14–5D). The zooplankton biomass reaches a peak in June and continues at a relatively high level until winter darkness begins in October. In this region above 70°N latitude, there is continuous darkness for about 3 months of winter and continuous illumination for a period of 3 months during summer.

In the **Antarctic** region, particularly at the southern end of the Atlantic Ocean, productivity patterns are similar to those found in the Barents Sea, except that the seasons are reversed, and productivity is somewhat greater. The most likely explanation for this greater productivity in Antarctic waters is the continual upwelling of water that has sunk in the North Atlantic. Moving southward as a deep-water mass, the

A.   BARENTS SEA PRODUCTIVITY

B.   ANTARCTIC UPWELLING

C.   Depth m (ft)

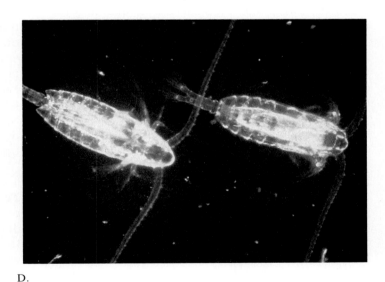

D.

FIGURE 14–5   Productivity in Polar Oceans.
A, diatom mass increases rapidly in the spring when the sun rises high enough in the sky to cause deep penetration of sunlight. As soon as this diatom food supply develops, the zooplankton population begins feeding on it, and the zooplankton biomass peaks early in the summer. *B,* the continuous upwelling of North Atlantic Deep Water keeps the Antarctic waters rich in nutrients. When the summer sun provides sufficient radiation, there is a huge period of biological productivity. *C,* polar water shows little temperature stratification. Nearly isothermal conditions from the surface to the bottom are common. *D,* copepods, *Calanus,* each about 8 mm in length. Photo courtesy of Scripps Institution of Oceanography, University of California, San Diego.

North Atlantic Deep Water surfaces hundreds of years later carrying high concentrations of nutrients (fig. 14–5B).

To illustrate the very great productivity that occurs during the short summer season in polar oceans, we can consider the growth rate of baby whales. The largest of all whales, the **blue whale** (fig. 15–13), migrates through the temperate and polar oceans at the times of maximum zooplankton productivity. As a result of this excellent timing, they are able to develop and support calves that during a gestation of 11 months reach lengths in excess of 7 m (23 ft) at birth. The mother suckles the calf for 6 months with a teat that actually pumps the youngster full of rich milk. By the time the calf is weaned, it is over 16 m (52 ft) in length and will in a period of 2 yr reach a length approaching 23 m (75 ft). In 3½ yr, a 60-ton blue whale has developed. This phenomenal growth rate gives some indication of the enormous biomass of copepods and krill that these large mammals feed upon.

In polar waters there is little density stratification to prevent the mixing of deeper water with shallow water. Polar waters are relatively isothermal (fig. 14–5C). In most polar areas, the surface waters freely mix with the deeper nutrient-rich water. There can be, however, some density segregation of water masses due to the summer melting of ice, a process that lays down a thin low-salinity layer that does not readily mix with the deeper waters.

There are usually high concentrations of phosphates and nitrates in the surface waters. Thus, photosynthetic productivity in the high latitudes is more commonly limited by the availability of solar energy than by the availability of nutrients. The productive season in these waters will be relatively short but will be characterized by an outstandingly high rate of production.

Tropical Region

In direct contrast with high productivity associated with the summer season in the polar seas, low productivity is the rule in the tropical regions of the open ocean. Light penetrates much deeper into the open tropical ocean than into the temperate and polar waters. This produces a very deep oxygen compensation depth. In the tropical ocean, however, a permanent thermocline produces a relatively permanent stratification of water masses and prevents mixing between the surface waters and the nutrient-rich deeper waters (fig. 14–6A).

At about 20° latitude we will commonly find the concentrations of phosphate and nitrate to be less than 1/100 of the concentrations of these nutrients in temperate oceans during winter. Nutrient-rich waters within the tropics lie for the most part below 150 m (492 ft), and the highest concentration of nutrients occurs between 500 and 1000 m (1640 and 3280 ft) depth.

There is generally a steady low rate of primary productivity in tropical oceans. Although the rate of tropical productivity is low, when we compare the total annual productivity of tropical oceans with that of the more productive temperate oceans, we find that the tropical productivity is generally at least half of that found in the temperate regions on an annual basis.

Within tropical regions there are three environments where productivity is unusually high—regions of equatorial upwelling, coastal upwelling, and coral reefs (fig. 14–6B, C). In areas where trade winds drive westerly equatorial currents on either side of the equator, surface water diverges as a result of the Ekman transport. This surface water that moves off toward higher latitudes is replaced by nutrient-rich water that surfaces from depths of up to 100 m (328 ft). This condition of equatorial upwelling is probably best developed in the eastern Pacific Ocean. The surface waters are driven away from the coast where the prevailing winds blow toward the equator and along the western margins of continents. They are replaced by nutrient-rich waters from depths of 200–900 m (656–2952 ft). As a result of this upwelling of nutrient-rich water, there is a high rate of primary productivity in these areas, and they support large fisheries. Such conditions exist along the southern coast of California and the southwest coast of Peru in the Pacific Ocean. Upwelling also occurs along the northwest coast of Morocco and the southwest coast of Africa in the Atlantic Ocean. The relatively high productivity of coral reef environments is not related to the upwelling process and is discussed in chapter 16.

Temperate Region

We have discussed the general productivity picture in the polar regions, where productivity is limited primarily by the availability of solar radiation, and in the tropical low-latitude areas, where the limiting factor is the availability of nutrients. We will next consider the temperate regions, where an alternation of these factors controls productivity in a pattern that is somewhat more complex.

Productivity in temperate oceans is at a very low level during the winter months, although the concentrations of nutrients is higher than at any time throughout the year. The limiting factor on productiv-

FIGURE 14–6   Productivity in Tropical Oceans.

In normal tropical regions *(A)* deep penetration of sunlight produces a deep compensation depth with a good supply of solar radiation available for photosynthesis. A permanent thermocline serves as an effective barrier to mixing of surface and deep water. As plants use nutrients in surface layer, productivity is retarded because the thermocline prevents replenishment of nutrients from deeper water. Tropical regions of high productivity (local areas where nutrients are brought to the surface) include areas of equatorial upwelling *(B)*, where nutrient-enriched water from 200 to 300 m (656 to 984 ft) replaces diverging equatorial surface water, and areas of coastal upwelling *(C)*, where surface water moved away from the shore is replaced by deep water that is rich in nutrients because it comes from below the euphotic layer.

ity during the winter season in the temperate ocean is the availability of solar energy. Since the sun is at its lowest elevation above the horizon during this season, a higher percentage of solar energy is reflected and a smaller percentage absorbed into the surface waters. The oxygen compensation depth for basic producers such as diatoms is so shallow that it does not allow growth of the diatom population (fig. 14–7).

As the sun rises higher in the sky during the spring season, the oxygen compensation depth deepens as the amount of solar radiation being absorbed by the surface water increases. Eventually, there is sufficient water volume included within the water column above the oxygen compensation depth to allow for the exponential growth of the diatom population, a **spring bloom**. This expanding population puts a tremendous demand on the nutrient supply in the euphotic zone. In most Northern Hemisphere areas, decreases in the diatom population will occur by May as a result of insufficient nutrient supply.

As the sun rises higher in the sky during the summer months, the surface waters in the temperate parts of the ocean are warmed and the water becomes separated from the deeper water masses by a seasonal thermocline that may develop at depths of

approximately 15 m (49 ft). As a result of this thermocline development, there is little or no exchange of water across this discontinuity, and the nutrients that are depleted from surface waters cannot be replaced by those available in the deep waters. Throughout the summer months the plant population will remain at a relatively low level but will again increase in some temperate areas during the autumn months.

The **autumn bloom** is much less spectacular than that of the spring because of the decreasing availability of solar radiation resulting from the sun dropping lower in the sky. This causes a decrease in surface temperature and the breakdown of the summer thermocline. A return of nutrients to the surface layer occurs as increased wind strength mixes it with the deeper water mass in which the nutrients have been trapped throughout the summer months. This bloom is very short-lived. The phytoplankton population begins to decrease rapidly. The limiting factor in this case is the opposite of that which reduced the population of the spring phytoplankton. In the case of the spring bloom, solar radiation was readily available, and the decrease in nutrient supply was the limiting factor.

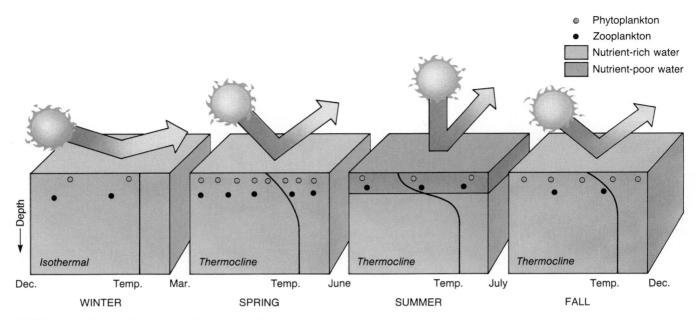

FIGURE 14–7   Productivity in Temperate Oceans.
In winter, the sun is low in the sky. Much solar energy is reflected with little being absorbed by ocean. The water column is *isothermal. Nutrients* are present throughout water column. The plant and animal populations are at low levels due to lack of solar energy for photosynthesis. In spring, the sun is higher in the sky. More solar energy is absorbed by ocean. A *thermocline* begins to develop. A "spring bloom" of plants occurs as a result of the availability of both solar energy and *nutrients*. In summer, the sun is high in the sky. Surface water warms and strong thermal stratification prevents mixing of surface and deep water. The *thermocline* reaches maximum development. When *nutrients* in the surface water are used up, the supply cannot be replenished from deep water. Plants become scarce in late summer. The increase in animal population that followed "spring bloom" is followed by a decrease. In fall, the sun is lower in the sky. Surface water cools. Fall winds aid in mixing surface and deep water. The *thermocline* begins to disappear. A small "fall bloom" of phytoplankton is terminated as too much solar energy is lost by reflection.

Additional Consideration Coastal waters with high nutrient levels are highly productive, or **eutrophic.** Most of the open ocean has a low level of nutrients and productivity. This low productivity is referred to as an **oligotrophic** condition. Figure 14–8 shows the general patterns of ocean productivity based on photosynthetic pigment concentrations as observed by the Coastal Zone Color Scanner (CZCS) aboard the *Nimbus 7* satellite.

Measurements of photosynthetic productivity of oligotrophic waters in the North Atlantic and North Pacific oceans indicate they may be from 2 to 7 (or even more) times as productive as C^{14} data indicate. Instead of using the small bottles used in the C^{14} measurements and suspending them for a short time in the ocean, some physical oceanographers have used a much larger "bottle." In their studies of the circulation within the subtropical gyres, physical oceanographers use "bottles" that are "capped" by the pycnoclines produced by the thermoclines beneath

the warm layers of surface water within the subtropical gyres of the open oceans. These "bottles" contain (depending on the design of the study) from tens to thousands of km^3 of water that record the average results of photosynthetic activity over periods of from a few months to decades.

The studies analyze the effect of photosynthesis on the oxygen concentrations (1) in the euphotic zone and (2) beneath the euphotic zone. In the North Pacific Ocean, oxygen saturations in **subsurface oxygen maximums (SOMs)** at depths between 50 and 100 m (164 and 328 ft) were from 110% to 120% at latitudes of from 30° to 40°N during summer months based on data obtained from 1962 through 1979. The excess oxygen (anything over 100%) may be due to trapped photosynthetically produced oxygen. The North Atlantic study determined the rate at which oxygen is used up beneath the euphotic zone (100 m) by decomposing organic matter falling toward the ocean floor. Oxygen utilization rates (OURs) of twice the accepted

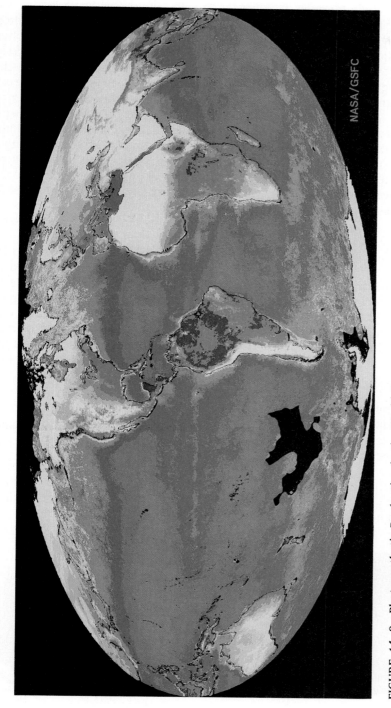

NASA/GSFC

FIGURE 14—8   Photosynthetic Production in the World Ocean.

Data obtained by the Coastal Zone Color Scanner (CZCS) aboard the satellite *Nimbus 7* between November 1978 and June 1981 were used to produce this false-color image. The CZCS senses changes in the color of ocean water caused by changing concentrations of photosynthetic pigment. Increasing photosynthetic productivity is correlated with increasing concentration of photosynthetic pigment and shows values below 0.1 mg/m³ (purple) in the oligotrophic open ocean and high values in excess of 10.0 mg/m³ (red) along eutrophic continental margins. Productivity of 1 mg/m³ is represented by the boundary between the yellow and green colors. The accuracy of these observations is 50%. Black represents areas of insufficient data. Image is courtesy of Jane A. Elrod and Gene Feldman, NASA/Goddard Space Flight Center.

mean oligotrophic rate indicate there may be much more organic matter produced in the oligotrophic euphotic zone than is indicated by C^{14} data.

The problem with the C^{14} method may be that by using such a small sample and exposing it for such a short time, periods of unusually high productivity may be missed and not averaged into the results. An interesting find may help explain how some of the biological productivity of oligotrophic water may have been missed. Mats of diatoms belonging to the genus *Rhizosolenia* have been found floating in the North Pacific Gyre and the Sargasso Sea. Averaging 7 cm (3 in.) in length, these mats possess symbiotic bacteria that fix molecular nitrogen (N_2) into nitrate (NO_3) usable as a nutrient by the diatoms (fig. 14–9). Also, bundles of nitrogen-fixing cyanobacteria, *Oscillatoria*, have been verified in the Sargasso Sea. Such mats are readily broken up by net tows and missed by standard bottle casts.

Recent studies in the Pacific and Atlantic oceans indicate that a large percentage (60% in one study) of oligotrophic photosynthesis is achieved by picoplankton (0.2–2.0 μm, or 0.00008–0.0008 in.) that pass through many filters used in productivity studies. Continued investigation of this problem may indicate the overall productivity of the oligotrophic ocean waters may have been greatly underestimated.

Chemosynthetic Productivity

Another source of potentially significant biological productivity in the oceans is occurring in the rift valleys of the oceanic spreading centers at depths of over 2500 m (8200 ft), where there is no light for photosynthesis. Interesting biological communities were first discovered in 1977 in association with the hydrothermal vents found within the rift valleys. A significant component of these communities is the vesti-

mentiferan worm population, a sample of which can be seen in figure 14–10. These 1 m (3.28 ft) tube worms and other chemosynthetic symbionts such as clams and mussels reach unusually large size in association with autotrophic bacteria that derive energy to produce their own food from the **hydrogen sulfide** (H_2S) gas dissolved in the water of the hydrothermal springs. By oxidizing the gas to form free sulfur (S) and sulfate (SO_4), the bacteria release chemical energy that they use much as plants use solar energy to carry on photosynthesis. Similar communities have been discovered near cold-water seeps at the base of the Florida Escarpment in the Gulf of Mexico, at hydrocarbon seeps and in association with trench deposits. The true significance of chemosynthetic bacterial productivity on the deep-ocean floor will not be fully understood until much more research on the phenomenon is conducted. It has the potential, however, of increasing our estimates of the biological productivity of the ocean.

Biochemists have recently discovered that bacteria can obtain the chemical energy needed for the synthesis of organic molecules through the oxidation of a large variety of compounds containing the metals iron, manganese, copper, nickel, and cobalt. These microorganisms may well be an important factor in the deposition of ore-quality deposits of the oxides of these metals on the ocean floor in the form of manganese nodules.

ENERGY TRANSFER

We have been discussing the general consideration related to availability of nutrients. Our attention will now be turned to the cycling of specific important classes of nutrients.

FIGURE 14–9 Nitrogen Fixation by Aggregates of Diatoms and Cyanobacteria in Oligotrophic Ocean Waters.

A, typical mat of *Rhizosolenia* about 5 cm long. It is composed of intertwining chains of *R castracanei* (wider cells) and *R. imbricata* (narrower cells). Within these cells, symbiotic bacteria fix nitrogen for uptake by the plant cells. Photo by James M. King. *B*, an aggregation of the nitrogen-fixing cyanobacteria *Oscillatoria* spp. with an O_2 probe inserted into it. The probe has a maximum diameter of about 5 μm. Photo courtesy of Hans W. Paerl, University of North Carolina.

B.

FIGURE 14-10 Chemosynthetic Life of the Galápagos Rift.

A, this photograph taken at the Galápagos Rift shows the large tube worms found there and at other deep-sea hydrothermal vents. These worms possess symbiont sulfur-oxidizing bacteria and chemosynthetically produce their own food by combining inorganic nutrients dissolved in the deep-ocean water. *B,* sample of sulfur-oxidizing bacteria from the Galápagos Rift with small particles of what is probably free sulfur attached to them. These and similar bacteria live symbiotically within the tissue of the tube worms, clams, and mussels found at the vents. They chemosynthetically produce food to support these organisms. Magnification 20,000×. Bar is 1 μm. Courtesy of Woods Hole Oceanographic Institution; *A* by Fred Grassle.

Marine Ecosystem

The term **biotic community** refers to the assemblage of organisms that live together within some definable area. An **ecosystem** includes the biotic community and **abiotic environment** with which it interacts in the exchange of energy and chemical substances. Within an ecosystem there are generally two basic categories of organisms—**producers and consumers**. Plants and some bacteria are the *autotrophic* producers and

have the capacity to nourish themselves through chemosynthetic and photosynthetic processes. The consumers are *heterotrophic* organisms that depend on the organic compounds produced by the autotrophs for their food supply.

Consumers may be divided into categories: **herbivores**, which feed directly on the plants; **carnivores**, which feed only on other animals; and **omnivores**, which feed on both. As the role of bacteria in the marine ecosystem becomes better understood, another category of consumers, the **bacteriovores**, which feed on bacteria, may be identified as an important component of the marine ecosystem. The **decomposers**, such as fungi and bacteria, break down the organic compounds of dead organisms and animals' excretions while taking some of these decomposition products for their own energy requirements. They release simple inorganic salts that are used by the plants as nutrients.

Energy Flow

Before considering the biogeochemical cycles that involve the transfer of organic and inorganic matter, we will consider the flow of energy in general. Most energy is put into a biotic community through plants. From the plants, it follows a unidirectional path (although cycles of energy exchange occur at many intermediate biological levels) that leads to a continual degradation of energy into forms less usable by organisms and culminating in increased **entropy**, or energy converted to a form where it is no longer available to do work. As can be seen in figure 14-11, which depicts the flow of energy through a plant-supported biotic community, energy enters the system as **radiant solar energy**, which is absorbed by the plants. Photosynthesis converts it to **chemical energy** that is used for plant respiration and growth. Plant mass is passed on to the animals to be used for growth and carrying on their various life-sustaining functions. Energy is expended by the animals as **mechanical** and **heat energy**. Finally, it becomes biologically useless as entropy increases.

Composition of Organic Matter

Having discussed the noncyclic nature of flow of energy through the biotic community, let us now consider the biogeochemical cycles involving matter that is not lost to the biotic community but is cycled by being converted from one chemical form to another by the various members of the community.

Biological mass is made up of compounds involving a number of elements that are abundant on earth.

COMPARISON OF SPRING BLOOMS IN SHELF, SLOPE, WARM-CORE RINGS, AND GULF STREAM WATERS

The "spring bloom" is a biological phenomenon that occurs as a biological response to changing physical conditions in the ocean. It represents, in the temperate latitudes, a period of rapidly growing phytoplankton populations during the spring season. The physical change that brings it on is the increased penetration of solar radiation in spring as compared to winter. During winter, there is no thermal stratification of the water column, so nutrients and phytoplankton are transported up into and down out of the thin euphotic zone so quickly that phytoplankton experience low levels of light exposure. This prevents the development of significant levels of primary productivity. As the sun rises higher and penetrates deeper into the ocean in spring, the surface waters heat up, and phytoplankton and nutrients are trapped in the warm surface layer of water that is separated from deeper waters by a deepening thermocline. The euphotic zone also deepens, and there is a rapid growth of phytoplankton, the "spring bloom," that lasts a short while—until the nutrients are depleted.

FIGURE 14A Temperature.
Processed AVHRR image of the U.S. east coast on April 24, 1982. The yellow (21°C) to reddish (26°C) band is the warm Gulf Stream. The greenish (16–20°C) Sargasso Sea is to the south of the Gulf Stream. Colder waters of the slope, shown in light blue (10–15°C), and shelf, shown in dark blue (less than 10°C), are to the north. The 200-m isobath is the boundary between the shelf and slope. Areas compared in the study—(a) Gulf Stream, (b) shelf, (c) slope, and (d) warm-core ring— are identified. Courtesy of Otis Brown and Robert Evans, University of Miami/RSMAS.

Ship studies have shown that the waters of the Gulf Stream and Sargasso Sea have low levels of biological productivity, with phytoplankton pigment concentrations less than 1 mg/m^3, whereas slope water may reach 3 mg/m^3 and shelf-water concentrations may rise to 7 mg/m^3. It is not possible to make such measurements simultaneously from a ship, so biological oceanographers turned to satellites in an effort to get productivity measurements from these areas simultaneously.

The Coastal Zone Color Scanner (CZCS) aboard the *Nimbus 7* satellite provides data that can be converted to phytoplankton pigment concentrations, and the Advanced Very High Resolution Radiometer (AVHRR) aboard the *NOAA 7* satellite can measure sea-surface temperature (SST). Measurements of these variables over a 28-day period from April 8 through May 5, 1982, were processed for this study. Figures 14A and 14B are images produced from AVHRR and CZCS data, respectively, obtained on April 24, 1982. Labeled on the images are the areas for which biological productivity comparisons were made. They are the (a) Gulf Stream, (b) shelf, (c) slope, and (d) warm-core ring.

FIGURE 14B Pigment Concentrations.

Processed CZCS image of the same area shown in *A* for the same day. Phytoplankton pigment concentrations, in mg/m^3, range from 0.05 (darkest blue) to over 3.0 (darkest brown). The cooler waters of the shelf and slope show the highest pigment concentrations—2.8 and 3.0, respectively. The warmer waters of the warm-core ring and Gulf Stream have lower pigment concentrations of 0.5 and 0.3, respectively. Courtesy of Otis Brown and Robert Evans, University of Miami/RSMAS.

Figure 14C shows how the temperatures and pigment concentrations changed over the 28-day period of the bloom. It is apparent that the shelf and slope waters experienced large blooms that lasted four weeks. The warm-core ring experienced two smaller bloom intervals of about ten days each separated by a storm event that broke down the thermal stratification, and the Gulf Stream area had a single short bloom of low magnitude late in April.

As would be expected from our understanding of what causes blooms to occur, the temperature of the shelf water shows a marked increase during the bloom, while the slope water warms up less dramatically. The AVHRR could sense no significant increase in temperature of the warm-core ring or Gulf Stream waters during the period. There may well have been a small increase in the water temperatures of these areas that was not detectable by the system.

The CZCS average pigment concentration for the shelf (2.8 mg/m^3) was, as would have been expected, 5.6 times greater than for the warm-core ring (0.5 mg/m^3) and 9.3 times greater than for the Gulf Stream (0.3 mg/m^3). The slope water value (3.0 mg/m^3) was, surprisingly, equivalent to that of the shelf water. When the relative depths of the euphotic zones are taken into account, the much deeper penetration of light in the clearer water of the warm-core ring and the Gulf Stream allows for a total primary productivity of the warm-core ring (529 mg/m^2/day) and the Gulf Stream (423 mg/m^2/day) that is about 50% of the productivity of the shelf (924 mg/m^2/day) and slope (917 mg/m^2/day) waters. Thus, we continue to accumulate evidence that the biological productivity of oligotrophic waters may well exceed previous estimates.

Reference: Brown, O. B., Evans, R. H., Brown, J. W., Gordon, H. R., Smith, R. C., and Baker, K. S. 1985. Phytoplankton blooming off the U.S. east coast: A satellite description. *Science* 229:4709, 163–67.

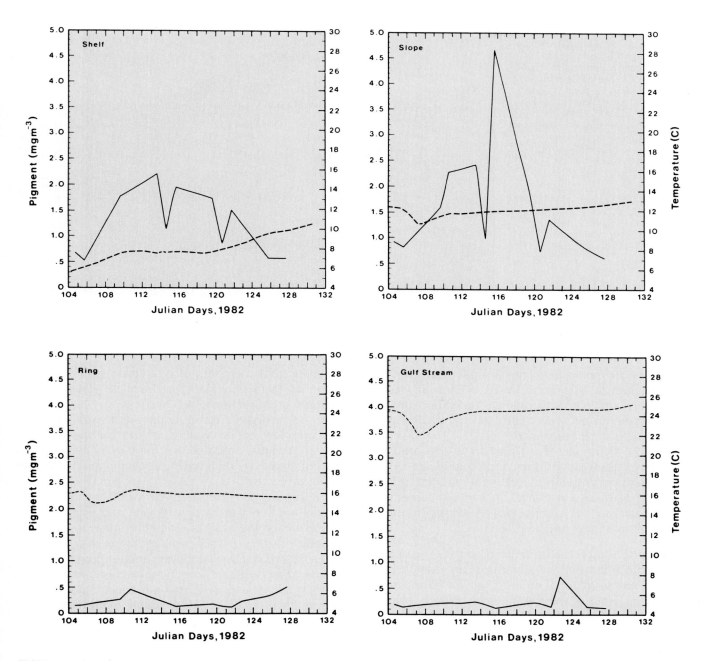

FIGURE 14C Changes in Temperature and Pigment Concentration over the 28-Day Period of the Study.

The plot of temperature (dashed line) and pigment concentration (solid line) for the warm waters of the Gulf Stream and warm-core ring at the bottom of the figure show very little change over the period of the study, whereas the cooler shelf- and slope-water plots at the top of the figure show a significant temperature increase over the period of study and pronounced "spring bloom" development. Courtesy of Otis Brown and Robert Evans, University of Miami/RSMAS.

FIGURE 14–11 Energy Flow through Photosynthetic Ecosystem.

Energy enters the system as radiant solar energy. It is converted to chemical energy through photosynthesis. Metabolism releases the chemical energy for conversion to mechanical energy and energy is lost from the biotic community as heat, which increases the entropy of the ecosystem.

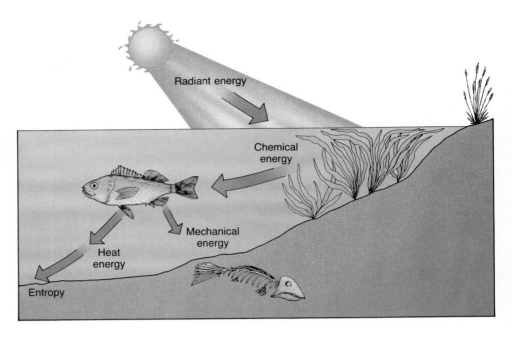

All of the elements that are naturally occurring can be assumed to be present in the oceans, at least in extremely small concentrations. However, some of these elements are in concentrations so small that we cannot detect their presence.

In view of the fact that living organisms are composed of carbohydrates, fats, and proteins, let us first concern ourselves with the elements that go into the makeup of these substances. About 20 elements in these basic organic compounds are considered to be essential to their production:

1. The **major constituents** of organic materials, those that individually make up at least 1% of organic material on a dry weight basis, are, in the order of their abundance, carbon, oxygen, hydrogen, nitrogen, and phosphorus.
2. **Secondary constituents** occurring in concentrations from 500 ppm to 10,000 ppm by weight are the elements sulfur, chlorine, potassium, sodium, calcium, magnesium, iron, and copper.
3. **Tertiary constituents** occurring in concentrations of less than 500 ppm of dry weight are boron, manganese, zinc, silicon, cobalt, iodine, and fluorine.

Considering the availability of these elements as solutes in ocean water, we find that the secondary constituents are in sufficient concentration to make it unlikely that members of this group would ever be limiting factors in plant productivity. However, the tertiary constituents, which occur in very low concentrations in organic compounds, are also found to be in very low concentrations in the marine environment. It seems probable that the tertiary constituents might create, by their absence, conditions that would limit productivity.

Returning to a consideration of the major constituents we find that carbon dioxide and water should certainly, on the basis of their concentrations, provide enough carbon, oxygen, and hydrogen to assure that these elements would never limit productivity. Nitrates and phosphates are the nutrients containing nitrogen and phosphorus that we need to consider in more detail since they do, under many conditions, limit marine productivity.

Biogeochemical Cycling

In biogeochemical cycles, elements will follow a pattern where an inorganic form is taken in by an autotrophic organism that synthesizes it into organic molecules. The organic matter follows a path through the biotic community that ends in bacterial decomposition of the organically produced compounds into the inorganic forms that may again be used as nutrients in plant production.

A New Role for Bacteria Although it has been widely accepted that zooplankton are the primary grazers of phytoplankton, new evidence indicates that free-living bacteria may consume up to 50% of the production of phytoplankton. The bacteria are thought to consume the **dissolved organic matter** that is lost from the conventional food web by three processes:

1. Phytoplankton exudate. As phytoplankton age they lose some of their cytoplasm directly into the ocean.
2. Phytoplankton "munchates." As phytoplankton are eaten by zooplankton, cytoplasm is "spilled" into the ocean.
3. Zooplankton excretions. The liquid excretions of zooplankton are dissolved in ocean water.

Free-living heterotrophic bacteria absorb this dissolved organic matter and re-enter the conventional food web primarily through the grazing of microscopic flagellates. Other larger zooplankton may be important in this process locally.

The details of the process by which free-living bacteria participate in the marine food web are still under investigation, but microbial ecologists studying the problem are convinced the evidence indicates bacteria have an important role in the transfer of energy through the marine ecosystem. Thus, bacteria may have more varied roles in the biogeochemical cycling of matter in the oceans than had previously been known.

Carbon, Nitrogen, and Phosphorus Although there is no concern that carbon concentrations in the marine environment are a limiting factor in marine productivity, we will examine the carbon cycle since this element is the basic component of organic compounds. To appreciate the excess of carbon dioxide in the world's oceans as it is related to the requirements for plant productivity, we should know that only about 1% of the total carbon present in the oceans is involved in plant productivity.

Comparatively, the soluble nitrogen compounds involved in plant productivity may be 10 times the total nitrogen compound concentration that can be measured as a yearly average. This level implies that the soluble nitrogen compounds must be cycled completely up to 10 times per year. Available phosphates may be turned over up to 4 times per year.

Comparing the ratio of carbon to nitrogen to phosphorus in dry weights of diatoms, we find that proportions are 41:7:1. This ratio is also observed in the zooplankton that feed on the diatoms, as well as in ocean water samples taken from the euphotic zone throughout the world. Thus, phytoplankton take up nutrients in the ratio in which they are available in the ocean water and pass them on to zooplankton in the same ratio. When these planktonic plants and animals die, carbon, nitrogen, and phosphorus are restored to the water in this ratio.

The Carbon Cycle This cycle involves the uptake of carbon dioxide by plants that use it in the photosynthetic process. Carbon dioxide is returned to the ocean water primarily through respiration of plants, animals, and bacteria and secondarily by autolytic breakdown of dissolved organic materials. **Autolytic decomposition** results from the action of enzymes present in the cells of organic tissue and does not require bacterial action (fig. 14–12a).

Carbon dioxide is present in the ocean in the molecular form (CO_2), as carbonic acid (H_2CO_3), as bicarbonate ions (HCO_3^-), and as carbonate ions (CO_3^{--}). In ocean waters of normal pH levels (8.1), bicarbonate ions are about 10 times more abundant than the carbonate ions, which are 10 times more abundant than the carbon dioxide molecules. Carbon dioxide is, in turn, about 100 times more abundant than carbonic acid.

A rapid rate of carbon dioxide consumption resulting from a phytoplankton increase will result in an increase in the pH as bicarbonate ions change to carbonate ions. A high rate of respiration, which produces unusually high concentrations of carbon dioxide, will drive the pH to lower levels, approaching a neutral value of 7. Extreme ranges of pH are rare in the ocean owing to the buffering activity of the carbon system that quickly restores the water to values within the 8.1 to 8.3 range. This buffering capacity of CO_2 was discussed in chapter 2.

The Nitrogen Cycle Nitrogen is primarily important in the production of **amino acids**, the building blocks of **proteins** that are synthesized by plants. These photosynthetic products are consumed by animals and free-living bacteria. They are then passed on to the saprobic (decomposing) bacteria, along with dead plant tissue, as dead animal tissue and excrement.

The **saprobic bacteria** gain energy from breaking down these compounds. This breakdown leads to the liberation of inorganic compounds, such as nitrates, that are the basic nutrient salts used by plants. Most nitrogen in the ocean is found as molecular nitrogen (N_2). The most common combined forms are **nitrite** (NO_2) and **nitrate** (NO_3). These are highly oxidized states of nitrogen. The most abundant reduced form is **ammonia** (NH_3).

Different bacteria involved in the nitrogen cycle make it somewhat complicated. Although most of the bacteria play the heterotrophic roles of consuming dissolved organic matter or converting organic compounds into inorganic salts, there are some which are able to fix molecular nitrogen into combined forms. These are **nitrogen-fixing bacteria. Denitrifying bacteria** make up another special group; their metabolism depends upon the breakdown of nitrates and the liberation of molecular nitrogen. The nitrogen cycle is graphically presented in figure 14–12B.

Autotrophic organisms of various types can use ammonium nitrogen and nitrites. However, the most important nutrient form of inorganic nitrogen, nitrate, can be utilized more efficiently by plants. Studies of nutrient cycles in various portions of the world's oceans indicate that the availability of nitrogen is clearly a limiting factor in productivity during summer months at mid- and low latitudes. This condition arises from the fact that the processes involved in converting particulate organic substance to nitrate salts through bacterial action may require up to 3 months. The conversion begins in the lower portions of the photosynthetic zone as the particular matter is sinking toward the ocean bottom.

There are three basic stages in the conversion of particulate organic matter to nitrate salts. The stages involve three distinct bacterial types. The first type reduces the particulate organic nitrogen to ammonium nitrogen, which is then acted upon by the second bacterial community, which converts it to nitrite. A third action is required to convert the nitrite to the prime organic salt species, nitrate. By the time this conversion is completed, the nitrogen compounds are usually below the euphotic zone and thus unavailable for photosynthesis. They cannot readily be returned to the euphotic zone during summer because of the strong thermostratification that exists throughout much of the ocean surface. If nitrates are to again become available to plants in these regions, their return must wait until the thermostratification disappears, allowing upwelling and mixing during the winter to effect the movement of the nutrients into the surface waters.

The Phosphorus Cycle The phosphorus cycle is simpler than the nitrogen cycle primarily because the bacterial action involved in breaking down the organic phosphorus compounds is simpler. This differ-

FIGURE 14–12 Biogeochemical Cycling.
The chemical components of organic matter enter the biological system through plant photosynthesis (or bacterial chemosynthesis). They are passed on to animal populations through feeding patterns. When the plants and animals die, their organic remains are converted to inorganic form by bacterial or other decomposition processes. In this form, they are again available for uptake by plants. _A_, carbon cycle. There is a large supply of inorganic carbon in the oceans, and it is not a limiting factor in biological productivity. Only about 1% of the carbon dissolved in the ocean is involved in plant productivity.

FIGURE 14–12 *continued.*

B, nitrogen cycle. The total amount of nitrogen fixed into organic molecules at any given time may be 10 times as great as the yearly average of soluble nitrogen compounds dissolved in ocean water. Therefore, each nitrogen atom must be recycled biogeochemically about 10 times per year. Also, the decomposition of organic nitrogen compounds back into the preferred inorganic form of nitrogen, nitrate (NO_3), requires three steps of bacterial decomposition. As a result, some of this process may not be completed until the molecules sink beneath the euphotic zone, where they are unavailable to plants. Nitrogen is considered to be the nutrient that is most likely to limit biological productivity as a result of its depletion. *C,* phosphorus cycle. Although each phosphorus atom may need to be recycled up to four times per year to maintain biological productivity in the oceans, it is seldom depleted to the point of limiting biological productivity. Two factors that help get organic phosphorus back into a form usable as nutrients by plants rather quickly are autolytic breakdown and the single step of bacterial decomposition required.

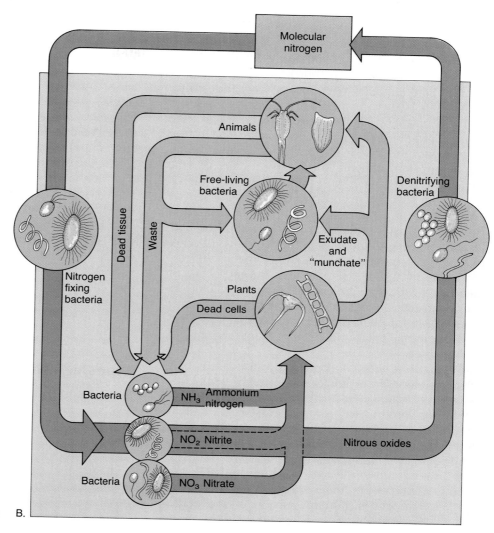

Inorganic nitrogen compounds

Organic nitrogen compounds

B.

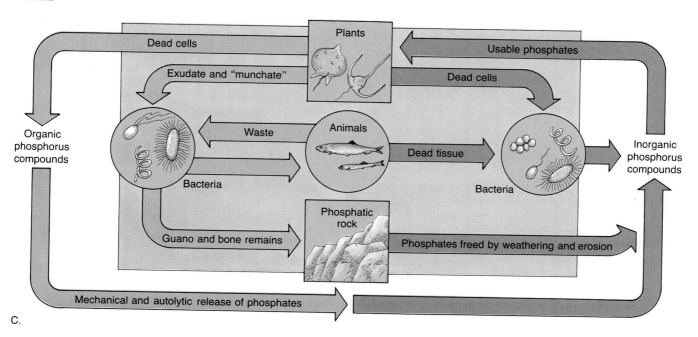

C.

ence can be studied by comparing figures 14–12B and 14–12C. The rate at which the organic compounds can be decomposed into inorganic **orthophosphates**, which are the phosphorus compounds primarily used by plants, is much more rapid than that of nitrogen breakdown. This quicker rate is due to autolytic breakdown of phosphatic organic material by enzymes and the single step required in the bacterial breakdown. As a result of the greater rate of breakdown for organic phosphorus, much of it can be completed above the oxygen compensation depth. It is therefore made available to plants within the photosynthetic zone. Although the concentration of phosphorus in the oceans is only about 1/7 that of the concentration of nitrogen, the fact that the recycling can take place within the photosynthetic zone usually allows adequate quantities of inorganic phosphorus to be available for plant productivity. The lack of phosphorus is rarely a limiting factor of plant productivity.

The Silicon Cycle As previously discussed, frustules of diatoms are composed of silica (SiO_2). Although the availability of silica can be a limiting factor in the productivity of diatoms, it is rarely a limiting factor of total primary productivity since not all phytoplankton require silica as a protective covering. There is very little likelihood that silicon will ever be a limiting nutrient in total productivity because it is very abundant in the rocks that make up the earth's crust and the small fraction of the earth's crustal silica that will eventually find its way into the sea is very great in relation to the silicon needs of the diatom population. Silicon concentrations range from unmeasurable up to 400 mg/m^3. Fluctuations in concentration roughly coincide with those observed in nitrogen and phosphorus. However, the fluctuation in silica concentrations displays a much greater amplitude than fluctuations for nitrogen and phosphorus. This condition is probably due to the fact that silica does not undergo bacterial decay and is taken directly into solution after undergoing autolytic breakdown.

TROPHIC LEVELS AND BIOMASS PYRAMIDS

Trophic Levels

The transfer of chemical energy stored in the mass of the ocean's plant population to the animal community occurs, in part, through the feeding process. Zooplankton feed as herbivores on the diatoms and other microscopic marine plants, while larger animals feed on the macroscopic algae and "grasses" found growing attached to the ocean bottom near shore. The herbivores will be fed upon by larger animals, carnivores, who in turn will be fed upon by another population of larger carnivores, and so on. Each of these feeding levels is a **trophic level**. As was discussed earlier in this chapter and illustrated in figure 14–12, an amount of secondary production equal to that of the zooplankton may be achieved by free-living bacteria.

In general, the individual members of a feeding population will be larger but not too much larger than the individual member of the population on which they feed. Although we can consider this to be generally true, there are outstanding exceptions to this condition. The blue whale, which is the largest animal known to have existed on earth, feeds upon the krill that grow to maximum lengths of 6 cm (2.4 in.).

Transfer Efficiency

In the transfer of energy between feeding, or trophic, levels we are greatly concerned with efficiency. There is a relatively high degree of variablility in the efficiency of various plant species under various laboratory conditions. As an average, we consider the percentage of light energy absorbed by plants and ultimately synthesized into the organic substances made available to herbivores to be about 2%.

The **gross ecological efficiency** at any trophic level is the ratio of energy passed on to the next higher trophic level divided by the energy received from the trophic level below. When we examine the transfer of energy from feeding population to feeding population within the ocean, we can readily see that of the biomass representing the food intake of a given population only a portion will be passed on to the next feeding level.

Figure 14–13 shows that some of the energy taken in as food by herbivores is passed by the animal as feces and the rest is assimilated by the animal. Of that portion of the energy that is assimilated, much is quickly converted through respiration to kinetic energy for maintaining life, while what remains is available for growth and reproduction. Only a portion of this mass is passed on to the next trophic level through feeding. Figure 14–14 represents diagrammatically the passage of energy between trophic levels through an entire ecosystem, from the solar energy assimilated by plants to the mass of the ultimate carnivore.

Many investigations have been conducted into the efficiency of energy transfer between trophic levels.

FIGURE 14–13 Passage of Energy through a Trophic Level.

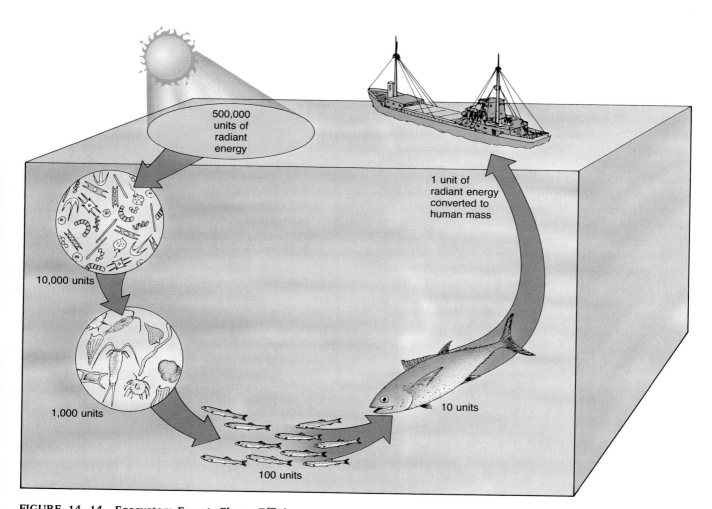

FIGURE 14–14 Ecosystem Energy Flow—Efficiency.
Based on probable efficiencies of energy transfer within the ecosystem, one unit of mass equivalent is added to the fifth trophic level (the tertiary carnivores) for every 500,000 units of radiant energy to the producers (plants). This value is based on a 2% efficiency of transfer by plants and 10% efficiency at all other levels.

There are many variables, such as the age of the animals involved, to be considered. Young animals display a higher growth efficiency than older animals. The availability of food can also alter efficiency. When food is plentiful, animals are observed to expend more energy in digestion and assimilation than when food is not readily available. Most efficiencies that have been determined range between 6% and 15%. It is widely accepted that ecological efficiencies in natural ecosystems average approximately 10%, but there is some evidence that in populations important to our present fisheries this efficiency may run as high as 20%. The true value of this efficiency is of practical importance to us because it determines the fish harvest we can anticipate from the oceans.

Biomass Pyramid

It is obvious from considering the energy losses that occur in each feeding population that there is some limit to the number of feeding populations in a food chain and that each feeding population must necessarily be smaller than the population upon which it feeds. It was previously mentioned that the individual members of a feeding population would be larger than their prey, but it seems they should also be less numerous.

Food Chains   Food chains represent a sequence of organisms through which energy is transferred from the primary producers through the herbivore and carnivore feeding populations until one feeding population does not have any predators, which marks the end of the chain. On the basis of our previous considerations of the efficiency of energy transfer between trophic levels, we see that for a population within a food chain it may be beneficial to feed as close to the primary producing population as possible. This arrangement increases the biomass available for food and the number of individuals in the population to be fed upon.

An example of an animal population that is an important fishery and usually represents the third trophic level in a food chain is the Newfoundland herring. Although some herring populations are involved in longer food chains, the Newfoundland herring feed primarily on a population of small crustaceans, copepods, that in turn feed upon diatoms (fig. 14–15).

Food Webs   It is, however, uncommon to see simple feeding relationships of the type described above in nature. More commonly the animals representing the last step in a linear food chain feed on a number of animals that have simple or complex feeding rela-

tionships, constituting a food web. The overall importance of food webs is not well understood, but one consequence for animals that feed through a web rather than a linear chain is their greater likelihood of survival if population extinctions or sharp decreases occurred within the web at or below their feeding level. Those animals involved in food webs, such as the North Sea herring illustrated in figure 14–15, are less likely to suffer from the extinction of one of the populations upon which they depend for food than the Newfoundland herring that feed only on copepods (fig. 14–5D). The extinction of the copepods in the latter case certainly would be expected to have a catastrophic effect on the herring population.

The Newfoundland herring population does, however, have an advantage over its relatives who feed through the broader-based food web. The Newfoundland herring are more likely to have a larger biomass to feed on since they are only two steps removed from the producers, while the North Sea herring represent the fourth level in some of the food chains within their web.

The ultimate effect of the transfer of energy between trophic levels can be seen in figure 14–16, which depicts the progressive decrease in numbers of individuals and total biomass at successive trophic levels as a result of decreased amounts of available energy.

A study of food webs in 55 marine ecosystems indicated an average of 4.45 trophic levels. Only two trophic levels were encountered at rocky shores in New England and Washington and at a Georgia salt marsh. Seven links were found in the longest food chains in Antarctic seas, while two tropical plankton communities in the Pacific Ocean were observed to have seven and ten trophic levels. Tropical epipelagic seas and Pacific Ocean upwelling zones had eight trophic levels. The study found that the amount of primary productivity and the degree of variability of the environment have little influence on the length of food chains within an ecosystem. The only variable that showed a clear relationship to the length of food chains was whether or not the environment was two- or three-dimensional. Three-dimensional (pelagic) marine ecosystems averaged 5.6 trophic levels compared to an average of 3.5 for two-dimensional (benthic) ecosystems.

SUMMARY

Organisms are divided into five kingdoms: Monera, single-celled organisms without a nucleus; Protista, single-celled organisms with a nucleus; Mycota, fungi; Metaphyta, many-celled plants; and Metazoa,

FIGURE 14–15 Food Chains—Food Webs.

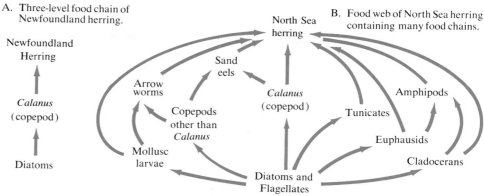

A. Three-level food chain of Newfoundland herring.

B. Food web of North Sea herring containing many food chains.

FIGURE 14–16 Biomass Pyramid.

The higher on the food chain an organism is found, the larger it is as an individual. However, the total biomass represented by a population high on the food chain is less than that for a population at a lower level.

many-celled animals. **Taxonomic classification** of organisms involves dividing the kingdoms into the increasingly specific groupings: **phylum, class, order, family, genus,** and **species.**

Protistan plants include the macroscopic algae **Phaeophyta** (kelp and *Sargassum*), **Chlorophyta** (green algae), and **Rhodophyta** (red algae). The microscopic algae include **Chrysophyta** (diatoms and coccolithophores) and **Pyrrophyta** (dinoflagellates). The more complex **Spermatophyta** are represented by a few genera of nearshore plants such as *Zostera* (eelgrass), *Phyllospadix* (surf grass), *Spartina* (marsh grass), and mangrove trees.

Plant productivity of the oceans is limited by the availability of solar radiation and of nutrients. The depth to which sufficient light penetrates to allow plants to produce only that amount of oxygen required for their respiration is the **oxygen compensation depth.** Plants cannot live successfully below this depth, which may occur at less than 20 m (65 ft) in turbid coastal waters or at a probable maximum of 150 m (492 ft) in the open ocean. Nutrients are most abundant in coastal areas owing to runoff and upwelling. In **high-latitude areas** thermoclines are generally absent, so upwelling can readily occur, and productivity is commonly limited more by the availability of solar radiation than lack of nutrients. In **low-latitude regions**, where a strong thermocline may exist year-round, productivity is limited except in areas of upwelling. The lack of nutrients is generally the limiting factor. In **temperate regions**, where distinct seasonal patterns are developed, productivity peaks in the spring and fall and is limited by lack of solar radiation in the winter and lack of nutrients in the summer.

In addition to the **primary productivity** through plant photosynthesis, organic biomass is produced through bacterial **chemosynthesis**. Chemosynthesis, recently observed on oceanic spreading centers in association with hydrothermal springs, is based on the release of energy by the oxidation of hydrogen sulfide.

Radiant energy captured by plants is converted to **chemical energy** and passed through the **biotic community**. It is expended as **mechanical** and **heat energy** and ultimately reaches a state of **entropy**, where it is biologically useless. There is, however, no loss of mass. The mass used as nutrients by plants is converted to biomass. Upon the death of organisms, the mass is decomposed to an inorganic form ready again for use as nutrients for plants. Of the nutrients required by plants, compounds of nitrogen are most likely to be depleted and restrict plant productivity. Since the total decomposition of organic nitrogen compounds to inorganic nutrients requires three stages of bacterial decomposition, these compounds may have sunk beneath the photosynthetic zone before decomposition is complete and therefore are unavailable to plants.

As energy is transferred from plants to **herbivores** and the various **carnivore** feeding levels, only about 10% of the mass taken in at one feeding level is passed on to the next. The ultimate effect of this decreased amount of energy that is passed between **trophic levels** higher in the food chain is a decrease in the number of individuals and total biomass of populations higher in the **food chain** or a combination of chains called a **food web**.

QUESTIONS AND EXERCISES

1. List the five kingdoms of organisms and the fundamental criteria used in assigning members to them.

2. Compare the macroscopic algae in terms of pigment, maximum depth in which they grow, common species, and size.

3. The Chrysophyta contains two classes of important phytoplankton. Compare their composition and the structure of their hard parts as well as their geological significance.

4. Discuss and compare the contributions of the Pyrrophyta genera *Ptychodiscus* and *Gonyaulax* to red tide development.

5. Define oxygen compensation depth, and explain the use of the dark and transparent bottle technique for its determination. Discuss how the quantity of oxygen produced by photosynthesis in each clear bottle is determined.

6. Compare the biological productivity of polar, temperate, and tropical regions of the oceans. Include a discussion of seasonal variables, thermal stratification of the water column, and the availability of nutrients and solar radiation.

7. Discuss chemosynthesis as a method of primary productivity. How does it differ from photosynthesis?

8. Describe the components of the marine ecosystem.

9. Describe the flow of energy through the biotic community, and include the forms to which solar radiation is converted. How does this flow differ from the manner in which mass is moved through the ecosystem?

10. What are the proportions by weight of carbon, nitrogen, and phosphorus in ocean water, phytoplankton, and zooplankton? Suggest how these amounts may support or refute the idea that life originated in the oceans.

11. Explain why nitrogen is much more likely than phosphorus to be a limiting factor in marine productivity.

12. How is the energy taken in by a feeding population lost so that only a small percentage is made available to the next feeding level? What is the average efficiency of energy transfer between trophic levels?

13. If a killer whale is a third-level carnivore, how much phytoplankton mass is required to add each gram of new mass to the whale? Assume 10% efficiency of energy transfer between trophic levels. Include a diagram.

14. Describe the probable advantage to the ultimate carnivore of the food web over a single food chain as a feeding strategy.

15. Discuss the effects of primary production, environmental variability, and environmental dimensionality on the length of food chains.

REFERENCES

Alldredge, A. L., and Cohen, Y. 1987. Can microscale chemical patches persist in the sea? Microelectrode study of marine snow, fecal pellets. *Science* 235:4789, 689–91.

Briand, F., and Cohen, J. E. 1987. Environmental correlates of food chain length. *Science* 238:4829, 956–59.

Craig, H., and Hayward, T. 1987. Oxygen supersaturation in the ocean: Biological versus physical contributions. *Science* 235:4785, 199–201.

George, D., and George, J. 1979. *Marine life: An illustrated encyclopedia of intervertebrates in the sea.* New York: Wiley-Interscience.

Grassle, J. F., et al. 1979. Galápagos '79: Initial findings of a deep-sea biological quest. *Oceanus* 22:2, 2–10.

Jenkins, W. J. 1982. Oxygen utilization rates in North Atlantic subtropical gyre and primary production in oligotrophic systems. *Nature* 300, 246–48.

Littler, M. M; Littler, D. S.; Blair, S. M.; and Norris, J. N. 1985. Deepest known plant life discovered on an uncharted seamount. *Science* 227:4683, 57–59.

Martinez, L.; Silver, M.; King, J.; and Alldredge, A. 1983. Nitrogen fixation by floating diatom mats: A source of new nitrogen to oligotrophic ocean waters. *Science* 221:4606, 152–54.

Parsons, T. R.; Takahashi, M.; and Hargrave, B. 1984. *Biological oceanographic processes,* 3rd ed. New York: Pergamon Press.

Pearl, H. W., and Begout, B. M. 1988. Direct measurement of O_2-depleted microzones in marine *Oscillatoria:* Relation to N_2 fixation. *Science* 241:4864, 442–45.

Platt, T., and Sathyendranath, S. 1988. Oceanic primary production: Estimation by remote sensing at local and regional scales. *Science* 241:4873, 1613–19.

Platt, T., Subba Rao, D. V., and Irwin, B. 1983. Photosynthesis of picoplankton in the oligotrophic ocean. *Nature* 301:5902, 702–4.

Rivkin, R.; Bosch, I.; Pearse, J.; and Lessard, E. 1986. Bacteriovory: A novel feeding mode for asteroid larvae. *Science* 233:4770, 1311–13.

Russell-Hunter, W. D. 1970. *Aquatic productivity.* New York: Macmillan.

Sherr, B. R., Sherr, E. B., and Hopkins, C. S. 1988. Trophic interactions within pelagic microbial communities; Indications of feedback regulation of carbon flow. *Hydrobiologia* 159:1, 19–26.

Shulenberger, E., and Reid, J. L. 1981. The Pacific shallow oxygen maximum, deep chlorophyll maximum, and primary productivity reconsidered. *Deep Sea Research* 28A:9, 901–919.

SUGGESTED READING

Sea Frontiers

Arehart, J. L. 1972. Diatoms and silicon. 18:2, 89–94.
A very readable description of the important role of silicon and other elements in the ecology of diatoms, including microphotographs showing the varied forms of diatoms.

Coleman, B. A., Doetsch, R. N., and Sjblad, R. D. 1986. Red tide: A recurrent marine phenomenon. 32:3, 184–91.
The problem of periodic red tides along the Florida, New England, and California coasts is discussed.

Ebert, C. H. V. 1978. El Niño: An unwanted visitor. 24:6, 347–51.
A description of the effects of El Niño and the attempts that are being made to understand its causes.

Idyll, C. P. 1971. The harvest of plankton. 17:5, 258–67.
An interesting discussion of the potential of zooplankton as a major fishery.

Jensen, A. C. 1973. WARNING—Red tide. 19:3, 164–75.
An informative discussion of what is known of the cause, nature, and effect of red tides.

Johnson, S. 1981. Crustacean symbiosis. 27:6, 351–60.
A description of various symbiotic relationships entered into by tropical shrimps and crabs.

McFadden, G. 1987. Not-so-naked ancestors. 33:1, 46–51.
The nature of the coverings of marine phytoplankton cells is revealed by the electron microscope.

Oremland, R. S. 1976. Microorganisms and marine ecology. 22:5, 305–10.
The role of such microorganisms as phytoplankton and bacteria in cycling matter in the oceans is discussed.

Philips, E. 1982. Biological sources of energy from the sea. 28:1, 36–46.
The potential for converting marine biomass to energy sources useful to society is discussed.

Scientific American

Benson, A. A. 1975. Role of wax in oceanic food chains. 232:3, 76–89.
   A report on the findings from observations made of the content of wax in the bodies of many marine animals, from copepods to small deep-water fishes, and their implications.

Childress, J. J., Feldback, H., and Somero, G. N. 1987. Symbiosis in the deep sea. 256:5, 114–21.
   Deep-sea hydrothermal vent animals have a symbiotic relationship with sulfur-oxidizing bacteria that allows them to live in the darkness of the deep ocean.

Coleman, G., and Coleman, W. J. 1990. How plants make oxygen. 262:2, 5–67.
   Photosynthesis is thought to be the source of atmospheric oxygen. The method by which plants produce oxygen is considered.

Levine, R. P. 1969. The mechanism of photosynthesis. 221:6, 58–71.
   Reveals what is known of the process by which energy is captured by plants and converted to useful forms of chemical energy while freeing oxygen to the atmosphere.

Pettitt, J.; Ducker, S.; and Knox, B. 1981. Submarine pollination. 244:3, 134–44.
   Discusses the pollination of sea grasses by wave action.

15
ANIMALS OF THE PELAGIC ENVIRONMENT

In the sunlit waters of the pelagic environment live the phytoplankton that provide the basis for nearly all of the marine biomass through their photosynthetic activities. This important marine population was discussed in chapter 14, so our emphasis here will be on the animal populations. The discussion will deal primarily with the adaptations that allow them to live successfully in the pelagic environment.

STAYING ABOVE THE OCEAN FLOOR

Since the tissues of the zooplankton and nekton (muscle, cartilage, scales, bone, and shell) are more dense than ocean water, pelagic animals can remain off the bottom only through the application of buoyancy or energy. Chapter 13 discussed how microplankton's decreased size aided in this process by an increased frictional resistance to sinking. However, the larger animals we will discuss in this chapter need additional means to remain off the ocean floor.

Gas Containers

Since air is approximately 0.001 the density of water at sea level, a small amount of it inside an organism significantly increases its buoyancy. Some cephalopods have rigid gas containers (fig. 15–1). The genus *Nautilus* has an external shell; the cuttlefish *Sepia* and deep-water squid *Spirula* have an internal chambered structure. These animals become neutrally buoyant and can maintain their positions in the water easily. Since the air pressure in their air chambers is always 1 atmosphere, they are limited in the depth to which they may venture. The *Nautilus* must stay above a depth of approximately 500 m (1640 ft), or its chambered shell will collapse as the external pressure approaches 50 atmospheres. The *Nautilus* is rarely observed below a depth of 250 m (820 ft).

Neutral buoyancy is achieved by some slow-moving bony fish through filling a gas bladder, or *swim bladder,* with gases. The swim bladder is normally not present in very active swimmers, such as the tuna, or in fish that live on the bottom. Some fish have a **pneumatic duct** that connects the swim bladder to the esophagus (fig. 15–2). These fish can add or remove air through this duct. In other fish, the gases of the swim bladder must be added or removed more slowly by an interchange with the blood. Since change in depth will cause the gas in the swim bladder to expand or contract, the fish removes or adds gas to the bladder to maintain a constant volume. Those fish without the pneumatic duct are limited in the rate at which they can make these adjustments and, therefore, cannot withstand rapid changes in depth.

The composition of gases in the swim bladders of shallow-water fishes is similar to that of the atmosphere. With increasing depth (fish with swim bladders have been taken from a depth of 7000 m, or 22,960 ft) the concentration of oxygen in the swim bladder gas increases from the 20% common near the surface to more than 90%. At the 700-atmosphere pressure that exists at a depth of 7000 m, gas is compressed to a density of 0.7 gm/cm^3. This is about the density of fat, so many deep-water fish have air bladders filled with fat instead of compressed gas.

Floating Forms

Floating at the surface are some relatively large members of the plankton, the evolutionarily primitive siphonophore and scyphozoan **coelenterates,** as well as the pelagic **tunicates.** Somewhat smaller are the **ctenophores** and **arrowworms.** They all have soft gelatinous bodies with little if any hard tissue. A major strategy of this body type is the replacement in body fluids of heavy sulfate ions with chloride ions to maintain osmotic equilibrium.

The **siphonophore** coelenterates are represented in all oceans by the genera *Physalia* (Portuguese man-of-war) and *Velella* (by-the-wind sailor). Their gas floats, pneumatophores, serve as floats and sails that allow the wind to push these colonial forms across the ocean surface (fig. 15–3). A colony of tiny sea-anemonelike **polyps** and jellyfishlike **medusae** (fig. 15–3B) are suspended beneath the float. The long tentacles of the polyps primarily capture food, and the medusae carry on the sexual reproduction required to create a new colony. The tentacles of the *Physalia* may be many meters long and possess **nematocysts** long enough to penetrate the skin of humans; they have been known to inflict a painful and occasionally dangerous neurotoxin poisoning. The colonies grow from the initial polyp through asexual budding.

The **scyphozoans,** or jellyfish, have a medusoid, bell-shaped body with a fringe of tentacles and a mouth at the end of a clapperlike extension hanging down under the bell-shaped float. Ranging in size from near microscopic to 2 m (6.6 ft) in diameter with 60-m (197-ft)

Preceding page: The humpback whale has aroused an especially large amount of interest with its "songs" and spectacular breaching behavior. Photo by Christopher Newbert.

380

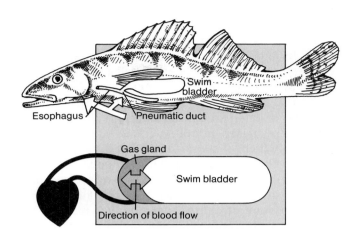

FIGURE 15–1 Gas Containers of the Cephalopods *Nautilus, Sepia,* and *Spirula*.

The *Nautilus* has an external chambered shell, and the *Sepia* and *Spirula* have internal chambered structures that can be filled with gas to provide buoyancy.

FIGURE 15–2 Swim Bladder of Some Bony Fishes.

Example of a swim bladder connected to the esophagus by the pneumatic duct, allowing air to be added or removed rapidly. In fish with no pneumatic duct, all gas must be added or removed through the blood. This requires more time and is achieved by a network of capillaries associated with the gas gland.

FIGURE 15–3 Planktonic Coelenterates.

A, Physalia and Jellyfish. *B,* medusa. Photo by Larry Ford.

tentacles, most jellyfish have bells with a diameter of less than 0.5 m (1.6 ft). Jellyfish are capable of movement through muscular contractions (*see* fig. 15–3A). Water enters the cavity under the bell and is forced out by contractions of muscles that circle the bell, jetting the animal ahead in short spurts. To allow the animal to swim generally in an upward direction, there are sensory organs spaced around the outer edge of the bell. These may be light-sensitive **ocelli** or gravity sensitive **otocysts**. This orientation ability is important because the jellyfish feed by swimming to the surface and sinking slowly through the life-rich surface waters.

Pelagic **tunicates** are generally transparent and barrel-shaped, with an anterior incurrent and posterior excurrent opening (fig. 15–4). They move by a feeble form of jet propulsion effected by the contraction of bands of muscles that force water out the excurrent opening. Reproduction of solitary forms is complicated, involving alternate sexual and asexual reproduction. The genus *Salpa* includes solitary forms reaching a length of 20 cm (8 in.) and smaller aggregate forms that produce new individuals by budding. Individual members of an aggregate chain may be 7 cm (2.8 in.) long, and the chain of newly budded members may reach great lengths (*see* fig. 15–4B). The genus *Pyrosoma* is luminescent and colonial. Individual members of the colony have their incurrent openings facing the outside surface of a tube-shaped colony that may be a few meters long. One end of the tube is closed, and the excurrent openings of the thousands of individuals all empty into the tube. Muscular contraction forces water out the open end to provide a means of propulsion.

Ctenophores are comb-bearing animals closely related to the coelenterates in that their radially symmetrical bodies develop around their mouths. (Some forms, however, also possess an anal opening from the digestive cavity. Often referred to as **comb jellies**, this group of animals is entirely pelagic and confined to the marine environment.) The body form of most members of the phylum Ctenophora is basically spherical, with eight rows of cilia spaced evenly around the sphere. It is from these structures that the name *comb-bearing* is derived; when magnified, they look like miniature combs. If a ctenophore possesses tentacles, there will be two that contain adhesive organs instead of stinging cells to capture prey. The genera *Pleurobranchia* (sea gooseberries) (fig. 15–4C) and the pink, elongated *Beroe* (fig. 15–4D) range in size from the gooseberry dimension of the former to over 15 cm (6 in.) in the latter.

The arrowworms (chaetognaths—bristle jawed) are transparent and difficult to see, although they may grow to more than 2.5 cm (1 in.) in length (fig. 15–5). The name *chaetognath* refers to the hairlike attachments around the mouth, which are used to grasp prey while they devour it. They are voracious feeders, eating primarily the small zooplankton; in turn, they are eaten by fishes and larger planktonic animals such as jellyfish. These exclusively marine, hermaphroditic animals are usually more abundant in the surface waters some distance from shore.

Swimming Forms

This section examines larger pelagic animals, the *nekton,* with substantial powers of locomotion. They include rapid-swimming invertebrate squids, as well as fish and marine mammals.

Swimming squid include the genera *Loligo* (common squid), *Ommastrephes* (flying squid), and the *Architeuthis* (giant squid). Active predators of small fish, the smaller varieties of squid have long slender bodies with paired fins (fig. 15–6). Unlike the less-active *Sepia* and *Spirula,* they have no hollow chambers in their bodies and therefore require more energy to remain in the upper water of the oceans. Trapping water in their mantle cavities and forcing it out through a siphon, these invertebrates can swim about as fast as any fish their size. Two long arms with pads containing suction cups at the ends are used to capture prey, and eight shorter arms containing suckers take the prey to the mouth, where it is crushed by a beaklike mouthpiece.

The consideration of locomotion in fish is more complex than that of the squid because of body motion and the role of fins. The basic movement of swimming fish is the passage of a wave of lateral body curvature from the front to the back of the fish. This is achieved by the alternate contraction and relaxation of muscle segments, **myomeres**, along the sides of the body. The backward pressure of the fish's body and fins produced by the movement of this wave provides the forward thrust (fig. 15–7).

Most active swimming fish have two sets of paired fins, **pelvic** and **pectoral**, used in maneuvers such as turning, braking, or balancing. When they are not in use, they can be folded against the body. Vertically oriented fins, **dorsal** and **anal**, serve primarily as stabilizers. The fin most important in propelling the high-speed fish is the tail, or **caudal**, fin. Caudal fins flare dorsally and ventrally to increase the surface area available to develop thrust. Increased surface area also increases frictional drag. The efficiency of the design of a caudal fin depends on its shape and can be expressed mathematically as an **aspect ratio**, calculated as follows:

$$\frac{(\text{fin height})^2}{\text{fin area}}$$

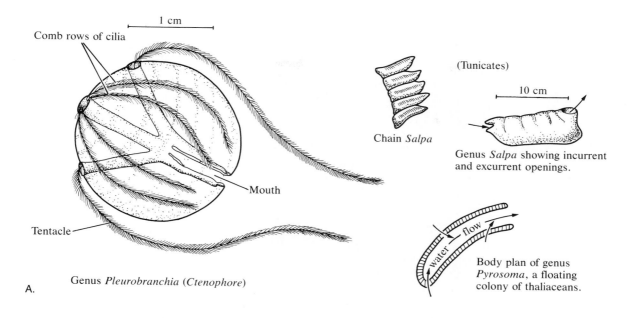

Genus *Pleurobranchia* (Ctenophore)

A.

Body plan of genus *Pyrosoma*, a floating colony of thaliaceans.

Genus *Salpa* showing incurrent and excurrent openings.

Chain *Salpa*

(Tunicates)

B.

C.

D.

FIGURE 15–4 Pelagic Tunicates and Ctenophores.

A, body structure. *B*, chain of salps. *C*, ctenophore, *Pleurobranchia*. *D*, ctenophore. *Beroe*. Photos *B*, *C*, and *D* by James M. King, Graphic Impressions.

A.

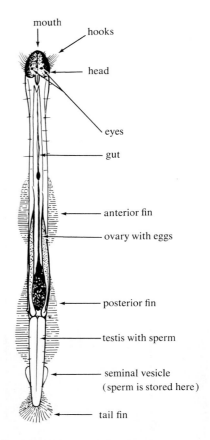

B. General Anatomy of an Arrowworm

FIGURE 15—5   Arrowworm.

A, head of chaetognath *Sagitta tenuis* from Gulf of Mexico, magnified 161 times. Photo courtesy of Howard J. Spero. *B,* adult chaetognaths reach lengths ranging from 1 to 5 cm (0.4 to 2 in.).

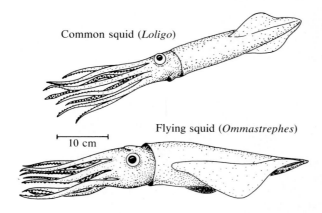

FIGURE 15—6   Squids.

There are five basic shapes of caudal fins: **rounded** (as on sculpin), **truncate** (bass), **forked** (yellowtail), **lunate** (tuna), and **heterocercal** (shark). The rounded fin, with an aspect ratio of approximately 1, is flexible (fig. 15—8) and useful in accelerating and maneuvering at slow speeds. The somewhat flexible truncate (ratio 3) and forked (ratio 5) tails will be found on

faster fish and may still be used for maneuvering. The lunate caudal fin, with an aspect ratio of up to 10, is found on the fast-cruising fishes such as tuna, marlin, and swordfish; it is very rigid and useless in maneuverability but very efficient in propelling. The heterocercal caudal fin is asymmetrical, with most of its mass and surface area in the dorsal lobe.

The heterocercal fin produces a significant lift to sharks as it is moved from side to side. This lift is important because sharks have no swim bladder and tend to sink when they stop moving. To aid this lifting, the pectoral fins are large and flat. Located on the anterior of the shark's body like the wings of an airplane, they serve as hydrofoils that lift the front of the shark's body to balance the posterior lift applied by the caudal fin. The shark loses a lot in maneuverability as a result of this modification of the pectoral fins.

Pectoral fins in many fish do more than just help maneuvering (fig. 15—9). Wrass and sculpins use them to "row" themselves through the water in jerky motions while the rest of the body remains motion-

**FIGURE 15–7 Fish—
Swimming Motions and Fins.**

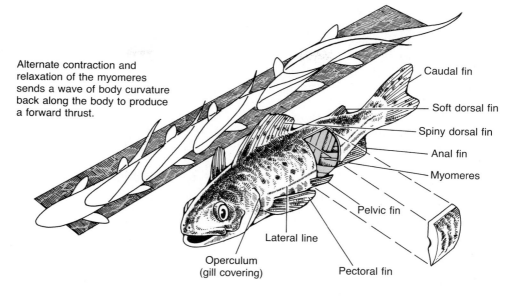

Alternate contraction and
relaxation of the myomeres
sends a wave of body curvature
back along the body to produce
a forward thrust.

Caudal fin

Soft dorsal fin

Spiny dorsal fin

Anal fin

Myomeres

Pelvic fin

Lateral line

Operculum
(gill covering)

Pectoral fin

**FIGURE 15–8 Caudal Fin
Shapes and Aspect Ratios.**

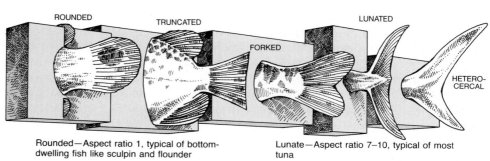

ROUNDED   TRUNCATED   LUNATED

FORKED

HETERO-
CERCAL

Rounded—Aspect ratio 1, typical of bottom-
dwelling fish like sculpin and flounder

Truncate—Aspect ratio 3, typical of salmon
and bass

Forked—Aspect ratio 5, typical of herring and
yellowtail

Lunate—Aspect ratio 7–10, typical of most
tuna

Heterocercal—Aspect ratio varies, typical of
sharks that need an upward thrust from their
swimming motion

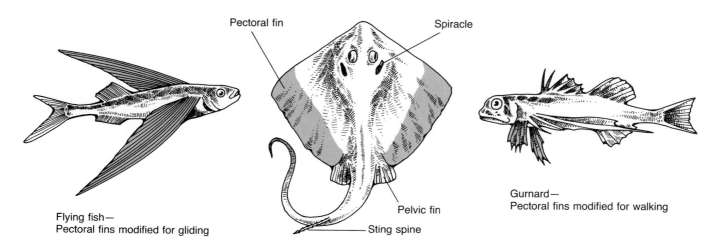

Pectoral fin

Spiracle

Pelvic fin

Sting spine

Flying fish—
Pectoral fins modified for gliding

Stingray—
Pectoral fins modified for swimming

Gurnard—
Pectoral fins modified for walking

FIGURE 15–9 Pectoral Fin Modifications.

less. Skates and rays swim by sending undulating motions across the greatly modified pectoral fins, and manta rays flap them like wings. The flying fish, *Exocoetus,* uses greatly enlarged pectoral fins to glide for up to 400 m (1312 ft) across the ocean surface after propelling itself with the enlarged ventral lobe of its caudal fin into the air to escape dolphins and other predators. Other fish have pectoral fins modified into fingerlike structures used to walk on the ocean floor.

Many fish in no great hurry propel themselves by undulation of the dorsal fin. Examples are triggerfish, sea horses, ocean sunfish, and trunkfish. The sunfish also uses its anal fin and stubby caudal fin to aid in the process, still without impressive results.

Modifications in Swimming Behavior

Some fish spend most of their time waiting patiently for prey and exert themselves only in short bursts as they lunge at the prey; others cruise relentlessly through the water seeking out prey. There is a marked difference in the nature of the musculature of fishes with these different styles of seeking out food. A grouper, representative of the **lungers**, has a truncate caudal fin (aspect ratio 3), and almost all its muscle tissue is *white*. Tuna, on the other hand, are **cruisers**, and less than half their muscle tissue is white; it is mostly of a variety referred to as *red* (fig. 15–10).

Red muscle fibers are much smaller in diameter (25–50 μm, or 0.01–0.02 in.) than **white muscle fibers** (135 μm, or 0.05 in.) and contain higher concentrations of **myoglobin**, a red pigment with an affinity for oxygen (O_2). These factors allow the red fibers to obtain a much greater oxygen supply than is possible for white fibers and therefore achieve a metabolic rate six times that of white fibers. The red muscle tissue is abundant in cruisers that constantly swim. The lungers can get along quite well with little red tissue because they need not constantly move. White tissue, which fatigues much more rapidly than red tissue, is used by the tuna for short periods of acceleration while on the attack. It is also quite adequate for propelling the grouper and other lungers during their quick passes at prey.

The **speed of fish** is thought to be closely related to body length. For tuna, well-adapted for sustained cruising and short bursts of high-speed swimming, cruising speed averages about 3 body lengths/second; maximum speed of about 10 body lengths/second can be maintained for only one second. A yellowfin tuna, *Thunnus albacares,* has been clocked at 74.6 km/hr (46 mi/h), or more than 20 body lengths/second, over a period of 0.19 second. Theoretically, a 4-m (13-ft) bluefin tuna, *Thunnus thynnus,* could reach speeds of up to 143.9 km/h (89.2 m/h). Many of the toothed whales are known to be capable of high rates of speed. The porpoise *Stenella* has been clocked at 40 km/h (25 mi/h), and it is believed that the top speed of killer whales may be in excess of 55 km/h (34 mi/h).

As is true with many chemical reactions, metabolic processes can proceed at greater rates at higher temperatures. Although fish are generally considered to be **poikilothermic** (cold-blooded), having body temperatures that conform to the temperature of their environment, there are some fast swimmers with body temperatures above that of the surrounding water. Some are only slightly above the temperature of the water: the mackerel (*Scomber*), yellowtail (*Seriola*), and bonito (*Sarda*) have body temperature elevations of 1.3°, 1.4°, and 1.8°C (2.3°, 2.5°, and 3.2°F), respectively. Other fast swimmers that have much higher elevations are members of the mackerel shark genera, *Lamna* and *Isurus,* as well as the tuna, *Thunnus.* **Homeothermic** bluefin tuna have been observed to maintain a body temperature of 30°–32°C (86°–90°F), regardless of the temperature of the water in which they are swimming. Although it is more commonly found in warmer water where the temperature difference is no more than 5°C (9°F), body temperatures of 30°C (86°F) have been measured in bluefin tuna swimming in 7°C (45°F) water.

Why do these fish exert so much energy to maintain their body temperatures at high levels when other fishes do quite well with ambient body temperatures? Their mode of behavior is that of a cruiser, and any adaptation (high temperature and metabolic rate) that can increase the power output of their muscle tissue helps them search out and capture prey.

The mackerel sharks and tuna are aided in conserving the heat energy needed to maintain their high body heats by a modification of the circulatory system (fig. 15–11). While most fish have a **dorsal aorta**, just beneath the vertebral column, that provides blood to the swimming muscles, the mackerel sharks and tuna have additional **cutaneous arteries** just beneath the skin on either side of the body. As cool blood flows into red muscle tissue, its temperature is increased by heat generated by muscle metabolism (muscle contractions). A fine network of tiny blood vessels within the muscle tissue is designed to minimize heat loss. The vessels that return the blood to the **cutaneous vein**, parallel to the cutaneous artery

A.

B.

FIGURE 15–10   Feeding Styles—Lunger and Cruiser.

A, lungers like the grouper sit patiently on the bottom and capture prey with quick, short lunges. Their muscle tissue is predominantly white. Photo courtesy of Scripps Institution of Oceanography, University of California, San Diego. *B, cruisers* like the tuna swim constantly in search of prey and capture it with short periods of high-speed swimming. Their muscle tissue is predominantly red. Photo courtesy of National Marine Fisheries Service.

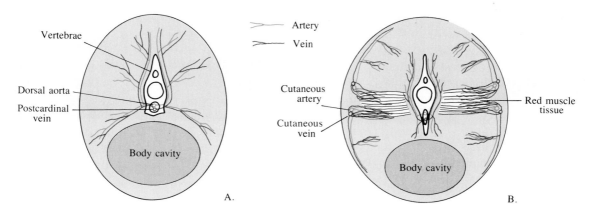

FIGURE 15–11   Circulatory Modifications in Warm-Blooded Fishes.

A, most cold-blooded (poikilothermic) fishes have major blood vessels arranged in the pattern represented on the left. All blood flows to muscle tissue from the dorsal aorta and returns to the postcardinal vein beneath the vertebral column. *B,* warm-blooded (homeothermic) fishes like the tuna have cutaneous arteries and veins that help maintain high blood temperature by using heat energy generated by contracting muscle tissue. Its blood vessel pattern is represented on the right.

A. Manatee

B. Elephant seals

C. Fur seal

D. Sea otter

FIGURE 15—12   Marine Mammals—Manatee, Pennipeds, and Sea Otter.
A, photo by Eric Hanaver. *B* and *C,* photos by Tim Ragen. *D,* photo from Scripps Institution of Oceanography, University of California, San Diego.

along the side of the fish, are all paired with small vessels carrying blood into the muscle tissue. In this way, the warm blood leaving the tissue helps to heat the cooler blood entering from the cutaneous artery.

MARINE MAMMALS

Ocean mammals include the coast-dwelling herbivores, the **sirenans** (sea cows—dugongs and manatees), and a variety of carnivorous forms (fig. 15—12). Spending all or most of their lives in coastal waters and coming ashore to breed are the **sea otter** and **pinnipeds** (sea lions, seals, and walruses). The truly oceanic mammals are the **cetaceans**—whales, porpoises, and dolphins (fig. 15—13). Since mammals evolved from reptiles on land some 200 million years ago and no marine forms are known earlier than 60 million years ago, it is believed that all marine forms evolved from some ancient land-dwellers.

Cetaceans

Well-known as the mammals best adapted to life in the oceans, the 75-plus species of cetaceans are of two basic types. The toothed whales, the **odontoceti,** include **sperm whales, porpoises,** and **dolphins.** The baleen whales, the **mysticeti,** probably evolved from the toothed whales some 30 million years ago. In place of teeth, they have baleen, plates of horny material, that hang down from the upper jaw to serve as a sieve. The body of cetaceans is more or less cigar-shaped, nearly hairless, and insulated with a thick layer of blubber. The forelimbs are modified into flippers that may be moved only at the "shoulder" joint. The hind limbs are vestigial, not attached to the rest of the skeleton and not externally visible. The skull is highly modified, with one nasal opening (toothed) or two openings (baleen) near the top. They propel themselves by vertical movements of a horizontal fluke (tail fin).

The baleen whales include four families. The **gray whale** has short, coarse baleen, no dorsal fin, and only two to five ventral grooves beneath the lower jaw. The **rorqual whales** have short baleen and many ventral grooves and are divided into two subfamilies. The **balaenopterids**, with long, slender bodies, small sickle-shaped dorsal fins, and flukes with smooth edges, include the **minke, Baird's, Bryde's, sei, fin,** and **blue whales**. The **megapterids,** or **humpback whales,** have a more robust body, long flippers, flukes with uneven trailing edges, tiny dorsal fins, and tubercles on the head. The fourth family includes the **right whales,** with long, fine baleen, broad triangular flukes, no dorsal fin, and no ventral grooves. The **"northern" right whale** is the baleen whale that is most threatened by extinction. The **southern right whale** of the Southern Hemisphere and the **bowhead whale** that remains near the Arctic pack-ice edge are the other members of this family. The smallest and least known of the baleen whales is the **pigmy right whale**. It is rare and found only in the cooler temperate waters of the Southern Hemisphere.

The toothed whales are predators that feed mostly on smaller fish and squid, although the killer whale is known to feed on a variety of larger animals, including other whales. The baleen whales feed primarily by filtering crustaceans at depths ranging from the surface down to and including the sediment of shallow ocean basins.

Modifications to Increase Swimming Speed

Cetaceans' muscles are not vastly more powerful than those of other mammals, so it is believed that their ability to swim at high speed must result from modifications that reduce frictional drag. To illustrate the importance of streamlining in reducing the energy requirements of swimmers, a small dolphin would require muscles five times more powerful than it has to swim at 40 km/h (25 mi/h) in turbulent flow than in streamlined laminar flow. In addition to having a streamlined body, cetaceans are believed to achieve near-laminar flow of water across their bodies with the aid of a specialized skin structure. The skin is composed of two layers: a soft outer layer that is 80% water and has narrow canals filled with spongy material and a stiffer inner layer composed mostly of tough connective tissue. The soft layer tends to reduce the pressure differences at the skin–water interface by compressing under regions of higher pressure and expanding in regions of low pressure.

Modifications to Allow Deep Diving

Where humans can free-dive to a maximum recorded depth of about 100 m (328 ft) and hold their breaths in rare instances for 6 minutes, the sperm whale, *Physeter,* is known to dive deeper than 2200 m (7216 ft), and the North Atlantic bottle-nosed dolphin, *Hyperoodon ampullatus,* can stay submerged for up to two hours.

One reason why whales can make deep dives and stay submerged for long periods is their ability to alternate between periods of **eupnea** (normal breathing) and **apnea** (cessation of breathing). The periods of apnea occur while the animal is submerged. To understand how some cetaceans are able to go for long periods without breathing requires a general knowledge of the lungs and their associated structures. Air that is inhaled passes through the **trachea,** into the **bronchi,** and through a series of smaller **bronchioles** and **alveolar ducts,** to tiny terminal chambers, the **alveoli** (fig. 15–14). The alveoli are lined by a thin **alveolar membrane** that is in contact with a dense bed of capillaries; the exchange of gases between the inspired air and the blood occurs across the alveolar membrane. Some cetaceans have an exceptionally large concentration of capillaries surrounding the alveoli, which have muscles that move air against the membrane by repeatedly contracting and expanding. Cetaceans take from 1 to 3 breaths per minute while resting, compared to about 15 in humans. Because of the longer time inhaled breath is held, the large capillary mass in contact with the alveolar membrane, and the circulation of the air by muscular action, many cetaceans can extract almost 90% of the O_2 in each breath, compared to the 4% to 20% extracted by terrestrial mammals.

To use this large amount of O_2, which can be taken into the blood efficiently during long periods under water, cetaceans may apply two strategies. They (1) store large amounts of O_2 and (2) can reduce oxygen use. The storage of so much O_2 is possible because prolonged divers have a much greater blood volume per unit of body mass than those that dive for only short periods. Compared to terrestrial animals, some cetaceans have twice as many red blood cells per unit of blood volume and up to 9 times as much myoglo-

FIGURE 15–13 Baleen and Toothed Whales.
No creature in the sea captures the human imagination like the whales, huge mammals that returned to the sea about 60 million years ago. Two suborders are present in today's oceans, the toothed whales (Odontoceti) and baleen whales (Mysticeti). Represented here are members of each group drawn to scale. The toothed whales—bottle-nosed dolphin, narwhal, killer whale, and sperm whale—are active predators and may track down their prey by use of echolocation. The baleen whales—humpback whale, right whale, and blue whale—fill their mouths with water and force it

out between long baleen slats that hang down from the upper jaw. Small fish, krill, and other plankton are trapped behind the baleen. The Pacific gray whale has short baleen slats and feeds by sucking in sediment from the shallow bottom of its North Pacific feeding grounds. Thus it strains out benthic amphipods and other invertebrates. Although many great whale populations were pushed to near extinction by nineteenth- and twentieth-century whaling efforts, it appears all will survive this threat. The major threat to their survival may now be the disruption of their feeding and breeding grounds.

FIGURE 15-14 Cetacean Modifications to Allow Prolonged Submergence.
A, basic lung design. Air enters the lung through the trachea, and oxygen is absorbed into the
blood through the walls of the alveoli. *B,* oxygen exchange in the alveolus. A dense mat of capillaries receives oxygen through the alveolar membrane. Because air is normally held in the lungs of
cetaceans for up to 1 minute before exhaling, as much as 90% of the oxygen can be extracted.
Whales have up to twice as many red blood cells and 9 times as much myoglobin in their muscle
tissue as terrestrial mammals in which to store this oxygen.

bin in the muscle tissue. Thus large supplies of O_2 can be chemically stored in the **hemoglobin** of the red blood cells and the myoglobin.

Additionally, the muscles that start a dive with a significant O_2 supply can continue to function through **anaerobic glycolysis** when the O_2 is used up. The muscle tissue is relatively insensitive to high levels of carbon dioxide, a product of aerobic respiration, and lactic acid, a product of anaerobic respiration.

Because the swimming muscles can function without oxygen during a dive, they and other organs, such as the digestive tract and kidneys, may be sealed off from the circulatory system by constriction of key arteries. The circulatory system services primarily the heart and brain. Because of the decreased circulatory requirements, the heart rate can be reduced by 20% to 50% of its normal rate. However, recent research has shown that no such reduction in heart rate occurs when the common dolphin *(Delphinus delphis),* the white whale *(Delphinapterus leucas),* and the bottle-nosed dolphin *(Tursiops truncatus)* dive.

Another problem with deep and prolonged dives is the absorption of compressed gases into the blood. When humans make dives using compressed air, the prolonged breathing of compressed air, particularly N_2, can result in **nitrogen narcosis, or decompression sickness** (the bends). Nitrogen narcosis is an effect similar to drunkenness, and it can occur when a diver either goes too deep or stays too long at depths greater than 30 m (98 ft). If a diver surfaces too rapidly, the lungs cannot remove the excess gases fast enough, and the reduced pressure may cause small bubbles to form in the blood and tissue. The bubbles interfere with blood circulation, and the resulting decompression sickness can cause excruciating pain, severe physical debilitation, or even death.

Cetaceans and other marine mammals do not suffer from these difficulties. Their main defense against absorbing too much N_2 seems to be a more flexible rib cage. By the time a cetacean has reached a depth of 70 m (230 ft), the rib cage has collapsed under the 8 atmospheres of pressure. The lungs within the rib cage also collapse, removing all air from the alveoli. Since most absorption of gases by the blood occurs across the alveolar membrane, the blood cannot absorb the compressed gases, and the problem of nitrogen narcosis is avoided.

It is, however, possible that the collapsible rib cage is not the main defense against the bends. A recent research effort designed to increase the nitrogen concentration in the lungs of a trained dolphin and give it the bends put enough nitrogen into the tissue of the animal to give a human a severe case of the bends. The dolphin suffered no ill effects. This whale, as well as other species, may have simply evolved an insensitivity to this gas not present in other mammals.

Use of Sound It has long been known that cetaceans, none of which possess vocal chords, make a variety of sounds. Speculations that the sounds may represent aspects of behavior from echolocation to a highly developed language have been presented. To the present, what we do know about the use of sound by cetaceans is limited.

The baleen whales produce sounds at frequencies generally below 5,000 Hz. They are believed to be produced by air passing through the trachea. Gray whales produce pulses that could conceivably be used for echolocation and moans that may serve to maintain contact with other gray whales.

The rorqual whales produce moans that last from one to many seconds. These sounds are in the 10 to 20 Hz range and are probably used to communicate over distances of up to 50 km (31 mi). There is some evidence that the moans of fin whales may be related to reproductive behavior, but much more observation is required before much can be said about the message content of the full range of sounds. The repertoire of sounds produced by humpback whales ranges from chirps to moans. Repeated sequences of sounds termed "songs" are produced primarily by adult males on breeding grounds and possibly communicate the males' readiness to mate and challenge other males.

Right whales and bowhead whales, the "chatterboxes" of the whale population, generate complex suites of moans during feeding and migration and on breeding grounds. Some sounds have been correlated with specific social behavior patterns. For example, the bowhead males produce "songs" that may be similar in function to those of the humpback males.

High-pitched and more-varied sounds are produced by toothed whales. The sounds fall into two categories: clicks and whistles. Both are used for communication, and some clicks are used for echolocation in which the animal's brain involuntarily determines the distance to an object by multiplying the velocity with which a sound signal travels and returns from the object by the time it takes to reach it and dividing this product by 2 (fig. 15–15).

$$D = \frac{V \times T}{2}$$

Although all marine mammals have good vision, conditions within the marine environment often limit its effectiveness. In coastal waters, where suspended sediment and dense plankton blooms make the water turbid, and in the deeper waters, where light is limited or absent, echolocation would surely be beneficial in pursuit of prey or the location of objects in the water. Using lower-frequency clicks at great distance and higher frequency at closer range, the bottlenose dolphin, *Tursiops truncatus*, can detect a school of fish at distances greater than 100 m (328 ft) and an individual 13.5-cm (5.3-in.) fish at a distance of 9 m (30 ft).

Sperm whales produce intense clicks in series that are called codas, which may serve as individual signatures. The "signature whistles" of some dolphins may serve a similar purpose. Very high intensity clicks are produced by some dolphins and sperm whales and may be used to stun prey. There is, however, no direct evidence of such behavior. Although there is no doubt that many species of toothed whales communicate using sound, there is no evidence of language in the sense of an organized structure and sequences of sound corresponding to the human use of sentence structure. The clicks of the bottlenose dolphin are of a frequency range that is partly audible to humans; however, some are of ultrasonic frequencies that can be repeated up to 800 times per second. Large toothed whales emit lower-frequency clicks at lower repetition rates. The **killer whale,** *Orcinus orca,* produces from 6 to 18 clicks per second, and the **sperm whale,** *Physeter catodon,* may produce from less than one to more than 40 clicks per second. Each click of the sperm whale contains 9 pulses, or **clickettes.** It has been estimated that the sperm whales can detect their main prey, squid, from a distance of up to 400 m (1312 ft) by use of their low-frequency scanning clicks.

How the pulses of clicks are produced and how the returning sound is received are not fully understood. However, there are some theories that attempt to answer both questions for certain species of toothed whales. Two of these theories of sound production are discussed in figure 15–16.

Whereas the human ear is sensitive to frequencies from 16 Hz to 20 kHz, some toothed whales respond to frequencies as high as 150 kHz. In most mammals, the bony housing of the inner ear structure is fused to the skull, so that, when submerged, sounds transmitted through the water are first picked up by the skull and travel to the hearing structure from many directions. This structure makes it impossible for such mammals to locate accurately the source of the sounds. Obviously such an arrangement for the hearing structure would not suit an animal that depended on echolocation to accurately determine the positions of objects in the water. All cetaceans have evolved structural features that insulate the inner ear hous-

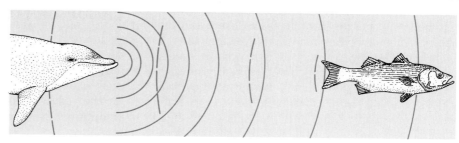

FIGURE 15–15 Echolocation. Clicking sound signals are generated by cetaceans and bounced off objects in the ocean to determine their size, shape, and distance.

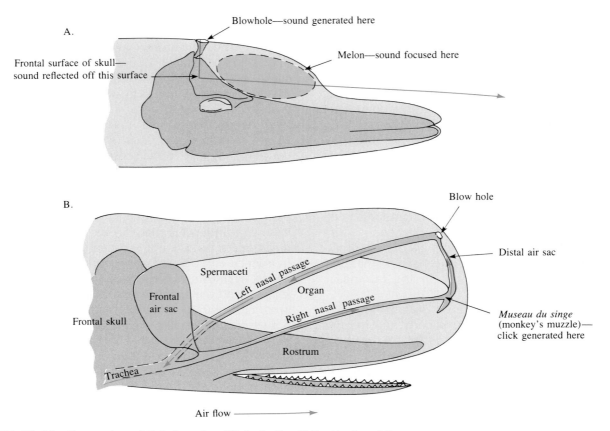

A.

Blowhole—sound generated here

Melon—sound focused here

Frontal surface of skull—
sound reflected off this surface

B.

Blow hole

Distal air sac

Spermaceti

Left nasal passage

Organ

Frontal
air sac

Right nasal passage

Frontal skull

Museau du singe
(monkey's muzzle)—
click generated here

Rostrum

Trachea

Air flow

FIGURE 15–16   Generation of Echolocation Clicks in Small-Toothed and Sperm Whales.

A, in small-toothed whales, the clicks may be generated within the blowhole mechanism, reflected off the frontal surface of the skull, and focused by the fatty melon into a forward-directed beam. *B,* structures that may be related to the generation of clicks by sperm whales. Air may pass from the trachea through the right nasal passage across the *museau du singe,* where clicks are generated. It passes along the distal air sac past the closed blow hole and returns to the lungs along the left nasal passage. The initial click is followed by 8 progressively weaker clicks that may result from sound energy bouncing back and forth between the frontal air sac and the distal air sac, being transmitted through the oil-filled spermaceti organ.

ing, the **tympanic bulla,** from the rest of the skull. In the toothed whales, the tympanic bullae are separated from the rest of the skull by an appreciable distance and surrounded by an extensive system of air sinuses. The sinuses are filled with an insulating emulsion of oil, mucus, and air and are surrounded by fibrous connective tissue and venous networks.

There is some disagreement on how sound reaches the inner ear to be processed. Some investigators believe it passes through the external auditory canal as it does in other mammals, yet this canal is usually blocked by a plug of ear wax in baleen whales. In toothed whales it may be nearly or completely covered by skin. A thin flaring of the lower jaw, less than 0.1 mm (0.004 in.) thick, is directed toward the tympanic bulla on either side of the head and connected with it by a fat- or oil-containing body.

It is believed that sound is picked up by the thin, flaring jawbone and passed to the inner ear via the connecting fat body. Tests of acoustical sensitivity have indicated the lower jaw is up to 6 times as sensitive as the external auditory canal.

GROUP BEHAVIOR

Depending on the size, feeding behavior, nature of the reproductive process, and other requirements of maintaining the species, many animals inhabiting the open ocean have developed patterns of group behavior that allow them to most efficiently exploit their environment. Two of these patterns are **schooling** and **migration.**

Schooling

Obtaining food occupies most of the time of many inhabitants of the open ocean. It is achieved by the fast and agile animals through active predation; other animals move more leisurely while filtering small food particles from the water. Examples of **predators** and **filter feeders** can be found in populations of pelagic animals from the tiny zooplankton to the massive whales. Patches of zooplankton may occur simply because the nutrients that support the phytoplankton they feed on may be in high concentrations in certain coastal waters. We usually do not refer to these patches as schools. The term **school** is usually reserved for well-defined social organizations of fish, squid, and crustaceans.

The number of individuals in a school can vary from a few larger predaceous fish to hundreds of thousands of small filter feeders. Within the school, individuals of the same size move in the same direction with equal spacing between each. This spacing is probably maintained through visual contact, and in the case of fish, by use of the **lateral line system** that detects vibrations of swimming neighbors. The school can turn abruptly or reverse direction as individuals at the flank or the rear of the school assume leadership positions (fig. 15–17).

The advantage of schooling seems obvious from the reproductive point of view. During spawning, it assures that there will be males to release sperm to fertilize the eggs shed into the water or deposited on the bottom by females. However, most investigators believe the most important function of schooling in small fish is **protection from predators**. On first consideration, it seems illogical that the formation of schools would serve such a purpose. Aren't the smaller fish making it easier for the predator by form-

ing a large mass? This is a difficult question; in fact, no experimental evidence provides the basis for an unequivocal answer. However, the consensus, based primarily on conjecture, is *no*.

The belief that small fish are safer from predators if they form schools is based in part on the following reasoning: Over two thousand species of fish are known to form schools. This fact alone indicates that this behavior has evolved and is so pervasive today among fish populations because, for fish with no other means of defense, it somehow provides a better chance of survival than swimming alone. How schooling is protective may grow out of the following considerations: (1) If members of a species form schools, they reduce the percentage of ocean volume in which a cruising predator might find one of their kind. (2) Should a predator encounter a large school, it is less likely to consume the entire unit than if it encounters a small school or an individual. (3) The school may appear as a single large and dangerous opponent to the potential predator and prevent some attacks. (4) Predators may find the continually changing position and direction of movement of fish within the school confusing, making attack particularly difficult for predators that can attack only one fish at a time. It is surely possible that there are other more subtle reasons for schooling. Whatever they might be, this gregarious behavior must enhance species survival because it is widely observed among pelagic animals.

Migration

Many oceanic animals undertake migrations of varying magnitudes. This behavior is observed among sea turtles, fish, and mammals. Some animals, for instance the sea turtles, return to dry beach deposits to lay their eggs. Other populations, including many of the baleen whales, migrate because the physical environment and nature of available food suitable for adults do not meet the needs of young members of their species.

Migratory routes of commercially important baleen whales have been well known since the mid-nineteenth century. The paths of these and other air-breathing mammals are easy to observe. However, the migratory paths of many less visible fish, even though they may also have high commercial value, have been more difficult to identify. Tagging studies have helped us understand the patterns of movement of these fish. Sampling the distribution patterns of eggs, larvae, young, and adult populations and radio-tracking individuals have also helped to identify migratory routes.

FIGURE 15–17 Schooling.
School of Pacific sardines, typical of the many small herringlike fishes that form large schools.

One aspect of migratory behavior that is of interest to investigators is **orientation**: How do migratory species orient themselves in time and space? To put the problem another way, How do they know where they are in relationship to where they want to go? How do they know when to leave their present location to get where they are going at the proper time?

Although these answers are still being sought, it is believed that all of the migratory species have an innate sense of time referred to as a **biological clock**. External manifestations of the biological clock of a species, called **circadian rhythms**, are physiological variations that occur independent of changes in the external environment. Circadian rhythms include cyclic changes in respiration rate and body temperature. However, some changes in the external environment can alter the circadian rhythms. It is possible that changes in food availability, water temperature, and length of daylight periods may serve to trigger seasonal migrations.

It seems that orientation in space must be a more complex problem than orientation to time. Animals such as mammals and turtles that migrate at the surface could well use their sight as a means of orientation. Gray whales that migrate close to land over a large part of their migratory route could possibly

identify landmarks along the shore. When out of sight of land, mammals and turtles could use the relative positions of the sun, moon, and stars to guide them on their way. Fish that migrate beneath the surface are thought to use smell, relative movement within ocean currents, and small induced currents (generated as the ions dissolved in ocean water are carried through the earth's magnetic field by ocean currents) to help them orient themselves in ocean space.

Many of the important food fish of the high northern latitudes deposit pelagic egg masses that are transported by currents. Such fish will actively migrate upcurrent during spring or summer and spawn in a location that will assure the current will carry the eggs back to a nursery ground where the food supply will be appropriate for newly hatched fry. Migrations of this type are usually over distances of a few tens or hundreds of kilometers. This behavior is observed in the population of **Atlantic cod**, *Gadus morhua*, that spawns along the south shore of Iceland. Adults that occupy feeding grounds along the north and east coasts of Iceland and around the southern tip of Greenland migrate to the spawning grounds in the late winter and early spring. This migration involves swimming against the East Greenland and Irminger currents and includes only mature adults at least 8

FIGURE 15—18 Migration Routes.

A, migration routes of the Icelandic cod, bluefin tuna, and Atlantic eel *(Anguilla). B,* migration routes of the Chinook salmon and the California gray whale.

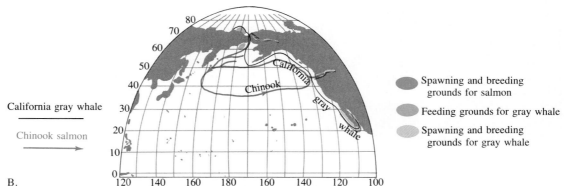

years of age. These spawning migrations are undertaken annually until the adults die at an age of 18 to 20 years (fig. 15–18A).

The spawning grounds are bathed in the relatively warm water of the Irminger Current, a northward-flowing branch of the Gulf Stream. Each female releases up to 15 million eggs, which float in the surface current. The drifting eggs, carried by the East Greenland and Irminger currents, hatch in about two weeks. The larvae feed at midwater depths while the currents are carrying them to the adult feeding grounds. They will remain there, undertaking only short onshore-offshore migrations, until they mature and make their first spawning migration.

Longer migrations are undertaken by two **Atlantic bluefin tuna** populations that spawn in tropical waters near the Azores in the eastern Atlantic and the Bahamas in the western Atlantic. Little is known about where the newly hatched tuna feed, but the adults are known to move north along the seaward edge of the Gulf Stream in May and June and feed on the herring and mackerel populations off the coasts of Newfoundland and Nova Scotia. They may move even farther north, but the route used by them to return to the spawning grounds is unknown.

Anguilla, the **Atlantic eel,** undertakes what appears to be a single round trip migration during its lifetime. Its behavior is **catadromous:** after being spawned in the ocean, it enters freshwater streams to spend its adult life and returns near the end of its life to the ocean spawning site.

Spawning is thought to occur in the Sargasso Sea southeast of Bermuda. The spawning has not been observed, and the earliest stage of *Anguilla* that has been observed is the leaf-shaped, transparent **leptocephalus larva.** The 5-mm (0.2-in.)-long larvae are carried into the Gulf Stream and north along the North American coast. After one year in the ocean, some of the leptocephali metamorphose into young eels, **elvers,** and move into the streams of North America. Others spend another year migrating to the European coast before metamorphosing and moving into European streams. A third population spends another year moving into the Mediterranean Sea before undergoing metamorphosis. The American and European populations are considered by some to be different species—*A. rostrata* and *A. anguilla,* respectively—based primarily on the fact that the American eel has 8 to 10 fewer vertebrae than the European eel.

After spending up to ten years in the freshwater environment, the mature eels undergo yet another change. They develop a silvery color pattern and enlarged eyes that are typical of fish inhabiting the me-

FIGURE 15–19 Varieties of Salmon.
Top, chum *(Oncorhynchus keta);* center, coho *(O. kisutch),* bottom, chinook *(O. tschawyscha).* Courtesy U.S. Fish and Wildlife Service.

sopelagic zone. They swim downriver to the ocean and presumably back to the spawning ground, where they are thought to die.

An opposite behavior pattern is characteristic of **salmon** of the north Atlantic and Pacific oceans. This **anadromous** behavior includes spawning in freshwater streams and spending most of their adult lives in the ocean. Pacific species die after spawning; Atlantic salmon return to the ocean.

Six species of **Pacific salmon,** *Oncorhynchus,* follow similar paths in their migrations (fig. 15–19). Spawning occurs during the late summer and early fall. Eggs are deposited and fertilized in gravel beds far upstream. The **chinook salmon** spawns from as far south as central California to Alaska. The young chinook hatching in the spring may head downstream as a silvery, filter-feeding smolt during its first or second year. By the time it reaches the ocean, the young chinook has become a predator, feeding on smaller fish. After spending about four years in the North Pacific, mature salmon of more than 10 kg (22 lb) in weight and 1 m (3.28 ft) in length return to their home streams to spawn (fig. 15–18B). Although it is not known how salmon achieve this homing, most investigators think the most likely sources of information processed by the salmon during their return trip to spawning grounds are odors and currents.

Some of the longest migrations known to occur in the open ocean are the seasonal migrations of baleen whales. Many baleen species feed in the colder high-latitude waters and breed and calve in warm tropical waters. Feeding occurs during summer while the long hours of sunlight illuminating the nutrient-rich waters produce a vast population of crustaceans to be fed upon. Only the vast supplies of food available in these waters makes it possible for the whales to maintain themselves during the long migrations and mating and calving season when feeding is minimal.

Although it is widely believed that many baleen whales calve in warm waters because the young could not survive in the cold waters of the feeding grounds, this may not be so. Bowhead whales live year-round at the Arctic pack-ice edge, and the smaller narwhal lives much of the year within the ice field. The fact that these whales can successfully calve in Arctic waters makes it difficult to understand why the larger calves of the species that are born in tropical waters could not also survive in the colder waters. The long migrations baleen whales now undertake may have resulted from productive feeding grounds that were once near the tropical calving

A.

FIGURE 15–20 California Gray Whale Feeding.

A, side-scan sonograph showing pits on the floor of the northern Bering Sea created by California gray whales feeding on bottom-dwelling amphipods and other invertebrates. The smaller 2- to 3-m (6.5- to 10-ft)-long pits produced by the initial feeding do not show up well at this scale. All of the readily visible pits have been enlarged by currents that flow in a northerly direction. The pits show this north-south alignment. Sonograph courtesy of William K. Sacco, Yale University. *B*, amphipod mat. A mat of mucus-lined burrows is created by ampeliscid amphipods that are the preferred prey of California gray whales. Photo courtesy of Kirk R. Johnson, Yale University.

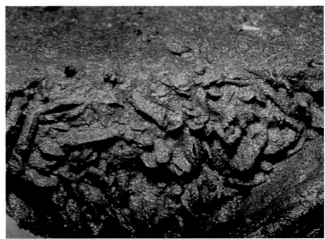

B.

grounds progressively moving poleward because of climate change. If this is the case, the breeding and calving grounds must be very important locations of social aggregation. Otherwise, they would surely have been abandoned for locations closer to the present feeding grounds. An alternate explanation for leaving the colder waters to calve is to avoid the killer whales that are more numerous in colder waters. They are a major threat to young whales. Because of the energy demands faced by female baleen whales in producing large offspring (the gestation period is up to 1 year providing them with fat-rich milk for several months, it is not uncommon for them to mate only once every two or three years.

The migration route of the *California gray whale* demonstrates how the conditions discussed above are met by whale migration patterns. Gray whales are moderately large whales, reaching lengths of 15 m (49 ft) and weighing over 30 metric tons. Populations of gray whales feed during the summer months in the extreme north Pacific Ocean, Sea of Okhotsk, Bering Sea, and Chukchi Sea. They are unique among baleen whales in that they do not feed by straining pelagic crustaceans and small fish from the water; instead they stir up bottom sediment with their snouts and feed on bottom-dwelling amphipods (fig. 15–20).

The western populations winter along the coast of Korea, but the population known as the California gray whale migrates from the Chukchi and Bering seas to winter in lagoons of the Pacific coast of Baja California and the Mexican mainland coast near the southern end of the Gulf of California (fig. 15–18B). The migration usually begins in September when pack ice begins to form over the continental shelf areas that are their feeding grounds. This migration, which is the longest known migration undertaken by any mammal, may involve a round trip distance of 22,000 km (13,662 mi). First to leave are the pregnant females. They are followed by a procession of mature females that are not pregnant, immature females, mature males, and immature males. After cutting through the Aleutian Islands by way of Unimak Pass, they follow the coast throughout their southern journey. Traveling at an average rate of about 200 km/day (124 mi/day), most of the whales reach the lagoons of Baja California by the end of January.

In these warm-water lagoons, the pregnant females give birth to 2-ton calves. The calves nurse and put on weight quickly during the next two months. While the calves are nursing, the mature males breed with the mature females that did not bear calves. Late in March, the return to the feeding grounds begins, with the procession order reversed. Most of the whales are back in the feeding grounds by the end of June, and they feed on prodigious amounts of amphipods to replenish their depleted store of fat and blubber before the next trip south.

Reproduction

As in the examples just described, reproduction of pelagic animals involves bringing the males and females of the species together for this purpose on some periodic basis. These gatherings usually occur during the warmer spring or summer months, when water temperatures are higher and primary productivity is at its peak.

Most of the **invertebrates** and fish that inhabit the open ocean are **oviparous**: they lay eggs that hatch in the open water or on the bottom into larval forms that become meroplankters. Animals that reproduce in this way usually produce enormous numbers of eggs—15 million eggs per season in the migrating **Icelandic cod**—because most will be lost to predators before they hatch or the larvae will be eaten by other members of the plankton community. However, some oviparous fish may produce only one or a few well-protected eggs per season; the cartilaginous sharks, skates, and rays include many such species (fig. 15–21).

The females of other fish species keep their fertilized eggs in their reproductive tract until they hatch. Although the young come into the world live, the process by which the embryos develop is the same as for those that develop from eggs that are laid. This behavior is called **ovoviviparous**, describing a method of reproduction in which the eggs are incubated inter-

FIGURE 15–21 Skate Egg Case.
This horny mermaid's purse protects the egg that develops within it until it is ready to hatch.

nally. The **sea horse** and **pipefish** exemplify a special twist to internal incubation where the female deposits the eggs in a ventral pouch of the male (fig. 15–22). The male carries the eggs until they hatch, and the young fish enter the ocean directly from this pouch. The additional protection against loss of eggs through pelagic predation results in the reduction of the number of eggs that must be produced; for example, the **dogfish shark**, *Squalus,* usually produces fewer than a dozen eggs at a time.

Viviparous animals give birth to live young, but the behavior is considered by most to require also that the young be given more than space in the mother's reproductive tract to incubate. Part of the behavior includes providing embryonic nutrient in addition to the yolk of the egg. Some sharks and rays have projections from the wall of the uterus called **villi** that secrete a rich milk. With this additional nutrition, the **stringray**, *Pteroplatea,* produces young that enter the ocean with a mass up to 50 times that of the mass of the yolk of the egg from which they came. In the case of the **white-tip shark**, *Carcharhinus,* this additional nutrition passes to the embryo through a yolk sac attached to the wall of the uterus. **Mammals** exhibit the highest level of viviparous behavior in that the embryo is encased in the placental sac. The mother's blood flows through the placenta, and the embryo receives nutrition from the mother through an umbilical cord by which it is attached to the placenta; essentially all nutrition is provided by the mother. This

behavior provides a high degree of protection to the developing young but puts a high energy demand on the mother. Thus mammalian births typically involve only one or a few young who will remain dependent on their mothers for protection and food for some time after their birth.

SUMMARY

Frictional resistance to sinking—a major factor in helping the tiny plankton stay near the surface—is not a major factor in keeping the larger nekton from sinking; they depend primarily on **buoyancy** or **swimming**. **Rigid gas containers** found in some cephalopods and **expandable swim bladders** of many bony fishes are adaptations that help increase buoyancy. Many invertebrates such as the jellyfish, tunicates, and arrowworms have soft, **gelatinous bodies** that reduce their density. The Portuguese man-of-war also has a **gas-filled float** that supports this pelagic colony. Many of these invertebrate forms are weak swimmers and depend primarily on buoyancy to maintain their positions near the surface.

Strong **swimmers**, the nekton—squid, fish, and mammals—depend on this expenditure of energy to obtain food in the water column. **Squid** swim by trapping water in their mantle cavities and forcing it out through a **siphon**. Most **fish** swim by creating a **wave of body curvature** that passes from the front to the back of the body and provides a forward thrust. The **caudal fin** is the most important in providing thrust, while the **paired pelvic** and **pectoral fins** are used for **maneuvering**. The vertically oriented **dorsal** and **anal fins** serve primarily as **stabilizers**. The efficiency with which a caudal fin produces thrust is indicated by its **aspect ratio**, equal to the square of the fin height divided by the fin area. The **rounded** caudal fin, found on the sculpin, is flexible, has a low aspect ratio of about 1, and can be used for maneuvering at slow speeds. The **lunate** fin is rigid, has a high aspect ratio of about 10, and is of little use in maneuvering. It is very efficient in producing thrust for fast swimmers such as the tuna.

Fish such as groupers are **lungers** that sit motionless and make short, quick passes at passing prey. They have mostly **white muscle tissue**. Tuna and other **cruisers**, constantly swimming in search of prey, possess mostly **red muscle tissue**. Red fibers have a greater affinity for oxygen and tire less rapidly than white fibers, enabling tuna to maintain a cruising speed of about 3 body lengths/s. Using their white muscles to help in short periods of rapid swimming, they can reach speeds of up to 10 body lengths/s. Al-

FIGURE 15–22 Ovoviviparous Behavior in Sea Horses.
The female deposits the eggs in the ventral pouch of the male, where they are protected during incubation. Photo by Larry Ford.

though most fish are **poikilothermic (cold-blooded)**, the tuna, *Thunnus,* is **homeothermic**; that is, it maintains a body temperature well above the temperature of the water.

The best-adapted mammals for life in the open ocean are the Cetacea. The **toothed whales** (Odontoceti) and **baleen whales** (Mysticeti) propel their highly streamlined bodies through the water with vertical movement of their **horizontal flukes**.

Whales are able to dive deep and stay submerged for unusually long periods of time because of modifications that allow them to absorb 90% of the O_2 in the air they inhale, store large quantities of oxygen, possibly reduce oxygen use, and collapse their lungs below depths of 100 m (328 ft).

Most whales and many other marine mammals are thought to use **echolocation** in finding their way through the ocean and locating prey. The clicking sounds emitted by whales are bounced off objects, and the animal can determine the size, shape, and distance of the objects by the nature of the returning signals and the time elapsed.

Schooling of active swimmers such as fish, squid, and crustaceans is not fully understood, but it has obvious advantages as a reproductive strategy. Schooling more likely serves a protective function, although the ways in which it may meet this end are speculative.

Migrations are observed among sea turtles, fish, and mammals and are related to the needs of reproduction and finding food. It is believed that **orientation** is maintained during migrations through use of visible landmarks, smell, and the earth's magnetic field. **Baleen whales** may migrate from their cold-water summer feeding grounds to warm low-latitude lagoons in winter so their young can be born into warm water; they might not be able to survive the cold water of higher latitudes at birth. Fishes such as the Atlantic cod swim upcurrent to deposit their eggs so that when they hatch the cod fry will be in water where suitable food is available.

The North Atlantic eel, *Anguilla,* is **catadromous**: it spawns in the deep waters of the Sargasso Sea and spends its adult life in freshwater streams of North America and Europe. The **Atlantic** and **Pacific salmon** are **anadromous**: they spawn in the freshwater streams of North America, Europe, and Asia and spend their adult life in the open ocean.

Most fish are **oviparous**, depositing their eggs in the ocean. Some sharks and rays maintain their eggs in a body cavity until they hatch; this **ovoviviparous** behavior provides a greater protection of the eggs. Where oviparous fish may produce millions of eggs, ovoviviparous fish may produce fewer than a dozen. The stingray, white-tip shark, and mammals are **viviparous**, not only providing space in their bodies for the eggs to develop but also providing nutrition in addition to the egg yolk. In mammals, the young will still be dependent on their mothers for nutrition for some time after birth.

QUESTIONS AND EXERCISES

1. Discuss how the rigid gas chambers in cephalopods may be more effective in limiting the depth to which they can descend than the flexible swim bladders of bony fish.

2. Describe the body form and lifestyle of the following plankters: *Physalia,* jellyfish, tunicates, ctenophorans, and arrowworms.

3. What are the major structural and physiological differences between the fast-swimming cruisers and the lungers that patiently lay in wait for their prey?

4. List the modifications that are thought to allow some cetaceans to (1) dive to great depths without suffering the bends and (2) stay submerged for long periods of time.

5. Describe the process by which the sperm whale may produce echolocation clicks.

6. Although there is disagreement about how it is achieved, discuss what most investigators believe to be the method by which sound reaches the inner ear of toothed whales.

7. Summarize the reasons some investigators believe schooling increases the safety of fishes from predators.

8. What are the methods believed to be used by migrating animals to maintain their orientation?

9. Why do fishes such as Icelandic cod swim upcurrent to spawn?

10. How are the migrations of the North Atlantic eels and Pacific salmon fundamentally different?

11. Why don't the California gray whales remain in the cold-water feeding grounds during the winter season?

12. Compare the reproductive behavior of oviparous, ovoviviparous, and viviparous animals.

REFERENCES

Carey, F. G. 1973. Fishes with warm bodies. *Scientific American* 228:2, 36–44.

Denton, E. J., and Shaw, T. I. 1962. The buoyancy of gelatinous marine animals. *Journal of Physiology* 161: 14P–15P.

George, D., and George, J. 1979. *Marine life: An illustrated encyclopedia of invertebrates in the sea.* New York: Wiley-Interscience.

Kanwisher, J. W., and Ridgway, S. H. 1983. The physiological ecology of whales and porpoises. *Scientific American* 248:6, 110–21.

MacGinitie, G. E., and MacGinitie, N. 1968. *Natural history of marine animals.* 2nd ed. New York: McGraw-Hill.

Mrosovsky, N.; Hopkins-Murphy, S. R.; and Richardson, J. I. 1984. Sex ratio of sea turtles: Seasonal changes. *Science* 225:4663, 739–40.

Norris, K. S., and Harvey, G. W. 1972. A theory for the function of the spermaceti organ of the sperm whale (*Physeter catodon* L). Animal orientation and navigation, 397–417. Washington, D.C.: National Aeronautics and Space Administration.

Pike, G. C. 1962. Migration and feeding of the gray whale (*Eschrichtius gibbosus*). *Journal of the Fisheries Research Board of Canada* 19:815–38.

Royce, W.; Smith, L. S.; and Hartt, A. C. 1968. Models of oceanic migrations of Pacific salmon and comments on guidance mechanisms. *Fishery Bulletin* 66:441–62.

Thorson, G. 1971. *Life in the sea.* New York: McGraw-Hill.

Würsig, B. 1989. Cetaceans. *Science* 244:4912, 1550–57.

SUGGESTED READING

Sea Frontiers

Bachand, R. G. 1985. Vision in marine animals. 31:2, 68–74. An overview of the types of eyes possessed by marine animals.

Bleecker, S. E. 1975. Fishes with electric know-how. 21:3, 142–48. A survey of fishes that use electrical fields to navigate, to capture prey, and to defend themselves.

Klimley, A. P. 1976. The white shark—a matter of size. 22:1, 2–8. Describes procedure used by Dr. John E. Randall for determining the size of sharks from the perimeter of the upper jaw and height of teeth.

Lineaweaver, T. H., III. 1971. The hotbloods. 17:2, 66–71. Discusses the physiology of the bluefin tuna and sharks that have high body temperatures.

Maranto, G. 1988. The Pacific walrus. 34:3, 152–59. A summary of the natural history of walruses of the Bering Sea and the role of humans as their predators.

Netboy, A. 1976. The mysterious eels. 22:3, 172–82. An informative discussion describing what is known of the migrations of the catadromous eels and their importance as a fishery.

O'Feldman, R. 1980. The dolphin project. 26:2, 114–18. A description of the response of Atlantic spotted dolphin to musical sounds.

Reeve, M. 1971. The deadly arrowworm. 17:3, 175–83. This important group of plankton is described in terms of body structure and life style.

Scientific American

Denton, E. 1960. The buoyancy of marine animals. 303:11, 118–28. The methods used by various marine animals to use buoyancy to maintain their position in the water column.

Donaldson, L. R. and Joyner, T. 1983. The salmonid fishes as a natural livestock. 249:1, 50–69. The genetic adaptability of the salmonid fishes may help them adapt to "ranching" operations.

Gosline, J. M., and DeMont, M. E. 1985. Jet propelled swimming in squids. 252:1, 96–103. By using jet propulsion resulting from expelling water through their siphons, squids can move as fast as the speediest fishes.

Gray, J. 1957. How fishes swim. 197:2, 48–54. The roles of musculature and fins in the swimming of fishes.

Kooyman, G. L. 1969. The Weddell seal. 221:2, 100–107. The life style and problems related to this mammal's living in water permanently covered with ice.

Leggett, W. C. 1973. Migration of shad. 228:3, 92–100. The routes of the anadromous shad in the rivers and Atlantic waters are described, along with factors that may control the migrations.

Rudd, J. T. 1956. The blue whale. 195:6, 46–65. A description of the ecology of the largest animal that ever lived.

Shaw, E. 1962. The schooling of fishes. 206:6, 128–36. A consideration of theories seeking to explain the schooling behavior of fishes.

Whitehead, H. 1985. Why whales leap. 252:3, 84–93.
  Whales seem to communicate with their spectacular lunges above the oceans surface.

Würsig, B. 1988. The behavior of baleen whales. 258:4, 102–107.
  A summary of the facts concerning the behavior of baleen whales.

Zapol, W. M. 1987. Diving adaptations of the Weddell seal. 256:6, 100–107.
  The physiological adaptations that enable the Weddell seal to make deep, long dives are discussed.

16
ANIMALS OF THE BENTHIC ENVIRONMENT

More than 90% of the species of animals found in the ocean live on or in the ocean floor. Ranging from the rocky, sandy, and muddy environments of the intertidal zone to the muddy deposits of deep-ocean trenches more than 11 km (6.8 mi) in depth, the ocean floor provides a varied environment that is home for a diverse benthic community.

Because it lives at or near the interface of the ocean floor and the ocean water, a benthic organism's success is closely tied to its ability to cope with physical conditions of the water, the ocean floor, and the other members of the biological community. The vast majority of benthic species live on the shallow bottom of the continental shelf, while only about 400 of the known 157,000 benthic species are found in the hadal zone of the deep-ocean trenches.

One of the most prominent variables affecting species diversification is temperature. We have previously discussed its effect on species diversity on a latitudinal basis; however, even at the same latitude, a significant difference in the number of benthic species is found on opposite sides of an ocean basin because of the effect of ocean currents on the coastal water temperature. Along the European coast where the Gulf Stream warms the water from the northern tip of Norway to the Spanish coast, over three times more species of benthos exist than are found along the similar latitudinal range of the Atlantic coast of North America, where the Labrador Current cools the water as far south as Cape Cod.

Figure 16–1 shows that the distribution of benthic biomass is patterned after the distribution of photosynthetic productivity in the surface waters shown in figure 14–8. This tells us that life on the ocean floor is very much dependent upon the primary photosynthetic productivity of the ocean-surface waters. We will nonetheless discuss some special benthic communities that are an exception to this general condition.

ROCKY SHORES

The rocky shore, especially the supralittoral and littoral zones, is the most readily observable of marine habitats. Along most shores, the littoral zone can be divided into the high-tide zone, mostly dry—covered

Preceding page: crown-of-thorns sea star, which threatened to destroy many coral reefs in the 1960s, being attacked by one of its few predators, the Pacific triton. Photo by Christopher Newbert.

by the highest high tide but not the lowest high tides; the middle-tide zone, exposed and covered equally—covered by all high tides and exposed during all low tides; and the low-tide zone, mostly wet—covered during the highest low tides and exposed during the lowest low tides (fig. 16–2A).

Along rocky shores these divisions of the littoral zone are often obvious because of the sharp boundaries between populations of attached organisms. Because each square centimeter of the littoral rocky shore is significantly different from the square centimeter above and below it, evolution has been able to produce organisms with the abilities to withstand very specific degrees of exposure to the atmosphere. This condition results in the most finely defined biozones known in the marine environment.

Supralittoral (Spray) Zone

Throughout the world the most obvious inhabitant of the rocky supralittoral zone is the periwinkle snail. The supralittoral zone can be easily identified if one considers the middle of the periwinkle belt to be the boundary between the littoral and the supralittoral. The genus Littorina includes species able to breathe air, like land snails. They need the spray of breaking waves as their only tie to the sea since they are viviparous (give birth to live young) rather than depositing their eggs into the water as do more-marine periwinkles and other snails (fig. 16–2B).

The periwinkles feed by rasping algae and lichens off the rocks of the spray zone with their filelike radulae.

Hiding among the cobbles and boulders covering the floors of sea caves well above the high-tide line are isopods belonging to the genus Ligia; they are commonly called rock lice or sea roaches (fig. 16–2C). Neither name is particularly flattering to these little scavengers that reach lengths of 3 cm (1.2 in.) and scurry about at night feeding on any type of organic debris.

Another indicator of the spray zone is a distant relative of the periwinkle snails—limpets belonging to the genus Acmaea (fig. 16–2D). Having a flattened conical shell, the limpets feed in a manner similar to that of the periwinkles. At low tide, limpets will remain motionless with their shells pulled down tight against the rock on which they make their homes. Some limpets are known to leave their home spots on the rock during high tide and feed for as long as the tide will allow. They will then return to their homes, marked by a depression or discoloration on the rock, to wait out the dry period of low tide.

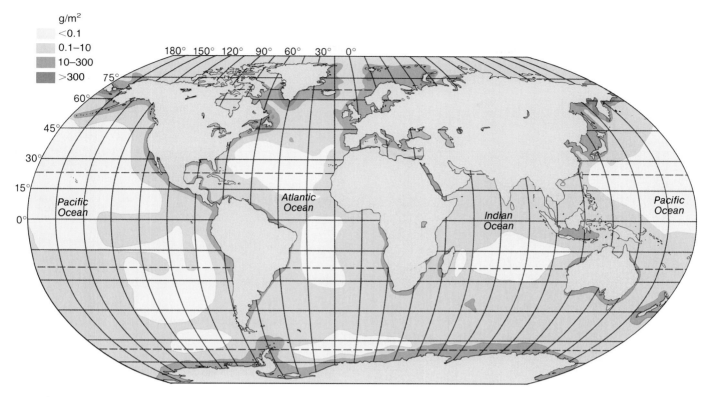

FIGURE 16–1   Distribution of Benthic Biomass of the World Ocean (g/m², wet weight).

Although the distribution pattern shown here is defensible in concept, it is based on too few samples to be regarded as accurate. The pattern is similar to that of photosynthetic pigment distribution shown in figure 14–8. This suggests that most of the benthic community is directly dependent on the photosynthetic production in the surface waters of the oceans for its nutritional support. High concentrations of biomass associated with hydrothermal vents and cold seeps are not reflected in this distribution. The concentrations beneath the centers of the subtropical gyre waters are 20,000 times less than those of the continental-shelf benthos. Up to 60% of the organic matter reaching the deep-ocean floor may enter the food chain through bacteria, and deep-sea bacteria may metabolize organic matter at a rate at least 10 times slower than bacteria on the continental shelf. This could reduce the rates of biological productivity on the deep-ocean floor to levels of 1/200,000 that occurring on the continental shelves. After Zenkevitch et al., 1971.

High-Tide Zone

Unlike the periwinkles, some of whom give birth to live young, the **barnacles** cannot abandon the sea for dry land. Their limit is the high-tide shoreline. This **hermaphroditic** animal fertilizes its neighbor's eggs by inserting a penis into a neighboring barnacle. In addition to their need of water for the planktonic existence of their larval forms, barnacles also are confined to the ocean by their method of feeding.

At the highest intertidal level exists the most prominent inhabitant of this zone, the **buckshot barnacle** (fig. 16–2E), so called because of its small size, which seldom exceeds 0.5 cm (0.2 in). This unusual crustacean attaches itself to a rock by the back of its neck,

secretes a protective calcium carbonate housing that looks like a tiny volcano, and extends its "legs," modified into featherlike **cirri**, to comb microscopic plankton from the water.

The most conspicuous plants in the high-tide zone are members of the genera *Fucus* (fig. 16–2F) in colder latitudes and *Pelvetia* (fig. 16–2G) in warmer latitudes. Both have thick cell walls to reduce water loss during periods of low tide. These plants continue to flourish in the middle-tide zone, where the variety of life forms is much greater than in the high-tide zone. Not only does the variety increase, but also the total biomass is much greater; there is, thus a greater competition for rock space by **sessile** (attached) forms (fig. 16–2F–I).

FIGURE 16–2 The Rocky Shore.

A, a typical rocky intertidal zone and some organisms typically found in its subzones (not to scale). *B,* periwinkles *(Littorina)* nestled in a depression near the upper limit of the high-tide zone. This behavior helps reduce exposure to direct sunlight. *C,* rock louse *(Ligia). D,* limpet *(Acmaea). E,* buckshot barnacles *(Chthamalus). F, G,* rock weeds, *Fucus* and *Pelvetia.* Sharing the middle-tide zone with these rock weeds are *H,* acorn barnacles *(Balanus)* and *I,* goose barnacles *(Pollicipes)* and mussels. Photo *D* by Greg West.

Spray zone

Highest high tide

High tide zone

Lowest high tide

Middle tide zone

Highest low tide

Low tide zone

Lowest low tide

Rock louse (*Ligia*)

Periwinkle (*Littorina*)

Limpet (*Acmaea*)

Buckshot barnacle (*chthamalus*) (*Balanus*)

Periwinkle (*Littorina*)

Limpet (*Acmaea*)   Chiton (*Nuttalina*)

Mussel (*Mytilus*) (*Madiolus*)

Chitons and Limpets

Hermit crab (*Pagurus*)

Goose barnacles (*Pollicipes*)

Acorn barnacle (*Balanus*)

Sea Star (*Asterias-Pisaster*)

Many species of animals and plants

Sea anemone (*Authopleura*)

A.

B.

C.

D.

E.

F.

Fucus

Pelvetia

G.

H.

I.

409

Middle-Tide Zone

Although the smaller acorn barnacles usually found in the upper-tide zone can occasionally be found in the upper reaches of the middle-tide zone, larger species of the genus *Balanus* (fig. 16–2H) are the more common barnacles of the middle-tide zone. The barnacle most characteristic of this zone is the **goose barnacle**, genus *Pollicipes* (fig. 16–2I), with two valves composed of numerous plates that house the animal, which attaches itself to the rock surface by a long muscular neck. The total length of this structure may exceed 10 cm (4 in.). Even more successful than barnacles in the competition for space in the middle-tide zone are the various species of **mussels** (fig. 16–2I) belonging to the genera *Mytilus* and *Modiolus;* they will attach to bare rock, algae, or barnacles. Settling on these surfaces as larval forms, they attach themselves by tough proteinous **byssal threads**. Given time, mussels would eventually overgrow all sessile forms.

But life in the intertidal zone is not simple. Mussels are fed upon by predators such as **sea stars** and **carnivorous snails**. Two common genera of sea stars are *Pisaster* and *Asterias*.

A sea star feeds by attaching its tube feet to its prey. There are five double rows, which radiate from its mouth at the center of its bottom side. The tube feet are in an oral groove beneath each ray, or arm. To get to the mussel tissue, protected by calcium carbonate bivalve covering, the sea star exerts a continuous pull on the valves by alternating the use of the tube feet so that some are always pulling while others are resting. The mussel is eventually fatigued and can no longer hold the valves closed. The valves open ever so slightly, and the sea star everts its stomach, slips it through the crack, and digests the mussel inside the mussel's shell (fig. 16–3).

Boring carnivorous snails feed on both barnacles and mussels. Their preferred food, barnacles, can be consumed after the snail envelops it with its fleshy foot and forces the valves open. A narcotic secretion called *purpurin* may also be secreted to make the opening of the valves easier. To feed on the mussels, the snail must use its radula to drill a hole through the shell, through which it can rasp away the flesh. The more widely distributed carnivorous snails are the **dogwhelk**, *Nucella,* and the similar Pacific coast species *Thais lamellosa*.

The dominant feature of the middle-tide zone along most rocky coasts is a **mussel bed** that thickens toward the bottom until it reaches an abrupt bottom limit. This may be so pronounced that it appears as though an invisible horizontal plane has prevented the mussels from growing below this depth. Protruding from the mussel bed will be numerous goose barnacles, and concentrated in the lower levels of the bed will be the sea stars browsing on the mussels. Less conspicuous forms common to the mussel beds are varieties of algae, hydroids, worms, clams, and crustaceans.

Where the rock surface flattens out within the middle-tide zone, **tide pools** trapping water as the tide ebbs serve as interesting microecosystems containing a wide variety of organisms. The largest member of this community will often be the sedentary relative of the jellyfish, the **sea anemone**. Shaped like a sack, the anemone has a flat **pedal disc** that provides a suction attachment to the rock surface. Directed upward, the open end of the sack has the only opening to the gut cavity, the mouth, surrounded by rows of tentacles. Unlike their pelagic relatives, the tide pool coelenterates do not have nematocysts long enough to penetrate human skin and therefore are not dangerous (fig. 16–4).

Swimming in the tide pools are a variety of small fish. The **opaleye** reaches a length of 5 cm (2 in.) in the pools and grows larger in deeper water offshore. It is easily identified by the tiny white spots on its back on either side of the dorsal fin. The **woolly sculpin** is a permanent resident of tide pools and grows longer than 15 cm (6 in.). It is covered with hairlike cirri and usually is resting on the bottom or walking on its large pectoral fins. Readily identified by its blunt head and continuous dorsal fin is the 15-cm **rockpool blenny** (fig. 16–5).

FIGURE 16–3   Sea Star Feeding on a Clam.

FIGURE 16–4 Sea Anemone.
A, sea anemone. Photo by Shane Anderson. *B*, basic body plan of an anemone and detail of the stinging mechanism.

A.

B.   Sea anemone
                 (polyp body form)

Before activation   Activated

Nematocyst

Trigger   Lid   Surface of
                          tentacle

Cnidoblast

Mouth   Tentacle

Ectoderm

Endoderm   Gut

Polyp-benthic body form
that reproduces asexually

FIGURE 16–5 Rockpool Blenny.

Photograph by Marty Snyderman.

The most interesting inhabitant of the tide pool is the **hermit crab**, *Pagurus* sp. (fig. 16–6A and B). With a well-armored pair of claws and cephalothorax, hermit crabs have a soft abdomen that is protected by some hard protective container, usually an empty snail shell. The abdomen has even developed a curl to the right to make it fit properly into snail shells. Once in the snail shell, the opening can be closed off by the crab's large claws. However, one primary enemy is the **octopus**, which is not bothered by the shell that the hermit crab has adopted. The octopus can cover the crab with the web surrounding its mouth and release a poisonous secretion that paralyzes the crab so that it can be pulled from the shell.

In tide pools near the lower limit of the middle-tide zone, **sea urchins** (fig. 16–6C and D), with their five-toothed structure centered on the bottom side of a hard spherical covering that supports many spines, may be found feeding on algae. The hard protective covering is called a **test** and is composed of fused calcium carbonate plates that are perforated to allow tube feet and dermal gills to extend through it.

Low-Tide Zone

Unlike the high- and middle-tide zones, the low-tide zone is dominated by plants rather than animals. A diverse community of animals exists, but its members are less obvious because they are hidden by the great variety of **seaweeds** and **surf grass**, *Phylospadix*. The encrusting red alga *Lithothamnion*, which is also seen in middle-zone tide pools, becomes very abundant in the lower-tide pools. In temperate latitudes, moderate-sized red and brown algae provide a drooping **algal canopy** beneath which much of the animal life is found (fig. 16–7A).

Nudibranchs (fig. 16–7B), snails with no shells and exposed gills on their backs, are common in quiet tide pools. They feed on sea anemones, sponges, and hydroids, delicate relatives of the sea anemones. Actually, almost all the phyla of animals are represented in the low-tide zone, where life is less rigorous than in the zones previously discussed. There is no great distinction between the fauna and flora immediately above and below the spring low-tide shoreline.

Scampering from crevice to crevice and in and out of tide pools across the full range of the intertidal zone are various species of **shore crabs** (fig. 16–8). Scavengers that help keep the shore clean, shore crabs spend most of the daylight hours hiding in cracks or beneath overhangs. At night they engage in most of their eating activity, shoveling in algae as rapidly as they can tear it from the rock surface with their **chelae**, large front claws. Shore crabs need to return to ocean water only periodically to wet their gills.

A.

B.

C.

D.

FIGURE 16—6 Hermit Crab and Sea Urchin.

A, hermit crab *(Pagurus)* in *Maxwellia gemma* shell. Photo by James McCullagh. *B,* hermit crab out of its shell; note the soft curved abdomen. *C,* sea urchins burrowed into the bottom of a lower middle-tide zone tide pool. *D,* the bottom side of this dead sea urchin shows the mouth at the center and the hard calcium carbonate test beneath the spiny exterior.

A.

B.

FIGURE 16–7   Inhabitants of the Low-Tide Zone.
A, algal canopy of the low-tide zone exposed during extremely low tide. *B,* hidden beneath the algal canopy may be this sea slug *(Caloria)* shown with its egg case. It is common in rocky low-tide and subtidal areas. The white-tipped projections called certata contain stinging cells from hydroids they have eaten. The stinging cells help protect them from predators. Photo *B* courtesy of Scripps Institution of Oceanography.

A.

B.

C.

FIGURE 16–8   Striped Shore Crab.
A, striped shore crab, *Pachygrapsus,* backed into a rock crevasse. *B,* dorsal view of a preserved *Pachygrapsus. C,* preserved female *Pachygrapsus* showing eggs under the broad abdomen.

SEDIMENT-COVERED SHORE

The Sediment

Sediment-covered shore includes **beaches, salt marshes,** or **mud flats** that represent lower-energy environments. As the energy level decreases, the particle size and sediment slope decrease, and sediment stability increases; the energy level refers to the strength of the waves and the longshore current. Another characteristic of sediments closely related to particle size is **permeability,** the ease with which fluid can move through the deposit. Because of the strength of cohesive attraction between flat clay-sized particles and the greater angularity of small silt- and fine sand-sized particles compared to the more rounded shape of coarser sand-sized particles, the permeability of sediments increases with increased particle size. Therefore, breaking waves can more readily percolate down through coarse sands and replace the oxygen used up by animals buried in the sediment. This readily available oxygen also enhances the process of bacterial decomposition of dead tissue.

Life in the Sediment

Fragments of kelp broken from the rocky substrate by heavy wave action are commonly piled high on beaches during storms. Thus, some of the **suspension feeders** that filter plankton from the clear water of the rocky shore are replaced by **deposit feeders** that eat detritus on many sandy beaches and ingest sediment in mud flats.

The sandy beach supports fewer species than the rocky shore, and mud flats fewer still; however, the number of individuals may be as high. In the low-tide zones of some beaches and on mud flats, as many as 5000 to 8000 burrowing clams have been counted in one square meter (10.8 ft^2). Burrowing is the most successful adaptation for life in the sediment-covered shore, so life is less visible. By burrowing a few centimeters beneath the surface, organisms can find a stable environment where they are not bothered by fluctuations of temperature and salinity and the threat of desiccation.

Burrowing in the sediment does not prevent animals from straining plankton from the clear water above the sediment as do many of the animals of the rocky shore. By the use of various techniques, the water above the sediment surface can be stripped of its plankton content by buried animals. Deposit feeders collect debris from the sediment surface and eat their way through the sediment, digesting its organic content and excreting inorganic particles. Burrowing carnivores chemically sense the presence of their prey in the darkness of the infaunal habitat (fig. 16–9).

The Sandy Beach

Bivalve Mollusks Of all the animals that live in the sediment-covered shore, the bivalve mollusks are surely among the best adapted. The greatest variety of clams are found buried beneath the low-tide regions of sandy beaches; their numbers decrease as the sands become muddier. The **pelecypods**, or bivalve mollusks, possess a soft body, a portion of which is called the **mantle**; this part secretes the calcium carbonate lateral valves that hinge together on the dorsal region of the body. The foot extends anteriorly to dig into the sediment and pull the valves and posterior **siphons** down after it. The tip of the foot is then pumped full of blood to make it swell and anchor the mollusk, while retractor muscles running the length of the foot contract and pull the animal down. How deep the bivalve can bury itself depends on the length of its siphons; they must reach above the sediment surface to pull in water from which plankton will be filtered. Oxygen is also extracted in the gill chamber before the water is expelled through the excurrent siphon. Indigestible particulate matter will be forced back out the incurrent siphon periodically by quick muscular contractions (fig. 16–10).

Annelid Worms A variety of **annelids**, or **segmented worms**, are also well adapted for life in the sediment. Most common of the sand worms is *Arenicola* sp., or **lugworm**. It lives in a U-shaped burrow, the walls of which are strengthened by mucus. With its finely hooked **parapods** gripping the sides of the tube, it moves forward and backward. The worm

FIGURE 16–9 Modes of Feeding along the Sediment-Covered Shore.

A, suspension feeding. This method is used by clams that bury themselves in sediment and extend siphons through the surface to pump in overlying water. They feed by filtering plankton and other organic matter that is suspended in the water. *B*, deposit feeding. Some deposit feeders such as the segmented worm *Arenicola* feed by ingesting the sediment and extracting organic matter contained within it. Others, like the amphipod *Orchestoidea*, feed on more concentrated deposits of organic matter, detritus, found on the sediment surface. *C*, carnivorous feeding. The sand star, *Astropecten*, cannot climb rocks like its sea star relatives, but it can burrow rapidly into the sand, where it feeds voraciously on crustaceans, mollusks, worms, and other echinoderms.

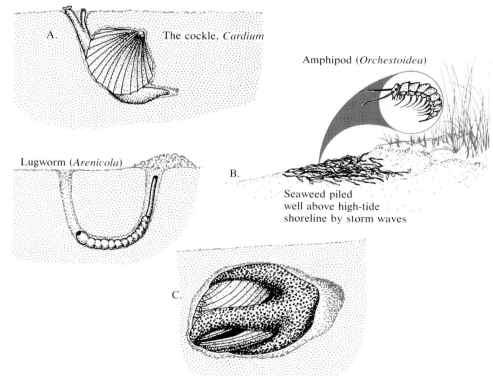

A. The cockle, *Cardium*

Amphipod (*Orchestoidea*)

Lugworm (*Arenicola*)

B.

Seaweed piled well above high-tide shoreline by storm waves

C.

FIGURE 16–10   How a Clam Burrows.
Clams exposed at the sediment surface will quickly burrow into the sediment by *(A)* extending the
point-shaped foot into the sediment and *(B)* forcing the foot deeper into the sediment and using
this increasing leverage to bring the exposed, shell-clad body toward vertical. When the foot has
penetrated deep enough, a bulbous anchor forms at the tip *(C)*, and a quick muscular contraction
pulls the entire animal into the sediment *(D)*. The siphons are then pushed up above the sediment
to pump in water from which the clam will extract food and oxygen *(E)*.

moves forward to feed and extends its proboscis up
into the head shaft of the burrow to loosen sand with
quick pulsing movements. A cone-shaped depression
forms at the surface over the head end of the burrow
as sand continually slides into the burrow and is ingested by the worm. As the sand passes through the
digestive tract, the organic content is digested, and
the processed sand is deposited as castings at the
surface surrounding the opening or openings to the
tail shaft (fig. 16–9B).

Crustaceans   Staying high on the beach and feeding
on kelp cast up by storm or high-tide waves, numerous
amphipod crustaceans called **sand hoppers** are found
(fig. 16–9B). They are known to jump distances more
than 2 m (6.6 ft). A common genus is *Orchestoidea*,
whose members usually range from 2 to 3 cm (0.8 to
1.2 in.) in length. Laterally flattened, they usually spend
the day buried in the sand or hidden in the kelp on
which they feed. They become active at night and may
form large clouds above the masses of seaweed on
which they are feeding.
   A larger crustacean that may be harder to find on
sandy beaches is a member of the variety of **sand crabs**

belonging to the genera *Blepharipoda, Emerita,* and
Lepidopa. Ranging in length from 2.5 to 8 cm (1 to 3 in.),
they move up and down the beach near the shoreline.
Burying their bodies into the sand and leaving their
long curved V-shaped antennae pointing up the beach
slope, these little crabs filter food particles out of the
water (fig. 16–11).

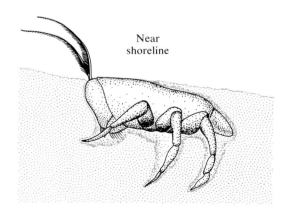

FIGURE 16–11   Sand Crab, *Emerita*.

Echinoderms Echinoderms are represented in the beach deposits by the **sand star**, *Astropecten*, and **heart urchins**. Members of the genus *Astropecten* are called *sand stars* because they are well adapted to prey on invertebrates buried in the low-tide region of sandy beaches. With five rays and a smooth dorsal surface, sand stars have sturdy spines along the margins of the rays and tube feet tapered to a point to aid them in moving through the sediment (fig. 16–9C).

The most widely distributed heart urchins belong to the genus *Echinocardium* (fig. 16–12). More flattened and elongated than the sea urchins of the rocky shore, the heart urchins live buried in the sand near the low-tide line. They gather sand grains into the mouth, where the coating of organic matter is scraped off and ingested.

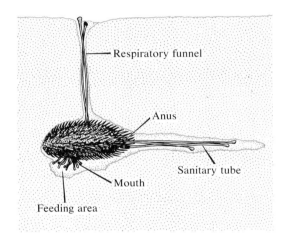

FIGURE 16–12 Heart Urchin, *Echinocardium*.
The chamber housing the body of the heart urchin, the respiratory funnel, and the sanitary tube are maintained by mucus distributed by the tube feet. The sand grains within the feeding area are processed very quickly, and the urchin must move a distance of 15 cm (6 in.), or two body lengths, about every half-hour.

Meiofauna Living in the spaces between sediment particles are tiny organisms ranging in size from 0.1 to 2 mm (0.004 to 0.08 in.) in length (fig. 16–13). They feed primarily on bacteria removed from the surface of sediment particles. The meiofauna population, composed primarily of polychaetes, mollusks, arthropods, and nematodes, is found in sediment from the intertidal zone to the deep-ocean trenches.

Intertidal Zonation Although it is difficult to see, there exists a characteristic faunal distribution across the intertidal range of sediment-covered shores that is similar to that observed on rocky shores. Although different species of animals are found in each of the corresponding intertidal zones, the distribution of the life forms across the intertidal rocky and sediment-covered shores is similar in two ways: the maximum number of species and the greatest biomass is found near the low-tide shoreline, and both decrease toward the high-tide shoreline. This zonation, shown in figure 16–14, is best developed on steeply sloping, coarse sand beaches and is less readily identified on the gentler-sloping fine sand beaches. The tiny clay-sized particles of the mud flat produce a deposit with essentially no slope, thus eliminating the possibility of zonation in this protected, low-energy environment.

The Mud Flat

Two widely distributed plants of the mud flat are **eelgrass**, *Zostera* (fig. 16–15), and **turtle grass**, *Thalassia*, which occupy the low-tide zone and adjacent shallow sublittoral regions bordering the flats. Numerous openings at the surface of a mud flat attest to a large population of bivalve mollusks and other invertebrates.

A.

B.

C.

FIGURE 16–13 Meiofauna.
Scanning electron micrographs of *A*, nematode head (×804). The projections and pit on right side are possibly sensory structures. *B*, amphipod (×20). This organism builds a burrow of cemented sand grains. *C*, polychaete (×55) with its proboscis extended. Courtesy of Howard J. Spero, University of South Carolina.

FIGURE 16–14   Intertidal Zonation on the Sediment-Covered Shore.
Zonation is best displayed on coarse sand beaches with steep slopes. As the sediment becomes finer and the beach slope decreases, zonation becomes less distinct. It disappears entirely on flat mud flats.

FIGURE 16–15   Life of the Mud Flat.

Zostera, eelgrass, and *Uca,* fiddler crab.

Quite visible and interesting inhabitants of the mud flats are the fiddler crabs, *Uca* (fig. 16–15), living in burrows that may be more than 1 m (3 ft) deep. Relatives of the shore crabs, they usually measure no more than 2 cm (0.8 in.) across the carapace. Fiddler crabs get their name from one outsized claw, up to 4 cm (1.6 in.) long, on the male. This large claw is waved around in such a manner the crab seems to be fiddling. The females have two normal-sized claws. The large claw of the male is used to court females and fight off competing males. They feed by extracting organic matter from the mud.

THE SHALLOW OFFSHORE OCEAN FLOOR (SUBLITTORAL)

Extending from the spring low-tide shoreline to the seaward edge of the continental shelf is an environment that is mainly sediment-covered, although bare rock exposures may be found locally. This region conforms roughly with what has been described as the sublittoral zone, the continental shelf.

The Rocky Bottom

A rocky bottom within the shallow inner sublittoral region is usually covered with algae. Along the Pacific Coast of North America, the giant bladder kelp, *Macrocystis,* is known to attach to rocks at depths as great as 30 m (98 ft) if the water is clear enough to allow sunlight to support plant growth. The giant bladder kelp and another fast-growing kelp, *Nereocystis,* often form bands of kelp forest along the Pacific Coast. Smaller tufts of red and brown algae are found on the bottom and living epiphytically (one plant living nonparasitically on another plant) on the kelp fronds. Epifauna commonly found growing along with the algal tufts on the fronds are hydroids and bryozoan colonies. All of these smaller life forms serve as food for many of the animals found living within the kelp forest community. Nudibranchs are important as predators on the hydroids and bryozoan colonies. Surprisingly, very few animals feed directly on the living kelp plant. Among those that do are the large sea hare, *Aplysia,* and sea urchins (fig. 16–16).

Large crustaceans called lobsters are common to rocky bottoms. They are a somewhat varied group with robust external skeletons. The palinuran spiny lobster is named for its spiny carapace and is characterized by two very large and spiny antennae that have noise-making devices near their bases. The genus *Palinurus* is found at depths greater than 20 m (66 ft) along the coast of Europe, reaches lengths up to 50 cm (20 in.), and is considered a delicacy. The species *Panulirus argus,* found in the Caribbean, is sometimes observed to migrate single file for unknown reasons over distances of several kilometers. *Panulirus interruptus* is the spiny lobster of the American west coast. All of the spiny lobsters are taken for food, but none are as highly regarded as the so-called true lobsters belonging to the genus *Homarus.* Although they are scavengers like their spiny relatives, the true lobsters—which include the American lobster, *Homarus americanus*—also feed on live animals, including mollusks, crustaceans, and members of their own species. Migrating into deep water during winter and returning to nearshore waters in summer, the American lobster is nocturnal and is found from Labrador to Cape Hatteras (fig. 16–17).

Oysters are sessile bivalve mollusks found in estuarine environments. They prefer locations where there is a steady flow of clean water to provide plankton and oxygen. Having great commercial importance throughout the world, they have been closely studied. Oyster beds consist of the empty shells of generation after generation cemented to rock bottom or one another, with the living generation on top. Each female will produce many millions of eggs each year which become planktonic larvae when fertilized. After a few weeks as plankton, the larvae settle to attach themselves to the bottom. The larvae prefer live oyster shells, dead oyster shells, and rock, in that order, as a substrate for attachment. Oysters serve as prey for a variety of sea stars, fishes, crabs, and boring snails that drill through the shell and rasp away the soft tissue of the oyster (fig. 16–18).

Coral Reefs

Although corals are found throughout the ocean, coral accumulations that might be classified as reefs are restricted to the warmer-water regions where the average monthly temperature exceeds 18°C (64°F) throughout the year (fig. 16–19). Such temperature conditions are found primarily between the tropics, although reefs grow to latitudes approaching 35°N and S on the western margins of ocean basins where warm-water masses move into the high-latitude areas and raise average temperatures.

Because of changes in wave energy, salinity, water depth, temperature, and other less obvious factors, there is a well-developed vertical and horizontal zonation of the reef slope. These zones are readily identified by the assemblages of plant and animal life found in and near them (fig. 16–20).

**FIGURE 16–16 Life of the
Sublittoral Zone.**
A, giant bladder kelp, *Macrocystis,*
and *B,* sea hares, *Aplysia californica,*
and sea urchins in a kelp forest.
C, purple sea urchin, *Strongylocen-
trotus purpuratus.* Photos *A* and *C*
by Mia Tegner; *B* by James Mc-
Cullagh.

A.

B.

C.

A.

B.

FIGURE 16–17   Spiny and American Lobsters.
A, spiny lobster. Photo by B. Kiwala. *B,* American lobster. Photo by Harold W. Pratt/Biological Photo Service.

FIGURE 16–18   An Oyster Drill Feeding on an Oyster.

Not only coral, but algae, mollusks, and foraminifers make important contributions to the reef structure. Reef-building corals are **hermatypic.** They have a symbiotic relationship with the algae zooxanthellae that live within the tissue of the coral polyp. Not only reef-building corals, but other reef animals have a similar symbiotic relationship with algae. These heterotrophs that derive part of their nutrition from algal symbionts are called **mixotrophs.** They include corals, foraminifers, sponges, and mollusks (fig. 16–21). Because light is essential for algal photosynthesis, reef-building corals are restricted to clear, shallow bottom waters. The algae contribute not only to nourishing the coral, but could contribute to the calcification capability of corals by extracting carbon dioxide from the animals' body fluids. This would increase the concentration of the carbonate ion needed for the precipitation of calcium carbonate. The relationship is representative of **mutualism,** in which both parties benefit. The corals contribute to the relationship by providing a supply of nutrients to the zooxanthellae.

Reef growth also requires that the water have a relatively normal salinity and that it be free from particulate matter. Therefore, we see very little coral reef growth near the mouths of rivers that lower the salinity and carry large quantities of suspended material that would choke the reef colony.

Coral reefs actually contain up to three times as much plant as animal biomass. The zooxanthellae account for less than 5% of the reef's plant mass; most of the rest is filamentous green algae. However, zooxanthellae account for up to 75% of the biomass of reef-building corals and provide the coral with up to 90% of its nutrition.

The greatest depth to which active coral growth extends is 150 m (492 ft), below which there is not enough sunlight to support hermatypic mutualism. Water motions are not great at these depths, so relatively delicate **plate corals** can live in the outer slope of the reef from 150 m (492 ft) to about 50 m (164 ft), or to depths where light intensity exceeds 4% of the surface intensity. From 50 m to about 20 m (66 ft), the strength of water motion from breaking waves increases on the side of the reef facing into the prevailing current flow. Correspondingly, the mass of coral growth and the strength of the coral structure supporting it increases toward the top of this zone,

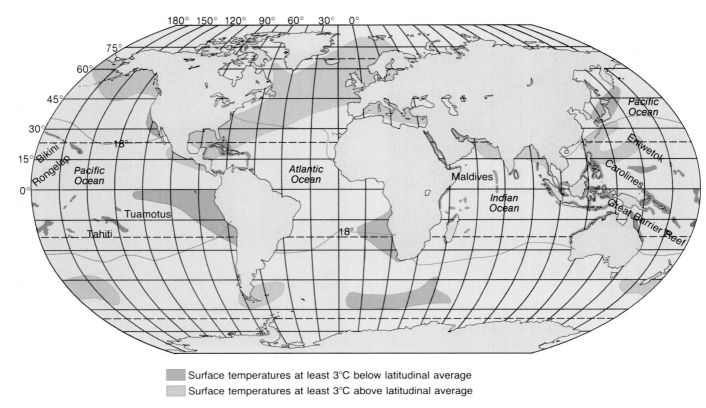

Surface temperatures at least 3°C below latitudinal average

Surface temperatures at least 3°C above latitudinal average

FIGURE 16—19 Coral Reef Distribution.

Minimum water temperatures of 18°C (64°F) in the surface waters of the Northern and Southern hemispheres occur in February and August, respectively. Coral reef development is restricted to the low-latitude area between the two 18°C temperature lines shown on the map. Observe that in each ocean basin, the width of the coral reef belt is wider on the west side because here boundary currents carry warm water away from the equator, while cold currents converge on the equator on the east side of the oceans.

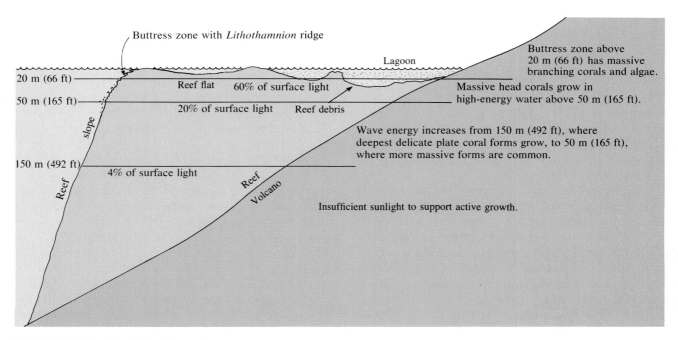

FIGURE 16—20 Coral Reef Zonation.

FIGURE 16–21 Coral Reef Inhabitants That Depend on Algal Symbionts.

A, polyps of a reef coral extend their tentacles to capture tiny planktonic organisms from the surrounding water. However, for most reef-building corals, most of their nourishment is provided by symbiotic algae, zooxanthellae, that live in the coral tissue. *B,* the blue-gray sponge on the left is *Niphates digitalis,* a totally heterotrophic sponge. On the right is *Angelas* sp., which contains some cyanobacterial symbionts. It is, like most reef-building corals, a mixotroph. Photo was taken in 20 m (60 ft) of water at Carrie Bow Cay, Beliz. *C,* a giant clam, *Tridacna gigas.* These suspension feeders also depend on symbiotic algae living in the mantle tissue. Photos *A* by Christopher Newbert, *B* by C.R. Wilkerson, Australian Institute of Marine Science, *C* by Ken Lucas/Biological Photo Service.

A.

B.

C.

where light intensity exceeds 20% of surface value. **Head corals** are dominant. The **buttress zone** extends from 20 m to the low-tide line; within it more-massive **branching corals**, *Acropora*, and encrusting algae such as the red alga, *Lithothamnion*, withstand the crashing waves. Light intensity exceeds 60% of surface intensity in the buttress zone. The waves cut **surge channels** across the algal ridge, which is inhabited by only a few animals—such as snails, limpets, and the slate pencil urchin, *Heterocentrotus*—that can withstand the constant beating of the surf. The surge channels extend down the reef slope as **debris channels** that carry the products of wave erosion. Small reef fish find protection from larger predators such as sharks, barracuda, and jacks in the debris grooves extending between the buttress ridges. The reef flat extends across the reef beyond the lagoons of atolls and barrier reefs. Here the reef may be under a few centimeters to a few meters of water at low tide. A variety of beautiful reef fish inhabit this shallow water. The sand of reef debris and foraminifer tests fills in the deeper holes and provides a home for sea cucumbers, worms, and a variety of mollusks. In the protected water behind the *Lithothamnion* ridge lagoonal reefs may form, where species of *Porites* and *Acropora* grow into beautiful large colonies. Gorgonian coral, anemones, crustaceans, mollusks, and echinoderms of great variety also will be found in the lagoon reef (fig. 16–22).

An obvious example of **commensalism** (one organism benefits while the other is unaffected) is the behavior of the **shrimpfish**, which swims head down among the long slender spines of the reef sea urchins. The urchin's spines are a significant deterrent to any predator, and the sea urchin is neither hindered nor aided by the presence of the little fish. The **clown fish** receives a similar protection by swimming among the tentacles of two species of sea anemones. This relationship is believed to be mutual since the anemones benefit by the clown fish serving as bait to draw other fish within reach of anemone tentacles and actually carrying food to them (fig. 16–23). A variety of **cleaner shrimp** and **cleaner fishes** that set up cleaning stations on coral reefs are known to be essential to the well-being of many of the larger fishes on the reef by removing parasites and infected tissue.

Coral Reefs and Nutrient Levels When human populations increase in lands adjacent to coral reefs, the reefs deteriorate. There are many aspects of human behavior that would obviously damage the reef, but one of the more subtle is the inevitable increase in the nutrient levels of the reef waters. As nutrient levels increase in reef waters, the dominant benthic community changes. At low nutrient levels the hermatypic corals and other reef animals that contain algal symbionts thrive. As nutrient levels in the water increase, conditions favor development of fleshy benthic plants at moderate nutrient levels and heterotrophic suspension feeders at high nutrient levels. At high nutrient levels, phytoplankton mass exceeds the benthic plant mass, and benthic populations tied to the phytoplankton food web dominate. The clarity of water is reduced by increased phytoplankton biomass, and the fast-growing members of the phytoplankton-based ecosystem destroy the reef structure by overgrowing the slow-growing coral and by **bioerosion**. Bioerosion by sea urchins and sponges is particularly effective in damaging the reef.

Bleaching of Coral Reef Communities The bleaching or loss of color by coral reef organisms has occurred on a local basis numerous times in the past. However, a mass mortality of at least 70% of the corals along the Pacific Central American coast occurred as a result of a bleaching episode associated with the severe El Niño of 1982–83. Although the cause is not known for certain, this eastern Pacific bleaching may have been caused by the increased water temperatures associated with the El Niño event. Whatever the cause, it will take many years for these reefs to recover.

One thing is clear, the direct cause of the bleaching is the expulsion of symbiont algae, zooxanthellae. Essentially all reef-building corals and some other reef mixotrophs are nourished by these algae that live within their tissue. The loss of this source of nourishment can be life threatening to the reef dwellers.

Figure 16–24 shows the bleaching of a round starlet coral, *Siderastrea siderea*, on Enrique Reef, Puerto Rico. The photograph was taken in November 1987. Normally a rusty brown color, this coral has been bleached on its lower left half.

A husband and wife team of coral reef biologists, Drs. Lucy Bunkley Williams and Ernest H. Williams, Jr., of the University of Puerto Rico, have provided the following statement to give us an expert view concerning the possible causes of coral reef bleaching and implications for the future:

Extensive bleaching (loss of zooxanthellae) of 60 species of stoney corals, fire corals, stylaster corals, sea anemones, coral-like anemones, zoanthids, gorgonians, and at least 3 orders of sponges, began in the northern Caribbean, western Colombia, Bahamas, Florida and

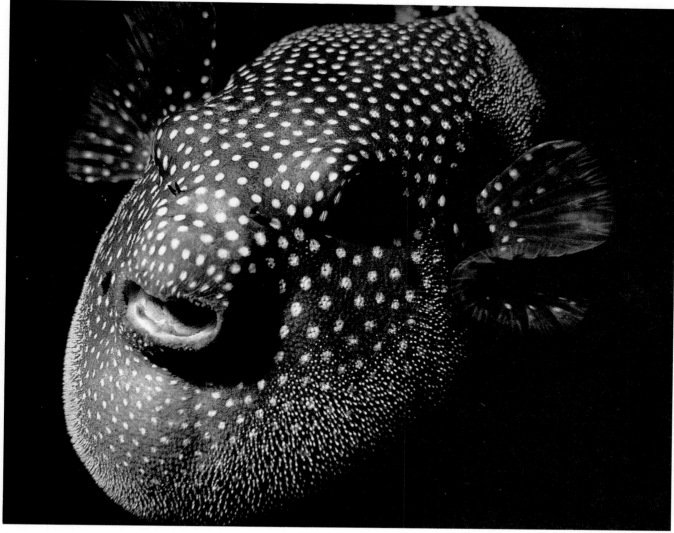

A.

FIGURE 16–22   Nonreefbuilding Inhabitants.

A, one of the more unusual of the great variety of colorful and strange fishes associated with coral reef and other shallow-water, tropical environments are the puffers. When a potential predator comes too close, they gulp water to fill a ventral stomach pouch, expanding their loose, scaleless skin to produce a spherical shape. Actually, puffers are seldom bothered by predators as they have viscera and skin that often contain deadly neurotoxin. Though they are feeble swimmers, puffers like this swelled-up member of the genus *Arothron* are virtually free from attack. *B,* many members of the coral family do not secrete the calcium carbonate of the reef builders. An example is a soft gorgonian coral, shown with feeding polyps extending from its branches. Photos by Christopher Newbert.

B.

FIGURE 16–23 Mutualism.
The clown fish, which lives unharmed among the stinging tentacles of the sea anemone, may pay for this protection by bringing food to the anemone. This symbiotic relationship that benefits both participants is called mutualism. Photo by Christopher Newbert.

FIGURE 16–24 Bleached Coral.
This round starlet coral, *Siderastrea siderea,* on Enrique Reef, Puerto Rico, has been bleached across the lower left side of the image. Photo by Dr. Lucy Bunkley Williams.

Texas from summer to early fall of 1987. Recovery of many animals began in late fall to early winter, while some bleaching continued and new bleaching began in other parts of the Caribbean at that time. The disturbance occurred from near the surface down to the approximate limit where zooxanthellae occur. Similar events occurred in the Pacific in Australia (January-June), Galápagos Islands (February-April), and possibly Hawaii, Thailand, and Indonesia. Some animals have died, but no massive die-offs as in the eastern Pacific during the 1982–83 bleaching event, have occurred. Higher than normal water temperatures, light-related effects, or a combination of both, have been suggested as the most likely causes of the event. Bleaching of coral reef animals, while very important, may have a larger meaning as an indicator of a more widespread and enduring problem.

The closing sentence may well be a reference to the broad scope of human-induced changes throughout the earth. Increased concentrations of greenhouse gases in the lower atmosphere; depletion of ozone in the upper atmosphere; pollution of the oceans with petroleum, plastics, sewage, and so on; and the bleaching of coral reefs are all symptoms of a worldwide pathology that will require major modifications in human behavior before it can be cured.

DEEP-OCEAN FLOOR

The Physical Environment

The deep-ocean floor includes the bathyal, abyssal, and hadal zones described in chapter 13. Light is present in only the lowest concentrations above 1000 m (3280 ft) and absent below this depth. Everywhere the temperature is low, rarely exceeding 3°C (37°F) and falling as low as −1.8°C (28.8°F) in the high latitudes. Pressure exceeds 200 atmospheres on the oceanic ridges, ranges between 300 and 500 atmospheres on the deep-ocean abyssal plains, and climbs to over 1000 atmospheres in the deepest trenches. (The pressure is one atmosphere at the ocean surface and increases by one atmosphere for each 10 m, or 33 ft, of depth.) Much of the deep-ocean floor is covered by at least a thin layer of sediment ranging from the mudlike abyssal clay deposits of the abyssal plains and deep trenches, through the oozes of the oceanic ridges and rises, to some coarse sediment deposited as turbidites on the continental rise. Near the crests of the oceanic ridges and rises and down the slopes of seamounts and oceanic islands, sediment is absent and basaltic ocean crust is the substrate.

Deep-Sea Hydrothermal Vent Biocommunities

In 1977 the first active hydrothermal vent field was discovered at a depth below 2500 m (8200 ft) in the Galápagos Rift, near the equator in the Eastern Pacific Ocean (figs. 16–25 and 16–26). Water temperatures in the immediate area of the vents ranged between 8° and 12°C (46° and 54°F), while the normal bottom water temperature at these depths is about 2°C (36°F). These vents were found to support the first known **hydrothermal vent biocommunities**, consisting of unusually large organisms for these depths. The more prominent members of this community are **tube worms** with tubes over 1 m long, giant **clams** up to 25 cm (10 in.) in length, large **mussels**, and two varieties of white crabs.

At 21°N latitude on the East Pacific Rise, south of the tip of Baja California, tall chimneys were found to belch out 350°C (662°F) vent water so rich in metal sulfides that it was black. These chimney vents, first observed in 1979, were composed primarily of sulfides of copper, zinc, and silver and came to be called **black smokers**. A new species of **vent fish** associated with some vent fields was first observed here.

The most important members of these biocommunities are the **sulfur-oxidizing bacteria** that synthesize, through chemosynthesis, the organic molecules that feed the community. The bacteria use the chemical energy released by sulfur oxidation much as marine plants use the sun's energy to carry on photosynthesis. Although some larger animals feed on the bacteria, most of them depend primarily on a symbiotic relationship with the bacteria. For instance, the tube worms and giant clams depend entirely on sulfur-oxidizing bacteria that live symbiotically in their tissues.

In 1981 the **Juan de Fuca Ridge** biocommunity was observed. Although vent fauna at this site is less robust than at the Galápagos Rift and on the East Pacific Rise, the **metallic sulfide deposits** associated with the vents aroused much interest because they are so near the U.S. coast.

The **Guaymas Basin**, in the Gulf of California, provided the first observation, in 1982, of hydrothermal vents beneath a thick layer of sediment. Of particular interest is the fact that sediment and sulfide samples recovered were saturated with hydrocarbons, which may have entered the food chain through bacterial uptake. The abundance and diversity of life may exceed that of the rocky bottom vents. A phenomenon similar to that of the Guaymas Basin was observed in 1987 at the **Mariana Basin**, a small spreading-center system beneath a sediment-filled back-arc basin.

FIGURE 16–25 Alvin Approaches a Hydrothermal Vent Community Typical of That Found on the East Pacific Rise.

In the lower left corner are a sea anemone and three octacorals, which are close relatives of the anemone. They may feed primarily on sulfur-oxidizing bacteria suspended in the water near the vents. Grenadier, or rattail fish, are common in the deep ocean and are usually the first to arrive at bait placed on the deep-ocean floor. Vestimentiferan tube worms *(Riftia)* and giant clams *(Calypotogena)* do not possess guts; they are nourished by chemosynthetic bacteria that live in their tissues. White brachyuran crabs swarm over the lava pillows, and a "black smoker" spews hot (350°C) sulfide-rich water from its metallic sulfide chimney.

Subsequent exporation has revealed numerous hydrothermal vent biocommunities in the Pacific Ocean, including one in the Southern Hemisphere on the East Pacific Rise at 20°S. The first active hydrothermal vents with associated biocommunities in the Atlantic Ocean were discovered in 1985 at depths below 3600 m (11,800 ft) near the axis of the Mid-Atlantic Ridge, which divides the Atlantic Ocean in half. The predominant fauna of these vents consists of shrimps that have no eye lens but can detect levels

of light emitted by the black smoker chimneys that is not visible to the human eye (fig. 16–27). After these Atlantic discoveries, at 23°N, and 26°N latitude, it now seems likely that similar biocommunities will be found in significant numbers in all the world's oceans.

A relict vent area found on the Galápagos spreading center indicates vent communities may have a relatively short life span. This inactive vent was identified by an accumulation of dead clams. It appears

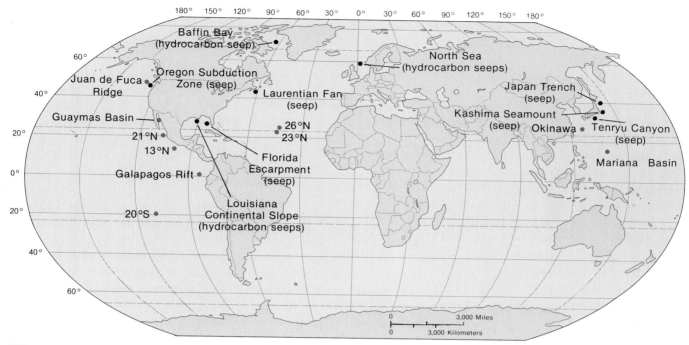

FIGURE 16–26 Hydrothermal Vents, Cold Seeps, and Hydrocarbon Seeps Known to Support Biocommunities.

FIGURE 16–27 Atlantic Ocean Hydrothermal Vent Organisms.

Swarm of particulate-feeding shrimp, the predominant animals observed at hydrothermal vents near 23°N and 26°N on the Mid-Atlantic Ridge. This swarm was photographed at 26°N. Courtesy of Peter A. Rona, NOAA.

LIFE ON THE DEEP-OCEAN FLOOR

FIGURE 16A   Dr. Robert R. Hessler.

Interview with Dr. Robert R. Hessler (fig. 16A), professor of oceanography, Scripps Institution of Oceanography, University of California at San Diego.

Q. Dr. Hessler, when does the evidence tell us life first appeared on the deep-ocean floor?

A. There is no absolute evidence for an answer to that question. The conditions for life on the deep-sea floor have probably existed for as long as the deep sea has existed. There has probably always been oxygen and food. The present ocean sediments give us evidence only back to the Cretaceous period (65 to 136 million years ago).

Q. How about the fossil record? Are there any fossils found in rocks that appear to have been deposited in an ancient deep-ocean environment?

A. I have never pursued this much myself. I think there are some. Some of the black shales are thought to have accumulated in deep-basin conditions.

Q. Do you have any ideas about, in an evolutionary sense, where life in the deep ocean came from?

A. My guess would be that ultimately life must have begun in shallow water and penetrated the deep sea from there. One would expect animal life to have evolved closest to the source of primary production, and that can only be in shallow water. Once you get deeper than a few hundred meters there isn't enough light to support primary productivity. Furthermore, considering the variety of physical circumstances in shallow water, the multitude of biogeographic barriers, and so on, I would have expected that to be the area where evolution was the most active. Influx into the deep sea has probably been continuous.

Q. What groups seem to be most effective in achieving this—moving from the shallow water into the deep benthic environment?

A. Some groups have done particularly well. All of the phyla make it into the deep sea. As you go to lower taxonomic levels, the deep-sea fauna is more and more unusual, so that at the level of species, hardly any exist that are common to both the deep sea and shallow water. The **percarid crustaceans,** macrofauna that brood their young—amphipods (fig. 16–13B), isopods, tanaids, and cumaceans—have been extraordinarily successful in the deep sea. You can get as many as 100 species of **isopod** in a single sample. **Bivalve mollusks** and polychaete worms are also very abundant and diverse. In certain environments, other groups tend to be important. Sometimes you find large quantities of sipunculids, sometimes large quantities of pogonophorans. Among the very small animals, another group that is very important is the nematodes. There are incredible quantities of **nematodes** in the deep sea. The tiny harpacticoid copepods (deposit-feeding copepods less than 2 mm, or 0.08 in., in length) are also very small and abundant. **Foraminifers** are an interesting group. One doesn't think of the foraminifers as being an important group of benthic organisms, but in the deep sea they are dominant, certainly in terms of biomass and probably also in terms of species diversity.

If you look at the megafauna (the animals that can be photographed), **holothurians** are fairly important. Glass sponges, tunicates, gorgonians, and certain kinds of decapods are also common. There are a variety of deep-sea families of fishes that are fairly characteristic. The **grenadiers** are almost always seen.

Q. Which of the physical factors such as pressure, temperature, and the absence of light most limits the vertical migration of life in the deep ocean?

A. **Temperature** seems to be the most important factor that prevents deep-sea animals from migrating upward. The deep sea is characterized by being isothermal; that is, at any one spot the temperature hardly varies by a measurable amount, and deep-sea animals have become accustomed to this. This means that even a slight change in temperature is likely to have an enormous effect. They can't break through this temperature barrier. However, at north and south high latitudes, you find portions of the deep-sea fauna in shallow water as well. I think the reason is that the temperature barrier isn't there. It's equally cold in the surface waters. With animals penetrating the deep sea from shallow water, the issue is more complex. My guess is that the most important factor in preventing them from penetrating the deep sea is not physical barriers, but **food**. In order to survive in the deep sea, animals have to be adapted to conditions in which both the rate of supply and the concentration of food are very low. An organism that does not have a life style that can cope with this is not going to make it in the deep sea. Temperature adaptation for a shallow-water animal isn't that difficult. Of the shallow-water species that live through the summer and winter, some are capable of adapting to 20° to 25°C (36° to 45°F) temperature changes on an annual basis. Shallow-water animals could get, or may already be, used to low temperature, but becoming used to low food supply requires major changes in behavior, morphology, and physiology. That's harder to achieve.

Q. The overall consensus is that food is in low supply in general on the deep-ocean floor?

A. That is correct. The reason we think that is so is because one finds so little life there. Any one sample has orders-of-magnitude less life in it than one finds in shallow water. There is only one sensible explanation for that. There's not much food.

Q. We are talking here in terms of biomass?

A. Total biomass—but it is also true in terms of numerical abundance. These two, of course, tend to be correlated.

Q. How about the occasional availability of large supplies of food? Does it have any significant bearing on the distribution of life in the deep ocean?

A. This is called the dead whale hypothesis. The first thing I have to mention is that I don't believe that anyone has written a completely convincing budget for the rate of food supply to the deep sea. So I don't think anyone knows, with any degree of confidence, the rate at which small particles of food are descending to the deep sea—and nobody knows anything at all about the big ones. What we do know about large organic falls is that there seems to be a group of animals that are very well adapted for taking advantage of them. These are the fish and amphipods. When a large organic fall hits the bottom, it isn't long before these animals are there in large quantities consuming it at a fairly rapid rate, and secondarily dispersing it to the rest of the fauna. This dispersal is achieved through their feces, through breaking the food parcel into small particles which are scattered about, and by eating there and dying elsewhere.

Q. What evidence is there that deep-ocean life is moving out of the deep sea and back to the shallow shelf environment from which you think it came?

A. There's not very much. You have to ask yourself, What kind of evidence can I rely on to detect such a phenomenon? It is difficult to come up with really strong criteria, but there is one which isn't bad. It is the absence of eyes. In a group I work on, the isopods, there's a large variety of families that live in the deep sea, but are also found in shallow water at high latitudes. What's interesting about them is that they lack eyes. I won't go into all of the details of the argument, but it looks as though the reason they lack eyes in shallow water is because they originally lost them in the deep sea. When they reinvaded shallow water, they couldn't evolve them again. This is an indication that there are some groups that have come back into shallow water in high latitudes. This apparently happens in other groups as well, but not with every group.

Q. Is there anything else that is of general importance to the nature of life on the deep-ocean floor that you might discuss?

A. Well, there are some very important issues. Let me mention some. First of all, let me emphasize that everything that has been done in the deep sea has been done on soft, mud bottoms. The reason for that is simply that oceanographers have gotten tired of losing their gear while sampling rocky bottoms. But now with the discovery of the **deep-sea hydrothermal vents** and their wonderful fauna, more and more people are getting interested in studying that hard bottom environment. I think that is going to be the impetus for studying hard bottom environments in general, of which there are an awful lot in the deep sea.

Staying with the hydrothermal vents for a second, I should emphasize that their fascinating fauna is really very different from the normal deep-sea fauna. Not the least of the differences is the fact it is so abundant. This is a nice thing to have found because it seems to be good documentation of the fact that food really is the limiting factor. Here is a source of large quantities of food in the deep sea, and, lo and behold, there is a large amount of biomass taking advantage of it. And these individuals are relatively large, as opposed to most of the animals of the deep sea.

Another point about the deep sea which deserves mention is that almost all deep-sea animals are **deposit feeders**. Here, I am talking about mud bottom, where suspension feeders and carnivores are rare. This introduces a rather interesting problem. For years now, there has been a debate as to how there can be so many deposit feeders in the same space at the same time. It is a very difficult issue, and probably never will be solved in the deep sea.

Q. With the discovery of large biomass concentrations at the vents, does this give us any hope that further study of the deep-ocean biology will lead to an increase in the food that will be available to us from the ocean?

A. Only to the most limited extent. Except for the vent community, which is inedible because of ambient water chemistry, life is too sparse on the ocean floor to support a fishery. There is a limited fishery for deep, bottom-associated fish, but the turnover rate is so slow in the deep sea that areas could be easily fished out.

that when the vent becomes inactive and the hydrogen sulfide that serves as the source of energy for the community is no longer available, the community dies.

Low-Temperature Seep Biocommunities

Three additional submarine spring environments have also been found to have chemosynthetically supported biocommunities. During the summer of 1984, an investigation of an ambient-temperature, hypersaline (46.2‰) seep at the base of the Florida Escarpment in the Gulf of Mexico (fig. 16–28) revealed a biocommunity similar in many respects to the hydrothermal vent communities. The seeping water appears to flow from joints at the base of the limestone escarpment and move out across the clay deposits of the abyssal plain at a depth of about 3200 m (10,496 ft).

The hydrogen sulfide–rich waters support a number of white bacterial mats that carry on chemosynthesis in a fashion similar to that of the bacteria of the hydrothermal vents. These and other chemosynthetic bacteria may provide most of the support for a diverse community of animals. The community includes holothurians, sea stars, shrimp, snails, limpets, brittle stars, anemones, vestimentiferan worms, galatheid crabs, clams, mussels, and zoarcid fish.

FIGURE 16–28 Hypersaline Seep Biocommunity at Base of Florida Escarpment.

A, rectangle is the location of the seep and biocommunity. *B,* seismic reflection profile of limestone Florida Escarpment and abyssal bedded sediments at its base. Arrow marks location of seep. *C,* the Florida Escarpment seep biocommunity consists of dense mussel beds over much of the image. White dots are small gastropods on mussel shells. At lower right are tube worms covered with hydrozoans and galatheid crabs. Fractures in the escarpment limestone are visible along top of the image. Courtesy of C. K. Paull, Scripps Institution of Oceanography, University of California, San Diego.

FIGURE 16–29 Hydrocarbon Seep Biocommunities in Gulf of Mexico.

Locations of trawls that recovered members of biocommunities living in vicinity of hydrocarbon seeps are shown. Courtesy of M. C. Kennicutt II, Texas A&M University.

Drilling into the limestone rocks of the platform east of the escarpment revealed fluids with temperatures of up to 115°C (239°F) and salinities of 250‰. This dense water apparently seeps deep into the jointed rocks of the platform and is trapped by the impermeable clay deposited on the ocean floor at the base of the escarpment. It then flows out on to the surface clays, mixes with sea water, and produces the spring water with salinity intermediate between that of the platform fluids and ocean water.

Also observed in 1984 were dense biological communities associated with oil and gas seeps on the Gulf of Mexico continental slope (fig. 16–29). Trawls at depths of between 600 and 700 m (1968 and 2296 ft) recovered epifauna and infauna similar to those observed at the hydrothermal vents and the hypersaline seep at the base of the Florida Escarpment. Carbon isotope analysis indicates the hydrocarbon-seep fauna is based on a chemosynthetic productivity that derives its energy from hydrogen sulfide and/or methane. Bacterial oxidation of methane results in the production of $CaCO_3$ slabs found here and at other hydrocarbon seeps shown in figure 16–26.

Finally, a third environment of vent communities was observed from *Alvin* during the summer of 1984. It is located at a **subduction zone seep** of the Juan de Fuca Plate at the base of the continental slope off the coast of Oregon (fig. 16–30). Here, the trench is filled with sediments. At the seaward edge of the slope,

FIGURE 16–30 Locations of Vent Communities off the Coast of Oregon.

The communities are associated with the subduction zone of the Juan de Fuca Plate. Sediment filling the trench is folded into a ridge with vents at its crest. Courtesy of L. D. Klum, Oregon State University.

clastic sediments are folded into a ridge. At the crest of the ridge, pore water escapes from the two-million-year-old folded sedimentary rocks into a thin overlying layer of soft sediment. Occurring at a depth of 2036 m (6678 ft) the seeps produce water that is only slightly warmer (about 0.3°C, or 0.5°F) than ambient conditions. The vent water contains CH_4 (methane) that is probably produced by decomposition of organic material in the sedimentary rocks. The methane serves as the source of energy for bacteria that oxidize it and chemosynthetically produce food for themselves and the rest of the community, which contains many of the same genera found at other vent and seep sites. During 1985, similar communities were located in subduction zones of the Japan Trench and Peru-Chile Trench. All the seeps are located on the landward side of the trenches at depths from 1300 to 5640 m (4264 to 18,499 ft).

SUMMARY

Over 98% of the approximately 160,000 species of marine animals live on the ocean floor. Most of these live on the continental shelf, with only 400-odd species found in deep-ocean trenches. The importance of temperature on the development of species diversity is reflected in the fact that there are three times as many species of benthos along the European coast warmed by the Gulf Stream than are along a similar length of North American coast cooled by the Labrador Current.

Because of tidal motions, the littoral (intertidal) zone can be divided into the high-tide zone (mostly dry), middle-tide zone (equally wet and dry), and the low-tide zone (mostly wet). The littoral zone is bounded by the supralittoral zone (covered only by storm waves) and the sublittoral zone, which extends below the low-tide shoreline.

The supralittoral zone along rocky coasts is characterized by the presence of the periwinkle snail, the rock louse, and the limpet.

Within the intertidal zone are found the barnacles, crustaceans that attach themselves to the rocks and comb plankton from the water. The barnacles most characteristic of the high-tide zone are the tiny buckshot barnacles. The most conspicuous plants of the high tide zone are *Fucus* along colder shores and *Pelvetia* in warmer latitudes. These plants become more abundant in the middle-tide zone, and the diversity and abundance of the flora and fauna in general increase toward the lower intertidal zone.

Larger species of acorn barnacles are found in the middle-tide zone. A common assemblage of rocky middle-tide zones includes the goose barnacle, the mussels, and sea stars. Joining the sea stars as predators on the barnacles and mussels are carnivorous snails.

Tide pools within the middle-tide zone will commonly house sea anemones, fishes such as the opaleye, woolly sculpin, and rockpool blenny, and hermit crabs. At the lower limit of the middle-tide zone sea urchins become numerous.

The dominance of animals in the upper- and middle-tide zones is ended in the low-tide zone, where plants become dominant.

Scampering across the entire intertidal zone are the scavenging shore crabs.

The only sediments usually found along rocky shores are a few cobbles and boulders. Moving into more protected segments of the shore, lower levels of wave energy allow deposition of sand and mud. Owing to the higher permeability of sand, sand deposits are usually well oxygenated.

Compared to the rocky shore, the diversity of species is reduced on and in beach deposits and is quite restricted in mud deposits. This does not mean that life is less abundant. Although life is less visible, up to 8000 burrowing clams have been recovered from one square meter (10.8 ft^2) of mud flat. Suspension feeders (filter feeders) characteristic of the rocky shore are still found in the sediment, but there is a great increase in the relative abundance of deposit feeders that ingest sediment and detritus.

Bivalve mollusks are well suited for sediment-covered shores. Clams that burrow into the sediment pump water through siphons and extract food and oxygen. The lugworm is a deposit feeder that ingests sand and extracts whatever organic content is available.

Amphipods, called sand hoppers, can be found feeding on the kelp deposited high on the beach by storm waves. Sand crabs filter food from the water with their long, curved antennae while buried in the sand near the shoreline.

Echinoderms are represented in the sandy beach by the sand star, which feeds on buried invertebrates, and the heart urchin, which scrapes organic matter from sand grains.

As is true for the rocky shore, the diversity of species and abundance of life on the sediment-covered shore increases toward the low-tide shoreline. Although it is more difficult to observe on sediment-covered shores, the high-, middle-, and low-tide zonation is also present. It is best developed on steep, coarse sand beaches and is probably missing on mud flats with little or no slope.

Although many burrowing forms inhabit the **mud flats,** the more visible life forms are the **eelgrass,** *Zostera,* and the **turtle grass,** *Thalassia,* found at its low-tide margin, and the **fiddler crabs,** *Uca,* that leave their burrows at low tide.

Attached to the rocky sublittoral bottom just beyond the shoreline is a band of algae including large **kelp** plants. Growing on the large fronds of the kelp are small varieties of **algae, hydroids,** and **bryozoans. Nudibranchs** feed on the small hydroids and bryozoans, and the sea hare and sea urchin feed on the kelp and other algae. **Spiny lobsters** are common to rocky bottoms in the Caribbean and along the U.S. west coast, and the **American lobster** is found from Labrador to Cape Hatteras.

Oyster beds found in estuarine environments consist of individuals that attach themselves to the bottom or the empty shells of previous generations. Living within these beds is a community consisting of suspension feeders such as **sponges, coelenterates, annelids, crustaceans,** and **tunicates. Sea stars, fishes, crabs,** and **boring snails** prey on the oysters.

Above a depth of 150 m (492 ft) in clear, nutrient-poor tropical waters, living **coral reef** can be found off the shores of islands and continents. Reef-building corals and other **mixotrophs** are **hermatypic,** containing algal symbionts, **zooxanthellae,** in their tissues. Delicate varieties are found at 150 m, and they become more massive near the surface, where wave energy is higher. The top 20 m (66 ft), the **buttress zone,** is reinforced with calcium carbonate deposited by algae such as *Lithothamnion,* producing an **algal ridge.** Waves cut **surge channels** across the ridge, and **debris channels** extend down the reef front below the surge channels. Many varieties of **commensalism** and **mutualism** are found within the coral reef biological community. There is some evidence that the potentially lethal "bleaching" of coral reefs may result from the expulsion of symbiont algae under stress of elevated temperature.

Although little is known of the **deep-ocean benthos,** it is clear that it is much more varied than previously thought. There seems to be a decreased variety of sea stars and mollusks and increased diversity of other echinoderms, isopods, amphipods, polychaetes, pycnogonids, and pogonophorans.

An important discovery of the **hydrothermal vent communities** made in 1977 on the Galápagos Rift has shown that, at least locally, **chemosynthesis** is an important means of primary productivity at these vents and at cold-water and hydrocarbon seeps.

QUESTIONS AND EXERCISES

1. Discuss the general distribution of life in the ocean. Include biomass and species diversity (variety of types of organisms) between the pelagic and benthic environments and within the benthic environment.

2. Diagram the intertidal zones of the rocky shore and list organisms characteristic of each zone.

3. Describe the dominant feature of the middle-tide zone along rocky coasts, the mussel bed. Include a discussion of other organisms found with the mussels.

4. List and describe crabs found within the rocky shore intertidal zone.

5. Discuss how sediment stability and permeability of sandy and muddy shores differ.

6. Other than predation, discuss the two types of feeding styles that are characteristic of the rocky, sandy, and muddy shores. One of these feeding styles is rather well-represented in all of these environments; name it, and give an example of an organism that uses it in each environment.

7. How does the diversity of species on sediment-covered shores compare with that of the rocky shore? Can you think of any reasons why this should be so? If you can, discuss them.

8. In which intertidal zone of a steeply sloping coarse sand beach would you find the following: clams, sand hoppers, ghost shrimp, sand crabs, and heart urchins?

9. What relationships exist among intertidal zonation, sediment particle size, and beach slope?

10. Discuss the dominant species of kelp, their epifauna, and animals that feed on kelp in the Pacific Coast kelp forest.

11. Where are spiny lobsters and American lobsters found?

12. Discuss the preferred environment, reproduction, and threats to survival of larval and adult forms of oysters.

13. Describe the environment suited to development of coral reefs.

14. Describe the zones of the reef slope, the characteristic coral types, and the physical factors related to zonation.

15. As one moves from the shoreline to the deep-ocean floor, what changes in the physical nature of the ocean floor can be expected?

16. Discuss the most important barriers to the migration of shallow-water benthos into the deep ocean and to the reverse migration.

17. What are the major differences between the conditions and biocommunities of the hydrothermal vents and the cold seeps? How are they similar?

REFERENCES

Childress, J. J.; Fisher, C. R.; Brooks, J. M.; Kennicutt, M. C., II; Bidigare, R.; and Anderson, A. E. 1986. A methanotrophic marine molluscan (Bivalvia, Mytilidae) symbiosis: Mussels fueled by gas. *Science* 233:4770, 1306–8.

George, D., and George, J. 1979. *Marine life: An illustrated encyclopedia of invertebrates in the sea.* New York: Wiley-Interscience.

Hallock, P., and Schlager, W. 1986. Nutrient excess and the demise of coral reefs and carbonate platforms. *Palaios* 1:389–98.

Hessler, R. R.; Ingram, C. L.; Yayanos, A. A.; and Burnett, B. R. 1978. Scavenging amphipods from the floor of the Philippine Trench. *Deep-Sea Research* 25:1029–47.

MacGinitie, G. E., and MacGinitie, N. 1968. *Natural history of marine animals.* 2nd ed. New York: McGraw-Hill.

Ricketts, E. F.; Calvin, J.; and Hedgpeth, J. 1968. *Between Pacific tides.* Stanford, Calif.: Stanford University Press.

Rona, P. A.; Klinkhammer, G.; Nelson, T. A.; Trefry, J. H.; and Elderfield, H. 1986. Black smokers, massive sulphides and vent biota at the Mid-Atlantic Ridge. *Nature* 321:6065, 33–37.

Williams, E. H., Jr.; Goenaga, C.; and Vicente, V. 1987. Mass bleaching on Atlantic coral reefs. *Science* 238:4830, 877–78.

Yonge, C. M. 1963. *The sea shore.* New York: Atheneum.

Zenkevitch, L. A.; Filatove, A.; Belyaev, G. M.; Lukyanove, T. S.; and Suetove, I. A. 1971. Quantitative distribution of zoobenthos in the world ocean. *Bulletin der Moskauer Gen der naturforscher, Abt. Biol.* 76, 27–33.

ZoBell, C. E. 1968. Bacterial life in the deep sea. In Proceedings of the U.S.-Japan Seminar on Marine Microbiology, August 1966, Tokyo. *Bulletin Misaki Marine Biology, Kyoto Inst. Univ.* 12:77–96.

SUGGESTED READING

Sea Frontiers

Coleman, N. 1974. Shell-less molluscs. 20:6, 338–42.
A description of nudibranchs, gastropods without shells.

George, J. D. 1970. The curious bristle-worms. 16:5, 291–300.
The variety of worms belonging to the class *Polychaeta* of phylum Annelida is described. The discussion includes locomotion, feeding, and reproductive habits of the various members of the class.

Gibson, M. E. 1981. The plight of *Allopora*. 27:4, 211–18.
The reason the author believes the unusual California hydrocoral is headed for the endangered species list is the topic of this article.

Ruggiero, G. 1985. The giant clam: Friend or foe? 31:1, 4–9.
The ecology and behavior of the giant clam *(Tridacna gigas)* is discussed in reference to whether it is a danger to divers.

Shinn, E. A. 1981. Time capsules in the sea. 27:6, 364–74.
The method by which geologists determine past climatic and environmental conditions by studying coral reefs is discussed.

Viola, F. J. 1989. Looking for exotic marine life? Don't leave the dock. 35:6, 336–341.
A description of marine life living on and near pilings for a boat dock are described along with good color photos.

Winston, J. E. 1990. Intertidal space wars. 33:1, 47–51.
Florida intertidal life is discussed.

Scientific American

Caldwell, R. L., and Dingle, H. 1976. Stomatopods. 234:1, 80–89.
Presents the ecology of these interesting crustaceans that have appendages specialized for spearing and smashing prey.

Feder, H. A. 1972. Escape responses in marine inverte-
    brates. 227:1, 92–100.
    Discusses the surprisingly rapid movements and other
    responses made by invertebrates to the presence of
    predators. Some interesting photographs accompany the
    text, which describes the escape responses of limpets,
    snails, clams, scallops, sea urchins, and sea anemones.

Wicksten, M. K. 1980. Decorator crabs. 242:2, 146–57.
    Describes how species of spider crabs use materials
    from their environment to camouflage themselves.

Yonge, C. M. 1975. Giant clams. 232:4, 96–105.
    The distribution and general ecology of the tridacnid
    clams, some of which grow to lengths well over 1 m (3.3
    ft), are investigated.

17
EXPLOITATION OF MARINE RESOURCES

In the austral spring of 1985, a hole in the stratospheric ozone layer was discovered over Antarctica (fig. 17–1). It was later proven to be caused by the formation of ozone-destroying chlorine monoxide (ClO). The chlorine was made available from chlorofluorocarbons (CFCs). In 1987, most industrial nations signed the Montreal Protocol, which calls for a 20% reduction in CFC production by 1993 and a 50% reduction by 1998. E. I. du Pont de Nemours and Co., a major manufacturer, then announced plans for a total phase-out of CFCs. Our ecosystem was beginning to get our attention.

Then from June 1987 through February 1988, more than 750 bottlenose dolphins died along the U.S east coast, and in the summer of 1988 thousands of seals died in the North Sea. The conclusions were that the dolphins died from red tide toxins and that canine distemper killed the seals. But the question was asked, "Could both populations have been stricken because their immune systems had been rendered dysfunctional by toxic chemicals?"

As 1988 continued, the Environmental Protection Agency announced that **acid rain**, already known to be a threat to freshwater systems, was significantly degrading Chesapeake Bay, and NASA's director of Goddard Institute of Space Studies told Congress that greenhouse warming is here. Although many scientists attempted to refute this claim, Gilbert, the largest hurricane in the history of the Western Hemisphere, came along, and we began to see articles in the media blaming it on the increased greenhouse effect (fig. 6D, page 147). Everyone was getting the message. Something is going wrong.

Although the average temperature of the earth has risen by 0.5°C (0.9°F) in the last 100 years, there is no proof it is the result of the 25% increase in atmospheric CO_2 and the increase in other greenhouse gases (methane, CFCs, etc.) that began with the industrial revolution. Some climate modelers believe the greenhouse warming will produce more cloud cover to block out the sun's rays and significantly reduce the warming effect. What is very clear is that our agricultural and industrial activities are producing changes in the environment. Although we cannot

Preceding page: False color image of the Pacific off the coast of Peru (right) to the Galápagos Archipelago (left). Reds and oranges indicate highest concentrations of marine phytoplankton, with yellows, greens, and blues indicating decreasing levels. Wind-driven coastal upwelling off Peru is reflected in a 100-km-wide chlorophyll-rich band along the coast and plumes of productive water extending some 500 to 1000 km offshore. *Nimbus 7* satellite CZCS image. Courtesy of Jane A. Elrod and Gene Feldman, NASA/Goddard Space Flight Center.

FIGURE 17–1   Map of Ozone Levels in the Atmosphere of the Southern Hemisphere.
The ozone concentration in the ozone "hole" observed over Antarctica on October 5, 1987, is about half of what it was in 1977. The data are from the Total Ozone Mapping Spectrometer (TOMS) on board the NASA satellite *Nimbus 7.* Courtesy of NASA.

eliminate modifications of the environment by human activities, we can begin a serious attempt to reduce the magnitude of combustion of fossil fuels and the spread of agriculture. We must preserve as much of the earth's plant communities as possible and replace much of what has been removed.

We must also increase our understanding of the basic cycles of materials within the earth's ecosystems. One of the critical problems in this area is the exchange of gases between the atmosphere and the ocean. But we cannot wait until we have proof that our activities are having a negative effect on the earth ecosystem before we act. Our activities that are known to change the environment the most must be modified while we continue to strive for understanding of the details of how the system works. Otherwise we are basing our future on a game of Russian roulette.

Rapid advances in technology have brought our capability to exploit the oceans to such a high level that great harm can and has been done in relatively short periods of time. Marine fisheries and petroleum found in marine sediments are two areas of the exploitation of marine resources where problems are well known. In this chapter, we will discuss the international effort to develop a legal framework for expanded exploitation of the oceans and some of the more important marine resources. The pollution that has resulted from the exploitation of resources and the disposal of wastes into the oceans is addressed in the following chapter.

LAW OF THE SEA

Because of the imminence of extensive exploitation of the ocean floor beyond the jurisdiction of the nations that border the seas and the problems developing from international fisheries and pollution, one of the most urgent problems awaiting international agreement has been the establishment of laws by which these activities may be regulated.

Before describing the activities needing regulation, we will discuss some history and some problems facing those who have worked to establish a new law of the sea. In 1609 **Hugo Grotius**, the Dutch jurist and statesman whose writings helped form international law, established in *Mare Liberum* the doctrine that the ocean is free to all nations. Although controversy continued over whether nations could control portions of the oceans, **Cornelius van Bynkershonk** produced a satisfactory solution to this problem in *De Dominio Maris,* published in 1702. It provided for national domain over the sea for the distance that could be protected by cannons on the shore. The first official dimension of this zone was established in 1772 when the British determined that they would exercise control over a strip of water one league (3 nautical miles) wide along the shores of the Commonwealth—the **territorial sea**.

Because technology was rapidly developing for drilling beneath the ocean, the first United Nations Conference on the Law of the Sea was held in Geneva in 1958, producing a treaty relating to the continental shelf. It stated that prospecting and mining minerals on the shelf are under the control of the country that owns the nearest land. Unfortunately, the seaward limit of the continental shelf was not well defined in the treaty.

A second, less productive United Nations Conference on the Law of the Sea was held in Geneva in 1960; the third had its first meeting in New York in 1973. Subsequent meetings were held at least once per year.

The third conference came to a disappointing conclusion in April of 1982, when a large majority—130 to 4, with 17 abstentions—voted in favor of a new Law of the Sea (LOS) convention. Most of the developing nations who could benefit significantly from the provisions of the treaty voted for its adoption. The opposition was led by the United States, who along with Turkey, Israel, and Venezuela, voted against it. These countries support private companies planning seabed mining operations and believe certain provisions of the treaty would make it unprofitable. Among the abstainers were the Soviet Union, Britain, Belgium, the Netherlands, Italy, and West Germany, all of whom are interested in seabed mining. As of October 31, 1987, 140 nations had signed the final act, but only 35 nations had ratified it. Because it is unlikely that all dissenting and abstaining nations will ratify the treaty, its ability to facilitate the exploitation of seabed resources will be severely hampered.

The primary features of the treaty are as follows:

Coastal Nations Jurisdiction: A uniform 12-mi (19.2-km) territorial sea and a 200-mi (322-km) **Exclusive Economic Zone (EEZ)** were established. The coastal nation has jurisdiction over mineral resources, fishing, and pollution within the EEZ and beyond it, up to 350 mi (564 km) from shore, if the continental shelf (defined geologically) extends beyond the EEZ (fig. 17–2).

FIGURE 17–2 The Exclusive Economic Zone (EEZ) of the Law of the Sea Treaty.
Whether or not the treaty is adopted, individual nations will declare their own EEZs and 42% of the world's oceans will come under the control of coastal nations. Source U.S. State Department.

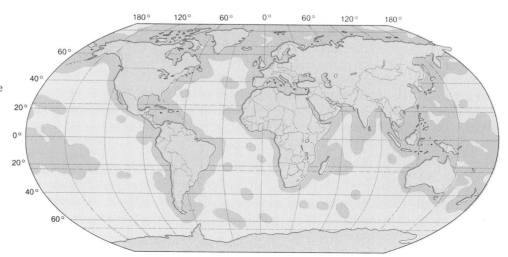

Ship-Passage: The right of free passage on the high seas is maintained; it is also provided for within territorial seas through straits used for international navigation.

Deep-Ocean Mineral Resources: Private exploitation of seabed resources may proceed under the regulation of the **International Seabed Authority (ISA)**, within which will be a mining company chartered by the United Nations, called **Enterprise.** Mining sites will be operated in pairs: one chosen by the private mining concern, the other by Enterprise, and both approved by the ISA. Revenues from the private concern's chosen site will go to that concern; revenues from the other site will be divided among the developing nations.

Arbitration of Disputes: A Law of the Sea tribunal will perform this function.

On March 10, 1983, President Reagan unilaterally declared the establishment of a 200-nautical-mile EEZ for the United States. This declared zone contained characteristics identical to those of the EEZ defined in the LOS convention, and President Reagan declared that the U.S. would honor all provisions of the convention related to the EEZ in its relations with all nations.

FISHERIES

Since well before the beginning of recorded history, humans have used the sea as a source of food. Although food from the sea has in recent years provided only about 5% of the protein consumed by the world's population, this seemingly small percentage has meant the difference between starvation and an adequate diet for millions of the world's inhabitants. If we are to continue to provide even this small percentage of the world's food supply through our marine fishery, we must continue to increase our understanding of all aspects of the fishery. Not only is it important that we understand the feeding and other ecological relationships that exist among the ocean's organisms, but also that we understand the effects of our attempts to exploit the biological resources of the oceans as a source of human food. We seek to increase the food resources we remove from the oceans, but it is clear that the greatest negative impact humans have had on the ecology of the oceans is through overfishing.

It is very disturbing that the marine fishery of fish, shellfish, and seaweeds from the ocean has increased only from 59.7 metric tons (MT) in 1970 to 76.9 MT in 1985, the latest year for which complete information is available—despite a significantly increased

TABLE 17–1   World Commercial Fish Catch, 1951–1985 (million metric tons).

Year	Fresh Water	Peruvian Anchovy	Other Marine	Total Marine	Total
1951	2.6	NA*	20.9	20.9	23.5
1952	2.8	NA	22.3	22.3	25.1
1052	3.0	NA	22.9	22.9	25.9
1954	3.2	NA	24.4	24.4	27.6
1955	3.4	NA	25.5	24.5	28.9
1956	3.5	0.1	27.2	27.3	30.8
1957	3.9	0.3	27.5	27.8	31.7
1958	4.5	0.8	28.0	28.8	33.3
1959	5.1	2.0	29.8	31.8	36.9
1960	5.6	3.5	31.1	34.6	40.2
1961	5.7	5.3	32.6	37.9	43.6
1962	5.8	7.1	31.9	39.0	44.8
1963	5.9	7.2	33.5	40.7	46.6
1964	6.2	9.8	35.9	45.7	51.9
1965	7.0	7.7	38.5	46.2	53.2
1966	7.3	9.6	40.4	50.0	57.3
1967	7.2	10.5	42.7	53.2	60.4
1968	7.4	11.3	45.2	56.2	63.9
1969	7.6	9.7	45.4	55.1	62.7
1970	8.4	13.1	46.6	59.7	68.1
1971	9.0	11.2	48.3	59.5	68.5
1972	5.7	4.8	53.7	58.5	64.2
1973	5.8	1.7	55.3	57.0	62.8
1974	5.8	4.0	56.8	60.7	66.5
1975	6.2	3.3	57.0	60.2	66.4
1976	5.9	4.3	59.7	63.9	69.8
1977	6.1	0.8	62.3	62.8	68.9
1978	5.8	1.4	63.3	64.6	70.4
1979	5.9	1.4	63.8	65.2	71.1
1980	6.2	0.8	65.4	65.8	72.0
1981	6.6	1.5	66.7	68.2	74.8
1982	6.8	1.8	67.9	69.7	76.5
1983	7.2	0.1	69.2	69.3	76.5
1984	9.7	0.1	73.0	73.1	82.8
1985	8.0	0.1	76.8	76.9	84.9

*NA–Not available

SOURCE: U.S Department of Commerce, N.O.A.A. publication. Fisheries of the United States, 1986.

fishery effort (table 17–1). It is also certain that the cost per unit of production is rapidly increasing.

The Japanese and the Soviet Union fishing industries are by far the largest in the world. The U.S. ranks fifth, with a fishery of less than half the size of that of the leading nations (table 17–2). The largest fisheries are for the small herringlike fish, including herring, sardines, and anchovies. The next most important fishery is for demersal codlike fish. They include the cod, hakes, and haddocks (table 17–3).

TABLE 17–2 Countries with the Largest Fishing Industries in 1985.

Country	Catch (million metric tons)
Japan	11.44
Soviet Union	10.52
China	6.78
Chile	4.80
USA	4.77
Peru	4.17
India	2.81
Republic of Korea	2.65
Thailand	2.12
Norway	2.11
World	84.9

SOURCE: Data from U.S. Department of Commerce.

TABLE 17–3 World Catch by Fish Type in 1985.

Fish Type	Catch (million metric tons)
Herrings, sardines, anchovies	21.2
Flounders, halibuts, soles	1.4
Cods, hakes, haddocks	12.4
Salmons, trouts, smelts	1.1
Miscellaneous marine fishes	8.9
Jacks, mullets, sauries	8.0
Miscellaneous freshwater fishes	8.4
Red fishes, basses, congers	5.3
Mackerels, snooks, cutlass fishes	3.7
Tunas, bonitos, billfishes	3.2
Crustaceans	3.4
Molluscs	6.1

SOURCE: Data from U.S. Department of Commerce.

Fishing peoples have always understood that fish must be permitted to reproduce if a fishery is to be maintained. However, it has become increasingly difficult to regulate fisheries in a manner that would assure sufficient reproduction to maintain the fishery. This has been primarily the result of the great demand that accompanies an increasing human population. Fisheries such as the anchovy, cod, flounder, haddock, herring, and sardine apparently are suffering from overfishing; that is, they are being harvested at a rate above the natural rate of reproduction in such numbers that their populations are driven below the size necessary to produce a **maximum sustainable yield**. International efforts are now being made to provide worldwide ocean fisheries management.

Fisheries and Marine Biology

Over the years, huge fluctuations in fish stocks that were not clearly related to fishing pressures convinced many fisheries biologists that a detailed study of fisheries ecology was needed.

Recruitment: The addition of young adult fish to the fishery was found to depend on survival at critical stages in the life of a fish. The mortality rate is very high for eggs and larvae. Mortality decreases in juvenile fish and reaches its lowest level in adults.

Larval survival: After the larva uses up its yolk reserves, it must find appropriate food or die. For instance, anchovy larvae require at least 20 phytoplankton cells/mL of water within 2½ days, and the minimum cell size must be at least 40 μm in diameter.

Juvenile survival: Once a fish survives the larval stage, it still must survive a juvenile existence before it can be recruited into the fishery. Juvenile fish may die from predation, disease, parasitism, genetic defects, and the effects of pollution. Availability of adequate food is still important because rapid growth rate is the best insurance against predators. If the juvenile fish grows faster than its predator, it will quickly become too large to be suitable prey. Assuming availability of suitable food, the juvenile fish will grow more rapidly in warmer water. Their feeding history is complex, as they may feed on a variety of plankton over a range of sizes as they grow larger. Even at a given size, juvenile fish may feed on a large size-range of plankton. However, there is usually a narrow size-range of plankton on which they can feed with high efficiency.

The Ecosystem Approach to Fishery Assessment

In an effort to understand the role of the physical environment and the many populations of organisms that affect fishery recruitment, fishery scientists have adopted an ecosystem approach to fishery assessment.

In most ecosystems, nitrogen is the nutrient that becomes so limited that it curtails primary productivity. Assessing the (1) *cycling of nitrogen within an ecosystem* and the (2) *influx of new nitrogen* into the system can lead to an estimate of the maximum amount of protein that can be removed from the ecosystem by fishing and still maintain a healthy fishery.

Less than 10% of the primary productivity in oligotrophic waters results from an influx of new nitrogen; about 50% of the primary production results from the influx of new nitrogen in areas of upwelling

FIGURE 17–3 Potential Fishery and Influx of Nitrogen.

All marine ecosystems are supported by nutrients that are recycled within the euphotic zone that contains their plant population and nutrients added by influx from surrounding and deeper waters. The biomass of fish taken from the ecosystem cannot exceed that which is supported by the influx of nutrients or the ecosystem will become stressed. The small arrows in *A* represent a small amount of nutrient flux into an oliogotrophic open-ocean ecosystem. The much larger arrows in *B* represent a large influx of nutrients into the euphotic zone of an upwelling region. This greater influx of nutrients makes possible a much greater fishery that can be extracted from the upwelling region than from the oligotrophic ecosystem.

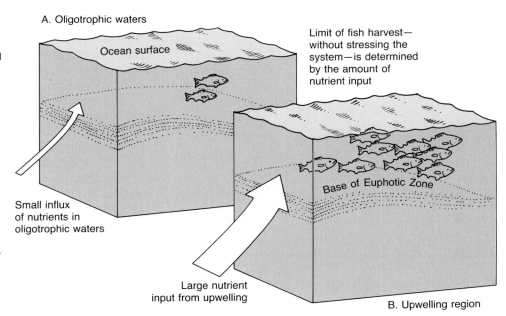

A. Oligotrophic waters

Ocean surface

Limit of fish harvest—without stressing the system—is determined by the amount of nutrient input

Small influx of nutrients in oligotrophic waters

Base of Euphotic Zone

Large nutrient input from upwelling

B. Upwelling region

such as occur in coastal Peruvian waters (see chapter opener photo and fig. 17–3). In both of the examples cited, the balance of the production is supported by nitrogen recycled within the ecosystem.

If, on a worldwide basis, slightly less than 20% of primary production results from the flux of new nitrogen into the ecosystem, we can estimate the **potential world fishery** on this basis. Such estimates must be considered tentative, but if 0.1% of this primary productivity sustained by new nitrogen found its way into the world fishery, the potential world fishery would be about 100 million MT. (metric tons). The present catch is just about 76 million MT. The key point here is that the fishery must not remove more fish than can be supported by the influx of new nitrogen; otherwise, the ecosystem supporting the fishery will be damaged.

What the fishery managers want to know is the **standing stock,** the biomass, of the fishery. Previous methods of fishery assessment were directed at making this determination on the basis of statistics of landings and scientific sampling of fish populations. It was discovered that these methods were inaccurate. It is much easier to sample the phytoplankton and zooplankton populations, and the productivity of these populations can be determined with a much greater degree of accuracy. There are still many problems in moving from phytoplankton standing stock to fishery standing stock.

The methods used to estimate the potential fishery involve knowledge of size grades, standing stocks, and size-doubling times of each component of the

food chain from the phytoplankton to the fishery species. Where such methods have been applied to a number of fisheries, the conclusion was that the fisheries were already taking an amount equal to the potential fishery.

It has become clear that the physical resources of the marine environment determine the level of biological productivity, and therefore the fishery standing stock. Any significant change in the physical resources will have a corresponding effect on the fishery.

Upwelling and Fisheries

Because so much of the productivity of upwelling areas is supported by the influx of nutrients, it is not surprising to find that these areas, which represent about 0.1% of the ocean surface area, account for around 50% of the world fishery. Within upwelling areas, research has identified two variables that significantly affect the fisheries associated with them. These variables are the **duration** and **rate** of upwelling (fig. 17–4). Generally, upwelling of long duration (more than 250 days) produces ideal conditions for fishery development. However, even in areas of shorter duration, high levels of productivity may still result if the rate of upwelling is moderate. Regardless of the duration, extremely low or high rates of upwelling produce a decreased fishery compared to one produced by moderate rates. Waters that are too stratified or too well mixed—where upwelling rates are too low or high, respectively—tend to develop smaller-sized phytoplankton species that cannot be grazed efficiently by fishes such as sardine and anchovy. This decrease in phytoplankton size is thought to result from the decreased concentration of nutrients that results from either condition. Under conditions of small-sized phytoplankton, herbivorous zooplankton must first feed on the tiny plant cells and subsequently be grazed by the fish.

Where upwelling rates are moderate, and the water column less turbulent, higher concentrations of nutrients support large species of diatoms that can be grazed directly by the fish. This elimination of the herbivorous zooplankton greatly increases the energy available to the fish and thus increases the potential fishery. Therefore, it appears that areas of upwelling of long duration and moderate rates provide the optimum environment to produce a large fishery.

New Technologies Useful in Studying Fisheries Ecology

The use of **side-directed sonar** has become increasingly effective in assessing the masses of populations within the ecosystem associated with fisheries. When these signals are reflected off schools of fish, squid, or other animals, the return signals can be analyzed to determine the volume and mass of the aggregations. Careful analysis of the frequencies that return with the greater resonance can tell investigators the species and even the age of organisms in the school (fig. 17–5).

Using a version of this sonar equipment attached to the manned submersible *Johnson Sea Link,* investigators added significantly to the knowledge of the ecosystem that supports the important fisheries of Georges Bank off the coast of New England (fig. 17–6).

FIGURE 17–4 Classification of Upwelling Systems in Terms of Duration and Rate of Upwelling.
Duration determines the absolute amount of production, and rate of upwelling determines the algal cell size. Moderate rates of upwelling for short or long duration (blocks 2 and 5) produce large enough concentrations of nutrients in nonturbulent water to support large algal cells that can be grazed directly by sardines and anchovies. These conditions will support relatively large fisheries. Whether the duration is short or long, low or high rates of upwelling will not allow high enough concentrations of nutrients to produce large algal cells for the fish to graze directly. Therefore, zooplankton must graze the tiny algal cells and be grazed in turn by the fish. This increases the length of the food chain and greatly reduces the amount of energy available to the fish. The result is a much smaller fishery biomass. After Wyatt 1980.

FIGURE 17–5. Fish Assessment with Sonar.

A, plan view, *B,* vertical section. Side-scan signal is used to determine area and abundance of fish schools. *C,* a record of a 115-km (71-mi) sonar transect conducted in Southern California Bight from Malibu to Oceanside. Species of fish in schools were not determined, but the major fisheries in this region are northern anchovy and jack mackerel. Courtesy of Paul E. Smith, National Marine Fisheries Service.

It had been a problem to explain how the benthic community on Georges Bank could support the large demersal fish (fish that live on or near the bottom) and squid fisheries there. The benthic and demersal zooplankton communities did not appear to be large enough to support these fisheries.

Sonar investigations in the submarine canyons along the southeast flank of Georges Bank revealed that large masses of the krill *Meganyctiphanes norvegica* (fig. 17–7) are present on or near the bottom day and night. In the fifty meters adjacent to the bottom, densities of krill averaged 813 animals/m^3. It is known that these krill are an important component of the diet of squid, cod, flounder, haddock, hake, pollock, and redfish. It is clear that if they descend into the canyons and deep waters surrounding Georges Bank to feed on this previously unknown food supply, the krill could easily provide the needed support for the existence of the large fisheries represented by these animals.

The Future of Fisheries Management

A disturbing trend in fisheries management developed during the recent past that reduced the priority of preserving the ecology of the fishery in making management decisions. Increased emphasis was given to social benefits such as occupational opportunity. It is the ecosystem, however, that makes the social benefits possible, so it was surely shortsighted to relegate the welfare of the ecosystem to any position less than the number one priority in fisheries management. The fact that this philosophy is being abandoned will surely increase the likelihood of successful fisheries management in the future.

Regulation

The effects of inadequate fisheries management can be seen in history of the fisheries in the northwest Atlantic. In this area, regulated by the International Commission for the Northwest Atlantic Fisheries, the

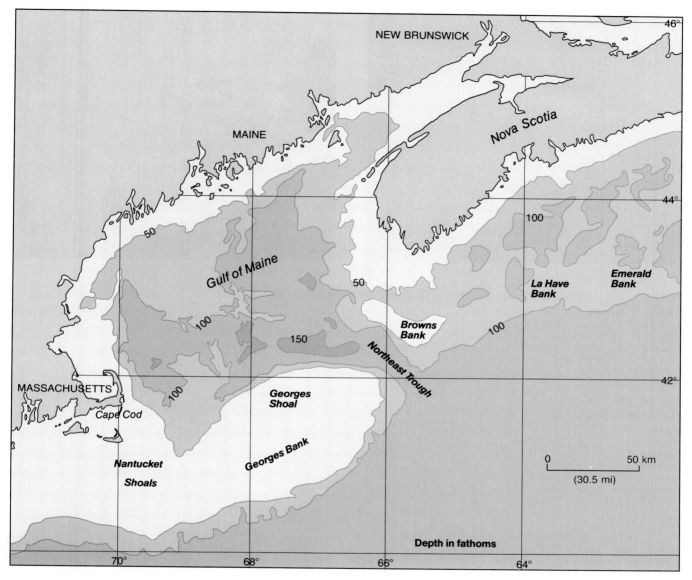

FIGURE 17–6 Georges Bank: A Major Fishing Grounds off the New England Coast.
The discovery, using side-directed sonar, of large krill populations in the submarine canyons of the southeast flank of Georges Bank may explain how the large demersal fishery of this region is supported. Previously, the benthic food resources did not seem sufficient to support such a large fishery.

fishing capacity of the international fleet increased 500% from 1966 to 1976; the total catch, however, rose by only 15%. This was a significant decrease in the catch per unit of effort, a good indication that the fishing stocks of the Newfoundland–Grand Banks area are being overexploited. Biologists who had been setting quotas for the major species within this region complained that enforcement was extremely difficult and that the quotas had become international currency used in political bargaining. Politicians were accused of allocating quotas for fish that exceeded

the total allowable catch set by the commission, thus depleting the reserve and preventing stock recovery.

Occurrences like these make it seem that the only answer for regulation with adequate enforcement may be to have fisheries controlled by the nation in whose coastal waters they exist. The difficulty of enforcing regulations by the international commission was largely responsible for Canada's unilateral decision to extend its right to control fish stocks for a distance of 200 mi (322 km) from its shores beginning January 1, 1977. The United States followed with a

A.

B.

FIGURE 17–7   High-Density Near-Bottom Krill Aggregations in Submarine Canyons off Georges Bank.

A, close-up view of the krill *Meganyctiphanes norvegica.* It is about 3.8 cm (1.5 in.) long. *B,* A large aggregation of the krill observed in the submarine canyons off the southeast flank of Georges Bank. Photos *A* by Pam Blades-Eckelbarger, Harbor Branch Oceanographic Institution, Inc. *B* by Marsh Youngbluth, Harbor Branch Oceanographic Institution, Inc.

FIGURE 17–8   Exclusive Economic Zone of the United States.

The boundaries of this region were established by the March 1, 1977, declaration of U.S. control of fisheries within 200 nautical miles of its coastal waters. Map courtesy of National Marine Fisheries Service.

FIGURE 17–9 Spotted Dolphin (Stenella attenuata).
Photo by W. High, courtesy of National Marine Fisheries Service.

similar action on March 1, 1977. This was a first step toward the declaration of a full-blown EEZ in 1983 (fig. 17–8).

In April 1990, the U.S. tuna canning industry responded to a boycott of tuna products initiated by the environmental group Earth Island Institute by declaring it would not buy or sell tuna caught by methods that kill or injure dolphins. The nature of the problem of dolphin deaths was graphically presented to the world in the video footage taken by biologist Samuel F. La Budde, and it was undoubtedly the impact of this footage that turned the tide. It was first aired in March 1988 and in just over two years accomplished

what the U.S. government has been unwilling to do since the Marine Mammal Protection Act was passed in 1972. It will greatly reduce the practice of using dophins to spot yellowtail tuna in the eastern Pacific Ocean. Spotted (Stenella attenuata) (fig. 17–9) and spinner (S. longirostris) dolphins have been used by the industry to spot yellowfin tuna schools that are for unknown reasons commonly found beneath the dolphins.

A measure designed in the 1970s to reduce the number of deaths was the development of purse seines of the type shown in figure 17–10. Along with the net change, a technique for its use called *back-*

FIGURE 17–10 Backdown Using Apron-Chute Rescue System.

A technique used to reduce the number of porpoises killed by U.S. tuna purse seiners of the East Pacific tuna fleet is referred to as *backdown*. Nets with small mesh in an apron-chute area prevent porpoises from being entangled as net is pulled under them. Courtesy of National Marine Fisheries Service.

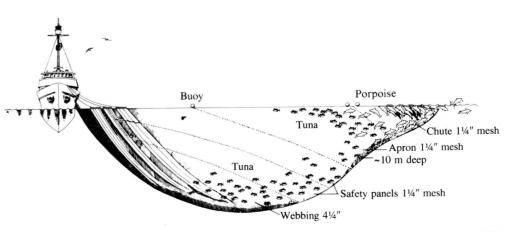

Buoy   Porpoise

Tuna

Chute 1¼" mesh

Apron 1¼" mesh

~10 m deep

Tuna

Safety panels 1¼" mesh

Webbing 4¼"

449

down helps dolphins escape over a fine-meshed apron as the net is pulled in. It was this process that was the focal point of La Budde's video, which showed dolphins being caught in the net and pulled over and crushed by the block pulley used to land the net.

Another means of netting tuna that will be affected by this decision is the use of *drift nets* or *gill nets,* which are made of monofilament line that is virtually invisible and cannot be detected by dolphin echolocation. Japan, Korea, and Taiwan are the nations with the largest drift net fleets. They are sending 1500 fishing vessels into the North Pacific and setting over 37,000 km (23,000 mi) of nets every day. They are said to be fishing for squid, but are illegally taking large quantities of salmon and steelhead trout. Drift netters are moving into the South Pacific and targeting immature 2-year-old tuna. There is considerable concern that this practice will destroy the South Pacific tuna fishery in a very short time. In addition to dolphins, birds and turtles are killed in these nets. At least 20% of a drift net catch is thrown away because it consists of unsalable species or is badly damaged. This most efficient means of stripping animals from the sea is extremely inefficient as a method of maintaining the health of the world's fisheries for future years, and it has become a focus of international negotiations. The U.S. and Japan have already agreed to a program of observers on Japanese vessels to monitor their locations and types of catch.

Fishing fleets from the Soviet Union, Poland, Japan, and Italy fish off the U.S. coasts, interfering at times with the U.S. lobster fishery. Occasionally, fishing vessels are boarded, seized, and escorted to port for violation of the fishing regulations established by the United States within its 200-mi (322-km) limit. Such violations can result in severe fines, imprisonment of the captain, and confiscation of the ship.

The U.S. fishing vessels, particularly from the tuna fleet, occasionally are seized for fishing within the EEZs of Latin American nations, where they have also been subjected to extremely large fines. In 1988, Japanese fishing fleets were accused of illegally taking at least 300,000 MT of pollock and cod from U.S. EEZ waters in the Bering Sea. It is a great irritation to many that the U.S. imports two-thirds of its fish products each year and that most of these imports are caught in U.S. waters. Although establishment of the EEZs throughout the world may aid in protecting fisheries, it will certainly create new political and economic problems. The U.S. fishing industry, happy with the establishment of a 200-mi (322-km) EEZ by our government, has discovered it too will be limited by the accompanying regulations.

PERUVIAN ANCHOVETA FISHERY

The magnificent **anchoveta** fishery that existed at the north end of the Peru Current can be used as a case study to show the importance of placing the understanding of the ecology of a fishery above all other considerations in fisheries management. The unfortunate expansion of this fishery was based in large part on the consideration of the jobs that were provided. Many of those jobs are gone forever.

Physical Conditions

Southeasterly winds blow along the coast of Peru and, owing to the earth's rotation, drive the surface waters away from the coast. The surface water is replaced by an upwelling of cold water from depths of 200–300 m (660–990 ft).

The Peru Current (fig. 17–11A) is the northward-flowing eastern boundary of the South Pacific current gyre. Speed seldom exceeds 0.5 km/h (0.3 mi/h), and the current extends to a depth of 700 m (2300 ft) near its seaward edge, some 200 km (124 mi) from the coast. The **Peru Countercurrent** knifes its way between the coastal and oceanic regions of the Peru Current near the equator. This southbound surface flow of warm equatorial water seldom reaches more than 2° or 3° south of the equator, except during the summer when it reaches its maximum strength. Beneath the surface current is the southward-flowing Peru Undercurrent.

Biological Conditions

Biologically, the significance of the cold water and nutrients provided by upwelling is great. During the years 1964–1970, fully 20% of the world fishery was represented by anchoveta taken off the coast of Peru.

The food web starts with diatoms and other phytoplankton that support a vast zooplankton population of copepods, arrowworms, fish larvae, and other small animals. Energy is passed on through a series of predators, but most stops with the anchoveta (fig. 17–11B), which once attained a biomass of 20 million MT. Natural predators that feed on the anchoveta are marine birds—boobies, cormorants, and pelicans—and large fishes and squid.

A Natural Hazard

Periodically during the summer, the southeasterly trade winds weaken. The northward-flowing current and upwelling slow down, and the Peru Countercur-

B.

FIGURE 17–11   The Peruvian Anchoveta Fishery.

A, the Peru Countercurrent is a warm-water mass that spreads over the surface and separates the Peru Current into coastal and oceanic regions. *B*, these small fish represented one of the world's major fisheries until the collapse of 1972. An active fish meal industry based on anchovy made Peru, for a while, the world's leading fishing nation. Photo courtesy of National Marine Fisheries Service. *C*, a catch of Peruvian anchovy during the early 1960s when the fishery was undergoing rapid development. Photo courtesy of FAO. *D*, after a peak catch of 12.3 million MT in 1970, the fishery declined.

C.

D.

rent may move as far as 1000 km (621 mi) south of the equator, laying down a warm low-salinity layer over the coastal water. Because this event commonly occurs around Christmas, it is called El Niño, The Child.

The increase in water temperature and decrease in nutrients brought about by the slowdown or complete halt of upwelling cause a severe decrease in the anchoveta population, the remnants of which seek deeper, cooler water. Marine birds that feed primarily on the anchoveta and produce the enormous **guano** (excrement rich in phosphorus and nitrogen compounds) deposits on the islands off the Peruvian coast have their populations cut severely by the decrease in the anchoveta.

In 1972, a severe El Niño occurred. The total biomass of anchoveta decreased from an estimated 20 million MT in 1971 to an estimated 2 million MT in 1973. But before blaming this decrease solely on the natural event of El Niño, with which the anchoveta have contended for thousands of years, another factor must be evaluated.

An Artificial Hazard

The Peruvian anchoveta fishery began to develop into a major industry in 1957. In 1960, the Instituto del Mar del Peru was established with the aid of the United Nations to study the fishery and recommend a management program to the government. It was hoped that early study would prevent the fishery from going the way of the Japan herring and California sardine fisheries that had collapsed as a result of overexploitation.

Biologists estimated that 10 million MT was the maximum sustainable yield of the anchoveta. Figure 17–11D shows that this limit was exceeded in 1968, 1970, and 1971. The 1970 catch of 12.3 million MT would probably reach 14 million MT if processing losses and spoilage were included.

Dr. Paul Smith, who is studying the anchoveta fishery problem for the National Marine Fisheries Service in the Southwest Fisheries Center at La Jolla, California, believes, as do other biologists familiar with the problem, that the yearly anchoveta catches may have been much larger than the official figures indicate. This belief grows out of the fact that many catches contained large numbers of **pelladilla**, which are young anchoveta that are low in protein. This underreporting would result from buyers reducing the total tonnage of a catch they are purchasing by a percentage determined by the proportion of the catch that was accounted for by high-water-content and low-oil-content pelladilla. For example, a 100-MT

catch may be recorded as a 70-MT catch because it contained a large number of pelladilla.

Dr. Smith feels the crisis may have resulted from a combination of the effects of El Niño and the industry's lack of understanding of the anchoveta fishery. This is reflected in the fact that the January 1972 allotment was set at 1.2 million MT. This figure was reached easily by the industry with about half the expected effort. Because the fishing vessels were able to meet their allotment so easily without going far from shore, the fishery was thought to have developed handsomely, and the February allotment was raised to 1.8 million MT. Much more effort was required to meet this allotment, and in March the fishery collapsed. What had happened?

Apparently, El Niño, or the Peru Countercurrent, had come south in January, but it had remained offshore. The anchovy were concentrated in the nearshore portion of the Peru Current that had not been affected (fig. 17–11A). In January the nets were filled easily near shore, and it was natural to think the high concentration of anchovy extended over the normal fishery area as well. This one event cannot explain the collapse of the fishery, but it is a symptom common to fisheries that are poorly understood.

It may be that even with upwelling, ammonia (NH_3) from decomposition of anchoveta excrement is required to support enough phytoplankton to allow the anchoveta population to expand. Because anchoveta populations are now small and scattered, they may swim through great expanses of water without adding much nutrient enrichment. When the population was larger, the whole region received that enrichment. It may require a long time, with a gradual increase in the population, for this nutrient supplement to be made available to the phytoplankton throughout the entire region.

The anchoveta are gradually being replaced by **sardines**, a more valuable variety of fish. The entire anchoveta catch was converted to fish meal and exported as poultry and hog feed, but the sardine fishery will be directed at producing canned and frozen products for direct human consumption.

With anchoveta fishing banned since 1982 and an emphasis on species conservation in Peru's future fishing activity, the anchoveta fishmeal fishery may be viable in the future. However, it will never reach previous levels. Although it is much smaller in biomass, the sardine fishery may eventually develop into a major source of revenue because of the higher value of the products that can be produced from it. To improve the present state of affairs, the Ministry of Fisheries will have to conduct a prudent program of fishery assessment and management.

MARICULTURE

Marine aquaculture, or **mariculture**, has been conducted for years throughout the Far East, making major contributions to the available food supply in this part of the world. Most attempts at commercial mariculture in the United States have not been successful because most of the projects were based on intensive production tainted with overambition and impatience. Although aquaculture accounts for about 13 million MT, or 15% of the world fishery, mariculture accounts for only about 10% of the aquaculture total. Freshwater projects have generally been more successful than those involving marine organisms.

Organisms chosen for mariculture should be popular marine products that command a high price, are easy and inexpensive to grow, and will reach marketable size within a year or less. Candidates should be hardy and resistant to disease and parasites. They should also be capable of feeding with a high degree of growth per unit of intake. Severe economic problems can result if the chosen organism is not able to reproduce or cannot be brought to sexual maturity in captivity.

Algae

Algae, in the form of certain seaweeds, are grown and marketed as luxury food. The life cycle of most of these algae is very complicated, and most of the successful operations are conducted in Japan, where spore-producing plants are cultivated in laboratories. When the spore-producing plants release their spores into the water, local growers come to the laboratories to dip their nets, ropes, or whatever devices they use for attachment into the water and allow the spores to attach themselves to these devices. They are next submerged in estuarine waters, where the spores grow into the desired seaweed form by using naturally available nutrients.

Bivalves

Cultivation of oysters and mussels is probably the most successful form of mariculture. Commercially, no bivalves have been reared from the larval stage to a marketing size in a controlled environment because of the problems related to producing sufficient phytoplankton to feed them. Usually hatchery-produced juveniles are reared to maturity in natural environments such as bays, estuaries, and other protected coastal waters. They may be set on the bottom or in trays or other suspension devices above the bottom (fig. 17–12). This simple procedure results in yields ranging from 10 to over 1000 MT/acre of edible meat per year. There is no attempt at artificial feeding, and the mollusks are fed by the phytoplankton carried to them by the tides and other water movements.

An interesting outcome of placing oil platforms in the coastal ocean has developed off the Santa Barbara, California, coast. Mussels grew so rapidly on the substructure of the platforms that it cost thousands of dollars per year to remove them from each platform.

In a deal that has worked out well for Bob Meek and his Ecomar Marine Consulting firm, the oil companies now get their platforms cleaned free and Ecomar harvests 3500 pounds of mussels per day. The mussels are sold along the west coast and as far inland as Chicago. The main market is the restaurant industry, which claims these mussels are as fine as can be found. Ecomar also cultures up to 50,000 scal-

FIGURE 17–12 Bamboo Raft Used in Culturing Oysters.

Oysters are suspended from bamboo rafts in nutrient-rich coastal waters off South Korea. This method of culturing off the bottom helps protect oysters from predators such as boring snails and sea stars. Photo from M. Grant Gross.

lops and oysters annually on trays that are suspended from the platforms.

Crustaceans

Crustaceans such as shrimp and lobster have not yet been raised in the United States with any great success, even though the price for such organisms is relatively high. The problem with most shrimp operations is that the animals must be fed extraneously, and this adds greatly to the cost of production. Worldwide, the production of shrimp exceeds 450,000 MT per year, which is about 22% of the world supply of this delicacy. Despite the modest state of development in the United States, shrimp are, on a worldwide basis, the most valuable commodity produced by mariculture. Asian nations lead the way, with China being responsible for 22% of world production. Other major producing nations are Ecuador, Taiwan, Indonesia, Thailand, and the Philippines. Also, Mexico is beginning to develop projects. Shrimp-rearing projects in the United States are found in South Carolina, Texas, and Hawaii.

The brightest spot in shrimp mariculture in the Western Hemisphere is Ecaudor, where a very profitable operation is producing 45,000 MT per year. This success in Ecuador, where the cost of labor and dealing with government regulation is low, contrasts with high costs in the United States, where a similar operation may have to cope with over 35 regulating agencies along with high labor costs. However, it is clear that shrimp can be produced in mariculture operations at much lower cost than they can be provided by fishing. This fact is beginning to be felt by the world shrimp fishing fleet. The largest problem the shrimp mariculture industry faces is the production of larvae, which are susceptible to bacterial and viral infections.

American lobsters *(Homarus americanus)* can easily be carried through the complete life cycle in captivity, but they have a cannibalistic nature. To be successfully grown, these lobsters would have to be compartmentalized and maintained at an optimum temperature of about 20°C (68°F) for two years. Artificial feeding would also be necessary. Encouraging results have been achieved in rearing American lobster in the warm effluent from an electrical generating plant in Bodega Bay, California. Similar projects will be tried in the coastal waters of Maine. The spiny lobster is not so easily reared through its long, complicated larva development, and less effort has been exerted toward developing commercial mariculture with these animals.

Fish

Mariculture with fish is also difficult, and only in Japan and Norway have significant marine programs been developed. Salmon, yellowtail, tuna, puffers, and a few other fishes are being raised in Japan on an experimental basis.

Norway has pioneered the rearing of **Atlantic salmon** in **suspended cages** with success far greater than that achieved throughout the rest of the world in salmon-ranching efforts. Norwegian salmon production reached 100,000 MT in 1990, and this technology is now being used in the Pacific Northwest. In Washington, Oregon, and British Columbia it has replaced

FIGURE 17–13 Milkfish Ponds.

Milkfish are harvested using a gill net. Brackish-water fish-culture techniques applied to ponds such as these in the Philippines are proving successful. FAO photo by P. Boonserm.

the failed salmon ranching, which released young salmon into the ocean in hope of their return. The returns were very poor. This new activity is controversial in the Pacific Northwest because of the introduction of Atlantic salmon into Pacific waters. Although the technology does not involve releasing the salmon from the cages, some have escaped and created a worrisome situation. There is concern about the effect the escaped Atlantic salmon might have on indigenous populations.

Good returns of salmon released by the Alaska Fish and Game hatcheries are being reported. In Canada, the addition of nitrates and phosphates to lakes provided a larger biomass of plankton for young salmon to feed on before going to sea. This rather simple modification, which had no ill effects on the ecology of the lake, increased salmon production in the lake and salmon returns from the open ocean. It is hoped that when the management of commercial operations is refined as knowledge is gained, we will be hearing success stories from the Pacific Northwest.

Some estuary fish farming involving **mullet** and **milkfish** has been practiced successfully in the Far East for centuries. These fishes are hardy and tolerate salinity ranging from that of fresh water to full sea water, and they do not have to be artificially fed. However, because they cannot be spawned and raised to sexual maturity in an artificial environment, the fry must be collected from a natural nursery environment. The costs of operating the farms are relatively low, and yields of about 1 MT/acre are common (fig. 17–13).

WHAT LIES AHEAD?

Mariculture requiring intensive artificial feeding and high labor cost cannot be expected to help alleviate the world shortage of animal protein. Actually, more food is consumed than is produced by such systems, which are typical of the enterprises that have been attempted in the United States. However, those farming enterprises that use natural feeding in estuarine environments have good potential for helping increase the food supply. Most of these efforts require juvenile animals that are obtained from natural nursery areas, which represent the greatest costs to the marine farmer. Potential for increased productivity using the 1 billion acres or so of coastal wetlands in the world would be quite significant. Assuming only 10% of the wetlands would be put into production, it could, based on the productivity that has been established for mullet and milkfish, result in 100 million MT of fish.

Many supporters of mariculture believe that this industry, which now accounts for less than 10% of the present food taken from the sea, can grow to a 60-million-MT/year business. Many scientists who have studied the field of mariculture are not as optimistic about the ability of the industry to significantly increase the world food supply. The new projects being developed in the United States are directed at luxury foods that will certainly be too expensive for the world's hungry poor. They think these projects may generate a profit but not much food.

The Food and Agriculture Organization (FAO) of the United Nations has estimated, on the basis of the yields of traditional fisheries, that the maximum catch that might be expected from the oceans is about 120 million MT. This estimate may be considered conservative, and could well be exceeded, particularly with the addition of new species to the significant fisheries. A candidate for addition to the list of major fisheries might be the **Antarctic krill**. **Grenadiers** and **lantern fishes**, found in deeper waters, and the **pelagic red crab** of the Eastern Pacific and subantarctic regions may also be added in the near future. Another significant addition would be **cephalopods** such as squid, cuttlefish, and octopus, which yield about 80% edible flesh compared to approximately 20–50% for the most common representative of our present fishery.

Surely improved management of existing fisheries, progress in mariculture, and the addition of new species to the world fishery will lead to a much needed increase in the supply of marine food.

NONLIVING RESOURCES

Petroleum

A significant event occurred in the United States during the first week of March 1976: for the first time the United States imported more oil than it produced. Plans were then made to explore and develop petroleum reserves off the Atlantic coast, which had until this time been free of such activities. Test wells were drilled 130 km (81 mi) east of Atlantic City, New Jersey, and 120 km (75 mi) off Cape Cod in the Georges Bank trough. Testing for petroleum off Baltimore Canyon and on Georges Bank has had discouraging results, and resistance to further exploration continues in the important fishery area of Georges Bank. Similar conflicts have resulted from the federal government's plans to develop petroleum reserves and drill exploratory wells along the California coast. Yet, the urgency to increase domestic petroleum produc-

FIGURE 17–14 Potential Petroleum Production Sites.
Areas of the U.S. continental shelf with the greatest potential for petroleum production. Courtesy of U.S. Department of Energy.

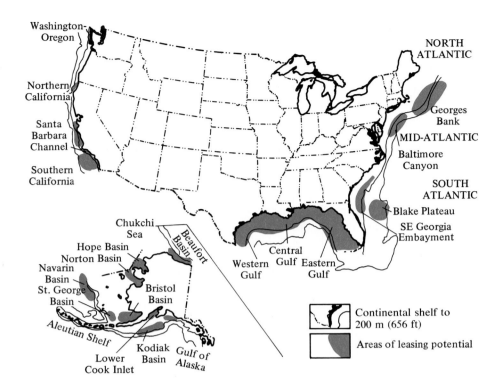

FIGURE 17–14 Potential Petroleum Production Sites.
Areas of the U.S. continental shelf with the greatest potential for petroleum production. Courtesy of U.S. Department of Energy.

tion remains. Figure 17–14 shows the areas of the U.S. continental margin with potential for oil and gas production.

After initial discoveries of North Sea oil and gas in the 1950s, new foreign discoveries continued, and additional fields were put into production. In the absence of international agreement, one of the more encouraging signs has been the relative ease with which the countries bordering the North Sea, a region of vast reserves, divided the petroleum rights among themselves. In April 1976, the United Kingdom recorded its first exportation of oil and is now receiving significant income from oil exports as a result of its North Sea reserves. Other offshore discoveries have been made near the Philippines and off the east and west coasts of Spain as well as on the continental shelf off the mouth of the Amazon River. In spite of the occasional glut of oil, the search for petroleum will be the major focus of exploration beneath the oceans in the immediate future.

The poor results of recent exploratory drilling in U.S. offshore waters has led to a major reduction in the estimated untapped reserves of the United States. Our total reserve of oil is estimated to be 35 billion barrels. This is a 16-year supply at the present rate of consumption and a 32-year supply if we import half of the oil we use. Our reserve of natural gas is believed to be about a 35-year supply, 263 trillion cubic

feet. Considering these facts along with the environmental damage that results from burning fossil fuel, maybe we will begin to think seriously about alternative sources of energy and lifestyle changes in the near future.

Minerals

The origin of metallic ores that contain rich deposits of copper, lead, zinc, and silver associated with the destructive margins of lithospheric plates is beginning to be understood. Their origin is related to the process of plate destruction (fig. 17–15). **Copper sulfide ores** associated with ophiolites—sections of oceanic crust that have been pushed up onto continental plates—are part of this process. A good example of this kind of ore is the **Troodos Massif** along the southern coast of Cyprus, which has been mined since early in the development of the Mediterranean civilizations (see "Ophiolites" on page 68).

The constructive margins of plates associated with oceanic ridges may be the source of this metallic enrichment. The metals found here are predominantly iron and manganese, with significant amounts of copper, nickel, cobalt, zinc, and barium. Analysis of such sediments in a 200-km^2 area of the **Atlantis Deep** in the **Red Sea** have shown that they contain more than 3 million MT of zinc, 1 million MT of copper, 80,000

MT of lead, and 5000 MT of silver. Similar iron- and manganese-rich deposits were recovered by submersibles diving along the Mid-Atlantic Ridge during the French-American Mid-Ocean Undersea Study (FAMOUS), as well as on the East Pacific Rise and the Juan de Fuca Ridge.

Figure 17–16 shows the general distribution of rich deposits of manganese nodules in the world ocean, and figure 5–14 shows the concentrations of manganese, cobalt, nickel, and copper. From the concentrations shown in figure 5–14, it appears that the area of the oceans with the greatest economic potential is in the north equatorial Pacific from Mexico to south of the Hawaiian Islands.

Deposits of phosphorite in nodules or bedded crusts may be found in water close to the edge of the continental shelf (fig. 17–16). These deposits are not likely to be mined in the near future because of the extensive deposits of phosphate on land.

Closer to shore and much easier to mine are **placer deposits** on the continental shelf. Such deposits are formed through erosion of minerals from the nearby landmasses and by concentration in old river channels or along beaches that were subsequently submerged by the rising ocean. These deposits are unconsolidated and can be easily lifted by dredging operations. They include concentrations of **diamonds** off southwest Africa, **gold** off Alaska, **tin** off Indonesia and southwest England, and titanium off Mozambique (fig. 17–16).

With the depletion of **sand and gravel deposits** on land or their unavailability through urban expansion and other land-use activities, it appears that the continental shelf will have to provide much of these aggregate materials in the near future (fig. 17–17). Economic studies have already shown that marine aggregate is competitive with land-derived sources in many areas. In fact, excluding petroleum, the mining of sand and gravel from the coastal ocean is the largest marine mining operation in existence today. Although environmental problems can result from dredging sand and gravel, they may be less serious than those created by mining land deposits in urban areas.

DESALINATION

With great expanses of the continents being arid and the available freshwater supplies throughout the world being used in greater volumes each year, it is becoming urgent that practical methods of **desalination**, or salt removal, be developed so we can take advantage of the greatest water supply on Earth—the oceans. Some of the methods that have been investigated to determine their potential usefulness toward this end are described below.

Distillation involves boiling salt water and passing the water vapor that is produced to a condenser,

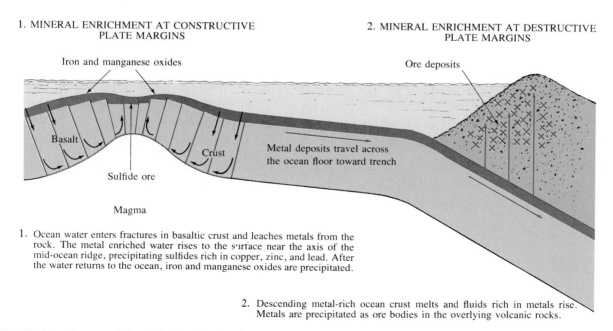

1. MINERAL ENRICHMENT AT CONSTRUCTIVE PLATE MARGINS

2. MINERAL ENRICHMENT AT DESTRUCTIVE PLATE MARGINS

Iron and manganese oxides

Ore deposits

Basalt

Crust

Sulfide ore

Metal deposits travel across the ocean floor toward trench

Magma

1. Ocean water enters fractures in basaltic crust and leaches metals from the rock. The metal enriched water rises to the surface near the axis of the mid-ocean ridge, precipitating sulfides rich in copper, zinc, and lead. After the water returns to the ocean, iron and manganese oxides are precipitated.

2. Descending metal-rich ocean crust melts and fluids rich in metals rise. Metals are precipitated as ore bodies in the overlying volcanic rocks.

FIGURE 17–15 Theory of Metallic Ore Production.

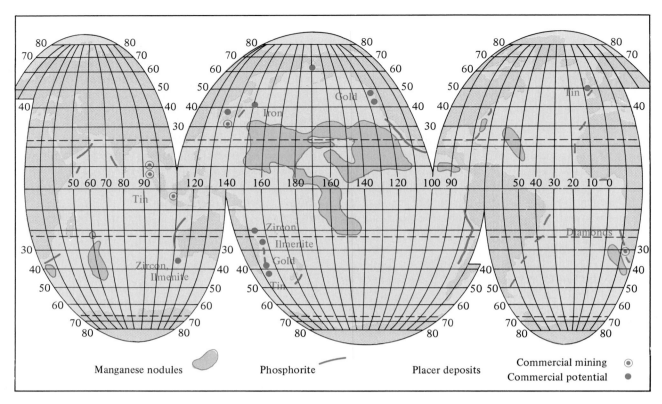

FIGURE 17–16   Manganese Nodule, Phosphorite, and Placer Deposits.
Base map courtesy of National Ocean Survey.

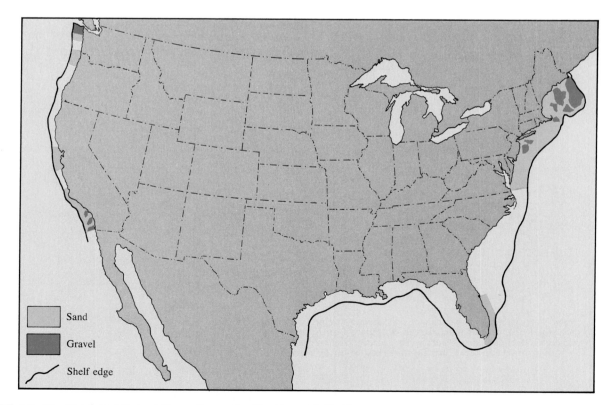

FIGURE 17–17   Sand and Gravel Deposits on the Continental Shelf.
From U.S. Department of Commerce Public Bulletin 188717.

where it is collected as fresh water. This simple procedure is expensive and requires the use of large amounts of heat energy. Increased efficiency will be required before it will be practical on a large-scale basis.

Another process that requires large amounts of energy is **electrolysis**. In this method, two volumes of fresh water—one containing a positive electrode and the other a negative electrode—are located on either side of a volume of sea water. The sea water is separated from each of the freshwater reservoirs by semipermeable membranes that are permeable to salt ions but not to water molecules. When an electrical current is applied, positive ions such as sodium ions are attracted to the negative electrode, and negative ions such as chlorine ions are attracted to the positive electrode. In time, enough ions are removed through the membranes to convert the sea water to fresh water.

Another application that is limited to small-scale use is **freezing**. This process that naturally produces sea ice may not be feasible on a large-scale basis, but imaginative thinkers have proposed the towing of icebergs to the coastal waters of arid lands. Here, the fresh water that is produced as the icebergs melt could be captured and pumped ashore.

Solar humidification does not require supplemental heating and has been successfully used in large-scale agricultural experiments in Israel, West Africa, and Peru. It involves the evaporation of water in a covered container placed in direct sunlight. Salt water in the container is subjected to evaporation, and the water vapor that condenses on the cover runs into trays from which it is collected.

A method that has much potential for large-scale projects is **reverse osmosis**. You will recall from our coverage of osmosis that a net flow of water molecules will pass through a water-permeable membrane from a freshwater solution into a saltwater solution. In reverse osmosis, pressure is applied to the saltwater solution, forcing it to flow through the water-permeable membrane into the freshwater solution (fig. 17–18). It is the reverse of natural osmosis. A significant problem with this method is that the membranes must be replaced frequently. At least 30 nations located in arid climates have built reverse osmosis units.

FIGURE 17–18   Reverse Osmosis Process Used to Produce Fresh Water from Salt Water.
This fundamentally simple process requires that pressure be applied to a reservoir of salt water to force it through a water-permeable membrane into a reservoir of fresh water, from which it is collected.

SUMMARY

As a result of a rapidly expanding human population and a growing need for food and minerals, knowledge and technology have developed to the point where extensive exploitation of the ocean's resources is near at hand. Already the ocean has been significantly affected by the search for food and minerals.

Fisheries management in coastal U.S. waters began with the prohibition against seining for mackerel imposed by the Plymouth Colony in 1684. Attempts at regulation of fisheries have not been successful to date. This failure may in large part be due to the fact that the ecology of most fisheries is not well understood. Present studies of the ecology of marine fisheries attempt to assess the factors that affect survival of fish eggs, larvae, and juveniles. Understanding these important factors helps fishery scientists make better predictions of **recruitment** of young adults to the fishery. A promising method of **fishery assessment** involves an **ecosystem approach**. This method requires knowledge of the annual nutrient flux into the ecosystem, **size grades** of organisms, **standing stock**, and **size-doubling times** of each component of the food chain.

About 50% of the world fishery comes from about 0.1% of the ocean surface area where **upwelling** occurs. Therefore, studies of **duration** and **rate of upwelling** in these regions have been undertaken. They reveal that moderate rates of upwelling over long periods of time—or even short durations—produce larger fisheries than high or low rates of upwelling.

The use of new technologies such as **side-directed sonar** have aided in fisheries assessment. Although fisheries management has not met with great success in the past, there is hope that increased knowledge gained from **ecological assessment programs** will make it possible to do better in the future. **Fisheries regulation** has been difficult also. From the overfishing of the northwest Atlantic to the collapse of the Peruvian anchoveta fishery, we see many examples of the difficulty of fisheries regulation. They will continue to be a challenge in the future.

Mariculture projects around the world have, in some cases, indicated potential for increasing the food taken from the ocean. The most successful mariculture to date has been directed at algae, **oysters, mussels, salmon, mullet,** and **milkfish**. These activities, combined with adding new species such as **krill, grenadiers, lantern fish, red crabs,** and **cephalopods** to our open-ocean fisheries, may help increase the amount of food we take from the ocean.

Inorganic resources such as petroleum, metal ores, phosphorite, and placer deposits of diamonds, gold, and other minerals and aggregates are targets of exploration and exploitation. **Petroleum** is already being extensively exploited, and plans are well developed for the mining of **manganese nodules**.

Desalination of ocean water to provide fresh water for domestic, home, and agricultural use may be achieved by **distillation, electrolysis, freezing, solar humidification,** and **reverse osmosis** of sea water. The last two methods show the greatest potential for practical applications.

QUESTIONS AND EXERCISES

1. Summarize the significant events in the development of international sea law and its main features.

2. What are the two critical survival stages that must occur before young adults can be recruited to a fishery, and what factor is most important in increasing the chance of survival at each stage?

3. Explain the relationship that exists between the influx of nutrients (nitrogen) into an ecosystem and the amount of the fishery that can safely be removed from it each year.

4. Which two characteristics of an upwelling system are related to the magnitude of the fishery it can produce? What combination of these factors appears to produce the maximum environment for the development of a large fishery?

5. What are the characteristics that the fisheries of the four major eastern boundary current upwellings of the subtropical gyres have in common?

6. Discuss how the discovery of large demersal aggregations of krill found in submarine canyons off the flank of Georges Bank relate to the ecology of the fisheries on the bank.

7. List and discuss some positive and negative effects that you think may result from the establishment of the 200-m (322-km) Exclusive Economic Zones.

8. Discuss the natural and artificial conditions that may have combined to cause the catastrophic decline of the Peruvian anchoveta fishery in 1972.

9. What characteristics are desirable in an organism chosen for a mariculture project?

10. What do algae and bivalve cultivation have in common that helps make those organisms good choices for mariculture operations?

11. Discuss the positive and troublesome aspects of mariculture of crustaceans.

12. Discuss how the mariculture of mullet and milkfish is similar to that of bivalves.

13. From which mariculture and natural fisheries do we have the greatest hope of obtaining an increased quantity of food from the sea?

14. Make a table listing regions of the (1) Atlantic Coast, (2) Gulf Coast, and (3) Pacific Coast of the conterminous United States and (4) Alaska where potential for petroleum production from the continental shelf is greatest.

15. Describe the relationship of metallic ore deposits in the lithosphere to the plate tectonics process.

16. Discuss the positive and negative aspects of mining gem and metal placer deposits and sand and gravel deposits on the continental shelves.

17. Briefly describe how the following methods of desalination would produce fresh water: distillation, electrolysis, freezing, solar humidification, and reverse osmosis.

REFERENCES

Backus, R. H., and Bourne, D. W. eds. 1987. *Georges Bank.* Cambridge, Massachusetts: The MIT Press.

Borgese, E. M., and Ginsburg, N. 1988. *Ocean Yearbook* 7. Chicago: University of Chicago Press.

Greene, C. H., Wiebe, P. H., Burczynski, J., and Youngbluth, M. J. 1988. Acoustical detection of high-density krill demersal layers in the sumbarine canyons off Georges Bank. *Science* 241:4863, 359–61.

Knauss, J. A. 1974. Marine science and the 1974 Law of the Sea Conference—Science faces a difficult future in changing Law of the Sea. *Science* 184, 1335–41.

Longhurst, A. R., and Pauly, D. 1987. *Ecology of tropical oceans.* San Diego: Academic Press.

Parsons, T. R., Takahashi, M., and Hargrave, B. 1984. *Biological oceanographic processes,* 3rd ed. New York: Pergamon Press.

Parrish, R., Bakun, A., Husby, D. M., and Nelson, C. S. 1983. Comparative climatology of selected environmental processes in relation to eastern boundary current pelagic fish reproduction. Unpublished F.A.O. study.

Royce, W. F. 1987. *Fishery development.* San Diego: Academic Press.

Smith, P. E. 1969. The horizontal dimensions and abundance of fish schools in the upper mixed layer as measured by sonar. La Jolla, Calif.: Bureau of Commercial Fisheries.

SUGGESTED READING

Sea Frontiers

Brown, M. 1984. Counting dolphins: New answers to an important question. 30:2, 68–75.
Describes a new method used to study the number of dolphins in the tropical Pacific Ocean.

Cooper, L. 1981. Seaweed with potential. 27:1, 24–27.
Describes how an increased demand for seaweeds as a source of food and natural chemical products has stimulated interest in the mariculture of macroalgae.

Cuyvers, L. 1984. Milkfish: Southeast Asia's protein machine. 30:3, 173–9.
The importance of milkfish mariculture operations to the food supply in Southeast Asia is discussed.

Iverson, E. S. 1984. Mariculture: Promise made, debt unpaid. 30:1, 20–29.
Progress in mariculture in North America is considered.

Iverson, E. S., and Iverson, J. A. 1987. Salmon-farming success in Norway. 33:5, 354–61.
The spectacular success of salmon farming of Atlantic salmon in Norway is described.

Iverson, E. S., and Jory, D. E. 1986. Shrimp culture in Ecuador: Farmers without seed. 32:6, 442–53.
The rapidly growing shrimp mariculture industry of Ecuador encounters some growing pains.

Katz, A. S. 1984. Herring fisheries: Comparing the two stories. 30:6, 335–41.
A discussion of how the San Francisco herring fishery management might use the history of the collapsed North Sea herring fishery to prevent their fishery from being overfished.

Maranto, G. 1988. Caught in conflict: Managers trapped in fisheries dilemma. 34:3, 144–51.
With the heavy pressure on fisheries and the conflict between recreational and commercial fishermen, fisheries managers are hard pressed to effectively regulate fisheries.

McCredie, S. 1990. Controversy travels with driftnet fleets. 36:1, 13–21.
The problems related to the development of driftnet fisheries in the Pacific Ocean are discussed.

Nicol, S. 1987. Krill: Food of the future? 33:1, 12–17.
Before krill can be considered as a viable fishery many questions about the cost of the fishery and processing as well as the fishery's effect on krill-eating mammal populations must be answered.

Rodhouse, P. G. 1989. Squid: The southern ocean's unknown quantity. 35:4, 206–11.
The increasing demand for squid has led to the development of an ecological assessment program of Antarctic squid populations.

Scientific American

Holt, S. J. 1969. The food resources of the ocean. 221:3, 178–97.
  A review of world fisheries and their potential for the future is presented. The critical problem of international regulation of fisheries is also explored.

Pinchot, G. B. 1970. Marine farming. 223:6, 14–21.
  Presents a brief history and description of efforts and potential for future marine farming, with an emphasis on the means by which nutrients may be economically introduced into low-nutrient waters.

18
MARINE POLLUTION

As society provides for an ever-increasing population, the demand placed on the oceans in meeting these needs is also increased. One of these roles has been disposal for a growing amount of waste materials, from sewage sludge to radioactive waste. An additional threat is the accidental contamination from transportation and production accidents associated with the efforts to meet present energy needs with petroleum.

The ocean is an important recreational resource in the United States because 80% of its population lives within easy access of the coasts of either the oceans or the Great Lakes. Well over 100 million Americans participate in marine recreational activities, spending almost $150 billion annually. It has not been unusual to see local areas of the shore closed to recreational activities such as fishing, boating, and swimming because of pollution.

We have little difficulty understanding the meaning of pollution. The **World Health Organization** defines pollution of the marine environment as

> The introduction by man, directly or indirectly, of substances or energy into the marine environment, including estuaries, which results or is likely to result in such deleterious effects as harm to living resources and marine life, hazards to human health, hindrance to marine activities, including fishing and other legitimate uses of the sea, impairment of quality for use of sea water and reduction of amenities.

But as far as the marine environment is concerned, there is great difficulty in establishing the degree to which pollution is occurring. In most cases, the ocean has not been studied sufficiently prior to society's introduction of pollutants. Therefore, we cannot tell how the marine environment has been altered by these activities.

CAPACITY OF THE OCEANS FOR ACCEPTING SOCIETY'S WASTES

Although, at present, a strong consensus is developing in the United States and in many other nations that we should not propose to dispose of any materials or energy in the ocean, a more realistic view is that we will have to consider the oceans as an option in disposing of some waste materials. An even more realistic view of this problem is that regardless of what we say we are going to do, polluting materials

Preceding page: The supertanker *Exxon Valdez* (foreground) aground in Prince William Sound near Valdez, Alaska. Photo copyright AP/Wide World Photos.

will continue to enter the oceans for the foreseeable future. In light of this probability, research into the capacity of the oceans to absorb various substances that could lead to the oceans becoming polluted continues.

One of the present goals of marine pollution research is to determine an elusive entity, the **assimilative capacity** of the oceans—the amount of a given material that can be contained within the ocean without causing an unacceptable degree of harm to the living or nonliving resources of the oceans. Determining at what concentrations an unacceptable degree of harm is done to the ocean's resources requires long-term monitoring of disposal activities. Although some researchers believe we have enough data to establish values for the assimilative capacity of the coastal ocean for certain substances, all agree that what we do not know greatly exceeds what we do know.

PREDICTING THE EFFECTS OF POLLUTION ON MARINE ORGANISMS

To date, the most widely used technique for determining the level of pollutant concentration that negatively affects the living resources of the ocean is the **standard laboratory bioassay**. Regulatory agencies such as the **Environmental Protection Agency** used a bioassay that determines the concentration of a pollutant that causes 50% mortality among the test organisms to set allowable concentration limits of pollutants in coastal water effluents. One shortcoming of the bioassay is that it does not predict the long-term effect of sublethal doses on the organisms.

It is desirable to be able to predict the effects of a given concentration of pollutants at four different levels of biological organization, some of which can be detected in a few minutes whereas others may require decades of observation. These levels of biological response, the time spans over which they occur, and examples of each follow.

1. **Biochemical-cellular**: *minutes to hours.* Exposure to organic pollutants such as hydrocarbons and PCBs produce dysfunction in metabolic processes, causing death.
2. **Organism**: *hours to months.* Pollution stress may create changes in physiological processes such as metabolism or digestive efficiency that will severely reduce the energy available for growth and reproduction.
3. **Population dynamics**: *months to decades.* Although it is difficult to tie long-term effects to a certain pollutant, short-term or chronic ef-

FIGURE 18–1 Controlled Ecosystem Pollution Experiment (CEPEX).
Views of CEPEX bags used in experiments to investigate the effects of pollutants on plankton communities. A scuba diver is shown for scale. Courtesy of the National Science Foundation.

fects may be tied to events such as oil spills or malfunctions at sewage plants. These responses occur as changes in the abundance and distribution of a species, population structure on an age-class basis, growth rates within age classes, fecundity, and incidence of disease.

4. Community dynamics and structure: *years to decades.* Communities, as assemblages of populations, become less diverse, and more-resistant populations increase in numbers under the stress of pollutants.

Much progress is needed before the identification of early warning signs of stress at each level of biological response is possible. Because the degree of system complexity and the time needed to measure a response increase exponentially from the biochemical-cellular level to the community level, the predictive difficulties also increase at each level.

The Controlled Ecosystem Pollution Experiment (CEPEX) used plastic experimental bags (fig. 18–1) to simulate communities of plankton ranging in size from 68 to 1700 m^3 (2400–60,000 ft^3) to investigate the effects of pollutants. Observations of the recovery rates of bacteria, phytoplankton, and zooplankton after pollution by heavy metals and hydrocarbons showed that the bacteria and phytoplankton, with their short generation times, recovered much more rapidly than the larger and more complex zooplankton.

MARINE POLLUTION CONTROL IN THE UNITED STATES

The United States has long been concerned with the legal protection of the marine environment. The first **Rivers and Harbors Act** of 1899 provided a straightforward prohibition against dumping any kind of refuse into the navigable waters of the United States. Most current regulations on waste disposal in the coastal ocean, the 200-mi (322-km) Exclusive Economic Zone, are contained in two major statutes. The **Federal Water Pollution Control Act (Clean Water Act)** of 1948 and its amendments deal primarily with problems of point sources of municipal and industrial waste and spills of oil and hazardous materials. The **Marine Protection, Research, and Sanctuaries Act (Ocean Dumping Act)** of 1972 controls dumping of wastes at sea, research, and the establishment of marine sanctuaries.

Although these regulations are quite complex, these laws basically (1) prohibit the dumping of materials known to be harmful and (2) specify the criteria under which other materials may be dumped. The basic premise of each seems to be that land disposal is preferable to marine disposal. If it cannot be proved to the satisfaction of the regulating agencies that dumping of material will not adversely affect human health, it will not be allowed in the ocean.

Despite the strong language of the laws, the coastal ocean has been adversely affected by the dumping of waste materials. Conversely, some waste materials that could possibly have been safely disposed of in the coastal ocean were not because of the difficulty in proving it would not have been harmful to dump them. It is very easy to prove the harmful effects of dangerous materials, but it is almost impossible to prove that other seemingly benign substances would not create a long-term harmful condition.

Because some scientists say that we can probably better predict the paths of materials dumped into the ocean than dumped on land, the option of ocean disposal received renewed attention. In the past, uncertainty led to a decision to prohibit dumping; then there was growing support for placing the obligation of proving harmful effect on the regulating agency in-

stead of requiring the dumping applicant to prove that no harmful effect will occur. With the public outcry in 1988 over littering of northeastern U.S. beaches with medical waste, the Ocean Dumping Ban Act, which bans sewage and industrial-waste dumping after 1991, was passed.

In light of recent advances in scientific knowledge, it is frustrating to have to admit the broad scope of our ignorance of the marine environment. Considering the great cost and amount of time required to remove uncertainty from the field of marine waste disposal in the near future, it would seem prudent to continue exercising a great deal of caution in pursuit of this option. Short-term political priorities point to stopping ocean dumping. Considering the rapid escalation of the cost of land-based disposal, economic reality tends to make ocean dumping more attractive. Clearly, politicians and their constituents need to have the best possible data available to proceed rationally. Some very difficult decisions will have to be made in the near future.

AREAS OF CONCERN

Plastics

During the last thirty years, the use of plastics has increased at a tremendous rate. The problem of disposing of this "throwaway" component of the western culture has already strained the capacity of our land-based disposal systems. Such waste is also an increasingly abundant component of oceanic flotsam.

Small pellets used in the production of essentially all plastic products are transported in bulk aboard commercial vessels. Found throughout the oceans, the pellets probably find their way into the oceans as a result of spillage at loading terminals. In coastal waters, plastic products used in fishing and thrown overboard by recreational and commercial vessels are common. They also find their way into the open ocean waters from careless dumping by commercial vessels.

The best documentation of negative effects of plastics in the ocean on marine organisms is found in the strangulation of seals and birds caught in plastic netting and packing straps (fig. 18–2). Marine turtles are known to mistake plastic bags for jellyfish or other transparent plankton on which they typically feed.

It is believed that all plastics that enter the ocean may eventually be removed by shorelines of islands and continents that "filter" out the particles as ocean currents wash against their shores. Beaches throughout the world are probably increasing their plastic

FIGURE 18–2 Elephant Seal with Plastic Packing Strap Tight Around Its Neck.

This animal was found on a beach of San Clemente Island off the Southern California coast. It was saved from strangulation by the removal of the plastic packing strap by scientists from the National Marine Fisheries Service. Photo by Wayne Perryman, NMFS.

pellet content as a result of the filtering process. Some Bermuda beaches have up to 10,000 pellets/m². The beaches of Menemsha Harbor on Martha's Vineyard yielded 16,000 plastic spherules/m². Thin plastic film products will eventually be broken into smaller particles by photochemical degradation, and others may sink as they become denser because of photochemical degradation and encrustation by epifauna such as bryozoans and hydroids. These data are from a survey conducted between 1984 and 1987; the survey also found more than 10,000 plastic pieces and 1500 pellets/km² in the northern Sargasso Sea between 28° and 40° north latitude. Compared to a survey conducted in 1972, the 1987 survey found that the concentration of plastic pellets had doubled.

The U.S. Congress is considering a bill to ban disposal of plastics in the 200-mile-wide Exclusive Economic Zone. Internationally, the Convention for the Prevention of Pollution from Ships prevents disposal of all plastics into the oceans. The acceptance of both of these regulations would aid greatly in turning around the increasing concentration of plastics in the world's oceans.

Petroleum

There have been a number of major oil spills since the *Torrey Canyon* ran aground at Seven Stones Rocks off the Cornwall coast on March 18, 1967. The spills have resulted from additional tanker accidents and the blowout of oil wells being drilled or produced in the coastal ocean. Although it is important to know how much oil enters the oceans and from what sources, our primary concern is the effect of hydrocarbon pollution on marine organisms and the marine environment in general.

Because hydrocarbons are organic substances biodegradable by microorganisms, they are considered by some to be among the least damaging pollutants of the ocean. Yet they are complex, containing molecules ranging in molecular mass from 16 to 20,000. Most crude oils will also contain, in addition to their hydrocarbon components, chemicals containing oxygen, nitrogen, sulfur, and trace metals. When this complex chemical mixture is combined with the ocean—another complex chemical mixture containing organisms—the result can indeed be negative to the organisms.

West Falmouth Harbor Because of the complexity of petroleum, it is difficult to answer the question, How long does it take for a shore to recover from an oil spill? The oldest well-studied oil spill in the United States occurred on September 16, 1969, near West Falmouth Harbor in Buzzards Bay, Mass. When the barge *Florida* came ashore and ruptured, currents carried the Number-2 fuel oil north into Wild Harbor, where the most severe damage occurred (fig. 18–3). The initial kill was almost total for intertidal and subtidal animals in the most severely oiled area. A severe reduction in species diversity was accompanied by rapid increases in the population of polychaete worms that were resistant to the oil. The species *Capitella capitata* accounted for up to 99.9% of the individuals taken in samples of the most severely oiled

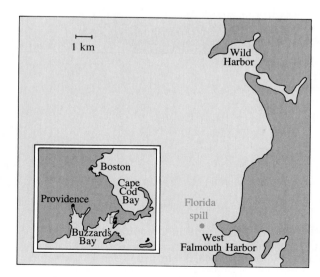

FIGURE 18–3 *Florida* **Oil Spill at West Falmouth Harbor.**

locations during the first year. Species diversity did not appreciably increase until well into the third year after the spill. Although conditions have returned to near normal, oil can still be found in the fine sediment of areas that underwent the most severe contamination.

Chedabucto Bay   A spill of heavy Bunker-C fuel oil in Chedabucto Bay, Nova Scotia, on February 4, 1970, by the vessel *Arrow* has provided some indication of

the recovery time. Based on changes in visible contamination, about one-third of the initially oiled 200 km (124 mi) of shore was clear after three months. This cleaning was primarily the result of wave action on the rocky shore. By 1973, mechanical cleaning by waves, evaporation, photochemical decomposition, and bacterial decomposition had cleaned all but the restricted low-energy estuarine marshes. Fully 75% of the shore was clean. Visible contamination had been reduced to 15% by 1976 (fig. 18–4).

A.

FIGURE 18–4   Oil Spill in Chedabucto Bay.
A, distribution of oil residues in the shore zone of Chedabucto Bay, Nova Scotia, Canada, determined by aerial inspection in 1970 and 1973 and by ground survey in 1976. *B,* oil-removal pattern of stranded Bunker-C oil from shorelines from 1970 through 1976. No oil is visible at the surface now, but it can be found by removing the surface sediments in some marshes.

B.

Although the oil disappeared rapidly from the rocky shore, the low-energy sediments of the marshes will not be cleaned so quickly. The half-life for removal of the visible oiling in the bay as a whole is about two years, but it appears that the half-life for removal from the fine sediment will be up to 25 years.

Dr. John H. Vandermeulen of the Bedford Institute of Oceanography is at present gathering data on the long-term effects of the Chedabucto Bay spill. He says that sandy and muddy sediments still have detectable hydrocarbons and that boulder beach deposits are still cemented with heavy tar deposits. Oil released from local commercial and recreational fishing activities over the past 18 years has made it impossible to tie the present contamination to the *Arrow* spill in 1970. In terms of total biomass, the area seems to be fully recovered.

The biological consequences of the Chedabucto Bay spill have been indicated by the recovery of kelp, *Fucus vesiculosus;* the salt marsh grass, *Spartina alterniflora;* and the soft-shell clam, *Mya arenaria.* Over half of the *Fucus* was destroyed, and a recovery half-life of four years is indicated. *Spartina* suffered a similar reduction but was much slower to start recovery. *Fucus* made a gradual recovery that began immediately, but *Spartina* showed no signs of recovery until two years after the spill. Once recovery started, however, it was rapid, and an estimate of the recovery half-life is about 3 years. The clam, *Mya,* does not show such an encouraging pattern of recovery. Severely stressed by the contamination, *Mya* has experienced a serious reduction in efficiency of food utilization. In some locations, *Mya* is barely surviving. Tissue and shell growth are greatly reduced, and the age-class structure has been altered because of the high degree of mortality that occurred with the spill. The recovery half-life of the *Mya* will surely be in excess of 10 years.

Argo Merchant When oil spills do not come ashore, the negative effects are not so obvious. The sinking of the *Argo Merchant* after it ran aground on Fishing Rip Shoals off Nantucket on December 15, 1976, provides the best-studied effects of such a spill. A full load of Number-6 fuel oil was dumped into the ocean 40 km (25 mi) southeast of Nantucket Island. Fortunately, the winds were such that no oil came ashore. The surface slick moved east out to sea and was gone by mid-January. Even though the oil was not visible for long, it did a significant amount of biological damage that was investigated by scientists from Woods Hole Oceanographic Institution, the National Marine Fisheries Service, and local institutions (fig. 18–5).

FIGURE 18–5 *Argo Merchant* Oil Spill, December 15, 1976.

A, location of the grounding site and oiled area of the ocean. *B, Argo Merchant,* after breaking up and spilling much of its cargo into the Atlantic Ocean southeast of Nantucket. *B,* photo courtesy of U.S. Environmental Protection Agency.

The primary observable effect of the *Argo Merchant* spill on marine organisms was in pelagic fish eggs (pollock and cod) in plankton samples taken shortly after the spill. Of the 49 pollock eggs recovered, 94% were oil-fouled, and 60% of the 60 cod eggs were in that state. Twenty percent of the cod eggs and 46% of the pollock eggs were dead or dying (fig. 18–6). Although little is known about the natural mortality rate of pelagic fish eggs, only 4% of a sample of cod eggs spawned in a laboratory were dead or dying at the same development stage.

Because most of the oil floated in a surface slick, contamination in subsurface water samples did not exceed more than 250 parts per billion. Other than the described damage to fish eggs and a significant amount of damage to other members of the plankton, little direct evidence of major biological damage was recovered. There were however, numerous oiled birds that washed ashore at Nantucket and Martha's Vineyard.

Because pollock females spawn about 225,000 and cod about 1 million eggs each season over a range extending from New Jersey to Greenland, it is unlikely that this single occurrence will have a major effect on those fisheries. However, it is easy to see why the fishing communities of the northwest Atlantic are so happy about the marginal results of oil exploration on Georges Bank. It may be that a fishery already severely stressed by overexploitation could not survive even the lowest level of pollution brought about by the establishment of oil production on the bank.

Prince William Sound The first major oil spill resulting from the development of the North Slope of Alaska petroleum reserves occurred March 23, 1989, when the California-bound *Exxon Valdez* went aground on rocks 40 km (25 mi) out of Valdez, Alaska (fig. 18–7). The 32,000 MT of oil spilled makes this the largest spill to occur in U.S. waters. This spill in the pristine waters of Prince William Sound spread to the Gulf of Alaska, and 1775 km (1000 mi) of shoreline were damaged. The U.S. **Fish and Wildlife Service** reported kthat at least 994 sea otters and 34,434 birds were killed by the spill. The actual kill could be ten times that amount by some official estimates. The total death toll will never be known. It was predicted that the Alaskan waters would have a long, slow recovery, but the fisheries closed in 1989 bounced back with record takes in 1990.

Encouraged by good results on small test plots, Exxon is spending $10 million to spread phosphorus- and nitrogen-rich fertilizers on Alaskan shorelines to

A.

B.

FIGURE 18–6 Pollock Eggs Affected by the *Argo Merchant* Spill.
A, these eggs, in the tail-bud and tail-free embryo stages, were taken from the edge of the oil slick. The outer membranes of the eggs at the upper left and lower right are contaminated with a tarlike oil; arrows point to some of the oil masses. The uncontaminated egg at the upper right has a malformed embryo; that at the lower left is collapsed and also has an abnormal embryo. The actual size of pollock eggs is about 1 mm (0.04 in.); of cod eggs, about 1.5 mm (0.06 in.). *B,* a portion of the surface of an oil-contaminated egg. The upper arrow points to one of many oil droplets; the lower arrow to one of the membrane pores. Scanning electron microscope: about 5000×. Courtesy of A. Crosby Longwell.

A.

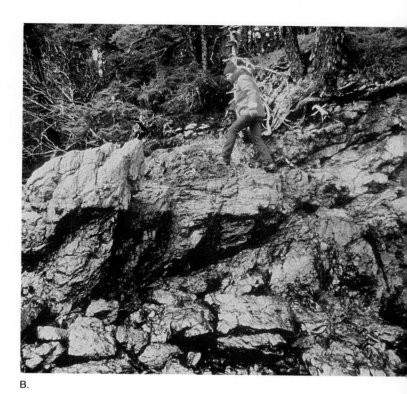

B.

FIGURE 18–7 *Exxon Valdez* Oil Spill.

A, location and extent of contamination resulting from the spill of heavy crude oil from the super-tanker *Exxon Valdez*. *B*, oil from the spill damaged shorelines and killed sea otters and birds. Photo courtesy of U.S. Coast Guard.

boost the development of indigenous oil-eating bacteria. It is hoped that this will result in a cleanup rate that will produce in two to three years what would require up to seven years under natural conditions.

Oil spills are a problem that will be with us for many years, and they may become more common as the petroleum reserves underlying the continental shelves of the world are increasingly exploited. The largest oil spill on record occurred June 3, 1979, as a result of this exploitation. The Petroleus Mexicanos (PIMEX) oil-drilling station **Ixtoc #1** in the Bay of Campeche off the Yucatán peninsula, Mexico, blew out and caught fire. Before it was capped on March 24, 1980, it spewed 468,000 MT of oil into the Gulf of Mexico (fig. 18–8).

Oil can be broken down by microorganisms such as bacteria, fungi, and yeast. Certain of these organisms are effective in breaking down only a particular variety of hydrocarbon, and none is effective against all forms. In 1980, **Dr. A. M. Chakrabarty**, now a microbiologist at the University of Illinois, worked at the General Electric Research and Development center in Schenectady, New York. There he produced a microorganism capable of breaking down nearly two-

thirds of the hydrocarbons in most crude oil spills. A similar strain of bacteria was released into the Gulf of Mexico to test its effectiveness in cleaning up the crude oil spilled after a June 8, 1990, explosion disabled the *Mega Borg*. Preliminary results indicate that the bacteria are effective. Carl Oppenheimer of the University of Texas, who conducted the test, reports no indication of negative effect on the ecology of the Gulf.

Sewage

Over 500,000 MT of sewage sludge containing a mixture of human waste, oil, zinc, copper, lead, silver, mercury, PCBs, and pesticides per year have been dumped through sewage outfalls into the coastal waters of southern California. A more massive 8 million MT have been dumped into the New York Bight each year. Although the Clean Water Act of 1972 prohibited dumping of sewage into the ocean after 1981, the high cost of treating and disposing of it on land resulted in extended waivers being granted to these areas. Nonbiodegradable debris was probably carried into the ocean through storm drains by heavy rains

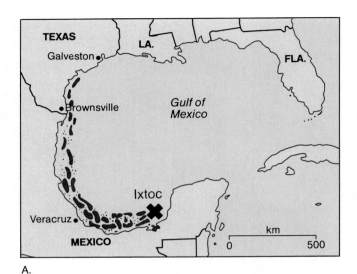

A.

FIGURE 18–8 Ixtoc #1 Blowout, June 3, 1979.
A, location of the Gulf of Campeche blowout and oil slick that
threatened the Texas coast during the summer of 1979. *B,* fire-
fighting equipment spraying water on the waters around the flam-
ing oil directly above the wellhead. Oil can be seen spreading
across the ocean surface. Photo courtesy of NOAA.

B.

and washed up on Atlantic coast beaches during the
summer of 1988. Ironically, this event, which was not
related to sewage disposal, hurt the tourist business
and was responsible for legislation again being
passed to terminate disposing of sewage into the
ocean.

It now appears that the political pressure to clean
up the coastal water is sufficient to assure that land
treatment and disposal will be the primary disposal
procedure of the future. For the present, Los Angeles
has stopped pumping sewage into the sea. If it is allowed to resume this practice, the sewage sludge
pumped through the lines will have had its toxic
chemicals and pathogens removed.

Off the east coast over 8 million MT of sewage
sludge was dumped by barge over dump sites total-
ing 150 km^2 (58 mi^2) each year during the past few
years. These dumps occurred at the New York Bight
Sludge Site and the Philadelphia Sludge Site shown
in figure 18–9. The New York Bight water depth is
about 29 mi (95 ft), and the Philadelphia's site is 40
m (130 ft) deep. In such shallow water, the water column displays relatively uniform characteristics from
top to bottom; even the smallest sludge particles
reach the bottom without having undergone much
horizontal transport, and the ecology of the dump site
can be totally destroyed. At the very least, the concentration of organic and inorganic nutrients seri-

**FIGURE 18–9 Previous and Present Atlantic Sewage
Sludge Disposal Sites.**

ously disrupts the biogeochemical cycling. Greatly reduced species diversity results, and in some locations the environment becomes anoxic.

The shallow-water sites have been abandoned, and the sewage is now being transported to a deep-water site 106 miles (171 kilometers) out (fig. 18–9). At the deep-water site, beyond the shelf break, there is usually a well-developed density gradient separating low-density, warmer surface water from high-density colder deep water. Internal waves moving along this density gradient can retard the sinking rate of particles and may allow a horizontal transport rate 100 times greater than the sinking rate.

Local fishermen are having none of the rosy scenario described above and were reporting adverse effects of the deep-water dumping on their fisheries soon after it began. As it stands now, this east coast dumping program for sewage is also doomed, and municipalities will have to find the money for land-based processing.

Radioactive Waste

As with other waste materials, a case can be made that oceans can accept a given amount of radioactive waste without doing significant harm to the marine environment. Since 1944, artificial **radionuclides** (radioactive atoms) have been reaching the oceans as **fallout** from the production and testing of nuclear weapons. This input has been about four times greater in the Northern Hemisphere than in the Southern Hemisphere. Because of the nature of the fallout process, no high concentration of any radionuclide has resulted. Table 18–1 shows the amount of radiation in thousands of curies that has reached the oceans via fallout. (A **curie** is a quantity of any radioactive nuclide in which exactly 3.7×10^{10} disintegrations occur per second.)

The second largest source of radionuclides is the **nuclear fuel cycle**. The release from this source occurs in local concentrations. Nuclear power generating plants release little radiation to the oceans, but nuclear fuel reprocessing plants contribute a significant quantity. Such plants are found at Sellafield and Dounreay in Great Britain and Cape de la Hague, France. The Irish Sea, which has been receiving reprocessing discharges since the early 1950s, is now the most radioactive area in the world ocean.

Another significant source of radioactive waste is a dumping operation being carried out over the past three decades under international agreement at various locations. Since 1967, the only significant dumping operation has been conducted in the northeast Atlantic under the coordination of the **Nuclear Energy Agency** of the **Organization for Economic Cooperation and Development**. The input of this operation is shown in table 18–1.

In order to determine safe input rates for various radionuclides, oceanographers must determine the average concentration of a radionuclide if mixed totally in the ocean. This is of primary importance for radionuclides with half-lifes much greater than the mixing time of the ocean; it is a major consideration for materials like plutonium, which has a half-life of 24,400 years. **Mixing time** is the time required to evenly distribute a substance added to the ocean. The scientific community's best guess is that the mixing time of the world ocean is several thousand years; that of the Atlantic Ocean may be less than 1000 years.

A second consideration is to determine the transfer mechanisms to critical populations; this is important for all radionuclides in areas near a dumping locality, regardless of the lengths of their half-lives.

The **International Atomic Energy Agency** has determined maximum dumping rates on the basis of these considerations. As an example, the approved dump-

TABLE 18–1 Artificial Radionuclides in the Oceans and Their Sources.

	Plutonium-239, -240 (kCi)	Cesium-137 (kCi)	Strontium-90 (kCi)	Carbon-14 (kCi)	Tritium (kCi)
Total worldwide fallout by early 70s	320	16,700	11,500	6,000	3,000,000
North Atlantic Ocean— early 70s	63	3,300	2,300		650,000
Windscale, 1957–78 discharge	14	830	130		370
	Total α-emitters	*Total β/γ-emitters (other than tritium)*			
The NEA dumpsite (1967–79)	8.3	258			262

ing rate for one of the most dangerous radionuclides, plutonium 239, is placed at 10,000 curies/yr. This limit would result in the plutonium concentration reaching the safe dose limit in 40,000 years.

Although research has been done on deep-sea disposal of high-level nuclear waste, it has now been terminated. The Nuclear Energy Agency has concluded that if this waste is to be disposed of in the thick sediments of abyssal plains, it will not occur for at least 50 years. Land-based disposal sites are now the top priority of research toward the solution of this problem. However, given the political opposition to the development of such facilities, we may see seabed disposal revived as an option in the future. See "Subsea Disposal of High-Level Nuclear Waste," on page 476 for a discussion of the proposed program.

Halogenated Hydrocarbons (DDT and PCBs)

DDT (dichlorodiphenyltrichlorethane) and PCBs (polychlorinated biphenyls) are found throughout the marine environment. They are persistent, biologically active chemicals that have been put into the oceans entirely as a result of human activities.

At the time of a near-total ban on U.S. production of the pesticide DDT in 1971, 2.0×10^9 kg (4.4×10^9 lb) had been manufactured, most of it by the United States. Its use in the Northern Hemisphere since 1972 has virtually ceased. The danger of excessive use of DDT and similar pesticides was first manifested in the marine environment in the effects on marine bird populations. During the 1960s, there was a serious decline in the **brown pelican** population of **Anacapa Island** off the coast of California; high concentrations of DDT in the fish eaten by the birds had caused the production of excessively thin egg shells. A decline in the **osprey** population of **Long Island Sound** that began in the late 1950s and continued throughout the 1960s was also caused by thinning egg shells brought on by DDT contamination. Studies showed a 1% increase in brown pelican and osprey egg shell thickness from 1970 to 1976, while the concentration of pesticide residue decreased. Because there was an increased egg hatching rate associated with these changes, it was concluded that DDT was the cause of the decline in the bird populations during the 1960s (fig. 18–10).

PCBs are industrial chemicals found in a variety of products from paint to plastics. They have been indicated as causes of spontaneous abortions in sea lions and the death of shrimp in **Escambia Bay, Florida.**

The main route by which DDT and PCBs enter the ocean is the atmosphere. They are concentrated ini-

FIGURE 18–10   Brown Pelican.
Pelican populations in the Channel Islands off Southern California are increasing, following a serious decline in the 1960s caused by DDT toxicity.

tially in the thin surface slick or organic chemicals at the ocean surface; then they gradually sink to the bottom attached to sinking particles. A study off the coast of Scotland indicated that open-ocean concentrations of DDT and PCBs are 10 and 12 times less, respectively, than in coastal waters. Long-term studies have shown that DDT residue content in mollusks along the U.S. coasts reached a peak in 1968. The pervasiveness of DDT and PCBs in the marine environment can best be demonstrated by the fact that Antarctic marine organisms contain measurable quantities of halogenated hydrocarbons. Obviously there has been no agriculture or industry on that continent to explain this presence; these substances have been transported from distant sources by winds and ocean currents.

Mercury

The stage was set for the first tragic occurrence of mercury poisoning with the establishment of a chemical factory on **Minamata Bay, Japan,** in 1938. One of the products of this plant was acetaldehyde, the production of which requires mercury as a catalyst. The first ecological changes in Minamata Bay were reported in 1950, human effects were noted in 1953, and the disease became epidemic in 1956. It was not until 1968 that the Japanese government declared mercury the cause of the disease, which involves a breakdown of the nervous system. The plant was immediately shut down, but by 1969 over 100 people were known to suffer from the disease. Almost half of these victims died. A second occurrence of mercury poisoning resulted from pollution by an acetaldehyde factory, shut down in 1965, that was located in Niigata, Japan. Between 1965 and 1970, 47 fishing families contracted the disease.

During the 1960s and 1970s, much attention was given to the problem of mercury contamination in seafood. Studies done on the amount of seafood consumed by various populations of humans have led to the establishment of safe levels of mercury content in fish to be marketed. To establish these levels for a given population, three variables must be considered:

1. fish consumption rate of the human population under consideration,
2. mercury concentration of the fish being consumed by that population, and
3. minimum ingestion rate of mercury that induces symptoms of disease.

If dependable data on these variables are available and a safety factor of 10 is applied to the determination, a maximum concentration level allowable for the contaminant (in this case mercury) can be established with a high degree of confidence that it will protect the health of the human population.

Figure 18–11 shows how such data can be used to arrive at a safe level of mercury concentration in fish

to be consumed by a population. For the general populations of Japan, Sweden, and the United States, the average individual fish consumption rates are 84, 56, and 17 g/day, respectively. By comparison, the members of the Minamata fishing community averaged 286 and 410 g/day during winter and summer seasons, respectively. Based on the minimum level of mercury consumption that is known to bring on symptoms of poisoning as determined by Swedish scientists using Japanese data—0.3 mg/day over a 200-day period—the three populations shown in figure 18–11 would begin to show symptoms if they ate fish with the following mercury concentrations: Japan, 4 ppm; Sweden, 6 ppm; and United States, 20 ppm. If a safety factor of 10 is applied, the maximum concentrations of mercury concentration in fish that could be safely consumed by these populations would be, for Japan, 0.4 ppm; for Sweden, 0.6 ppm; and for the United States, 2.0 ppm.

Although the U.S. Food and Drug Administration (FDA) initially established an extremely cautious limit of 0.5 mg/kg (ppm), the present limit of 1 mg/kg adequately protects the health of U.S. citizens. Essentially all tuna falls below this concentration, and most swordfish will be acceptable. The present limit amounts to the use of a safety factor of 20 instead of 10, so unless one eats an unusually large amount of tuna or swordfish, we should have little concern about consuming these fishes.

FIGURE 18–11 Mercury Concentrations in Fish vs. Consumption Rates for Various Populations.

Plotted curves show the safe ingestion level of mercury until the first signs of Minamata disease can be expected (0.3 mg/day), and the safe ingestion level of mercury after the safety factor of 10 is included (0.03 mg/day). For the United States, it appears that tuna and swordfish are safe to consume, although some swordfish will be banned because their mercury concentrations are above the present 1.0 mg/kg (ppm) FDA limit.

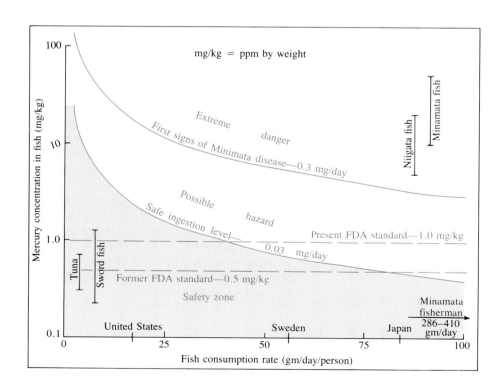

SUBSEA DISPOSAL OF HIGH-LEVEL NUCLEAR WASTE

For nearly four decades, high-level nuclear waste has been accumulating from the production of nuclear weapons and commercial power generation. The U.S. alone has more than 75 million gal of waste from weapons production and 12,000 MT of spent reactor fuel that must be safely disposed of until it is no longer dangerous.

This material will be radioactive for more than one million years, but a "safe" disposal system would likely require a shorter time of highly secure confinement. The wastes are composed of over 50 isotopes, each of which has different chemical, half-life, abundance, and radioactive emission characteristics. Considering such factors, investigators believe confinement that allows no more release of radiation into the atmosphere than that of the natural uranium ore from which it was generated may be considered safe and is possible with today's technology (fig. 18A).

The major emphasis at present is on disposal at land sites because the materials remain accessible. However, since 1974, research has been conducted into the possibility of disposal of high-level nuclear waste in the deep sea. This research was terminated by the United States in 1986 because of budget constraints and on an international basis in 1987 when international agreement was reached to pursue land-based disposal. However, the seabed disposal research did produce encouraging results. Given the political realities that may be faced as work proceeds toward land-based disposal, deep-sea disposal may be considered again in the future. The major features of the seabed disposal program are described below.

Given the requirements that the waste must be kept out of the way of human activities, safe from exposure by natural erosion, and away from seismically active regions, the centers of oceanic lithospheric plates might be ideal disposal sites. Such regions are beneath at least 5000 m of water far from the marine activities of humans. Fishing, petroleum production, and mining activities are now conducted primarily on the continental shelves. Only a potential area of manganese nodule mining between Mexico and the Hawaiian Islands would need to be avoided.

The lithospheric plates are covered with up to 1000 m of fine sediment that has accumulated uninterrupted for over 100 million years in some regions. This pattern of sediment accumulation indicates long periods of stability as the plates move across the earth's surface, and the pattern can be expected to continue for millions of years. In addition, the sediment could serve as an ideal medium in which to place the high-level nuclear waste. Many of the radionuclides, after being released from the canisters (which may fail within 1000 years), would naturally adhere to the clay particles in the sediment. This would slow diffusion of radioactivity away from the burial site.

A number of mid-plate, mid-gyre sites (MPGs) have been identified in the North Atlantic and North Pacific oceans (fig. 18B). The initial plan is to place the canisters under 30 to 100 m of sediment at intervals of at least 100 m. How

FIGURE 18A Radiation Doses from Various Sources in Millirems per Year.

The Environmental Protection Agency is responsible for setting the minimum performance requirements (levels of radiation emitted from high-level disposal sites). It is believed that they will be no higher than that of natural uranium ore.

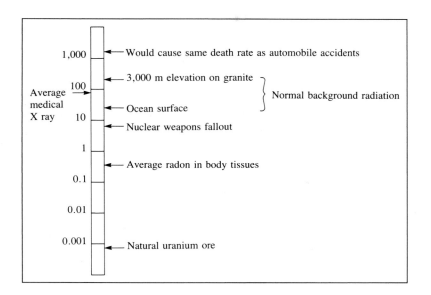

FIGURE 18B MPG Regions.

Mid-plate, mid-gyre (MPG) regions in northern oceans where the environment can be expected to be stable for millions of years.

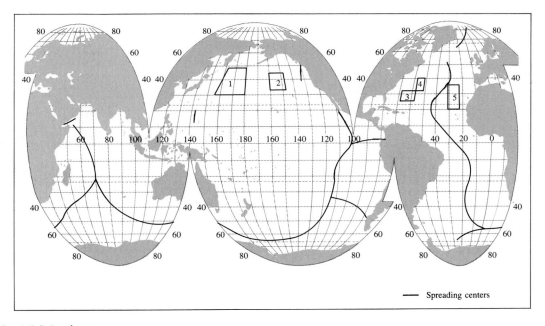

to emplace the canisters has not been decided yet, but it may be possible to simply drop them from an appropriate distance above the ocean floor (fig. 18C).

Models indicate that heat release from the waste will be transferred almost entirely by conduction, so upward convection of radiation carried by sediment pore water may be negligible. This is a crucial consideration. Although these results are promising, yet to be completed are models of how well the burial holes will reseal themselves and the means by which radiation will be transported through the water column once it reaches the sediment surface. Should this option ever be used, the present thinking is that it will not be for at least another 50 years. Much additional testing will be conducted before such a form of disposal would be put into operation.

FIGURE 18C Canister Placement.

Plan of canister emplacement shows them buried at least 30 m (98 ft) beneath ocean floor spaced 100 m (328 ft) apart. Because of the softness of the sediment, it may be possible to implant the canisters by simply dropping them from an appropriate distance above the sediment surface. (Figure not to scale.)

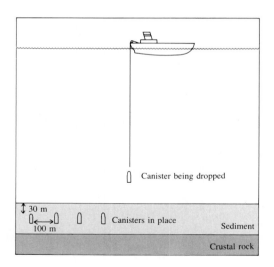

Canister being dropped

30 m

100 m

Canisters in place

Sediment

Crustal rock

INTERNATIONAL EFFORTS TO PROTECT THE MARINE ENVIRONMENT

Beginning with the *Action Plan for the Human Environment* established by the United Nations in 1972, a comprehensive plan for international environmental protection has evolved. It involves procedures for assessment of problem areas and monitoring activities that are potentially sources of pollution.

One of the most critically polluted areas identified was the Mediterranean Sea, polluted in many coastal regions from essentially every conceivable source—domestic sewage, industrial discharge, pesticides, and petroleum. To aid its recovery, an action plan was initiated in 1975, completed in 1976, and has been ratified by 13 of the 18 nations with Mediterranean coasts.

Other regions identified as environmental units with action plans in effect are the Red Sea, Gulf of Aden, East Asian Seas, East Africa, and Kuwait (fig. 18–12). Areas with action plans in preparation are the Caribbean, South Asian Seas, West and Central Africa, Southeast Pacific, and Southwest Pacific.

If such efforts proceed effectively, they can reverse the trend toward progressive degradation of the environment that accelerated throughout the 1960s. Recent efforts appear to have slowed this process and may well have reversed it in local coastal areas. The prognosis for the future health of the marine environment is much improved from a decade ago; this change serves as testimony to the human will to preserve the quality of the earth's environment. There is still much to do, and a continuing effort is required if we are to succeed.

SUMMARY

Pollution is the introduction of substances or energy into the marine environment that results in harm to the living resources of the oceans or humans that use these resources. Scientists are trying to determine the ocean's assimilative capacity—the amount of a substance it can absorb without becoming polluted.

The U.S. government has been concerned about coastal pollution since it passed the **Rivers and Harbors Act** in 1899. Later laws have been passed to prohibit the dumping of materials known to be harmful and to specify criteria under which other materials may be dumped.

The amount of **plastic** accumulating in the oceans has increased dramatically. Certain forms of plastic are known to be lethal to marine mammals and tur-

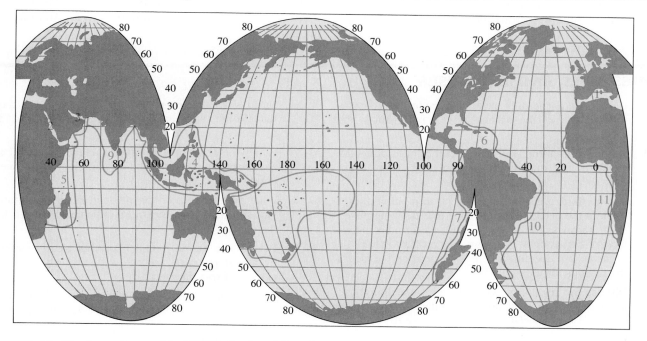

FIGURE 18–12   Areas Covered by UNEP's Regional Seas Program.
According to the United Nations Environment Program, the definition of boundaries is the responsibility of the governments concerned, and those shown here are only illustrative. Environmental units with action plans in effect or in preparation include (1) Mediterranean Sea, (2) Red Sea and Gulf of Aden, (3) Kuwait region, (4) East Asian Seas, (5) East Africa, (6) Caribbean Sea, (7) Southeast Pacific, (8) Southwest Pacific, (9) South Asian Seas, (10) Southwest Atlantic, and (11) West and Central Africa.

tles, and national and international legislation is being considered to ban the disposal of plastic in the oceans.

Study of areas such as **Wild Harbor**, Massachusetts, and **Chedabucto Bay**, Nova Scotia, have shown that **oil** pollution reduces species diversity and persists longer on muddy bottoms than on sandy or rocky bottoms. Oil spills that do not come ashore, like that of the *Argo Merchant*, do considerably less environmental damage than those that do. The greatest damage resulting from the *Argo Merchant* spill was probably to plankton, especially fish eggs.

Although 1972 legislation required an end to dumping of **sewage** in the coastal ocean by 1981, exceptions continued to be made. Especially off the **southern California** coast and in **New York Bight**, large amounts of sewage and sewage sludge continued to be dumped. Increased public concern resulted in new legislation that will prohibit sewage dumping in the ocean. Some scientists expected that the transfer of dump sites along the northeast Atlantic coast to deeper water farther offshore would make it safe to continue to dump sewage from Philadelphia and the New York area into the coastal ocean. However, fishermen have complained that the deep-water dumping has damaged fisheries less than one year after it began.

Most of the **radioactive waste** in the ocean comes from **fallout** resulting from the testing of nuclear explosives since 1944. However, the areas where the greatest concentration of **radionuclides** is entering the ocean are near the fuel-reprocessing plants in Great Britain and France.

DDT pollution produced a decline in the **Long Island osprey** population in the 1950s and the **brown pelican** population of the **California** coast in the 1960s. Virtual cessation of DDT use in the Northern Hemisphere in 1972 allowed the recovery of both populations. The DDT had thinned the egg shells and reduced the number of successful hatchings.

The first major human disaster resulting from ocean pollution reached epidemic proportions in 1956 within **Minamata Bay, Japan.** Of more than 100 people affected by **mercury** poisoning, almost half died from its effects. A similar episode occurred during the late 1960s at **Niigata, Japan.** Mercury contamination levels have been set on the basis of fish consumption of populations throughout the world, and they appear to have been effective in preventing further poisonings.

With its **Action Plan for the Human Environment**, initiated in 1972, the United Nations Environment Program has established eleven environmental units within which assessment and monitoring have begun or are planned. They are the Mediterranean Sea, Red Sea, Kuwait, Caribbean, West and Central Africa, East Asian Seas, Southeast Pacific, Southwest Pacific, East Africa, South Asian Seas, and Southwest Atlantic.

QUESTIONS AND EXERCISES

1. Discuss why it is difficult to determine the assimilative capacity of the oceans for accepting our waste materials.

2. List the acts of the U.S. Congress that have protected water quality, and describe restrictions contained in each.

3. Do you think it is a better policy (1) to prohibit dumping into the ocean any substance that can't be proven to be harmless to the ocean or (2) to consider the effects of disposing of the substance on land and in the ocean before deciding where to dispose of it? Discuss the reasons for your choice.

4. Discuss the means by which plastics are thought to enter the ocean, their known negative effects on organisms, and legislation being considered to outlaw dumping of plastics in the ocean.

5. Describe the effect of oil spills on species diversity and recovery in the benthos of Wild Harbor and Chedabucto Bay.

6. Discuss whether oil spills that wash ashore or those that do not, such as that of the *Argo Merchant*, are more destructive to marine life.

7. How would dumping sewage in deeper water off the east coast help reduce the degradation of the ocean bottom?

8. What components of the nuclear energy generation industry introduce the most radioactive wastes into the oceans?

9. Discuss the animal populations that clearly suffered from DDT pollution and the way in which this negative effect was manifested.

10. Refer to figure 18–11 and discuss the desirability of setting a mercury contamination level for fish at the 1.0 mg/kg level for the citizens of the United States, Sweden, and Japan.

REFERENCES

Bascom, W. 1978. Quantifying man's influence on coastal waters. *Journal of Ocean Engineering* 3:4.

Borgese, E. M., and Ginsburg, N., eds. *1988 Ocean Yearbook,* Vol 7. Chicago: University of Chicago Press.

Comptroller General of the United States. 1977. Problems and progress in regulating ocean dumping of sewage sludge and industrial wastes.

Clark, W. C. 1989. Managing planet earth. *Scientific American* 261:3, 47–54.

International Atomic Energy Agency. 1980. International symposium on the impacts of radionuclide releases into the marine environment.

Krendle, P. A., ed. 1975. Heavy metals in the aquatic environment. New York; Pergamon Press.

Lawrence, W. W. 1979. *Of acceptable risk: Science and the determination of safety.* Los Altos, Cal.: William Kaufman, Inc.

Mearns, A. J., and Word, J. Q. 1981. Forecasting effects of sewage solids on marine benthic communities. In *Ecological effects of environmental stress,* ed. J. O'Connor and G. Mayer. Estuarine Research Foundation, special publication.

National Academy of Sciences, National Research Council. 1979. Report on a multimedium management of municipal sludge, Vol. 9. National Advisory Committee on Oceans and Atmosphere. 1980. The role of the ocean in a waste management strategy. A Special Report to the President and Congress.

National Oceanic and Atmospheric Administration. 1979. Assimilative capacity of U.S. coastal waters for pollutants. Proceedings of Workshop at Crystal Mountain, Washington, July 29–August 4, 1979.

Oil spills and spills of hazardous substances. 1975. Washington, D.C.: U.S. Environmental Protection Agency.

Ramanathan, V. 1988. The greenhouse theory of climate change: A test by an inadvertent global experiment. *Science* 240:4850, 293–299.

Wilbur, R. J. 1987. Plastic in the North Atlantic. *Oceanus* 30:3, 61–68.

SUGGESTED READING

Sea Frontiers

Driessen, P. K. 1987. Oil rigs and sea life: A shortgun marriage that works. 33:5, 362–72.
Despite resistance to offshore drilling because of esthetic and environmental concerns, there is evidence that offshore drilling rigs enhance the ecology of many areas.

Hersch, S. L. 1988. Death of the dolphins: Investigating the east coast die-off. 34:4, 200–207.
A marine mammalogist investigating the cause of the deaths of dolphins along the east coast discusses the evidence developed since the deaths began in June 1987.

Hull, E. W. S. 1978. Oil spills: the causes and the cures. 24:6, 360–69.
A general discussion of oil spills, their frequency, and methods to clean them up.

Idyll, C. P. 1971. Mercury and fish. 17:4, 230–40.
An article written just after the mercury scare of 1970 discusses the effects of mercury on marine life.

Nowadnick, J. 1977. There is but one ocean. 23:3, 130–40.
Problems related to exploitation of the ocean in terms of environmental impact and international politics are discussed.

Parker, P. A. 1990. Clearing the oceans of plastics. 36:2, 18–27.
The history and scope of the problems of plastics in the ocean are covered.

Parry, J. 1983. Nations unite to fight pollution. 29:3, 143–50.
The efforts of the United Nations Environment Program are described.

Reynolds, J. E., and Wilcox, J. R. 1987. People, power plants and manatees. 33:4, 263–69.
The effects of water heating by power plants on the distribution of manatees is discussed.

White, I. C. 1985. An organization to Help Combat Oil Spills. 31:1, 15–21.
The efforts of the International Tanker Owners Pollution Federation Limited to increase the efficiency of cleanup operations and reduce the negative effects of oil spills have resulted in more-intelligent responses to oil spills.

Scientific American

Bascom, W. 1974. Disposal of waste in the ocean. 231:2, 16–25.
An informative overview of people's misconceptions concerning the disposal of waste in the oceans and the methods that may allow safe use of the oceans for such a purpose.

Butler, J. N. 1975. Pelagic tar. 232:6, 90–97.
Deals with the origin and distribution of lumps of tar found floating at the ocean surface and their possible implications to marine biology.

Houghton, R. A., and Woodwell, G. M. 1989. Global climatic change. 260:4, 36–44.
The increases in atmospheric carbon dioxide and methane caused by human activities will surely change the earth's climate. Drastic changes in human behavior are required to reduce the threat of change.

Theme Issue. 1989. Managing planet Earth. 261:3, 1–175.
Articles covering the broad scope of environmental problems are presented in this issue.

APPENDIX I
SCIENTIFIC NOTATION

To simplify writing very large and very small numbers, scientists indicate the number of zeros in them by scientific notation. One integer is placed to the left of the decimal, and a multiplication times a power of ten tells which direction and how far the decimal is moved to write the number out in its long form. For example:

$$2.13 \times 10^5 = 213,000$$
$$\text{or}$$
$$2.13 \times 10^{-5} = 0.0000213$$

Further examples showing numbers that are powers of ten are:

$$1,000,000,000 = 1.0 \times 10^9, \quad \text{or } 10^9$$
$$1,000,000 = 1.0 \times 10^6, \quad \text{or } 10^6$$
$$1,000 = 1.0 \times 10^3, \quad \text{or } 10^3$$
$$100 = 1.0 \times 10^2, \quad \text{or } 10^2$$
$$10 = 1.0 \times 10^1, \quad \text{or } 10^1$$
$$1 = 1.0 \times 10^0, \quad \text{or } 10^0$$
$$0.1 = 1.0 \times 10^{-1}, \text{or } 10^{-1}$$
$$0.01 = 1.0 \times 10^{-2}, \text{or } 10^{-2}$$
$$0.001 = 1.0 \times 10^{-3}, \text{or } 10^{-3}$$
$$0.000001 = 1.0 \times 10^{-6}, \text{or } 10^{-6}$$
$$0.000000001 = 1.0 \times 10^{-9}, \text{or } 10^{-9}$$

To add or subtract numbers written as powers of ten, they must be converted to the same power:

Addition		Subtraction	
2.1×10^3	0.021×10^5	3.4×10^4	3.4×10^4
$+1.0 \times 10^5 =$	$+1.000 \times 10^5$	$-2.0 \times 10^3 =$	-0.2×10^4
	1.021×10^5		3.2×10^4

To multiply or divide numbers written as powers of ten, the exponents are added or subtracted:

Multiplication	Division
6.04×10^2	3.0×10^3
$\times\ 2.1 \times 10^4$	$\div 1.5 \times 10^2$
12.684×10^6	2.0×10^1

APPENDIX II
THE METRIC SYSTEM
AND CONVERSION FACTORS

Units of Length
1 micrometer (μm) = 10^{-6} m = 0.000394 in.
1 millimeter (mm) = 10^{-3} m = 0.0394 in. = 10^3 μm
1 centimeter (cm) = 10^{-2} m = 0.394 in. = 10^4 μm
1 meter (m) = 10^2 cm = 39.4 in. = 3.28 ft = 1.09 yd = 0.547 fath
1 kilometer (km) = 10^3 m = 0.621 statute mi = 0.540 nautical mi

Units of Volume
1 liter = 10^3 cm^3 = 1.0567 liquid qt = 0.264 U.S. gal
1 cubic meter (m^3) = 10^6 cm^3 = 10^3 l = 35.3 ft^3 = 264 U.S. gal
1 cubic kilometer (km^3) = 10^9 m^3 = 10^{15} cm^3 = 0.24 statute mi^3

Units of Area
1 square centimeter (cm^2) = 0.155 in.^{2}
1 square meter (m^2) = 10.7 ft^2
1 square kilometer (km^2) = 0.292 nautical mi^2 = 0.386 statute mi^2

Units of Time
1 day = 8.64 $\times$ 10^4 s (mean solar day)
1 year = 8765.8 h = 3.156 $\times$ 10^7 s

Units of Mass
1 gram (g) = 0.035 oz
1 kilogram (kg) = 10^3 g = 2.205 lb
1 metric ton = 10^6 g = 2205 lb = 1.105 tons

Units of Speed
1 centimeter per second (cm/s) = 0.0328 ft/s
1 meter per second (m/s) = 2.24 statute mi/h = 1.94 kn
1 kilometer per hour (km/h) = 27.8 cm/s = 0.55 kn
1 knot (1 nautical mile per hour, kn) = 1.15 statute mi/h = 0.51 m/s

Units of Temperature

Celsius (°C)	Fahrenheit (°F)	Kelvin (K)	
−237.2	−459.7	0	Absolute zero (lowest possible temperature)
0	32	273.2	Freezing point of water
100	212	373.2	Boiling point of water

Conversions: °F = (1.8 $\times$ °C) + 32
°C = (°F − 32)/1.8

APPENDIX III
PERIODIC TABLE OF THE ELEMENTS

Light Metals

Transitional Elements

Heavy Metals

Nonmetals

Atomic number
Symbol of element
Atomic weight
Name of element

```
1
H
1.0080
Hydrogen
```

Inert gas
Gas
Liquid
Solid — all others

	I A	II A	III B	IV B	V B	VI B	VII B		VIII B			I B	II B	III A	IV A	V A	VI A	VII A	VIII A
1	1 H 1.0080 Hydrogen																		2 He 4.003 Helium
2	3 Li 6.939 Lithium	4 Be 9.012 Beryllium												5 B 10.81 Boron	6 C 12.011 Carbon	7 N 14.007 Nitrogen	8 O 15.9994 Oxygen	9 F 18.998 Fluorine	10 Ne 20.183 Neon
3	11 Na 22.990 Sodium	12 Mg 24.31 Magnesium												13 Al 26.98 Aluminum	14 Si 28.09 Silicon	15 P 30.974 Phosphorus	16 S 32.064 Sulfur	17 Cl 35.453 Chlorine	18 Ar 39.948 Argon
4	19 K 39.102 Potassium	20 Ca 40.08 Calcium	21 Sc 44.96 Scandium	22 Ti 47.90 Titanium	23 V 50.94 Vanadium	24 Cr 52.00 Chromium	25 Mn 54.94 Manganese	26 Fe 55.85 Iron	27 Co 58.93 Cobalt	28 Ni 58.71 Nickel	29 Cu 63.54 Copper	30 Zn 65.37 Zinc	31 Ga 69.72 Gallium	32 Ge 72.59 Germanium	33 As 74.92 Arsenic	34 Se 78.96 Selenium	35 Br 79.909 Bromine	36 Kr 83.80 Krypton
5	37 Rb 85.47 Rubidium	38 Sr 87.62 Strontium	39 Y 88.91 Yttrium	40 Zr 91.22 Zirconium	41 Nb 92.91 Niobium	42 Mo 95.94 Molybdenum	43 Tc (99) Technetium	44 Ru 101.1 Ruthenium	45 Rh 102.90 Rhodium	46 Pd 106.4 Palladium	47 Ag 107.870 Silver	48 Cd 112.40 Cadmium	49 In 114.82 Indium	50 Sn 118.69 Tin	51 Sb 121.75 Antimony	52 Te 127.60 Tellurium	53 I 126.90 Iodine	54 Xe 131.30 Xenon
6	55 Cs 132.91 Cesium	56 Ba 137.34 Barium	57 TO 71	72 Hf 178.49 Hafnium	73 Ta 180.95 Tantalum	74 W 183.85 Tungsten	75 Re 186.2 Rhenium	76 Os 190.2 Osmium	77 Ir 192.2 Iridium	78 Pt 195.09 Platinum	79 Au 197.0 Gold	80 Hg 200.59 Mercury	81 Tl 204.37 Thallium	82 Pb 207.19 Lead	83 Bi 208.98 Bismuth	84 Po (210) Polonium	85 At (210) Astatine	86 Rn (222) Radon
7	87 Fr (223) Francium	88 Ra 226.05 Radium	89 TO 103															

Lanthanide series

57 LA 138.91 Lanthanum	58 Ce 140.12 Cerium	59 Pr 140.91 Praseodymium	60 Nd 144.24 Neodymium	61 Pm (147) Promethium	62 Sm 150.35 Samarium	63 Eu 151.96 Europium	64 Gd 157.25 Gadolinium	65 Tb 158.92 Terbium	66 Dy 162.50 Dysprosium	67 Ho 164.93 Holmium	68 Er 167.26 Erbium	69 Tm 168.93 Thullium	70 Yb 173.04 Ytterbium	71 Lu 174.97 Lutetium

Actinide series

89 Ac (227) Actinium	90 Th 232.04 Thorium	91 Pa (231) Protactinium	92 U 238.03 Uranium	93 Np (237) Neptunium	94 Pu (242) Plutonium	95 Am (243) Americium	96 Cm (247) Curium	97 Bk (249) Berkelium	98 Cf (251) Californium	99 Es (254) Einsteinium	100 Fm (253) Fermium	101 Md (256) Mendelevium	102 No (254) Nobelium	103 Lw (257) Lawrencium

APPENDIX IV
THE GEOLOGIC TIME TABLE

Era	Period	Epoch & Age 10^6 Yr		Development of Life and Orogenic Events	
Cenozoic	Quaternary	Recent			
			0.01	Pacific Coast Ranges form	
		Pleistocene			
			1.6		
		Pliocene			
			5		
	Tertiary	Miocene			
			24		
		Oligocene			
			37		
		Eocene			
			58		
		Paleocene			
			66	First primates	Rocky Mts. form
Mesozoic	Cretaceous				
			144		
	Jurassic			First birds	Sierra Nevada Ranges form
			208	First mammals	
	Triassic			First flowering plants	
			245		
Paleozoic	Permian				
			286	First reptiles	
	Pennsylvanian				
			320	First insects	Appalachian Mts. form
	Mississippian				
			360	First amphibians	
	Devonian				
			408	First land plants	
	Silurian				
			438	First vertebrates (fish)	
	Ordovician				
			505	First abundant animal fossils	
	Cambrian				
			570		
	Eocambrian				
			640	First algae, bacteria	
Precambrian			4500		

Dates based on Geological Society of America Geological Time Scale, 1983.

APPENDIX V
THE PHYLOGENETIC TREE

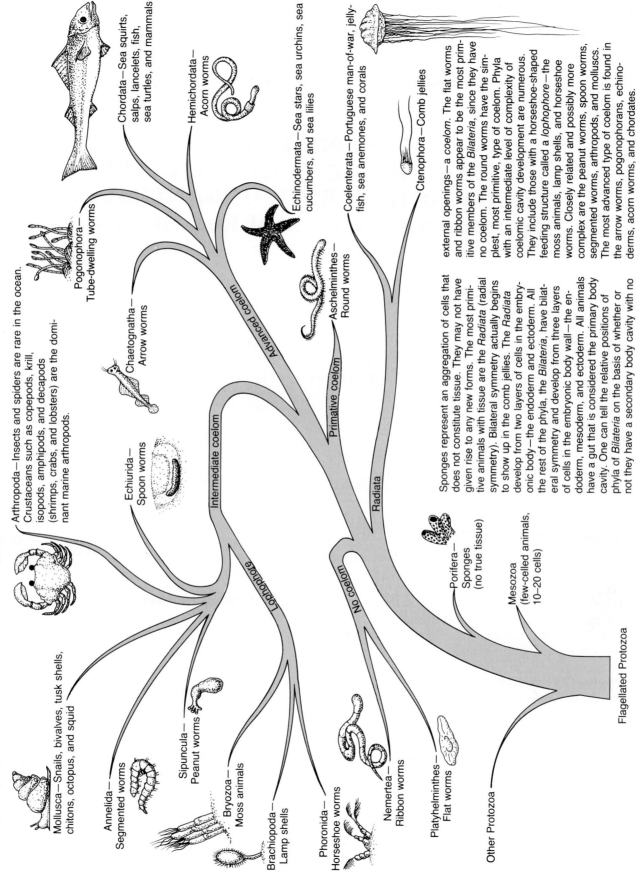

Mollusca—Snails, bivalves, tusk shells, chitons, octopus, and squid

Annelida—Segmented worms

Sipuncula—Peanut worms

Bryozoa—Moss animals

Brachiopoda—Lamp shells

Phoronida—Horseshoe worms

Nemertea—Ribbon worms

Platyhelminthes—Flat worms

Other Protozoa

Mesozoa (few-celled animals, 10–20 cells)

Porifera—Sponges (no true tissue)

Aschelminthes—Round worms

Coelenterata—Portuguese man-of-war, jellyfish, sea anemones, and corals

Ctenophora—Comb jellies

Echinodermata—Sea stars, sea urchins, sea cucumbers, and sea lilies

Hemichordata—Acorn worms

Chordata—Sea squirts, salps, lancelets, fish, sea turtles, and mammals

Pogonophora—Tube-dwelling worms

Chaetognatha—Arrow worms

Echiurida—Spoon worms

Arthropoda—Insects and spiders are rare in the ocean. Crustaceans such as copepods, krill, isopods, amphipods, and decapods (shrimps, crabs, and lobsters) are the dominant marine arthropods.

Lophophore

Intermediate coelom

Advanced coelom

Primitive coelom

Radiata

No coelom

Flagellated Protozoa

Sponges represent an aggregation of cells that does not constitute tissue. They may not have given rise to any new forms. The most primitive animals with tissue are the *Radiata* (radial symmetry). Bilateral symmetry actually begins to show up in the comb jellies. The *Radiata* develop from two layers of cells in the embryonic body—the endoderm and ectoderm. All the rest of the phyla, the *Bilateria*, have bilateral symmetry and develop from three layers of cells in the embryonic body wall—the endoderm, mesoderm, and ectoderm. All animals have a gut that is considered the primary body cavity. One can tell the relative positions of phyla of *Bilateria* on the basis of whether or not they have a secondary body cavity with no external openings—a *coelom*. The flat worms and ribbon worms appear to be the most primitive members of the *Bilateria*, since they have no coelom. The round worms have the simplest, most primitive, type of coelom. Phyla with an intermediate level of complexity of coelomic cavity development are numerous. They include those with a horseshoe-shaped feeding structure called a *lophophore*—the moss animals, lamp shells, and horseshoe worms. Closely related and possibly more complex are the peanut worms, spoon worms, segmented worms, arthropods, and molluscs. The most advanced type of coelom is found in the arrow worms, pogonophorans, echinoderms, acorn worms, and chordates.

APPENDIX VI
TAXONOMIC CLASSIFICATION
OF COMMON MARINE ORGANISMS (with appropriate number of known species)

Kingdom Monera
Organisms without nuclear membranes; nuclear material is spread throughout cell; predominantly unicellular.

Phylum Schizophyta.   Smallest known cells; bacteria (1500 species).

Phylum Cyanophyta.   Blue-green algae; chlorophyll *a,* carotene and phycobilin pigments (200 species).

Kingdom Protista
Organisms with nuclear material confined to nucleus by a membrane.

Phylum Chrysophyta.   Golden-brown algae; includes diatoms, coccolithophores, and silicoflagellates; chlorophyll *a* and *c,* xanthophyll, and carotene pigments (6000+ species).

Phylum Pyrrophyta.   Dinoflagellate algae; chlorophyll *a* and *c,* xanthophyll, and carotene pigments (1100 species).

Phylum Chlorophyta.   Green algae, chlorophyll *a* and *b,* and carotene pigments (7000 species).

Phylum Phaeophyta.   Brown algae; chlorophyll *a* and *c,* xanthophyll, and carotene pigments (1500 species).

Phylum Rhodophyta.   Red algae; chlorophyll *a,* carotene, and phycobilin pigments (4000 species).

Phylum Protozoa.   Nonphotosynthetic, heterotrophic protists (27,400 species).
Class Mastigophora.   Flagellated; dinoflagellates (5200 species).
Class Sarcodina.   Amoeboid; foraminifers and radiolarians (11,500 species).
Class Ciliophora.   Ciliated (6000 species).

Kingdom Fungi

Phylum Mycophyta.   Fungi, lichens; most fungi are decomposers found on the ocean floor, whereas lichens inhabit the upper intertidal zones (3000 species of fungi, 160 species of lichen).

Kingdom Metaphyta
Multicellular, complex plants.

Phylum Tracheophyta.   Vascular plants with roots, stems, and leaves that are serviced by special cells that carry food and fluids (287,200 species).
Class Angiospermae.   Flowering plants with seeds contained in a closed vessel (275,000 species).

Kingdom Metazoa
Multicellular animals.

Phylum Porifera.   Sponges; spicules are the only hard parts in these sessile animals that do not possess tissue (10,000 species).
Class Calcarea.   Calcium carbonate spicules (50 species).

Class Desmospongiae. Skeleton may be composed of siliceous spicules or spongin fibers or be nonexistent (9500 species).

Class Sclerospongiae. Coralline sponges; massive skeleton composed of calcium carbonate, siliceous spicules, and organic fibers (7 species).

Class Hexactinellida. Glass sponges; six-rayed with siliceous spicules (450 species).

Phylum Cnidaria (coelenterata). Radially symmetrical, two-cell, layered body wall with one opening to gut cavity; polyp (asexual, sexual, benthic) and medusa (sexual, pelagic) body forms (10,000 species).

Class Hydrozoa. Polypoid colonies such as pelagic Portuguese man-of-war and benthic *Obelia* common; medusa present in reproductive cycle but reduced in size (3000 species).

Class Scyphozoa. Jellyfish; medusa up to 1 m (3.3 ft) in diameter is dominant form; polyp is small if present (250 species).

Class Anthozoa. Corals and anemones possessing only polypoid body form and reproducing asexually and sexually (6500 species).

Phylum Ctenophora. Predominantly planktonic comb jellies; basic eight-sided radial symmetry modified by secondary bilateral symmetry (80 species).

Phylum Platyhelminthes. Flatworms; bilateral symmetry; hermaphroditic (25,000 species).

Phylum Nemertea. Ribbon worms; as long as 30 m (98 ft); benthic and pelagic (800 species).

Phylum Nematoda. Roundworms; marine forms are primarily free-living and benthic; most 1 to 3 mm (0.04 to 0.12 in.) in length (5000 marine species).

Phylum Rotifera. Ciliated, unsegmented forms less than 2 mm (0.08 in.) in length (1500 species, only a few marine).

Phylum Bryozoa. Moss animals; benthic, branching or encrusting colonies; lophophore feeding structure (4500 species).

Phylum Branchiopoda. Lamp shells; lophophorate benthic bivalves (300 species).

Phylum Phoronida. Horseshoe worms; 24-cm (9-in.)-long lophorate tube worms that live in sediment of shallow and temperate shallow waters (15 species).

Phylum Sipuncula. Peanut worms, benthic (325 species).

Phylum Echiura. Spoon worms; sausage-shaped with spoon-shaped proboscis; burrow in sediment or live under rocks (130 species).

Phylum Pogonophora. Tube-dwelling, gutless worms 5 to 80 cm (2 to 32 in.) long; absorb organic matter through body wall (100 species).

Phylum Tardigrada. Marine meiofauna that have the ability to survive long periods in a cryptobiotic state (diversity is poorly known).

Phylum Mollusca. Soft bodies possessing a muscular foot and mantle that usually secretes calcium carbonate shell (75,000 species).

Class Monoplacophora. Rare trench-dwelling forms with segmented bodies and limpetlike shells (10 species).

Class Polyplacophora. Chitons; oval, flattened body covered by eight overlapping plates (600 species).

Class Gastropoda. Large, diverse group of snails and their relatives; shell spiral if present (64,500 species).

Class Bivalvia. Bivalves; includes mostly filter-feeding clams, mussels, oysters, and scallops (7500 species).

Class Aplacophora. Tusk shells; sand-burrowing organisms that feed on small animals living in sand deposits (350 species).

Class Cephalopoda. Octopus, squid, and cuttlefish that possess no external shell except in the genus *Nautilus* (600 species).

Phylum Annelida. Segmented worms in which musculature, circulatory, nervous, excretory, and reproductive systems may be repeated in many segments; mostly benthic (10,000 marine species).

Phylum Arthropoda. Jointed-legged animals with segmented body covered by an exoskeleton (30,000 marine species).

Subphylum Crustacea. Calcareous exoskeleton, two pairs of antennae; cephalon, thorax, and abdomen body parts; includes copepods, ostracods, barnacles, shrimp, lobsters, and crabs (26,000 species).

Subphylum Chelicerata. Horseshoe crabs (4 species).
Class Merostomata.
Class Pycnogonida. Sea spiders.

Subphylum Uniramia. Insects; genus *Halobites* is the only truly marine insect.

Phylum Chaetognatha. Arrowworms; mostly planktonic, transparent and slender; up to 10 cm (4 in.) long (50 species).

Phylum Hemichordata. Acorn worms and pterobranchs; primitive nerve cord; gill slits; benthic (90 species).

Phylum Echinodermata. Spiny-skinned animals; benthic animals with secondary radial symmetry and water vascular system (6000 species).
Class Stelleroidea. Star-shaped animals.
Subclass Asteroidea. Starfishes; free-living, flattened body with five or more rays with tube feet used for locomotion; mouth down (1600 species).
Subclass Ophiuroidea. Brittle stars and basket stars; prominent central disc with slender rays; tube feet used for feeding; mouth down (200 species).
Class Concentricycloidea. Sea daisies; tiny disc-shaped form found in a sample taken from crevice in driftwood at a depth of 1000 m (3280 ft) near New Zealand (1 species).
Class Echinoidea. Sea urchins, sand dollars, heart urchins; free-living forms without rays; calcium carbonate test; mouth down or forward (860 species).
Class Holothuroidea. Sea cucumbers; soft bodies with radial symmetry obscured; mouth forward (900 species).
Class Crinoidea. Sea lilies, feather stars; cup-shaped body attached to bottom by a jointed stalk or appendages; mouth up (630 species).

Phylum Chordata. Notochord; dorsal nerve cord and gills or gill slits (55,000 species).

Subphylum Urochordata. Tunicates; chordate characteristics in larval stage only; benthic sea squirts and planktonic thaliaceans and larvaceans (1375 species).

Subphylum Cephalochordata. *Amphioxus,* or lancelets; live in coarse temperate and tropical sediment (25 species).

Subphylum Vertebrata. Internal skeleton; spinal column of vertebrae; brain (52,000 species).
Class Agnatha. Lampreys and hagfishes; most primitive vertebrates with cartilaginous skeleton, no jaws and no scales (50 species).
Class Chondrichthyes. Sharks, skates, and rays; cartilaginous skeleton; 5 to 7 gill openings; placoid scales (625 species).
Class Osteichthyes. Bony fishes; cycloid scales; covered gill opening; swim bladder common (30,000 species).
Class Amphibia. Frogs, toads, and salamanders; Asian mud flat frogs are the only amphibians that tolerate marine water (2600 species).
Class Reptilia. Snakes, turtles, lizards, and alligators; orders Squamata (snakes) and Chelonia (turtles) are major marine groups (6500 species).
Class Aves. Birds; many live on and in the ocean but all must return to land to breed (8600 species).
Class Mammalia. Warm-blooded; hair; mammary glands; bear live young; marine representatives found in the orders Sirenia (sea cows, dugong), Cetacea (whales) and Carnivora (sea otter, pinnipeds) (4100 species).

APPENDIX VII
CALCULATIONS RELATIVE TO
SOURCE OF OCEAN WATER

The Mantle Mass of the mantle = density $\times$ volume = $(4.5 \text{ g/cm}^3) \times (1.0 \times 10^{27} \text{ cm}^3)$
$$= 4.5 \times 10^{27} \text{ g}$$

Mass of the oceans = $(1.025 \text{ g/cm}^3) \times (1.4 \times 10^{24} \text{cm}^3) = 1.4 \times 10^{24} \text{ g}$
Percent of mass of mantle represented by mass of oceans:
$$\frac{\text{Mass of oceans}}{\text{Mass of mantle}} = \frac{1.4 \times 10^{24} \text{ g}}{4.5 \times 10^{27} \text{ g}} = 0.00031 = 0.031\%$$

Percent of mass of stony meteorites made up of water = 0.5%

Ratio of percent of meteorite mass composed of water to minimum percent of mantle mass that must be composed of water to allow mantle to be considered source of oceans:
$$\frac{\text{Meteorite}}{\text{Mantle}} = \frac{0.5\%}{0.031\%} = 16.1$$

If the mantle has the same water content as the meteorites, it contains 16 times the mass of water needed to account for the earth's oceans.

APPENDIX VIII
ROOTS, PREFIXES, AND SUFFIXES*

a- not, without
ab-, abs- off, away, from
abysso- deep
acanth- spine
acro- the top
aero- air, atmosphere
aigial- beach, shore
albi- white
alga- seaweed
alti- high, tall
alve- cavity, pit
amoebe- change
an- without, not
annel- ring
annu- year
anomal- irregular, uneven
antho- flower
apex- tip
aplysio- sponge, filthiness
aqua- water
arachno- spider
arena- sand
arthor- joint
asthen- weak, feeble
astro- star
auto- self
avi- bird
bacterio- bacteria
balaeno- whale
balano- acorn
barnaco- goose
batho- deep
bentho- the deep sea
bio- living

botryo- bunch of grapes
brachio- arm
branchio- gill
broncho- windpipe
bryo- moss
bysso- a fine thread
calci- limestone ($CaCO_3$)
calori- heat
capill- hair
cara- head
carno- flesh
cartilagi- gristle
caryo- nucleus
cat- down, downward
cen-, ceno- recent
cephalo- head
chaeto- bristle
chiton- tunic
chlor- green
choano- funnel, collar
chondri- cartilage
cilio- small hair
cirri- hair
clino- slope
cnido- nettle
cocco- berry
coelo- hollow
cope- oar
crusta- rind
cteno- comb
cyano- dark blue
cypri- Venus, lovely
-cyst bladder, bag
-cyte cell

deca- ten
delphi- dolphin
di- two, double
dino- whirling
diplo- double, two
dolio- barrel
dors- back
-drom- run, a race
echino- spiny
eco- house, abode
ecto- outside, outer
edrio- a seat
en- in, into
endo- inner, within
entero- gut
epi- upon, above
estuar- the sea
eu- good, well
exo- out, without
fec- dregs
fecund- fruitful
flacci- flake
flagell- whip
fluvi- river
fossili- dug up
fuco- red
geno- birth, race
geo- the earth
giga- very large
globo- ball, globe
gnatho- jaw
guano- dung
gymno- naked, bare

*Source: D. J. Borror, 1960. *Dictionary of Word Roots and Combining Forms.* Palo Alto, Calif.: National Press Books.

halo- salt
haplo- single
helio- the sun
helminth- worm
hemi- half
herbi- plant
herpeto- creeping
hetero- different
hexa- six
holo- whole
homo-, homeo- alike
hydro- water
hygro- wet
hyper- over, above, excess
hypo- under, beneath
ichthyo- fish
-idae members of the animal family of
infra- below, beneath
insecti- cut into
insula- an island
inter- between
involute- intricate
iso- equal
-ite rock
juven- young
juxta- near to
kera- horn
kilo- one thousand
lacto- milk
lamino- layer
larvi- ghost
latent hidden
latero- side
lati- broad, wide
lemni- water plant
limno- marshy lake
lipo- fat
litho- stone
litorial near the seashore
lopho- tuft
lorica- armor
luci- light
luna moon
lux light
macro- large
mala- jaw, cheek
mamilla- teat
mandibulo- jaw
mantle- cloak
mari- the sea
masti- chewing
mastigo- whip
masto- breast, nipple
maxillo- jaw

madi- middle
medus- a jellyfish
mega- great, large
meio- less
meridio- noon
meso- middle
meta- after
meteor- in the sky
-meter measure
-metry science of measuring
mid- middle
milli- thousandth
mio- less
moll- soft
mono- one, single
-morph form
nano- dwarf
necto- swimming
nemato- thread
neo- new
neph- cloud
nerito- sea nymph
noct- night
-nomy the science of
nucleo- nucleus
nutri- nourishing
o-, oo- egg
ob- reversed
octa- eight
oculo- eye
odonto- teeth
oiko- house, dwelling
oligo- few, scant
-ology science of
omni- all
ophi- a serpent
opto- the eye, vision
orni- a bird
-osis condition
ovo- egg
pan- all
para- beside, near
pari- equal
pecti- comb
pedi- foot
penta- five
peri- all around
phaeo- dusky
pholado- lurking in a hole
-phore carrier of
photo- light
phyto- plant
pinnati- feather
pisci- fish
plani- flat, level

plankto- wandering
pleisto- most
pleuro- side
plio- more
pluri- several
pneuma- air, breath
pod- foot
poikilo- variegated
poly- many
polyp many footed
poro- channel
post- behind, after
-pous foot
pre- before
pro- before, forward
procto- anus
proto- first
pseudo- false
ptero- wing
pulmo- lung
pycno- dense
quadra- four
quasi- almost
radi- radial
rhizo- root
rhodo- rose-colored
sali- salt
schizo- split, division
scyphi- cup
semi- half
septi- partition
sessil- sedentary
siphono- tube
-sis process
spiro- spiral, coil
stoma- mouth
strati- layer
sub- below
supra- above
symbio- living together
taxo- arrangement
tecto- covering
terra- earth
terti- third
thalasso- the sea
trocho- wheel
tropho- nourishment
tunic- cloak, covering
turbi- disturbed
un- not
vel- veil
ventro- underside
xantho- yellow
xipho- sword
zoo- animal

APPENDIX IX
LATITUDE AND LONGITUDE DETERMINATION

FIGURE 1   Determining Latitude.

The method of determining latitude used by Pytheas was to measure the angle between the horizon and the North Star, a star that is directly above the North Pole. Latitude north of the equator is the angle between the two sightings. The present North Star is Polaris. Similar determinations may be made in the Southern Hemisphere by using the Southern Cross as Polaris is used in the Northern Hemisphere.

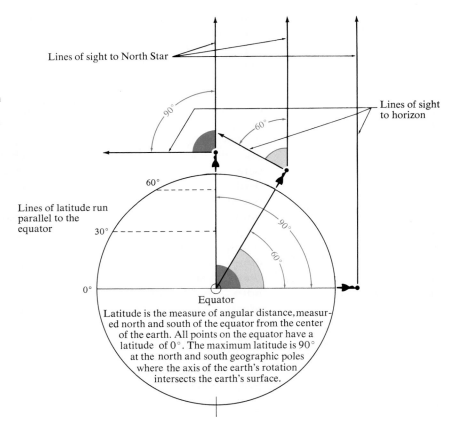

Lines of sight to North Star

Lines of sight to horizon

Lines of latitude run parallel to the equator

Latitude is the measure of angular distance, measured north and south of the equator from the center of the earth. All points on the equator have a latitude of 0°. The maximum latitude is 90° at the north and south geographic poles where the axis of the earth's rotation intersects the earth's surface.

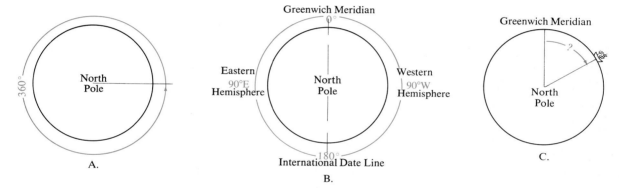

FIGURE 2   Determining Longitude.

View of earth from outer space, looking down on North Pole. *A*, as the earth turns on its rotational axis, it moves through 360° of angle per 24 h. *B*, a prime meridian that runs through Greenwich, England, was selected as the reference meridian, and the earth was divided into a Western and Eastern hemisphere. After John Harrison's chronometer was developed, many ships carried it showing the time on the Greenwich Meridian—Greenwich time. *C*, since the earth rotates through 15° of angle, longitude, per h (360° ÷ 24 h = 15°/h), a ship's captain could easily determine his longitude each day at noon. For example, a ship sets sail west across the Atlantic Ocean, checking its longitude each day at noon (when the sun crosses the meridian running directly overhead).

One day when the sun is at the noon position, the captain checks the chronometer. It reads 16:18 h (0:00 is midnight and 12:00 is noon). What is the ship's longitude? Longitude solution:

16:18 h = 4:18 P.M.

The earth rotates through ¼° (15′) of angle per minute of time.

(One degree of arc is divided into 60 min)

4 h × 15°/h = 60° of longitude

18 minutes of time × 15 minutes of angle/minute of time = 270 minutes of arc

270 min ÷ 60 min/degree (°) = 4.5° of longitude

60° + 4.5° = 64.5° W longitude

GLOSSARY

Abiotic Without life.

Abyssal clay Deep-ocean (oceanic) deposits containing less than 30% biogenous sediment.

Abyssal hill A volcanic extension above the ocean floor less than 1 km (0.62 mi).

Abyssal zone Benthic environment from 4000 to 6000 m (13,120 to 19,680 ft).

Abyssopelagic zone Open-ocean (oceanic) environment below 4000 m (13,120 ft) depth.

Acid A solution in which the H^+ concentration is greater than the OH^- concentration. In a water solution, the hydrogen ion concentration is greater than one part per 10^7.

Acoelomate Without a secondary body cavity (coelom).

Adiabatic Pertaining to a change in the temperature of a mass resulting from compression or expansion. Requires no addition or loss of heat from the mass.

Algae One-celled or many-celled plants that have no root, stem, or leaf systems; simple plants.

Alkaline Pertaining to a solution in which the OH^- concentration is greater than the H^+ concentration. In a water solution, the hydrogen ion concentration is less than one part per 10^7.

Amphidromic point A nodal or no-tide point in the ocean or sea around which the crest of the tide wave rotates during one tidal period.

Amphineura Class of mollusks with eight dorsal calcareous plates; includes chitons.

Amphipoda Crustacean order containing laterally compressed members such as the "sand hoppers."

Amplitude Height of wave crest above or trough below still water.

Anadromous Pertaining to a species of fish that spawns in fresh water, then migrates into the ocean to grow to maturity.

Anaerobic respiration Respiration carried on in the absence of free oxygen (O_2). Some bacteria and protozoans carry on respiration this way.

Anion A negatively charged ion.

Annelida Phylum of elongated, segmented worms.

Antinode Zone of maximum vertical particle movement in standing waves where crest and trough formation alternate.

Aphelion The point farthest from the sun in the orbit of a planet or comet.

Aphotic Without light.

Apogee The point farthest from the earth in the orbit of the moon or a human-made satellite.

Aschelminthes Phylum of wormlike pseudocoelomates.

Aspect ratio The index of propulsive efficiency obtained by dividing the square of fin height by the area of the fin.

Asthenosphere A plastic layer in the upper mantle from depths as shallow as 10 km (6.2 mi) to as deep as 800 km (497 mi) which may allow lateral movement of lithospheric plates and isostatic adjustments.

Atlantic-type margin The passive trailing edge of a continent that is subsiding because of lithospheric cooling and increasing sediment load.

Atom The smallest particle of an element that can combine with similar particles of other elements to produce compounds.

Atomic number A number representing the relative position of an element in the periodic table of elements. It is equal to the number of positive charges in the atom's nucleus.

Augite A dark mineral, usually black, rich in iron and magnesium. An important constituent of basalt, the rock that is characteristic of the ocean crust.

Authigenic minerals Minerals that form in place, such as phosphorite and those found in manganese nodules.

Autolytic decomposition The breakdown of organic matter by enzymes activated when an organism dies.

Autotroph Plant or bacterium that can synthesize organic compounds from inorganic nutrients.

Auxospore A diatom cell that has shed its frustule to allow growth.

Backshore The inner portion of the shore, lying landward of the mean spring-tide high-water line. Acted upon by the ocean only during exceptionally high tides and storms.

Bacterioplankton Bacteria that live as plankton.

Bacteriovore An organism that feeds primarily on bacteria.

Barnacle *See* Cirripedia.

Barrier island A long, narrow, wave-built island separated from the mainland by a lagoon.

Basalt A dark-colored volcanic rock characteristic of the ocean crust. Contains minerals with relatively high iron and magnesium content.

Base A compound that releases hydroxide ions in aqueous solution.

Bathyal Benthic environment from 200 to 4000 m (656 to 13,120 ft).

Bathymetry The study of ocean depth.

Bathypelagic Open-ocean environment of approximately 1000 to 4000 m (3280 to 13,120 ft).

Bay barrier A marine deposit attached to the mainland at both ends and extending entirely across the mouth of a bay, separating the bay from the open water.

Beach Sediment seaward of the coastline through the surf zone that is in transport along the shore and within the surf zone.

Benthic Pertaining to the ocean bottom.

Benthic nepheloid layer A layer of turbid water adjacent to the ocean floor; created by bottom currents.

Benthos The forms of marine life that live on the ocean bottom.

Biogenous sediment Sediment containing material produced by plants or animals; e.g., coral reefs, shell fragments, and housings of diatoms, radiolarians, foraminifers, and coccolithophores.

Biomass Total weight of the organisms in a particular habitat, species, or group of species.

Biotic community All the organisms that live within some definable area.

Bore A steep-fronted tide crest that moves up a river in association with high tide.

Breaker zone Region where waves break at the seaward margin of the surf zone.

Bryozoa Phylum of colonial animals that often share one coelomic cavity. Encrusting and branching forms secrete a protective housing (zooecium) of calcium carbonate or chitinous material. Possess lophophore feeding structure.

Buffering Of chemical or other processes, any process that reduces impact.

Calcareous Containing calcium carbonate.

Calcium carbonate compensation depth The depth at which ocean water dissolves calcium carbonate as rapidly as it is deposited from above. There is no calcium carbonate in sediment deposited beneath this depth (usually between 4000 and 5000 m, or 13,120 and 16,400 ft).

Calorie Unit of energy defined as the amount of heat required to raise the temperature of 1 g of water 1°C (1.8°F).

Capillary wave Ocean wave whose wavelength is less than 1.74 cm (0.7 in.). The dominant restoring force for such waves is surface tension.

Carapace Chitinous or calcareous shield that covers the cephalothorax of some crustaceans. Dorsal portion of a turtle shell.

Carbohydrate An organic compound consisting of carbon, hydrogen, and oxygen. Sugars and starches are examples.

Carbon cycle A process occurring in stars that releases energy. It involves the conversion of hydrogen to helium with carbon acting as a catalyst. During the process, a C^{12} atom grows to become a N^{15} atom. N^{15} combines with H^1 (proton) to produce $C^{12} + He^4 + \gamma$.

Carnivore An animal that depends solely or chiefly on other animals for its food supply.

Carotene (carotin) A red to yellow pigment found in plants.

Catadromous Pertaining to a species of fish which spawns at sea, then migrates into a freshwater stream or lake to grow to maturity.

Cation A positively charged ion.

Celsius temperature scale Scale in which 0°C = 273.16°K; 0°C = freezing point of water; 100°C = boiling point of water.

Center of gravity The point where the entire mass of a body may be considered to be concentrated.

Centrifugal force A force that seems to make an object move away from the center of a curved path it is following. It results from the application of a centripetal force acting against the inertia of the object.

Centripetal force A center-seeking force that tends to make rotating bodies move toward the center of rotation.

Cephalopoda A class of the phylum Mollusca with a well-developed pair of eyes and a ring of tentacles surrounding the mouth. The shell is absent or internal on most members. The class includes the squid, octopus, and *Nautilus.*

Cetacea An order of marine mammals that includes the whales.

Chaetognatha A phylum of elongate, transparent, worm-like pelagic animals commonly called *arrowworms.*

Chela Arthropod appendage modified to form a pincer.

Chemosynthesis The formation of organic compounds from inorganic substances using energy derived from oxidation.

Chiton Common name for any member of the Amphineura class of mollusks with eight dorsal plates.

Chlorinity The chloride content of seawater expressed in grams per kilogram (g/kg) or parts per thousand (‰) by weight. It includes all the halide ions (F^+, Cl^+, Br^+, I^+, and At^+).

Chlorophyll A group of green pigments that make it possible for plants to carry on photosynthesis.

Chlorophyta Green algae; characterized by the presence of chlorophyll and other pigments.

Chrysophyta An important phylum of planktonic algae, including the diatoms. The presence of chlorophyll is masked by the pigment carotin, which gives the plants a golden color.

Cilia Short hairlike structures common on lower animals. Beating in unison, they may be used for locomotion or to create water currents that carry food toward the mouth of the animal.

Cirripedia An order of crustaceans with up to six pairs of thoracic appendages that strain food from the water. They are barnacles that attach themselves to a substrate and secrete an external calcareous housing.

Clastic Pertaining to a rock or sediment composed of broken fragments of pre-existing rocks. Two common examples are beach deposits and sandstone.

Clay A term relating to particle size between silt and colloid. Clay minerals are hydrous aluminum silicates with plastic, expansive, and cation exchange properties.

Cnidoblast Stinging cell of the phylum Coelenterata; contains a stinging mechanism (nematocyst) used in defense and capturing prey.

Coast A strip of land that extends inland from the coastline as far as marine influence is evidenced in the landforms.

Coastline Landward limit of the highest storm waves' effect on the shore.

Coccolithophore A microscopic planktonic form of algae, encased by a covering composed of calcareous discs (coccoliths).

Coelenterata Phylum of radially symmetrical animals that includes two basic body forms, the medusa and the polyp. Includes jellyfish (medusoid) and sea anemones (polypoid).

Coelom Secondary body cavity (gut cavity is primary cavity). Forms within the mesoderm in higher animals, is lined with peritoneum, and contains vital organs.

Cold current A current carrying cold water into areas of warmer water. It generally travels toward the equator.

Colloid Substance having particles of a size smaller than clay.

Colonial animals Animals that live in groups of attached or separate individuals. Groups of individuals may serve special functions.

Comb jelly Common name for members of the phylum Ctenophora. (See Ctenophora.)

Commensalism A symbiotic relationship in which one organism benefits at no expense to its host.

Compensation depth, CaCO₃ The depth at which the amount of calcium carbonate ($CaCO_3$) produced by the organisms in the overlying water column is equal to the amount of $CaCO_3$ the water column can dissolve. There will be no $CaCO_3$ deposition below this depth.

Compensation depth, O₂ The depth where the oxygen produced by photosynthesis is equal to the oxygen requirements of plant respiration. A plant population cannot be sustained below this depth, which will be greater in the open ocean than near the shore owing to the relatively deeper light penetration in the open ocean.

Compound A substance containing two or more elements combined in fixed proportions.

Condensation The conversion of water from the vapor to the liquid state. When it occurs, the energy required to vaporize the water is released. This is about 585 cal/g of water at 20°C (68°F).

Conduction The transmission of heat by the passage of energy from particle to particle.

Conjunction An apparent closeness of two or more heavenly bodies. During the new moon, the sun and moon are in conjunction on the same side of the earth.

Conservative property A property of surface ocean water that is changed only by mixing and diffusion after the water sinks below the surface.

Consumer A heterotrophic organism that consumes an external supply of organic matter.

Continent About one-third of the earth's surface that rises above the deep-ocean floor to be exposed above sea level. Continents are composed primarily of granite, an igneous rock of lower density than the basaltic oceanic crust.

Continental borderland A highly irregular portion of the continental margin that is submerged beneath the ocean and is characterized by depths greater than those characteristic of the continental shelf.

Continental drift A term applied to early theories supporting the possibility that the continents are in motion over the earth's surface.

Continental rise A gently sloping depositional surface at the base of the continental slope.

Continental shelf A gently sloping depositional surface extending from the low-water line to a marked increase in slope around the margin of a continent.

Continental slope A relatively steeply sloping surface lying seaward of the continental shelf.

Convection Process by which, in a fluid being heated, the warmer part of the mass will rise and the cooler portions will sink. If the heat source is stationary, cells may develop as the rising warm water cools and sinks in regions on either side of the axis of rising.

Convergence Coming together of water masses in polar, tropical, and subtropical regions of the ocean. Along these lines of convergence, the denser mass will sink beneath the others.

Copepoda An order of microscopic to nearly microscopic crustaceans that are important members of the zooplankton in temperate and subpolar waters.

Coral A group of benthic coelenterates that exist as individuals or in colonies and may secrete external skeletons of calcium carbonate. Under the proper conditions, corals may produce reefs composed of their external skeletons and the $CaCO_3$ material secreted by varieties of algae associated with the reefs.

Corange lines Lines on a map joining points of equal tide range.

Coriolis effect An effect resulting from the earth's rotation that causes particles in motion to be deflected to the right in the Northern Hemisphere and to the left in the Southern Hemisphere.

Cosmogenous sediment All sediment derived from outer space.

Cotidal lines Lines connecting points where high tides occur simultaneously.

Crust Unit of the earth's structure that is composed of basaltic ocean crust and granitic continental crust. The total thickness of the crustal units may range from 5 km (3.1 mi) beneath the ocean to 50 km (31 mi) beneath the continents.

Crustacea A class of the phylum Arthropoda that includes barnacles, copepods, lobsters, crabs, and shrimp.

Crystalline rock Igneous or metamorphic rocks. These rocks are made up of crystalline particles with orderly molecular structures.

Ctenophora A phylum of gelatinous organisms that are more or less spheroidal with biradial symmetry. These exclusively marine animals have eight rows of ciliated combs for locomotion, and most have two tentacles for capturing prey.

Curie point The temperature above which thermal agitation prevents spontaneous alignment of magnetic particles with the earth's magnetic field.

Current A horizontal movement of water.

Cypris The advanced, free-swimming larval stage of barnacles. After attaching to a substrate, it metamorphoses into the adult.

Decapoda 1. An order of crustaceans with five pairs of thoracic "walking legs," including crabs, shrimp, and lobsters. 2. Suborder of cephalopod mollusks with ten arms that includes squids and cuttlefish.

Decomposers Primarily bacteria that break down nonliving organic material, extract some of the products of decomposition for their own needs, and make available the compounds needed for plant production.

Deep boundary current Relatively strong deep current flowing across the continental rise along the western margin of an ocean basin.

Deep scattering layer A layer of marine organisms in the open ocean that scatters signals from an echo sounder. The organisms migrate daily from near the surface at night to more than 800 m (2624 ft) during the day.

Deep sea system System that includes all benthic environments beneath the littoral (sublittoral, bathyal, abyssal, and hadal).

Deep water That portion of the water column from the base of the permanent thermocline or pycnocline to the ocean floor. Water temperatures are relatively uniform throughout.

Deep-water wave Ocean wave traveling in water that has a depth greater than one-half the average wavelength. Its velocity is independent of water depth.

Delta A low-lying deposit at the mouth of a river.

Density Mass per unit volume of a substance. Usually expressed as grams per cubic centimeter. For ocean water with a salinity of 35‰ at 0°C, the density is 1.028 g/cm^3. (*See* sigma.)

Desiccation Process of drying out.

Detritus Any loose material produced directly from rock disintegration. (Organic: material resulting from the disintegration of dead organic remains.)

Diatom Member of the class Bacillariophyceae of algae; possesses a wall of overlapping silica valves.

Diffraction Any bending of a wave around an obstacle that cannot be interpreted as refraction or reflection.

Diffusion The transfer of material or a property by random molecular movement. The movement is from a region in which the material or the property is high in concentration to regions of low concentration.

Dinoflagellates Single-celled microscopic organisms that may possess chlorophyll and belong to the plant phylum Pyrrophyta (autotrophic) or may ingest food and belong to the class Mastigophora of the animal phylum Protozoa (heterotrophic).

Diploblastic Pertaining to body structure composed of two cell layers: ectoderm and endoderm. Diploblastic phyla are Porifera, Coelenterata, and Ctenophora.

Discoaster A tiny star-shaped plate that may have formed on a coccolithophorelike algal cell which died out at the start of the Pleistocene epoch.

Discontinuity An abrupt change in a property, such as temperature or salinity, at a line or surface.

Disphotic zone The dimly lit zone corresponding approximately with the mesopelagic in which there is not enough light to carry on photosynthesis. Sometimes called the *twilight zone.*

Distributary A small stream flowing away from a main stream. Such streams are characteristic of deltas.

Diurnal inequality The difference in the heights of two successive high or low waters during a lunar (tidal) day.

Diurnal tide A tide with one high water and one low water during a tidal day. Tidal period is 24 hrs 50 min.

Divergence A horizontal flow of fluid from a central region, as occurs in upwelling.

Doldrums A belt of light, variable winds 10° to 15° north and south of the equator, resulting from the vertical flow of low-density air within this equatorial belt. Doldrums is the common name for the Intertropical Convergence Zone.

Dolomite A common rock-forming mineral; a calcium-magnesium carbonate [$CaMg(CO_3)_2$].

Dolphin 1. A brilliantly colored fish of the genus *Coryphaena*. 2. The name applied to the small, beaked members of the cetacean family Delphinidae.

Dorsal Pertaining to the back or upper surface of most animals.

Drifts Thick sediment deposits on the continental rise produced where the western boundary undercurrent slows and loses sediment as it changes direction to follow the base of the continental slope.

Dynamic topography A surface configuration resulting from the geopotential difference between a given surface and a reference surface of no motion. A contour map of this surface is useful in estimating the nature of geostrophic currents.

Earthquake A sudden motion or trembling in the earth caused by the sudden release of slowly accumulated strain by faulting (movement along a fracture in the earth's crust) or volcanic activity.

Ebb current Seaward-flowing current during a decrease in the height of the tide.

Echinodermata Phylum of animals that have bilateral symmetry in larval forms and usually a five-sided radial symmetry as adults. Benthic and possessing rigid or articulating exoskeletons of calcium carbonate with spines, this phylum includes sea stars, brittle stars, sea urchins, sand dollars, sea cucumbers, and sea lilies.

Echo sounding Determining the depth of water by measuring the time required for a sonic or ultrasonic signal to travel to the bottom and back to the ship that emitted the signal.

Ecological efficiency Efficiency with which energy is transferred from one trophic level to the next, or the ratio of the amount of protoplasm added to a trophic level to the amount of food required to produce it.

Ecosystem All the organisms in a biotic community and the abiotic environmental factors with which they interact.

Ectoderm Outermost layer of cells in an animal embryo. In vertebrates it gives rise to the skin, nervous system, sense organs, etc.

Eddy A current of any fluid forming on the side of or within a main current. It usually moves in a circular path and develops where currents encounter obstacles or flow past one another.

Edreobenthos Benthos that are attached to the bottom of the ocean.

Ekman spiral A theoretical consideration of the effect of a steady wind blowing over an ocean of unlimited depth and breadth and of uniform viscosity. The result is a surface flow at 45° to the right of the wind in the Northern Hemisphere. Water at increasing depth will drift in directions increasingly to the right until at about 100 m (328 ft) depth it is moving in a direction opposite to that of the wind. The net water transport is 90° to the wind, and velocity decreases with depth.

Ekman transport The net transport of surface water set in motion by wind. Owing to the Ekman spiral phenomenon, it is theoretically in a direction 90° to the right and 90° to the left of the wind direction in the Northern Hemisphere and Southern Hemisphere, respectively.

Electromagnetic energy Energy that travels as waves or particles with the speed of light. Different kinds possess different properties based on wavelength. The longest wavelengths (up to 10 km, or 6.21 mi, in length) belong to radio waves. At the other end of the spectrum are cosmic rays with great penetrating power and wavelengths of less than 0.000001 μm.

Electron A negatively charged particle in orbit around the nucleus of an atom.

Electron cloud The organized assemblage of electrons surrounding the nucleus of an atom.

Element One of a number of substances, each of which is composed entirely of like atoms.

El Niño A southerly-flowing warm current that generally develops off the coast of Ecuador shortly after Christmas. Occasionally it will move farther south into Peruvian coastal waters and cause the widespread death of plankton and fish.

Emergent shoreline A shoreline resulting from the emergence of the ocean floor relative to the ocean surface. It is usually rather straight and characterized by marine features usually found at a greater depth.

Endoderm Innermost cell layer of an embryo. Develops into the digestive and excretory systems and forms the lining for the respiratory system, etc., in vertebrates.

Endothermic reaction A chemical reaction that absorbs energy. For example, energy is stored in the organic products of the chemical reaction photosynthesis.

Entropy A quantity reflecting the degree of uniform distribution of heat energy in a system. It increases with time and represents a state in which energy is unrecoverable for work.

Environment The sum of all physical, chemical, and biological factors to which an organism or community is subjected.

Epicenter The point on the earth's surface that is directly above the focus of an earthquake.

Epifauna Animals that live on the ocean bottom, either attached or moving freely over it.

Epipelagic zone The upper region of the oceanic province, extending to a depth of 200 m (656 ft).

Equatorial tide A semimonthly tide occurring when the moon is over the equator. It displays a minimal diurnal inequality.

Equilibrium tide theory A tidal hypothesis that considers the ocean to be of uniform nature and depth throughout the earth's surface. It is further assumed that this ocean will respond instantly to the gravitational forces of the sun and moon.

Equinox The times when the sun is over the equator, making day and night of equal length throughout the earth. *Vernal equinox* occurs about March 21 as the sun is moving into the Northern Hemisphere. *Autumnal equinox* occurs about September 21 as the sun is moving into the Southern Hemisphere.

Estuarine circulation Circulation characteristic of an estuary and other bodies of water having restricted circulation with the ocean; this results from an excess of run-off and precipitation as compared to evaporation. Surface flow is toward the ocean with a subsurface counter flow.

Estuary The mouth of a river valley, or a bay or lagoon receiving fresh water, where marine influence is manifested as tidal effects and increased salinity of the fresh water.

Euphausiacea An order of planktonic crustaceans ranging in length from 5 to 30 cm (1.97 to 11.82 in.). Most possess luminous organs, and some are the principal food for baleen whales.

Euphotic zone The surface layer of the ocean that receives enough light to support photosynthesis. The bottom of this zone, which is marked by the oxygen compensation depth, varies and reaches a maximum value of around 150 m (492 ft) in the very clearest open ocean water.

Euryhaline Pertaining to the ability of a marine organism to tolerate a wide range of salinity.

Eurythermal   Pertaining to the ability of a marine organism to tolerate a wide range of temperature.

Eutrophic   Characterized by an abundance of nutrients.

Evaporation   The physical process of converting a liquid to a gas. Commonly considered to occur at a temperature below the boiling point of the liquid.

Excess volatiles   Volatile compounds found in the oceans, sediments, and atmosphere in quantities greater than the chemical weathering of crystalline rock could produce. They are considered to have been produced by volcanic action.

Exothermic reaction   A chemical reaction that liberates energy. For example, the energy stored in the products of photosynthesis is released by the chemical reaction respiration.

Fahrenheit temperature scale (°F)   Scale in which the freezing point of water is 32°; boiling point of water is 212°.

Fan   A gently sloping, fan-shaped feature normally located at the lower end of a canyon.

Fast ice   Sea ice that is attached to the shore and therefore remains stationary.

Fat   A colorless, odorless organic compound consisting of carbon, hydrogen, and oxygen; insoluble in water.

Fathom   A unit of ocean depth commonly used in countries using the English system of units. It is equal to 1.83 m, or 6 ft.

Fault   A fracture or fracture zone in the earth's crust along which displacement has occurred.

Fault block   A crustal block bounded on at least two sides by faults. Usually elongate; if it is down-dropped it produces a graben; if uplifted, it is a horst.

Fauna   The animal life of any particular area or of any particular time.

Fecal pellet   Organic excrement found in marine sediment and produced primarily by invertebrates. Usually of an ovoid form that is less than 1 mm (0.04 in.) in length.

Fetch   1. Area of the open ocean over which the wind blows with constant speed and direction, thereby creating a wave system. 2. The distance across the fetch (wave-generating area) measured in a direction parallel to the direction of the wind.

Fjord   A long, narrow, deep, U-shaped inlet that usually represents the seaward end of a glacial valley that has become partially submerged after the melting of the glacier.

Flagellum   A whiplike living process used by some cells for locomotion.

Floe   A piece of floating ice other than fast ice or icebergs. May range in dimension from about 20 cm (7.9 in.) across to more than a kilometer.

Flood basalt   A basalt plateau composed of from 1 to 2 million km^3 of basalt deposited in a period of from 1 to 2 million years.

Flood current   A tidal current associated with increasing height of the tide, generally moving toward the shore.

Flora   The plant life of any particular area or of any particular time.

Food chain   The passage of energy materials from producers through a sequence of herbivores and carnivores.

Food web   A group of interrelated food chains.

Foraminifera   An order of planktonic and benthic protozoans that possess protective coverings usually composed of calcium carbonate.

Forced standing wave   A wave that is generated and maintained by a periodic force, such as the gravitational attraction of the moon.

Foreshore   The portion of the shore lying between the normal high- and low-water marks—the intertidal zone.

Fortnight   Half a synodic month (29.5 days), or about 14.75 days. Normally used in reference to a period of time equal to two weeks. The time that elapses between the new moon and full moon.

Fossil   Any remains, trace, or imprint of an organism that has been preserved in rocks.

Fracture zone   An extensive linear zone of unusually irregular ocean floor topography, characterized by large seamounts, steep-sided or asymmetrical ridges, troughs, or long, steep slopes. Usually represents ancient, inactive transform fault zones.

Free standing wave   A wave, created by a sudden rather than continuous impulse, that continues to exist after the generating force is gone.

Freezing point   The temperature at which a liquid becomes a solid under any given set of conditions. The freezing point of water is 0°C (32°F) under atmospheric pressure.

Fringing reef   A reef that is directly attached to the shore of an island or continent. It may extend more than 1 km (0.62 mi) from shore. The outer margin is submerged and often consists of algal limestone, coral rock, and living coral.

Frustule   The siliceous covering of a diatom, consisting of two halves (epitheca and hypotheca).

Fucoxanthin   The reddish-brown pigment that gives brown algae its characteristic color.

Furrows   Parallel, troughlike structures cut into mud waves by bottom currents. They are aligned in the direction of current flow.

Gastropoda   A class of mollusks, most of which possess an asymmetrical spiral one-piece shell and a well-developed flattened foot. A well-developed head will usually have two eyes and one or two pairs of tentacles. Includes snails, limpets, abalone, cowries, sea hares, and sea slugs.

Geostrophic current   A current that develops out of the earth's rotation and is the result of a near balance between gravitational force and the Coriolis effect.

Gill   A thin-walled projection from some part of the external body or the digestive tract; used for respiration in a water environment.

Glacial Epoch   The Pleistocene Epoch, the earlier of two divisions of the Quaterary Period of geologic time. During this time, high-latitude continental areas now free of ice were covered by continental glaciers.

Glacier   A large mass of ice formed on land by the recrystallization of old compacted snow. It flows from an area of accumulation to an area of wasting, where ice is removed from the glacier by melting or calving.

Gondwanaland   A hypothetical protocontinent of the Southern Hemisphere named for the Gondwana region of India. It included the present continental masses of Africa, Antarctica, Australia, India, and South America.

Graben   An elongate, downdropped crustal block that is bounded by faults. The rift valleys along the axes of spreading centers are grabens.

Graded bedding   Stratification in which each layer displays a decrease in grain size from bottom to top.

Gradient   The rate of increase or decrease of one quantity or characteristic relative to a unit change in another. For example, the slope of the ocean floor is a change in elevation (a vertical linear measurement) per unit of horizontal distance covered. Commonly measured in m/km.

Granite   A light-colored igneous rock characteristic of the continental crust. Rich in nonferromagnesian minerals such as feldspar and quartz.

Gravitational stability   The degree to which segments of the water column tend to remain stationary. The portion of the water column with the greatest degree of gravitational stability is called the pycnocline, where water at the top is much less dense than water at the bottom. This zone of density stratification normally coincides with the thermocline.

Gravity wave   A wave for which the dominant restoring force is gravity. Such waves have a wavelength of more than 1.74 cm (0.7 in.), and their velocity of propagation is controlled mainly by gravity.

Groin   A low, artificial structure projecting into the ocean from the shore to interfere with longshore transportation of sediment. It usually has the purpose of trapping sand to cause the buildup of a beach.

Gross primary production   The total amount of organic material produced by autotrophs.

Guyot   A tablemount; a conical volcanic feature on the ocean floor that has had the top truncated to a relatively flat surface.

Gyre   A circular motion. Used mainly in reference to the circular motion of water in each of the major ocean basins centered in subtropical high-pressure regions.

Habitat   A place where a particular plant or animal lives. Generally refers to a smaller area than environment.

Hadal   Pertaining to the deepest ocean environment, specifically that of ocean trenches deeper than 6 km (3.7 mi).

Half-life   The time required for half the atoms of a radioactive isotope sample to decay to atoms of another element.

Halocline   A layer of water which has a high rate of salinity change in the vertical dimension.

Headland   A steep-faced irregularity of the coast that extends out into the ocean.

Heat   Energy moving from a high-temperature system to a lower-temperature system. The heat gained by the one system may be used to raise its temperature or to do work.

Heat budget (global)   The equilibrium that exists on the average between the amount of heat absorbed by the earth and its atmosphere in one year, and the amount of heat radiated back into space in one year.

Heat capacity   Usually defined as the amount of heat required to raise the temperature of 1 g of a substance 1°C.

Herbivore   An animal that relies chiefly or solely on plants for its food.

Hermaphroditic   Pertaining to the possession of both functional male and functional female reproductive organs by an animal. It is rare for both systems to function at the same time.

Hermatypic coral   Reef-building coral that has symbiotic algae in its ectodermal tissue. Cannot produce a reef structure below the euphotic zone.

Herpetobenthos   Benthos that walk or crawl across the ocean floor.

Heterotroph   Animals and bacteria that depend on the organic compounds produced by other animals and plants as food. Organisms not capable of producing their own food by photosynthesis or chemosynthesis.

Higher high water (HHW)   The higher of two high waters occurring during a tidal day where tides are mixed.

Higher low water (HLW)   The higher of two low waters occurring during a tidal day where tides are mixed.

High water (HW)   The highest level reached by the rising tide before it begins to recede.

Holoplankton   Organisms that spend their entire life as members of the plankton.

Homeotherm   An animal which maintains a precisely controlled internal body temperature using its own heating and cooling mechanisms.

Homologous   Pertaining to a basic similarity of structures in different organisms resulting from a similar embryonic origin and development; e.g., the foreflippers of a seal and the arms and hands of a man are homologous.

Hook   A spit or narrow cape of sand or gravel with an end that bends landward to form a "hook."

Horse latitudes   The latitude belts between 30° and 35° north and south where winds are light and variable since the principal movement of air masses at these latitudes is one of vertical descent.

Hurricane   A tropical cyclone in which winds reach velocities above 120 km/h (73 mi/h). Generally applied to such storms in the North Atlantic Ocean, eastern North Pacific Ocean, Caribbean Sea, and Gulf of Mexico. Such storms in the western Pacific Ocean are called *typhoons.*

Hydrocarbon   An organic compound consisting solely of hydrogen and carbon. Petroleum is a mixture of many hydrocarbon compounds.

Hydrogenous sediment   Sediment that forms from ocean water precipitation or ion exchange between existing sediment and ocean water. Examples are manganese nodules, phosphorite, glauconite, phillipsite, and montmorillonite.

Hydrologic cycle   The cycle of water exchange among the atmosphere, land, and ocean through the processes of evaporation, precipitation, runoff, and subsurface percolation.

Hydrophilic   Pertaining to the property of attracting water.

Hydrophobic   Pertaining to the property of being impossible or difficult to wet with water.

Hydrozoa   A class of coelenterates that characteristically exhibit alternation of generations, with a sessile polypoid colony giving rise to a pelagic medusoid form by asexual budding.

Hypertonic   Pertaining to the property of an aqueous solution having a higher osmotic pressure (salinity) than another aqueous solution separated by a semipermeable membrane allowing osmosis. The hypertonic fluid will gain water molecules through the membrane from the other fluid.

Hypotonic   Pertaining to the property of an aqueous solution having a lower osmotic pressure (salinity) than another aqueous solution separated by a semipermeable membrane allowing osmosis. The hypotonic fluid will lose water molecules through the membrane to the other fluid.

Iceberg   A massive piece of glacier ice that has broken from the front of the glacier (calved) into a body of water. It floats with its tip at least 5 m (16.4 ft) above the water's surface and with at least 4/5 of its mass submerged.

Ice floe   *See* floe.

Ice shelf   A thick layer of ice with a relatively flat surface which is attached to and nourished by a continental glacier from one side. The shelf, which is for the most part afloat, may extend above water level by more than 50 m (164 ft) along its seaward cliff (formed by the break-off of large tabular chunks of ice that become icebergs).

Igneous rock   One of the three main classes into which all rocks are divided, i.e., igneous, metamorphic, and sedimentary. Rock that forms from the solidification of molten or partly molten material (magma).

Inertia   Tendency that a body at rest will stay at rest and a body in motion will remain in a uniform motion in a straight line unless acted on by some external force. Stated in Newton's first law of motion.

Infauna   Animals that live buried in the soft substrate (sand or mud).

Infrared radiation   Electromagnetic radiation between the wavelengths of 0.8 μm (0.00032 in.) and about 1000 μm (0.394 in.). It is bounded on the shorter-wavelength side by the visible spectrum and on the long side by microwave radiation.

Inner sublittoral zone   The section of ocean floor from the low-tide shoreline to the point where attached plants stop growing.

In situ   In place.

In situ density (σ, sigma)   Density of a sample of water at its original depth.

Insolation   The rate at which solar radiation is received per unit of surface area at any point at or above the earth's surface.

Interface   A surface separating two substances of different properties, i.e., density, salinity, or temperature. In oceanography, it usually refers to a separation of two layers of water with different densities caused by significant differences in temperature and/or salinity.

Internal wave   A wave that develops below the surface of a fluid, the density of which changes with increased depth. This change may be gradual or occur abruptly at an interface.

Intertidal zone   Littoral zone; the foreshore. The ocean floor covered by the highest normal tides and exposed by the lowest normal tides, and the water environment of the tide pools within this region.

Intertropical Convergence Zone   Zone where northeast trade winds and southeast trade winds converge. Averages about 5°N in the Pacific and Atlantic oceans and 7°S in the Indian Ocean.

Ion   An atom that becomes electrically charged by gaining or losing one or more electrons. The loss of electrons produces a positively charged cation, and the gain of electrons produces a negatively charged anion.

Ionic bond   A chemical bond resulting from the electrical attraction that exists between cations and anions.

Island arc system   A linear arrangement of islands, many of which are volcanic, usually curved so the concave side faces a sea separating the islands from a continent. The convex side faces the open ocean and is bounded by a deep-ocean trench.

Isobar   Lines connecting values of equal pressure on a map or graph.

Isohaline   Of the same salinity.

Isopoda   An order of dorsoventrally flattened crustaceans that are mostly scavengers or parasites on other crustaceans or fish.

Isostasy   A condition of equilibrium, comparable to buoyancy, in which the rigid crustal units float on the underlying mantle.

Isotherm   A line connecting points of equal temperature.

Isothermal   Of the same temperature.

Isotonic   Pertaining to the property of having equal osmotic pressure. If two such fluids were separated by a semipermeable membrane that allows osmosis to occur, there would be no net transfer of water molecules across the membrane.

Isotope   One of several atoms of an element that has a different number of neutrons, and therefore a different atomic mass, than the other atoms, or isotopes, of the element.

Jellyfish   1. A free-swimming, umbrella-shaped medusoid member of the coelenterate class, Scyphozoa. 2. Also frequently applied to the medusoid forms of other coelenterates.

Jet stream   An easterly moving air mass at an elevation of about 10 km (6.2 mi). Moving at speeds that can exceed 300 km/h (186 mi/h), the jet stream follows a wavy path in the midlatitudes and influences how far polar air masses may extend into the lower latitudes.

Jetty A structure built from the shore into a body of water to protect a harbor or a navigable passage from being shoaled by deposition of longshore (littoral) drift material.

Juvenile water Water that is derived directly from magma, being released for the first time at the earth's surface as the magma crystallizes to igneous rock.

Kelp Large varieties of Phaeophyta (brown algae).

Kelvin temperature scale (K) Scale in which 0° K = −273.16°C. One degree on the Kelvin scale equals the same temperature range as one degree on the Celsius scale. 0 K is the lowest temperature possible.

Kelvin wave Wave that results when a progressive tide wave moves from the open ocean into and out of a relatively narrow body of water during a tidal cycle; the tidal range will be greater on the right side of the narrow body of water during flood tide. This results from the fact that the channel rotates in a counterclockwise direction as the earth rotates, while the wave tends to move in a straight line.

Key A low, flat island composed of sand or coral debris that accumulates on a reef flat.

Kinetic energy Energy of motion. It increases as the mass or velocity of the object in motion increases.

Knot (kt) Unit of speed equal to 1 nautical mile per hour (approximately 51 centimeters per second).

Krill A common name frequently applied to members of the crustacean order Euphausiacea (euphausids).

Lagoon A shallow stretch of seawater partly or completely separated from the open ocean by an elongate narrow strip of land, such as a reef or barrier island.

Lamina A layer.

Laminar flow Flow in which a fluid flows in parallel layers or sheets. The direction of flow at any point does not change with time; nonturbulent flow.

Langmuir circulation A cellular circulation set up by winds that blow consistently in one direction with velocities above 12 km/h (7.5 mi/h). Helical spirals running parallel to the wind direction are alternately clockwise and counterclockwise.

Larva An embryo that is on its own before it assumes the characteristics of the adult of the species.

Latent heat The quantity of heat gained or lost per unit of mass as a substance undergoes a change of state (liquid to solid, etc.) at a given temperature and pressure.

Lateritic soil A red subsoil rich in secondary iron and aluminum oxides. It is characteristic of intertropical soils, may contain much quartz and kaolinite, and can harden to a bricklike substance.

Latitude Location on the earth's surface based on angular distance north or south of the equator. Equator, 0°; North Pole, 90°N; South Pole, 90°S.

Laurasia A hypothetical protocontinent of the Northern Hemisphere. The name is derived from Laurentia, per-

taining to the Canadian Shield of North America, and Eurasia, of which it was composed.

Lava Fluid magma coming from an opening in the earth's surface, or the same material after it solidifies.

Leeward Direction toward which the wind is blowing or waves are moving.

Levee Natural (resulting from deposition during flooding) or man-made low ridges on either side of a river channel.

Lichen Organism involving a photosynthetic, mutualistic relationship between an alga and a fungus. The alga is protected by the fungus, which is dependent on the alga for photosynthetically produced food.

Limestone A class of sedimentary rock composed of at least 50% calcium or magnesium carbonate. Limestone may be either biogenous or hydrogenous.

Limpet A mollusk of the class Gastropoda that possesses a low conical shell exhibiting no spiraling in the adult form.

Lithogenous sediment Sediment composed of mineral grains derived from the rock of continents and islands and transported to the ocean by wind or running water.

Lithosphere The outer layer of the earth's structure, including the crust and the upper mantle to a depth of about 200 km (124 mi). It is this layer that breaks into the plates that are the major elements of global plate tectonics.

Lithothamnion ridge A feature common to the seaward edge of a reef structure, characterized by the presence of the red alga, _Lithothamnion._

Littoral zone The benthic zone between the highest seaward and lowest normal water marks; the intertidal zone.

Lobster Large marine crustacean used as food. _Homarus americanus_ (American lobster) possesses two large chelae (pincers) and is found off the New England coast. _Panulirus_ sp. (spiny lobsters or rock lobsters) have no chelae but possess long spiny antennae effective in warding off predators. _P. argus_ is found off the coast of Florida and in the West Indies, while _P. interruptus_ is common along the coast of southern California.

Longitude Location on the earth's surface based on angular distance east or west of the Greenwich Meridian (0° longitude). 180° longitude is the International Date Line.

Longitudinal wave A wave in which particle vibration is parallel to the direction of energy propagation.

Longshore current A current located in the surf zone and running parallel to the shore as a result of breaking waves.

Longshore drift The load of sediment transported along the beach from the breaker zone to the top of the swash line in association with the longshore current.

Lophophore Horseshoe-shaped feeding structure bearing ciliated tentacles characteristic of the phyla Bryozoa, Brachiopoda, and Phoronidea.

Lower high water (LHW) The lower of two high waters occurring during a tidal day where tides are mixed.

Lower low water (LLW) The lower of two low waters occurring during a tidal day where tides are mixed.

Low water (LW)   The lowest level reached by the water surface at low tide before the rise toward high tide begins.

Lunar day   The time interval between two successive transits of the moon over a meridian (approximately 24 hours and 50 minutes of solar time).

Lunar hour   One twenty-fourth of a lunar day (about 62.1 minutes).

Lunar tide   The part of the tide caused solely by the tide-producing force of the moon.

Magma   Fluid rock material which solidifies to form igneous rock.

Magnetic anomaly   Distortion of the regular pattern of the earth's magnetic field, resulting from the various magnetic properties of local concentrations of ferromagnetic minerals in the earth's crust.

Manganese nodules   Concretionary lumps containing oxides of iron, manganese, copper, and nickel found scattered over the ocean floor.

Mangrove swamp   A marshlike environment that is dominated by mangrove trees. They are restricted to latitudes below 30°.

Mantle   The zone between the core and crust of the earth; rich in ferromagnesium minerals. In pelecypods, the portion of the body that secretes shell material.

Marginal sea   A semienclosed body of water adjacent to a continent.

Marsh   An area of soft, wet land. Flat land periodically flooded by saltwater; common in portions of lagoons.

Maximum sustainable yield   The maximum fishery biomass that can be removed yearly and be sustained by the fishery ecosystem.

Meander   A sinuous curve, bend, or turn in the course of a current.

Mean high water (MHW)   The average height of all the high waters occurring over a 19-year period.

Mean low water (MLW)   The average height of the low waters occurring over a 19-year period.

Mean sea level (MSL)   The mean surface water level determined by averaging all stages of the tide over a 19-year period, usually determined from hourly height observations along an open coast.

Mean tidal range   The difference between mean high water and mean low water.

Mediterranean circulation   Circulation characteristic of bodies of water with restricted ocean circulation that results from an excess of evaporation as compared to precipitation and runoff. Surface flow is into the restricted body of water with a subsurface outflow, as exists between the Mediterranean Sea and the Atlantic Ocean.

Medusa   A free-swimming, bell-shaped coelenterate body form with a mouth at the end of a central projection and tentacles around the periphery. Reproduces sexually.

Meridian of longitude   Half a great circle terminating at the North and South poles.

Meroplankton   Planktonic larval forms of organisms that are members of the benthos or nekton as adults.

Mesoderm   A primitive cell layer of the embryo that develops between the endoderm and ectoderm. In vertebrates, it gives rise to the skeleton, muscles, circulatory and excretory systems, and most of the reproductive system.

Mesoglea   Jellylike substance found between the endoderm and ectoderm in coelenterates and ctenophores. Ranges from noncellular and nonliving to a highly cellular nature, which leads some to believe it represents a stage in the development of the mesoderm.

Mesopelagic zone   That portion of the oceanic province from about 200 to 1000 m (656 to 3280 ft) depth. Corresponds approximately with the disphotic (twilight) zone.

Metamorphic rock   Rock that has undergone recrystallization while in the solid state in response to changes of temperature, pressure, and chemical environment.

Metaphyta   Kingdom of many-celled plants.

Metazoa   Kingdom of many-celled animals.

Microcontinent   A submarine plateau that is an isolated fragment of the continental crust. It usually has linear features; found primarily in the Indian Ocean.

Microplankton   Net plankton. Plankton not easily seen by the unaided eye, but easily recovered from the ocean with the aid of a fine-mesh plankton net.

Mineral   An inorganic substance occurring naturally in the earth and having distinctive physical properties and a chemical composition that can be expressed by a chemical formula.

Mixed tide   A tide having two high and two low waters per tidal day with a marked diurnal inequality. Such a tide may also show alternating periods of diurnal and semidiurnal components.

Mohorovičić discontinuity   A sharp seismic discontinuity between the crust and mantle of the earth. It may be as shallow as 5 km (3.1 mi) below the ocean floor or as deep as 60 km (37 mi) beneath some continental mountain ranges. Also known as the Moho.

Mole   The weight of a substance in grams numerically equal to its molecular weight (gram molecule). 1 mole of water (H_2O) is 18 g.

Molecule   The smallest particle of an element or compound that, in the free state, retains the characteristics of the substance.

Mollusca   Phylum of soft, unsegmented animals usually protected by a calcareous shell and having a muscular foot for locomotion; includes snails, clams, chitons, and octopuses.

Molt   The periodic shedding of the exoskeleton by arthropods to permit growth.

Monera   The kingdom of organisms that do not have nuclear material confined within a sheath but spread throughout the cell, such as bacteria and blue-green algae.

Mononodal   Pertaining to a standing wave with only one nodal point or nodal line.

Monsoons   A name for seasonal winds derived from the Arabic word for season, *mausim*. The term was originally applied to winds over the Arabian Sea that blow from the southwest during summer and the northeast during winter.

Moraine Unsorted material deposited at the margins of glaciers. Many such deposits have become economically important as fishing banks after being submerged by the rising level of the ocean.

Mud Sediment consisting of silt- and clay-sized particles smaller than 0.06 mm (0.002 in.). Actually, small amounts of larger particles will also be present.

Mud waves Wave features with lengths of 2 to 3 km (1.2 to 1.9 mi) that bottom currents produce on the surfaces of drifts or ridges.

Mutualism A symbiotic relationship in which both partners benefit.

Mycota Kingdom of fungi. Marine fungi obtain nutrition primarily as decomposers, parasites, or in mutualistic relationships with algae.

Nadir The point on the celestial sphere directly opposite the zenith and directly beneath the observer.

Nanoplankton Plankton less than 50 μm (0.02 in.) in length that cannot be captured in a plankton net and must be removed from the water by centrifuge or special microfilters.

Nansen bottle A device used by oceanographers to obtain samples of ocean water from beneath the surface.

Nauplius A microscopic, free-swimming larval stage of crustaceans such as copepods, ostracods, and decapods. Typically has three pairs of appendages.

Neap tide Tide of minimal range occurring when the moon is in quadrature.

Nearshore zone The seaward zone from the shoreline to the line of breakers.

Nektobenthos Those members of the benthos that actively swim and spend much time off the bottom.

Nekton Pelagic animals such as adult squids, fish, and mammals that can determine their positions in the ocean by swimming.

Nematath A linear alignment of volcanic features produced by a hot spot as a tectonic plate moves over it.

Nepheloid layer Well-mixed, turbid layer of water at the base of the oceanic water column. It is particularly well developed in the western boundary undercurrent.

Neritic province That portion of the pelagic environment from the shoreline to a depth of 200 m (656 ft).

Neritic sediment That sediment composed primarily of lithogenous particles and deposited relatively rapidly on the continental shelf, continental slope, and continental rise.

Net primary production The remaining amount of organic material produced by autotrophs after they have met their respiration needs.

Neutron An electrically neutral particle found in the nucleus of most atoms. It has a mass approximately equal to that of a proton.

Niche The ecological role of an organism and its position in the ecosystem.

Nitrogen fixation Conversion by bacteria of atmospheric nitrogen (N_2) to oxides of nitrogen (NO_2, NO_3) usable by plants in primary production.

Node The point on a standing wave where vertical motion is lacking or minimal. If this condition extends across the surface of an oscillating body of water, the line of no vertical motion is a nodal line.

Nonconservative property A property of ocean water attained at the surface and changed after the water sinks below the surface by processes other than mixing and diffusion. For example, dissolved oxygen content will be altered by biological activity.

Nucleus 1. A central, membrane-bound mass in eukaryotic cells; containing chromosomes. 2. The central, positively charged part of an atom; containing protons and neutrons.

Nudibranch Sea slug. A member of the mollusk class Gastropoda that has no protective covering as an adult. Respiration is carried on by gills or other projections on the dorsal surface.

Nutrient Any organic or inorganic compound used by plants in primary production. Nitrogen and phosphorus compounds are important examples.

Obduction The reverse of subduction. In the case of ophiolites, the rock is pushed up on the continent instead of subducting beneath it.

Oceanic crust A mass of rock with basaltic composition that is about 5 km (3.1 mi) thick and that forms the earth's crust beneath ocean basins.

Oceanic province That division of the pelagic environment where the water depth is greater than 200 m (656 ft).

Oceanic ridge A linear, seismic mountain range that extends through all the major oceans, rising from 1 to 3 km (0.62 to 1.9 mi) above the deep-ocean basins. Averaging 1500 km (932 mi) in width, rift valleys are common along the central axis. A source of new oceanic crustal material.

Oceanic sediment The inorganic abyssal clays and the organic oozes that accumulate particle by particle on the deep-ocean floor.

Offshore The comparatively flat, submerged zone of variable width extending from the breaker line to the edge of the continental shelf.

Oligotrophic Characterized by a low level of nutrients.

Omnivore An animal that feeds on both plants and animals.

Ooze A pelagic sediment containing at least 30% skeletal remains of pelagic organisms, the balance being clay minerals. Oozes are further defined by the chemical composition of the organic remains (siliceous or calcareous) and by their characteristic organisms (diatom ooze, foraminiferan ooze, radiolarian ooze, pteropod ooze).

Ophiolite A suite of rocks representing what is believed to be a piece of oceanic lithosphere embedded in a continent.

Opposition The separation of two heavenly bodies by 180° relative to the earth. The sun and moon are in opposition during the full moon phase.

Organic chemistry The branch of chemistry dealing with carbon compounds.

Orthogonal lines Lines drawn perpendicularly to wave fronts and spaced uniformly so that equal amounts of energy are contained by the segments of the wave front lying between any two orthogonal lines in a series. The areas where energy is concentrated as the waves break on the shore can be identified by the convergence of the orthogonal lines.

Osmosis Passage of water molecules through a semipermeable membrane separating two aqueous solutions of different solute concentration. The water molecules pass from the solution of lower solute concentration to the higher.

Osmotic pressure A measure of the tendency for osmosis to occur. It is the pressure that must be applied to the more-concentrated solution to prevent the passage of water molecules from the less-concentrated solution.

Ostracoda An order of crustaceans that are minute and compressed within a bivalve shell.

Outer sublittoral zone The section of ocean floor from the seaward edge of the inner sublittoral zone to a depth of 200 m (656 ft). No attached plants grow here.

Outgassing A process, resulting from heating, by which gases and water vapor are released from molten rocks. It has produced the earth's atmosphere and oceans.

Oviparous Pertaining to an animal that releases eggs which develop and hatch outside its body.

Ovoviviparous Pertaining to an animal that incubates eggs inside the mother until they hatch.

Oxygen compensation depth The depth at which marine plants photosynthesize at a rate which exactly meets their respiration needs (the base of the euphotic zone).

Oxygen utilization rate (OUR) The rate of use of dissolved oxygen caused by the respiration of animals and bacterial decomposition of dead organic matter descending to the bottom. The OUR is highest just beneath the euphotic zone (from about 100 to 1000 m, or 328 to 2280 ft) and gradually decreases with depth as the mass of dead organic matter and living animals decreases. Beneath 1000 m, oxygen values may increase owing to the inflow of oxygen carried by deep- and bottom-water masses that descend from the high-latitude ocean surface.

Pacific-type margin Leading edge of a continent that undergoes tectonic uplift as a result of lithospheric plate convergence.

Pack ice Any area of sea ice other than fast ice. Less than 3 m (9.8 ft) thick, it covers the ocean sufficiently to make navigation possible only by icebreakers.

Paleoceanography Branch of oceanography pertaining to the biological and physical character of ancient oceans.

Paleogeography Branch of geography pertaining to the shapes and positions of ancient continents and oceans.

Pancake ice Circular pieces of newly formed sea ice from 30 cm to 3 m (0.98 to 9.8 ft) in diameter that form in early fall in polar regions.

Pangaea A hypothetical supercontinent of the geologic past that contained all the continental crust of the earth.

Panthalassa A hypothetical proto-ocean surrounding Pangaea.

Parapodia Flat protuberances on each side of most segments of polychaete worms. Most possess cirri and setae (bristlelike projections); may be modified for special functions such as feeding, locomotion, and respiration.

Parasite An organism that takes its nutrients from the tissues of another organism and benefits at the host's expense.

Parasitism A symbiotic relationship in which the parasite harms the host from which it takes its nutrition.

Partial tide One of the harmonic components comprising the tide at any location. The periods of the partial tides are derived from the various combinations of the angular velocities of the earth, sun, and moon relative to one another.

Pedicellariae Minute stalked or unstalked pincerlike structures around the base of spines and dermal branchiae in certain echinoderms, especially Asteroidea, Echinoidea, and Ophiuroidea. They snap shut on debris and small organisms to keep the surface of the echinoderm clean.

Pelagic environment The open-ocean environment which is divided into the neritic province (water depth 0 to 200 m, or 0 to 656 ft) and the oceanic province (water depth greater than 200 m, or 656 ft).

Pelecypoda A class of mollusks characterized by two fairly symmetrical lateral valves with a dorsal hinge. These filter feeders pump water through the filter system and over gills through posterior siphons. Many possess a hatchet-shaped foot used for locomotion and burrowing. Includes clams, oysters, mussels, and scallops.

Perigee The point on the orbit of an earth satellite (moon) that is nearest the earth.

Perihelion That point on the orbit of a planet or comet around the sun that is closest to the sun.

Peritoneum Thin membrane lining the coelom and covering all organs in the coelom.

Permeability A condition which allows the passage of liquids through a substance.

Phaeophyta Brown algae characterized by the carotinoid pigment fucoxanthin. Contains the largest members of the marine plant community.

Photosynthesis The process by which plants produce carbohydrate from carbon dioxide and water in the presence of chlorophyll, using light energy and releasing oxygen.

Phycoerythrin A red pigment characteristic of the Rhodophyta (red algae).

Phytoplankton Plant plankton. The most important community of primary producers in the ocean.

Plankton Passively drifting or weakly swimming organisms that are dependent on currents. Includes mostly microscopic algae, protozoans, and larval forms of higher animals.

Plankton bloom A very high concentration of phytoplankton, resulting from a rapid rate of reproduction as conditions become optimum during the spring in high-latitude areas. Less obvious causes produce blooms that may be destructive in other areas.

Plankton net Plankton-extracting device that is cone-shaped and typically of synthetic material. It is towed through the water or lifted vertically to extract plankton down to a size of 50 μm (0.02 in.).

Pneumatophore An air-containing float found on siphonophores.

Pogonophora A phylum of entirely marine tube worms that have no gut and are found only in water deeper than 20 m (65.6 ft).

Poikilotherm An organism whose body temperature varies with and is largely controlled by the temperature of its environment.

Polychaeta Class of annelid worms that includes most of the marine segmented worms.

Polynya A nonlinear opening in sea ice.

Polyp A single individual of a colony or a solitary attached coelenterate.

Population A group of individuals of one species living in an area.

Porifera Phylum of sponges. Supporting structure composed of $CaCO_3$ or SiO_2 spicules or fibrous spongin. Water currents created by flagella-waving choanocytes enter tiny pores, pass through canals, and exit through a larger osculum.

Potential density (σ_0, sigma-theta) The density of ocean water when the effect of the temperature increase resulting from compression is removed.

Primary productivity The amount of organic matter organisms synthesize from inorganic substances within a given volume of water or habitat in a unit of time.

Prime meridian The meridian of longitude 0° used as a reference for measuring longitude; the Greenwich Meridian.

Progressive wave A wave in which the waveform progressively moves.

Propagation The transmission of energy through a medium.

Protein A complex organic compound that contains nitrogen. It is composed principally of amino acids, and is an essential part of an organism's food requirement.

Protista A kingdom of organisms that includes all one-celled forms with nuclear material confined to a nuclear sheath. Includes the animal phylum Protozoa and the phyla of algal plants.

Proton A nuclear particle of all atoms, containing one unit of electrical charge.

Protoplasm The complicated self-perpetuating living material making up all organisms. The elements carbon, hydrogen, and oxygen constitute more than 95%; water and dissolved salts make up from 50% to 97% of most plants and animals, with carbohydrates, lipids (fats), and proteins constituting the remainder.

Protozoa Phylum of one-celled animals with nuclear material confined within a nuclear sheath.

Pseudocoelomate Phyla of animals with a coelom that forms by the separation of the endoderm and ectoderm. Coelom is not lined with peritoneum.

Pseudopodia Extensions of protoplasm in broad, flat, or long needlelike projections used for locomotion or feeding. Typical of amoeboid forms such as foraminifers and radiolarians.

Pteropoda An order of pelagic gastropods in which the foot is modified for swimming and the shell may be present or absent.

Pycnocline A layer of water in which a high rate of change in density in the vertical dimension is present.

Pycnogonid A spiderlike arthropod found on the ocean bottom at all depths. The more commonly observed nearshore varieties are usually less than 1 cm (0.4 in.) across, while deeper-water varieties may reach spreads of over 1 m (3.28 ft).

Pyrrophyta A phylum of microscopic algae that possesses flagella for locomotion—the dinoflagellates.

Quadrature Configuration in which two heavenly bodies are 90° apart when viewed from the center of a third body. During the 1st- and 3rd-quarter phases of the moon, the moon is in quadrature relative to the sun when viewed from the earth.

Radiata A grouping of phyla with primary radial symmetry—phyla Coelenterata and Ctenophora.

Radioactivity The spontaneous breakdown of the nucleus of an atom resulting in the emission of radiant energy in the form of particles or waves.

Radiolaria An order of planktonic and benthic protozoans that possess protective coverings usually composed of silica.

Ray A cartilaginous fish in which the body is dorso-ventrally flattened, eyes and spiracles are on the upper surface, and gill slits are on the bottom. The tail is reduced to a whiplike appendage. Includes electric rays, manta rays, and stingrays.

Red tide A reddish-brown discoloration of surface water, usually in coastal areas, caused by high concentrations of microscopic organisms, usually dinoflagellates. It probably results from increased availability of certain nutrients for various reasons. Toxins produced by the dinoflagellates may kill fish directly, or large populations of animal forms that spring up to feed on the plants, along with decaying plant and animal remains, may use up the oxygen in the surface water to cause asphyxiation of many animals.

Reef A consolidated rock (a hazard to navigation) with a depth of 20 m (65.6 ft) or less.

Reef flat A platform of coral fragments and sand that is relatively exposed at low tide.

Reef front The upper seaward face of a reef from the reef edge (seaward margin of reef flat) to the depth at which living coral and coralline algae become rare (16 to 30 m, or 52 to 98 ft).

Reflection The process in which a wave has part of its energy returned seaward by a reflecting surface.

Refraction The process by which the part of a wave in shallow water is slowed down to cause the wave to bend and tend to align itself with the underwater contours.

Relict beach A beach deposit laid down and submerged by a rise in sea level. It is still identifiable on the continental shelf, indicating no deposition is presently taking place at that location on the shelf.

Residence time The average length of time a particle of any substance spends in the ocean. It is calculated by dividing the total amount of the substance in the ocean by the rate of its introduction into the ocean or the rate at which it leaves the ocean.

Respiration The process by which organisms utilize organic materials (food) as a source of energy. As the energy is released, oxygen is used and carbon dioxide and water are produced.

Reversing current The tide current as it occurs at the margins of landmasses. The water flows in and out for approximately equal periods of time, separated by slack water when the water is still at high and low tidal extremes.

Rhodophyta Phylum of algae composed primarily of small encrusting, branching, or filamentous plants that receive their characteristic red color from the presence of the pigment phycoerythrin. With a worldwide distribution, they are found at greater depths than other algae.

Ridge *See* drifts; oceanic ridge.

Rip current A strong narrow surface or near-surface current of high velocity (up to 4 km/h, or 2.5 mi/h) flowing seaward through the breaker zone at nearly right angles to the shore. It represents the return to the ocean of water that has been piled up on the shore by incoming waves.

Ripples Capillary waves. Ten-to 15 cm (4 to 6 in.)-long waves. They are found on the sides of furrows cut into mud waves by current action.

Rise A long, broad elevation that rises gently and rather smoothly from the deep-ocean floor.

Rotary current Tidal current as observed in the open ocean. The tidal crest makes one complete rotation during a tidal period.

Sabellid A member of the annelid family Sabellidae that lives in a tube composed of shell fragments, sand, and agglutinous material. Featherlike gills and feeding structures filter food from the water above the tube opening.

Salinity A measure of the quantity of dissolved solids in ocean water. Formally, it is the total amount of dissolved solids in ocean water in parts per thousand by weight after all carbonate has been converted to oxide, the bromide and iodide to chloride, and all the organic matter oxidized. It is normally computed from conductivity, refractive index, or chlorinity.

Salinometer A conductance-measuring device that measures the salinity of ocean water to a precision of 0.003‰.

Salpa Genus of pelagic tunicates that are cylindrical, transparent, and found in all oceans.

Salt Any substance that yields ions other than hydrogen or hydroxyl. Salts are produced from acids by replacing the hydrogen with a metal.

Salt marsh A relatively flat area of the shore where fine sediment is deposited and salt-tolerant grasses grow. One of the most biologically productive regions on the earth's surface.

Sand Particle size ranging from 1/16 to 2 mm (0.0025 to 0.08 in.). It pertains to particles that lie between silt and granules on the Wentworth scale of grain size.

Sargasso Sea A region of convergence in the North Atlantic lying south and east of Bermuda where the water is a very clear deep blue in color and contains large quantities of floating *Sargassum.*

Sargassum A brown alga characterized by a bushy form, substantial holdfast when attached, and a yellow-brown, green-yellow, or orange color. Two species, *S. fluitans* and *S. natans,* make up most of the macroscopic vegetation in the Sargasso Sea.

Scaphopoda A class of mollusk commonly called *tusk shells.* The shell is an elongate cone open at both ends. The conical foot surrounded by threadlike tentacles extends from the larger end to aid the animal in burrowing.

Scarp A linear, steep slope on the ocean floor separating gently sloping or flat surfaces.

Scavenger An animal that feeds on dead organisms.

Schizocoelomate Animal that has a coelomic cavity called a *schizocoelom* that develops from a split in the mesoderm and is lined with peritoneum. It is present in members of the phyla Annelida, Arthropoda, Bryozoa, Mollusca, and Phoronida.

Scyphozoa A class of coelenterates that includes the true jellyfish, in which the medusoid body form predominates and the polyp is reduced or absent.

Sea 1. A subdivision of an ocean. Two types of seas are identifiable and defined. They are the *mediterranean seas,* where a number of seas are grouped together collectively as one sea, and *marginal seas* that are connected individually to the ocean. 2. A portion of the ocean where waves are being generated by wind.

Sea anemone A member of the class Anthozoa whose bright color, tentacles, and general appearance resemble a flower.

Sea arch An opening through a headland caused by wave erosion. Usually develops as sea caves are extended from one or both sides of the headland.

Sea cave A cavity at the base of a sea cliff; formed by wave erosion.

Sea cow An aquatic, herbivorous mammal of the order Sirenia that includes the dugong and manatee.

Sea cucumber A common name given to members of the echinoderm class Holotheuroidea.

Sea-floor spreading A process producing the lithosphere when convective upwelling of magma along the oceanic ridges moves away at rates of from 1 to 10 cm (0.4 to 4 in.) per year.

Sea ice Any form of ice originating from the freezing of ocean water.

Seamount An individual peak extending over 1000 m (3280 ft) above the ocean floor.

Sea snake A reptile belonging to the family Hydrophiidae with venom similar to that of cobras. Found primarily in the coastal waters of the Indian Ocean and the western Pacific Ocean.

Sea state A description of the ocean surface that includes the average height of the highest one-third of the waves observed in a wave train; referred to a numerical code.

Sea turtle Any turtle of the reptilian order Testudinata; widely found in warm water.

Sea urchin An echinoderm belonging to the class Echinoidea; possessing a fused test (external covering) and well-developed spines.

Sediment Particles of organic or inorganic origin that accumulate in loose form.

Sedimentary rock A rock resulting from the consolidation of loose sediment, or a rock resulting from chemical precipitation, e.g., sandstone and limestone.

Seiche A standing wave of an enclosed or semienclosed body of water that may have a period ranging from a few minutes to a few hours, depending on the dimensions of the basin. The wave motion continues after the initiating force has ceased.

Seismic Pertaining to an earthquake or earth vibration, including those that are artificially induced.

Seismic sea wave *See* tsunami.

Semidiurnal tide Tide having two high and low waters per tidal day, with small inequalities between successive highs and lows. Tidal period is about 12 hours and 25 minutes solar time; semidaily tide.

Serpulid A polychaete worm belonging to the family Serpulidae that builds a calcareous or leathery tube on a submerged surface.

Sessile Permanently attached to the substrate and not free to move about.

Setae Hairlike or needlelike projections on the exoskeletons of arthropods. Similar structures are found on some annelids.

Shallow-water wave A wave on the surface of the water whose wavelength is at least 20 times water depth. The bottom affects the orbit of water particles, and velocity is determined by water depth.

$$V(m/s) = 3.1\sqrt{\text{water depth (m)}}.$$

Shoal A shallow.

Shore The section of land seaward of the coast; extends from highest level of wave action during storms to the low-water line.

Shoreline The line marking the intersection of the water surface with the shore. Migrates up and down as the tide rises and falls.

Sialic Pertaining to the composition of the granitic continental crust, which is rich in silica and aluminum.

Sigma (σ) Symbol for *in situ* density.

Sigma-theta (σ_t) Symbol for potential density.

Silica Silicon dioxide (SiO_2).

Sill A submarine ridge partially separating bodies of water such as fjords and seas from one another or from the open ocean.

Silt A particle size ranging from 1/128 to 1/16 mm (0.003 to 0.0025 in.). It is intermediate between sand and clay.

Simatic Pertaining to the portion of the earth's crust underlying the oceans and the sialic continents; rich in magnesium and iron and therefore more dense than the sialic continental crust.

Siphonophora An order of hydrozoan coelenterates that forms pelagic colonies containing both polyps and medusae. Examples are *Physalia* and *Velella*.

Slack water Condition existing when a reversing tidal current changes direction at high or low water. Current velocity is zero.

Slick A smooth patch on an otherwise rippled surface caused by a monomolecular film of organic material that reduces surface tension.

Slope current Another name for the western boundary undercurrent; derived from the fact that these currents flow along the base of the continental slope.

Solar tide The partial tide caused by the tide-producing forces of the sun.

Solar wind The motion of interplanetary plasma or ionized particles away from the sun.

Solstice The time when the sun is directly over one of the tropics. In the Northern Hemisphere, the summer solstice occurs on June 21 or 22, when the sun is over the Tropic of Cancer, and the winter solstice occurs on December 21 or 22, when the sun is over the Tropic of Capricorn.

Solute A substance dissolved in a solution. Salts are the solute in salt water.

Solution A state in which a solute is homogeneously mixed with a liquid solvent. Water is the solvent in ocean water.

Sonar An acronym for *sound navigation and ranging*. A method by which objects may be located in the ocean.

Sounding Measuring the depth of water beneath a ship.

Specific gravity The density ratio of a given substance to pure water at 4°C (39.2°F) and at atmospheric pressure.

Specific heat The quantity of heat required to raise the temperature of 1 g of a given substance 1°C (1.8°F). For water it is 1 cal.

Spermatophyta Seed-bearing plants.

Spicule A minute needlelike calcareous or siliceous form found in sponges, radiolarians, chitons, and echinoderms that acts to support the tissue or provide a protective covering.

Spit A small point, low tongue, or narrow embankment of land having one end attached to the mainland and the other terminating in open water. It commonly consists of sand deposited by longshore currents.

Sponge *See* Porifera.

Spray zone A zone extending above the highest high-tide shoreline to the coastline. Water covers it only during storms.

Spring tide Tide of maximum range occurring every fortnight close to the times when the moon is new and full.

Stack An isolated mass of rock that has been detached by wave erosion from the headland and projects from the ocean.

Standing crop   The biomass of a population at any given time.

Standing wave   A wave whose form oscillates vertically without progressive movement. The region of maximum vertical motion is an *antinode*. On either side are *nodes* where there is no vertical motion but maximum horizontal motion.

Stenohaline   Pertaining to organisms that can withstand only a small range of salinity change.

Stenothermal   Pertaining to organisms that can withstand only a small range of temperature change.

Storm surge   A rise above normal water level resulting from wind stress and reduced atmospheric pressure during storms. Consequences can be more severe if it occurs in association with high tide.

Stromatolite   A calcium carbonate sedimentary structure in which algal assemblages trap sediment and bind it into forms that are often dome-shaped. They are known to form only in shallow-water environments.

Subduction   A process by which one lithospheric plate descends beneath another. The surface expression of such a process may be an island arc-trench system or a folded mountain range.

Sublimation   The transformation of the solid state of a substance to a vapor without going through the liquid phase, and vice versa.

Sublittoral zone   That portion of the benthic environment extending from low tide to a depth of 200 m (656 ft). Some consider it to be the surface of the continental shelf.

Submarine canyon   A steep V-shaped canyon cut into the continental shelf or slope.

Submergent shoreline   Shoreline on landforms developed under subaerial processes; formed by the relative submergence of a landmass. It is characterized by bays and promontories and is more irregular than a shoreline of emergence.

Subneritic province   The ocean floor underlying the neritic water column.

Suboceanic province   The ocean floor underlying the oceanic water column.

Substrate   The base on which an organism lives and grows.

Subsurface current   A current usually flowing below the pycnocline, generally at a slower speed and in a different direction than the surface current.

Subsurface oxygen maximum (SOM)   Due to the fact the oxygen produced by photosynthesis is trapped by thermal stratification, dissolved oxygen maximums occur in the low-latitude and temperate oceans. SOM dissolved-oxygen values at depths of 50 to 100 m (164 to 328 ft) are often 20% above saturation.

Supralittoral zone   The spray zone above the high-tide mark.

Surf zone   The region between the shoreline and the line of breakers, where most wave energy is released.

Sverdrup (sv)   A unit of volume transport equal to 1 million m^3/s.

Swash   The rush of water up on the beach following the breaking of a wave.

Swell   One of a series of regular long-period waves that travels out of a wave-generating area.

Swim bladder   An elongate, gas-filled sack dorsal to the digestive tract of most bony fishes. It serves chiefly as a hydrostatic organ to help achieve neutral buoyancy.

Symbiosis   A relationship between two species in which one or both benefit and neither or one is harmed. Examples are commensalism, mutualism, and parasitism.

Tablemount   A flat-topped seamount; a guyot.

Tectonics   The study of the origin and history of structural deformation of the earth's crust.

Temperature gradient   The rate of temperature change within the water column. It is steep within the thermocline, where water temperature changes rapidly with changing depth, but less steep in the deep water, where distribution patterns are more uniform.

Territorial sea   A zone extending seaward from the shore or internal waters of a nation for a distance of 12 mi (19.3 km) as defined by the UNCLOS. The coastal state has full authority over this zone but must allow rights of innocent passage.

Tethys   An oceanic body of water that separated the protocontinents Gondwanaland and Laurasia along the alignment of the present Alpine-Himalayan mountain belt.

Thermocline   A layer of water in which a rapid change in temperature can be measured in the vertical dimension.

Thermohaline circulation   The vertical movement of ocean water driven by density differences resulting from the combined effects of variations in temperature and salinity.

Tidal bore   *See* bore.

Tidal day   *See* lunar day.

Tidal period   Elapsed time between successive high or low waters.

Tidal range   The difference in height between consecutive high and low waters. The time frame of comparison may also be a day, month, or year.

Tide   Periodic rise and fall of the ocean surface and connected bodies of water resulting from the unequal gravitational attraction of the moon and sun on different parts of the earth.

Tide wave   The long-period gravity wave generated by the tide-generating forces described above and manifested in the rise and fall of the tide.

Tintinnid   A ciliate protozoan of the family Tintinnidae with a tubular to vase-shaped outer shell.

Tissue   An aggregate of cells and their products developed by organisms for the performance of a particular function.

Tombolo   A sand or gravel bar that connects an island with another island or the mainland.

Topography   The configuration of a surface. In oceanography, it refers to the ocean bottom or the surface of a mass of water.

Trade winds The air masses moving from subtropical high-pressure belts toward the equator. They are northeasterly in the Northern Hemisphere and southeasterly in the Southern Hemisphere.

Transform fault A fault characteristic of oceanic ridges along which the ridges are horizontally offset.

Transitional crust Thinned section of continental crust at the trailing edge of a continent; created by the breaking apart of an ancient continent over a newly formed spreading center.

Transitional wave A wave moving from deep water to shallow water that has a wavelength more than 2 but less than 20 times the water depth. Particle orbits are beginning to be influenced by the bottom.

Transverse wave A wave in which particle motion is at right angles to energy propagation.

Trawl A sturdy bag or net that can be dragged along the ocean bottom, or at various depths above the bottom, to catch fish.

Trench A long, narrow, and deep depression on the ocean floor with relatively steep sides.

Triploblastic Pertaining to the embryonic structure characterized by three layers of cell tissue—endoderm, mesoderm, and ectoderm.

Trophic level A nourishment level in a food chain. Plant producers constitute the lowest level, followed by herbivores and carnivores at the higher levels.

Tropical tide A tide occurring bimonthly when the moon is at its maximum declination north and south of the equator. It is during tropical tides that the greatest diurnal inequalities occur.

Tsunami Seismic sea wave. A long-period gravity wave generated by a submarine earthquake or volcanic event. Not noticeable on the open ocean but builds up to great heights in shallow water.

Tube worms. *See* sabellid and serpulid.

Tunicates Members of the chordate subphylum Urochordata, which includes sacklike animals. Some are sessile (sea squirts) while others are pelagic (salps).

Turbidite A sediment or rock formed from sediment deposited by turbidity currents; characterized by both horizontally and vertically graded bedding.

Turbidity A state of reduced clarity in a fluid; caused by the presence of suspended matter.

Turbidity current A gravity current resulting from a density increase brought about by increased water turbidity. Possibly initiated by some sudden force such as an earthquake, the turbid mass continues under the force of gravity down a submarine slope.

Turbulence A disorderly flow of ocean water.

Turbulent flow Flow in which the flow lines are confused owing to random velocity fluctuations.

Typhoon A severe tropical storm in the western Pacific.

Ultraplankton Plankton smaller than 5 μm (0.002 in.). Very difficult to separate from the water.

Ultrasonic Pertaining to sound frequencies above human range (above 20,000 cycles per second).

Ultraviolet radiation Electromagnetic radiation shorter than visible radiation and longer than X-rays. The approximate range is from 1 to 400 nanometers (nm).

Upper water That portion of the water column from the ocean surface to the base of the permanent thermocline or pycnocline. Surface temperatures are relatively high and uniform to the base of the mixed layer (100 to 200 m, or 328 to 656 ft) and decrease rapidly to the base of the thermocline (about 1000 m, or 3280 ft).

Upwelling The process by which deep, cold, nutrient-laden water is brought to the surface, usually by the wind divergence of equatorial currents or coastal winds push water away from the coast.

Valence The combining capacity of an element measured by the number of hydrogen atoms with which it will combine.

Van der Waals force Weak attractive force between molecules; a result of the interaction between the nuclear particles of one molecule and the electrons of another.

Vector A physical quantity that has magnitude and direction. Examples are force, acceleration, and velocity.

Veliger A planktonic larval stage of many gastropods with two ciliated lobes.

Ventral Pertaining to the lower or under surface.

Vertebrata Subphylum of chordates that includes those animals with a well-developed brain and a skeleton of bone or cartilage; includes fish, amphibians, reptiles, birds, and mammals.

Viscosity The property of a substance to offer resistance to flow; internal friction.

Viviparous Pertaining to an animal that gives birth to living young.

Wadati-Benioff seismic zone A plane descending from beneath oceanic trenches, at an angle of approximately 45°, in which earthquake foci are located. It is thought to represent the core of subducting lithosphere.

Warm current A current carrying warm water into areas of colder water. It generally moves poleward.

Water mass A body of water identifiable by its temperature, salinity, or chemical content.

Wave A disturbance that moves over or through a medium with a speed determined by the properties of the medium.

Wave-cut bench A gently sloping surface produced by wave erosion and extending from the base of the wave-cut cliff out under the offshore region.

Wave-cut cliff A cliff produced by wave erosion cutting landward.

Wave height Vertical distance between a crest and the preceding trough.

Wavelength Horizontal distance between two corresponding points on successive waves, such as from crest to crest.

Wave period The elapsed time between the passage of two successive wave crests past a fixed point.

Wave steepness Ratio of wave height to wavelength.

Wave train   A series of waves from the same direction.

Weathering   A process by which rocks are broken down by chemical and mechanical means.

Westerly winds   The air masses moving away from the subtropical high-pressure belts toward higher latitudes. They are southwesterly in the Northern Hemisphere and northwesterly in the Southern Hemisphere.

Western boundary undercurrent   Relatively strong bottom current flowing across the continental rise along the western boundary of ocean basins.

Windrows   Rows of floating debris aligned parallel to the direction of the wind; resulting from Langmuir circulation.

Windward   Pertaining to the direction from which the wind is blowing.

Zenith   That point on the celestial sphere directly over the observer.

Zooplankton   Animal plankton.

INDEX